This book provides the first comprehensive account of the proteins involved in blood coagulation, fibrinolysis and the complement system. A major section of the book is devoted to each of these three systems, with separate chapters dealing in detail with the structural aspects and the different functional processes. Topics covered in the blood coagulation section include the activation of factors IX and X and prothrombin, and the formation and stabilisation of fibrin. The fibrinolysis section includes the activation of plasminogen, the degradation of fibrin and the regulation of fibrinolysis. The complement system itself is covered in chapters dealing with classical activation, alternative activation, the lytic complex and the regulatory processes involved. In addition, there is a section which deals with special topics, including the kinin system, signal peptides, haemostasis, as well as the evolution of protein structure. This volume will be of use to researchers and advanced students in the fields of haematology, immunology and clinical chemistry.

Mechanisms in blood coagulation, fibrinolysis and the complement system

Mechanisms in blood coagulation fibrinolysis and the complement system

TORBEN HALKIER
Department of Molecular Biology and Plant Physiology
University of Aarhus

Translation by Paul Woolley

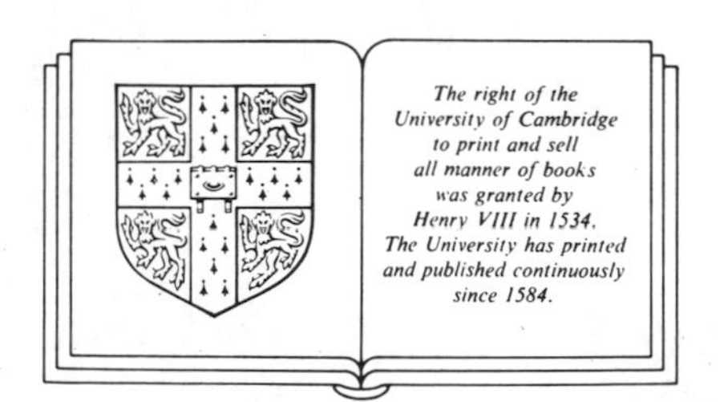
The right of the
University of Cambridge
to print and sell
all manner of books
was granted by
Henry VIII in 1534.
The University has printed
and published continuously
since 1584.

CAMBRIDGE UNIVERSITY PRESS
Cambridge
New York   Port Chester   Melbourne   Sydney

Published by the Press Syndicate of the University of Cambridge
The Pitt Building, Trumpington Street, Cambridge CB2 1RP
40 West 20th Street, New York, NY 10011-4211, USA
10 Stamford Road, Oakleigh, Victoria 3166, Australia

First published 1991

Printed in Great Britain at the University Press, Cambridge

British Library cataloguing in publication data

Halkier, Torben
Mechanisms in blood coagulation, fibrinolysis and the complement system.
1. Man. Blood. Coagulation
I. Title
612.115

Library of Congress cataloguing in publication data available

ISBN 0 521 38187 8 hardback

SE

Contents

Foreword

This book was originally prepared as teaching material and as a background compendium in Danish for the final-year undergraduate course 'Mechanisms in blood coagulation, fibrinolysis and the complement system' held at the Department of Molecular Biology and Plant Physiology at the University of Aarhus from 1987 to 1989.

It was initially motivated by the fact that merely reading selected articles in research journals does not provide the student with a sufficient background for the understanding of mechanisms in the three systems referred to in the title. It is no use understanding the fine mechanics of biochemistry, if the overall principles disappear in a mass of details.

Many good reasons can be thought of for compiling a book of this kind, which is both a compendium of components – proteins – and also a guide to the mechanisms by which they act and interact. One reason is that, in spite of the best intentions, a course on the subject in question can easily get out of perspective without some kind of summarising document – the theme is colossal, while the time at the disposal of most students is very limited. Another is that reading original articles can be confusing for one unacquainted with the field; the author hopes that this will not be the case with this book, where the material has been both selected and subjected to a certain amount of pre-digestion. A third reason is the chance to emphasize aspects of particular interest, and a fourth is the possibility of making comparisons and thus of seeing things in a better perspective.

It will, it is hoped, be unnecessary to state that the content of this book reflects more the author's interpretation than it does Completeness or Truth.

It is here appropriate for me to thank Staffan Magnusson* for having introduced me to the field of blood coagulation and for many wide-ranging discussions, just as I should like to thank my students for their patience and help. An especial word of thanks must be addressed to Paul Woolley for his assiduous and expert translation from the original Danish manuscript and to Vilhelm Tetens for first suggesting that this book should be addressed to a wider public.

Århus, January 1990.

* The news of the untimely death of Staffan Magnusson reached Århus just before the revised manuscript of this book went to the press.

Introduction

It is a matter of everyday experience that a scratch, a cut or a more serious wound leads to bleeding, and, likewise, that the flow of blood stops in time, as long as the wound was not too large. Blood coagulates; first a scab, or clot, is formed, and later scar tissue. The processes involved are summarised by the terms *coagulation*, *haemostasis* and *fibrinolysis*, and they involve a sizeable collection of proteins acting in an even more sizeable collection of biochemical processes.

The coagulation of blood, and events subsequent to it, can well be regarded as a kind of defence system for stopping holes in complex organisms so as to minimise penetration by foreign bodies and aggressive micro-organisms that later could cause severe damage. However, these do get through sometimes, and penetration by them can result in inflammation at the point of the original damage. A part of the system for combating foreign organisms is the *complement system*, which is the non-specific part of the immune system.

In the following pages, an attempt is made to give an up-to-date picture of our knowledge of the mechanisms of blood coagulation, fibrinolysis and the complement system. The amino-acid sequences of the proteins to be described are summarised in an appendix. The source of the proteins described is usually stated, but where this is not mentioned the proteins are, implicitly, of human origin. Likewise, the physiological systems described are, as far as possible, human.

Many of the proteins that we shall meet in this book have common domains, defined by their amino-acid sequences. In the figures in which the domain structures of individual proteins are shown, standard domain elements will be used, and these are defined in Table 1. For quantitative comparison of homology, the percentage of identical amino-acid residues in the aligned protein chains divided by the number of residues in the shorter chain. In the text this will be referred to as the *degree of identity*.

Table 1.

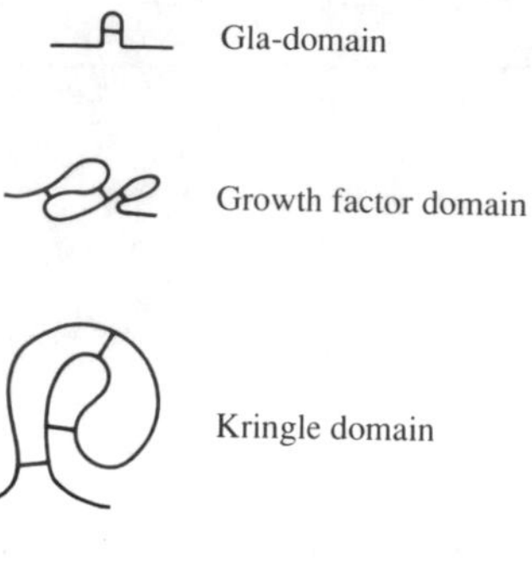

Gla-domain

Growth factor domain

Kringle domain

Short consensus repeat (SCR) domain

Prekallikein/factor XI domain

Fibronectin type I domain

Fibronectin type II domain

Serine proteinase domain

Part I Blood Coagulation

1.

The coagulation process: an introduction

Blood coagulation is a field of considerable research interest, so it has already been dealt with in a number of classical reviews (1–3). Recent years have seen a rapid increase in the number of protein sequences determined, by the sequencing of proteins, of cDNA or of genes, with the result that the sequences of all the proteins directly involved in the coagulation system are now known (4) (Table 2). This has led to a considerable increase in our understanding of the mechanism of blood coagulation, although of course merely knowing a sequence does not in itself enable us to understand the components of a system or the way in which they interact.

Traditionally, two pathways of blood coagulation have been distinguished (Fig. 1): the *intrinsic* and the *extrinsic* pathway. These two names allude to the fact that while the intrinsic pathway leads to thrombin formation, using only protein factors that are present in plasma, the extrinsic pathway requires also the presence of a lipid-dependent membrane glycoprotein that does not occur in plasma. However, it is becoming clear that the division of the coagulation process into two branches is somewhat artificial, since several of the components in the two branches appear to interact with one another.

The process of coagulation is classically regarded as a cascade reaction, as originally suggested (5, 6). But recent developments suggest that the cascade concept has outlived its usefulness, because a number of the enzymic reactions leading to the formation of thrombin require the presence of thrombin-activated co-factors if they are to proceed at a physiologically relevant speed. It therefore seems closer to the truth to describe the coagulation system as a cyclic reaction network.

Table 2. *The proteins of the blood coagulation system.*

Protein	Concentration in plasma	Physically-estimated molecular mass (kDa)	Number of amino-acid residues	Calculated molecular mass (kDa)	Gene size (kbp)	Chromosome (and position, if known)
Factor XII	500 nM	80	596	66.9	12	5q33-qter
Plasma prekallikrein	500 nM	85–88	619	69.2	unknown	unknown
Factor XI	35 nM	2×75	2×607	136	23	4
HMM-kininogen	1 μM	110	626	69.9	27	unknown
Factor IX	80 nM	55–57	415	47	34	Xq27.1
Factor VIII	0.5 nM	>250	2332	264.7	186	Xq27.3
Factor X	170 nM	$17+42$	$139+306$	49.6	27	13q34
Factor V	20 nM	>300	2196	249	unknown	1q21–25
Prothrombin	1.1 μM	72	579	65.7	20.2	11p11–q12
Factor VII	10 nM	50	406	45.5	12.8	13q34
Tissue factor	–	43–53	263	29.6	12.4	1q21–22
Fibrinogen $(A\alpha B\beta\gamma)_2$	6–13 μM	340	2×1482	329.6	–	–
$A\alpha$ chain	–	73	610	66.1	unknown	4q26–32
$B\beta$ chain	–	60	461	52.3	10	4q26–32
γ chain	–	53	411	46.4	10.5	4q26–32
Factor XIII $(\alpha_2 b_2)$	70 nM	320	2×1372	301	–	–
α-chain	–	80	731	80.5	>160	6p24–25
β-chain	130 nM	80	641	70	unknown	1q31–32.1
Antithrombin III	2.5 μM	54–68	432	49	19	1q23–25
$C\bar{1}$-inhibitor	2 μM	104	478	52.9	17	11p11.2–q13
α_1-proteinase inhibitor	20–50 μM	52	394	44.3	12	14q31–32.3
EPI/LACI[a]	2–3 nM	40	276	32	unknown	unknown
Protein C	60 nM	$21+41$	$155+262$	47.5	11.2	2q14–21
Protein S	150 nM[b]	69	635	70.7	>45	3
Thrombomodulin	–	78	557	58.7	3.7	20
Protein Ca inhibitor	90 nM	57	387	43.8	unknown	unknown

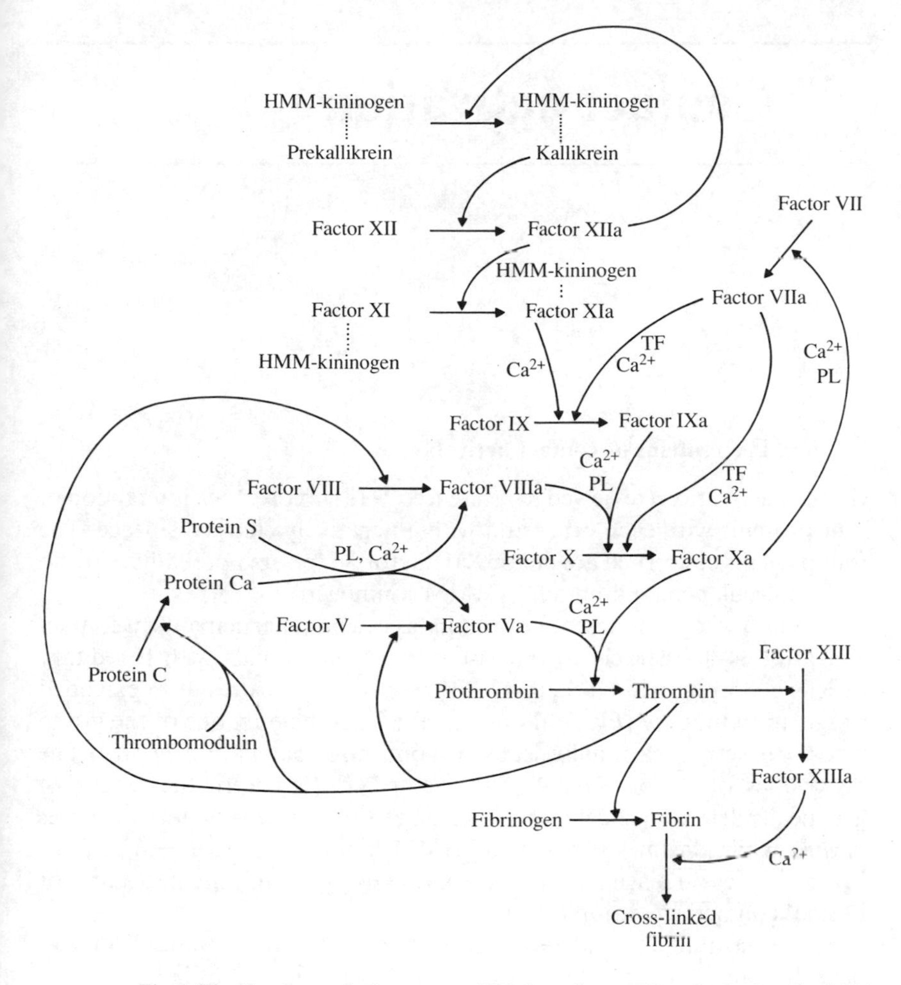

Fig. 1. The blood coagulation system. (TF, tissue factor; PL, anionic phospholipid.)

2.

Contact activation

2.1 The proteins of contact activation

Contact activation (reviewed in references 7–13) involves the interaction of four proteins with each other and with a negatively charged surface. The four proteins concerned are factor XII, factor XI, plasma prekallikrein and high-molecular-mass kininogen (HMM-kininogen).

Our knowledge of these four proteins as essential participants in contact activation is due largely to the discovery of individuals with hereditary deficiencies in individual factors. All these deficiencies result in extended coagulation times *in vitro*, although factor XI is the only one of the group whose absence occasionally leads to abnormal coagulation *in vivo*. The discoveries of factor XI (14) and factor XII (15) were the results of genetically determined deficiencies leading to lengthened coagulation times *in vitro*, while plasma kallikrein and HMM-kininogen were already known before studies on plasma lacking them revealed that they are necessary for normal contact activation (16–20).

It was inevitable that some of these coagulation factors should come to bear the names of the patients who were first recognised as lacking them. Factor XII is frequently referred to by its trivial name of Hageman factor (ironically, John Hageman eventually died of a pulmonary embolism (22)). Plasma prekallikrein is called Fletcher factor, while HMM-kininogen is variously called Fitzgerald, Williams, Flaujeac or Reid factor. Factor XI has not been named after a patient, but is occasionally referred to as 'plasma thromboplastin antecedent'.

Before discussing the actual mechanism of contact activation, we shall introduce the individual participant proteins in turn.

2.1.1. *Factor XII*

Factor XII is a single-chained glycoprotein. It circulates in the bloodstream as the inactive precursor (zymogen) of the active serine proteinase factor

Fig. 2. The domain structure of factor XII. → shows the position at which cleavage by plasma kallikrein leads to α-factor XIIa and ⇒ the positions at which extended cleavage by the same enzyme leads to β-factor XIIa. ▼ shows the intron positions, ◆ the N-bound carbohydrate, and ● the O-bound carbohydrate.

XIIa, which takes part in the initial steps of the intrinsic pathway of coagulation. Its molecular mass is estimated from SDS-polyacrylamide gels as 80 kDa, of which some 17% is carbohydrate (21). The concentration of factor XII in normal human plasma is about 40 μg/ml (≈500 nM) (24).

The amino-acid sequence of factor XII has been determined both by protein-sequencing (25, 26) and by cDNA-sequencing (27–29). The structure of the gene for factor XII is also known (30). This gene, located at position q33–qter of chromosome 5 (31), has 14 exons separated by 13 introns and covers about 12 kbp of genomic DNA. (Note: in general the number of introns in a gene is equal to the number of exons minus one. This is the case for all the genes whose structures are mentioned in this book, so the number of introns will not usually be stated explicitly.) The lengths of the exons of the gene for factor XII lie between 57 bp and 320 bp, while the introns are between 80 bp and 4.5 kbp long. Genomic sequencing and cDNA-sequencing have also shown that the signal peptide for factor XII is 19 amino acids in length (see also Chapter 20).

The amino-acid chain of human factor XII (Appendix, Fig. A1) contains 596 amino-acid residues with a calculated molecular mass of 66.9 kDa. If the above figure of 17% carbohydrate is included, then the calculated molecular mass is 80.4 kDa.

The structure of factor XII (Fig. 2) shows a feature also seen in many of the other proteins that we shall meet: a mosaic structure of multiple homology – that is, homology with several other proteins. (The possible evolutionary significance and cause of this are discussed in Chapter 27.) The positions of the introns in the factor XII gene and the relation of these to the amino-acid sequence are shown in Figure 2.

Starting at the N terminus of factor XII, we observe first a domain homologous to the 'type II internally homologous regions' originally found in fibronectin (32, 36), followed by a domain homologous to epidermal growth factor (EGF) and therefore known as the growth-factor domain (33, 34). These are followed in turn by a domain showing homology with type I homologies of fibronectin (32), another growth-factor domain, a domain homologous to the kringles originally described for prothrombin (35), a portion called the connecting strand, and finally the presumed catalytic region, homologous to known serine proteases.

Although the disulphide pattern of factor XII is not known, the disulphide bridges are assumed to be placed in the same manner as in homologous domains whose disulphide patterns have been determined. This means that the following disulphide pattern in factor XII can be assumed: in the fibronectin type II domain, Cys28–Cys54 and Cys42–Cys69; for the first growth-factor domain, Cys79–Cys91, Cys85–Cys100 and Cys102–Cys111; in the fibronectin type I domain, Cys115–Cys144 and Cys142–Cys151; in the second growth-factor domain, Cys159–Cys170, Cys164–Cys179 and Cys181–Cys190; in the kringle domain, Cys198–Cys276, Cys219–Cys258 and Cys247–Cys271; and in the serine-proteinase domain, Cys340–Cys467, Cys378–Cys394, Cys417–Cys420, Cys386–Cys456, Cys481–Cys550, Cys513–Cys529 and Cys540–Cys571.

The carbohydrate moieties bound to factor XII have been found experimentally at residues Asn230 and Asn414 (N-bound) and at Thr280, Thr286, Ser289, Thr309, Thr310 and Thr318 (O-bound) (25, 26).

The activation of factor XII to factor XIIa occurs by cleavage of the peptide bond between Arg353 and Val354 (26, 27). The resulting derivative of factor XII is termed α-factor XIIa, and it can be converted to β-factor XIIa by cleavage of the bonds Arg334–Asn335 and Arg343–Leu344 (26, 27). All of these three cleavages are performed by plasma kallikrein.

Factor XIIa thus has two chains, consisting of a catalytic serine-proteinase domain disulphide-bonded either to the whole N-terminal region (in α-factor XIIa, residues 1 to 353) or to a nonapeptide (in β-factor XIIa, residues 335 to 343). A covalent linkage between the chains is maintained by a disulphide bond between Cys340 and, as is believed, Cys467. Considerations of homology suggest that the catalytic apparatus in α-factor XIIa involves His40, Asp89 and Ser191 in the processed chain containing the catalytic site, corresponding to His393, Asp 442 and Ser 544 in unprocessed factor XII.

The genetic defect underlying deficiency of factor XII activity has been found in a single case. Here Cys571 was replaced by Ser (37). Since the complete amino-acid sequence of the variant molecule was not determined, it is possible – though unlikely – that other substitutions were also present.

2.1.2. *Plasma prekallikrein*

Plasma prekallikrein is a single-chained glycoprotein that circulates as the (inactive) zymogen precursor of an active serine proteinase. The active protein is plasma kallikrein, and it participates in the contact activation pathway of coagulation. The concentration of plasma prekallikrein in

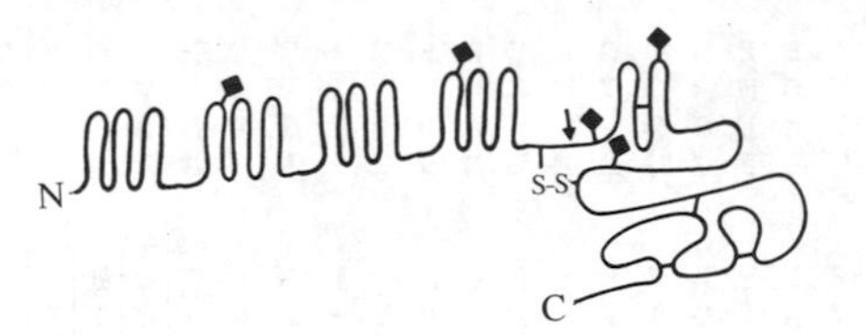

Fig. 3. The domain structure of plasma prekallikrein. → shows the position of cleavage by factor XIIa and ◆ the N-bound carbohydrate.

normal human plasma lies between 40 μg/ml and 55 μg/ml (≈ 500 nM) (38–40). Attempts to isolate it usually yield two forms, with molecular masses estimated from SDS-polyacrylamide gels to be 85 kDa and 88 kDa and with a carbohydrate content of about 15% (41). Isoelectric focussing shows seven bands – that is, plasma prekallikrein appears in seven forms, each of which has its characteristic isoelectric point (42). Four of these forms predominate. Each isoelectric variant contains both molecular-mass variants, but in differing proportions.

The amino-acid sequence of plasma prekallikrein has been determined partially by protein-sequencing (43) and completely by cDNA-sequencing (43) (Appendix, Fig. A2). Included in the cDNA sequence is a signal peptide of 19 amino acids. Neither the gene structure of plasma prekallikrein nor its chromosomal location is known.

The molecular masses calculated from the cDNA sequence are 69.2 kDa without carbohydrate and 80 kDa if the 15% carbohydrate is taken into account. This result is significantly lower than the experimental figures of 85 kDa and 88 kDa (above). However, determinations of molecular mass from SDS-polyacrylamide gels sometimes show deviations, and sedimentation-equilibrium studies indicate a molecular mass of 82 kDa for bovine plasma prekallikrein (44). The heterogeneity in molecular mass is assumed to be due to heterogeneity in the carbohydrate content or in the position of the C terminus, since the purified plasma kallikrein shows no heterogeneity in respect of its N-terminal sequence. Protein-chemical studies have shown that there is N-bound carbohydrate at the residues Asn108, Asn289, Asn377, Asn434 and Asn475.

The structure of plasma kallikrein (Fig. 3) is less complex than that of factor XII. Starting at the N terminus, we observe a succession of four internally homologous domains, of about 90 amino-acid residues each, followed by the serine-proteinase domain. A comparison of the internally homologous regions is given in Figure 4. These domains are almost unique; apart from plasma prekallikrein, they occur in only one other protein, factor XI, to be described below.

The activation of plasma prekallikrein to give plasma kallikrein is carried out by factor XIIa, which cleaves the bond Arg371–Ile372, resulting

```
  1 G C L T Q L Y E N A F F R G G D V A S M Y T P N A Q Y C Q M R
 91 A C H R D I Y K G V D M R G V N F N V S K V S S V E E C Q K R
181 G C H M N I F Q H L A F S D V D V A R V L T P D A F V C R T I
272 P C H S K I Y P G V D F G G E E L N V T F V K G V N V C Q E T

C T F H P R C L L F S F L P A S S I N D M E K R F G C F L K D S V
C T N N I R C Q F F S Y A T Q T F H K A E Y - R N N C L L K Y S P
C T Y H P N C L F F T F Y T N V W K I E S Q - R N V C L L K T S E
C T K M I R C Q F F T Y S L L P E D C K E E - K C K C F L R L S M

T G T L P K V - H R T G A V S G H S L K Q C G H Q I S 90
G G T P T A I K V L S N V E S G F S L K P C A L S E I 180
S G T P S S S T P Q E N T I S G Y S L L T C K R T L P E 271
D G S P T R I A Y G T Q G S S G Y S L R L C N T G D N S 362
```

Fig. 4. Internally homologous domains in plasma prekallikrein. Only the positions at which amino acids are the same in all four domains are outlined. Pairwise comparison of the four domains reveals the following degrees of amino-acid identity: domains 1 and 2, 29%; domains 1 and 3, 33%; domains 1 and 4, 26%; domains 2 and 3, 30%; domains 2 and 4, 40%; domains 3 and 4, 25%.

in a heavy chain from the N terminus (52 kDa) and a light chain from the C terminus (33 kDa and 36 kDa) (45). The heterogeneity in molecular mass can thus be attributed to the light chain, which also contains three bound carbohydrate chains. The disulphide-bond pattern has not been demonstrated experimentally, but in the catalytic domain it is assumed to be the same as in other serine proteases, and the position of the disulphide bond between the heavy and light chains can also be deduced from this. Thus the following disulphide bonds are taken to exist: Cys364–Cys484 (which connects the two chains after activation), Cys400–Cys416, Cys498–Cys565, Cys529–Cys544 and Cys555–Cys583. The four internally homologous domains, which together make up the heavy chain of plasma kallikrein, contain six cysteine residues each in homologous positions, and it is expected that these will prove to form three internal S–S connections in each domain. The fourth domain contains two 'extra' cysteines – Cys321 and Cys326 – and these are presumed to make up a fourth internal disulphide bridge in this domain.

Amino-acid-sequencing (43) has revealed a dimorphism at position 124, where the residue that is usually Asn can instead be Ser. On the basis of homology, the catalytic apparatus in plasma kallikrein is taken to involve the residues His44, Asp93 and Ser 188 in the light chain, corresponding to His415, Asp464 and Ser559 in plasma prekallikrein.

2.1.3. *Factor XI*

Factor XI is unique in being a zymogen for a serine proteinase and at the same time consisting of two identical polypeptide chains held together by

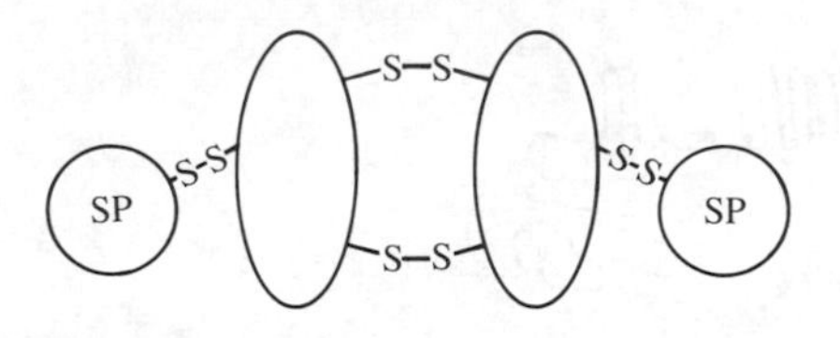

Fig. 5. The two-chain structure of factor XI. (SP, serine-proteinase domain.) The symmetry of the S–S bridged structure has not yet been determined with certainty.

one or more disulphide bridges (Fig. 5). The active enzyme – factor XIa – participates, as do factors XIIa and plasma kallikrein, in the contact activation of blood coagulation. Factor XI is a glycoprotein containing 5% carbohydrate, and its concentration in normal human plasma is about 6 μg/ml ($\approx$35 nM) (46, 47).

The molecular mass of the dimer that makes up the factor XI molecule is estimated by SDS-polyacrylamide gel electrophoresis to lie between 125 kDa and 160 kDa, and the corresponding estimate for the monomer lies between 60 kDa and 83 kDa (48, 49).

The amino-acid sequence of factor XI has been determined largely by cDNA-sequencing (50) (Appendix, Fig. A3); the cDNA sequence includes a signal peptide of 18 residues. The gene for factor XI, found on chromosome 4 (51), covers some 23 kbp of genomic DNA and is distributed among 15 exons of 56 bp to 332 bp that are separated by introns whose sizes range from 88 bp to 4.5 kbp (52).

The molecular mass of the monomer, as deduced from the cDNA sequence, is 68.0 kDa, giving 136.0 kDa for the dimer. The inclusion of 5% carbohydrate brings this value up to about 143 kDa. The carbohydrate is exclusively N-bound, and each monomer possesses five possible sites of N-glycosylation: Asn72, Asn108, Asn335, Asn432 and Asn473. (By 'possible' or 'potential' glycosylation sites we refer here, and at other points in this book, to sites exhibiting the known consensus sequence N-X-S/T for N-glycosylation; see Chapter 19.) The number of N-glycosylation sites used in prekallikrein is also five, but the mass of carbohydrate bound to factor XI is much less than that bound to prekallikrein, which means that either some of the attachment sites in factor XI are not used or else the attached carbohydrate units are much smaller.

The monomeric factor XI chain (Fig. 6) consists of 607 amino-acid residues. It shows homology with prekallikrein over its entire length, with 58% identical amino acids (Fig. 7). In the N-terminal region, the four internally homologous 90-residue domains are found (Fig. 8) and the C-terminal region contains the presumed catalytic serine-proteinase domain. The positions of the introns in the gene for factor XI, relative to the amino-acid sequence, are shown in Figure 6.

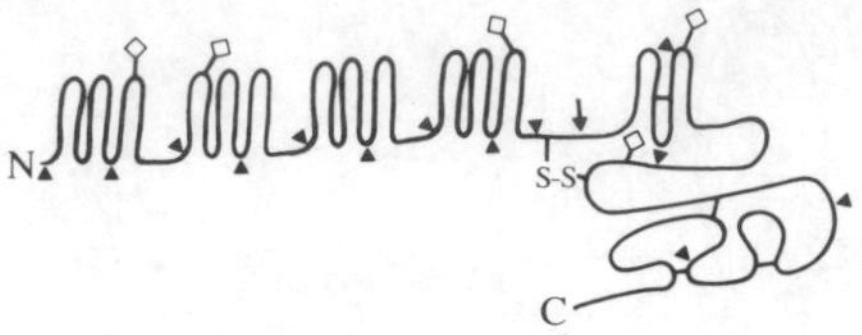

Fig. 6. The domain structure of factor XI, shown for one half-molecule. → shows the position of cleavage by factor XIIa, ▼ the positions of the introns and ◇ the potential attachment sites for N-bound carbohydrate.

The activation of factor XI to give factor XIa is performed by factor XIIa; it cleaves the bond between Arg369 and Ile370 in both monomers. The presence of two catalytically active domains per molecule of factor XIa has been demonstrated by titration with antithrombin III, which gave a stoichiometry of inhibition between enzyme and inhibitor of 1:2 (49). The result of the activation process is a four-chained serine proteinase held together by disulphide bridges, with two identical heavy chains (about 40 kDa) and two light ones (about 26 kDa).

The disulphide-bond pattern in factor XI has not been determined experimentally. In the serine-proteinase domain, it is assumed that this pattern corresponds to that of other, known serine proteases, and the same assumption is made for disulphide bridges connecting the light and heavy chains after activation. This leads to the following presumed pattern of disulphide bonds: Cys362–Cys482 (linking light and heavy chains), Cys398–Cys414, Cys496–Cys563, Cys527–Cys542 and Cys553–Cys581. In the four internally homologous domains, there are six cysteine residues, and these are presumed to make three S–S bridges within each domain. Unlike prekallikrein, whose fourth 90–residue domain contains two further Cys residues that suggest the existence of an additional intradomain bridge, the first and fourth 90–residue domains of factor XI contain one additional Cys each (Cys11 and Cys321), and these may be assumed to bind the monomeric chains to each other.

The catalytic mechanism of factor XIa, suggested by its homology with other serine proteases, is thought to involve His44, Asp93 and Ser188 in the light chain, corresponding to His413, Asp462 and Ser557 in the factor XI polypeptide.

The genetic basis for deficiency in factor XI activity has been determined in a few cases (53). From 12 alleles investigated from six individuals with factor XI deficiency, one allele was found with a mutation in the donor splicing site of intron 14, five alleles contained a mutation at Glu117 to give a stop codon and the remaining six alleles contained the substitution Phe283→Leu.

```
Plasma PK  1 G C L T Q L Y E N A F F R G G D V A S M Y T P N A Q
Factor XI  1 E C V T Q L L K D T C F E G G D I T T V F T P S A K

Y C Q M R C T F H P R C L L F S F L P A S S I N D M E K R F G C F
Y C Q V V C T Y H P R C L L F T F T A E S P S E D P T R W F T C V

L K D S V T G T L P K V H R T G A V S G H S L K Q C G H Q I S A C
L K D S V T E T L P R V N R T A A I S G Y S F K Q C S H Q I S A C

H R D I Y K G V D M R G V N F N V S K V S S V E E C Q K R C T N N
N K D I Y V D L D M K G I N Y N S S V A K S A Q E C Q E R C T D D

I R C Q F F S Y A T Q T F H K A E Y R N N C L L K Y S P G G T P T
V H C H F F T Y A T R Q F P S L E H R N I C L L K H T Q T G T P T

A I K V L S N V E S G F S L K P C A L S E I G C H M N I F Q H L A
R I T K L D K V V S G F S L K S C A L S N L A C I R D I F P N T V

F S D V D V A R V L T P D A F V C R T I C T Y H P N C L F F T F Y
F A D S N I D S V M A P D A F V C G R I C T H H P G C L F F T F F

T N V W K I E S Q R N V C L L K T S E S G T P S S S T P Q E N T I
S Q E W P K E S Q R N L C L L K T S E S G L P S T R I K K S K A L

S G Y S L L T C K R T L P E P C H S K I Y P G V D F G G E E L N V
S G F S L Q S C R H S I P V F C H S S F Y H D T D F L G E E L D I

T F V K G V N V C Q E T C T K M I R C Q F F T Y S L L P E D C K E
V A A K S H E A C Q K L C T N A V R C Q F F T Y T P A Q A S C N E

E K C K C F L R L S M D G S P T R I A Y G T Q G S S G Y S L R L C
G K G K C Y L K L S S N G S P T K I L H G R G G I S G Y T L R L C

N T G D N S V C T T K T S T R I V G G T N S S W G E W P W Q V S L
K M - D N E - C T T K I K P R I V G G T A S V R G E W P W Q V T L

Q V K L T A Q R H L C G G S L I G H Q W V L T A A H C F D G L P L
H T T S P T Q R H L C G G S I I G N Q W I L T A A H C F Y G V E S

Q D V W R I Y S G I L N L S D I T K D T P F S Q I K E I I I H Q N
P K I L R V Y S G I L N Q S E I K E D T S F F G V Q E I I I H D Q

Y K V S E G N H D I A L I K L Q A P L N Y T E F Q K P I C L P S K
Y K M A E S G Y D I A L L K L E T T V N Y T D S Q R P I C L P S K

G D S T T I Y T N C W V T G W G F S K E K G E I Q N I L Q K V N I
G D R N V I Y T D C W V T G W G Y R K L R D K I Q N T L Q K A K I

P L V T N E E C Q K R Y Q D Y K I T Q R M V C A G Y K E G G K D A
P L V T N E E C Q K R Y R G H K I T H K M I C A G Y R E G G K D A

C K G D S G G P L V C K H N G M W R L V G I T S W G E G C A R R E
C K G D S G G P L S C K H N E V W H L V G I T S W G E G C A Q R E

Q P G V Y T K V A E Y M D W I L E K T Q S S D G K A Q M Q S P A 619
R P G V Y T N V V E Y V D W I L E K T Q A V 607
```

Fig. 7. Comparison of the amino-acid sequences for plasma prekallikrein (PK) and factor XI. Only the positions at which amino acids are identical are outlined. The degree of identity between these is 58%.

```
  1 E C V T Q L L K D T C F E G G D I T T V F T P S A K Y C Q V V
 91 A C N K D I Y V D L D M K G I N Y N S S V A K S A Q E C Q E R
181 A C I R D I F P N T V F A D S N I D S V M A P D A F V C G R I
272 F C H S S F Y H D T D F L G E E L D I V A A K S H E A C Q K L

C T Y H P R C L L F T F T A E S P S E D P T R W F T C V L K D S V
C T D D V H C H F F T Y A T R Q F P S L E H R - N I C L L K H T Q
C T H H P G C L F F T F F S Q E W P K E S Q R - N L C L L K T S E
C T N A V R C Q F F T Y T P A Q A S C N E G K - G K C Y L K L S S

T E T L P R V - N R T A A I S G Y S F K Q C S H Q I S 90
T G T P T R I T K L D K V V S G F S L K S C A L S N L 180
S G L P S T R I K K S K A L S G F S L Q S C R H S I P V 271
N G S P T K I L H G R G G I S G Y T L R L C K M D N E 361
```

Fig. 8. Internally homologous domains in the factor XI monomer. Only the positions at which amino acids are the same in all four domains are outlined. Pairwise comparison of the four domains reveals the following degrees of amino-acid identity: domains 1 and 2, 27%; domains 1 and 3, 32%; domains 1 and 4, 29%; domains 2 and 3, 38%; domains 2 and 4, 36%; domains 3 and 4, 25%.

2.1.4. *High-molecular-mass kininogen*

We have now reached the first protein in the coagulation system that is not the zymogen for an enzyme. HMM-kininogen functions in contact activation as a co-factor, and in addition it has other rôles in other systems (e.g. in the kinin system, and in its function as an inhibitor of cysteine proteases).

HMM-kininogen is a single-chained glycoprotein containing about 13% carbohydrate (54). Its concentration in normal human plasma has been determined to be 70–90 μg/ml (≈ 1 μM) (40, 55). The molecular mass of HMM-kininogen, as determined by SDS-polyacrylamide gel electrophoresis, lies between 110 kDa and 120 kDa (56, 57), while sedimentation-equilibrium studies suggest a value of 108 kDa (56).

The amino-acid sequence of human HMM-kininogen has been determined both by protein-sequencing (58, 59) and by cDNA-sequencing (60) (Appendix, Fig. A4). The cDNA sequence includes an 18–residue signal peptide. The gene for human HMM-kininogen has also been investigated (61); it consists of 11 exons and is spread over some 27 kbp of genomic DNA. The exons are between 78 bp and 2093 bp long and the introns between 121 bp and 6.5 kbp.

The amino-acid sequence of the protein allows the molecular mass to be calculated as 69.9 kDa; however, even after allowance for glycosylation, this is much lower than the physically determined value. The attached carbohydrate is both N-bound (3 sites: Asn151, Asn187 and Asn276) and O-bound (9 sites: Thr383, Thr515, Thr524, Thr528, Thr539, Thr553, Ser559, Thr575 and Thr610) (58, 59).

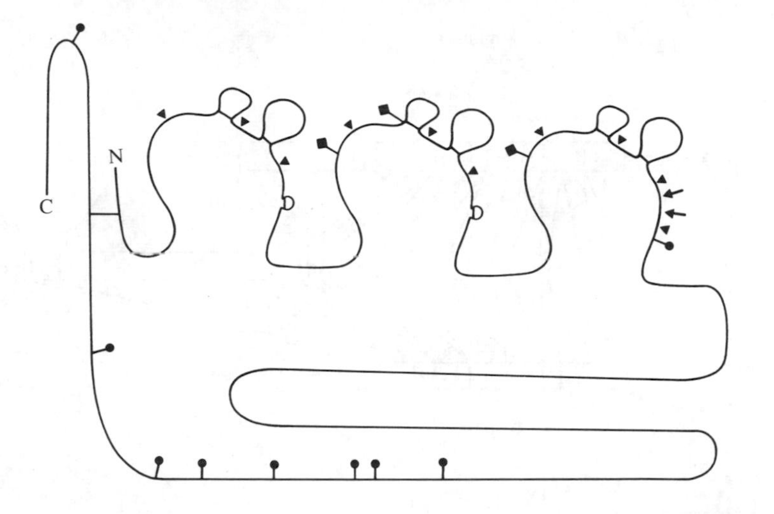

Fig. 9. The domain structure of HMM-kininogen. → shows the positions of cleavage by plasma kallikrein; cleavage at these positions results in the formation of bradykinin. ▼ shows the positions of the introns, ◆ the N-bound carbohydrate and ● the O-bound carbohydrate.

Arg-Pro-Pro-Gly-Phe-Ser-Pro-Phe-Arg

Fig. 10. The amino-acid sequence of bradykinin.

The 626 amino-acid residues of HMM-kininogen (Fig. 9) are organised into three structural elements. These are defined by the cleavage pattern of HMM-kininogen when it is digested with plasma kallikrein. Since HMM-kininogen is not an enzyme, there is no special terminology used to distinguish between the single-chain and the cloven forms.

The amino-acid sequence of HMM-kininogen contains an internal nonapeptide called bradykinin (Fig. 10). Bradykinin has proline at position 3, but this residue is sometimes hydroxylated to give 4–hydroxyproline (62–64).

Bradykinin can be released by proteolysis with kallikreins – in this case, with plasma kallikrein. Plasma kallikrein cleaves HMM-kininogen at two points, between Lys362 and Arg363 and between Arg371 and Ser372. Cleavage is followed by the release of the intermediate peptide bradykinin and the reorganisation of HMM-kininogen into a dimer with a heavy chain of 362 residues, from the N-terminal region of the precursor, and a light chain of 255 residues, from the C-terminal region. The two are held together by one disulphide bridge.

Before continuing, we should note two points. (i) The kinin system is a

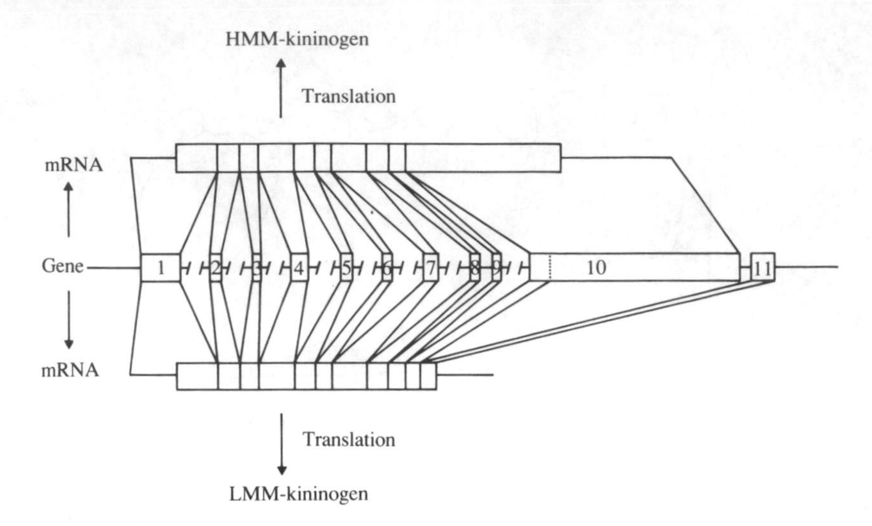

Fig. 11. The formation of HMM-kininogen and LMM-kininogen from the kininogen gene by alternative use of the exons. Exons are shown as numbered rectangles. (Modified from reference 61.)

biologically important system in its own right (see Chapter 18). (ii) In addition to HMM-kininogen, there exists a protein called low-molecular-mass kininogen (LMM-kininogen). LMM-kininogen plays no part in the blood coagulation system, but it is involved in the kinin system.

Chemically, LMM-kininogen and HMM-kininogen are identical through the entire heavy chain, the bradykinin part and the first 12 amino acids at the N terminus of the light chain. Thereafter the resemblance vanishes, and the light chain in LMM-kininogen consists of a mere 26 further residues, giving 38 residues in all. In fact, in humans there is only one kininogen gene (61), and the two kininogens are produced by different use of its exons (Fig. 11). The statement made earlier, that the gene for HMM-kininogen has 11 exons, must therefore be modified: it is the gene common to both kininogens that has 11 exons, and differential use of these gives rise to one or the other molecule, with the appropriate C terminus. HMM-kininogen is encoded by the first 10 exons, while LMM-kininogen is encoded by the first nine exons plus a part of exon 10 plus all of exon 11. Figure 9 shows schematically how the intron positions in the kininogen gene lie in relation to the amino-acid sequence of HMM-kininogen.

If we examine the heavy chain of HMM-kininogen (which is identical with the heavy chain in LMM-kininogen), we observe three internally homologous sequences (Fig. 12) in the N-terminal region (59, 61, 65). Parts of these sequences are homologous to mammalian cysteine protease inhibitors (Fig. 12), and in fact both HMM-kininogen and LMM-

(a)

```
              1 Q E S Q S E E I D C N D K D L F K A V D A A L K
117 V V T A Q Y D C L G C V H P I S T Q S P D L E P I L R H G I Q
239 V Q P P T K I C V G C P R D I P T N S P E L E E T L T H T I T

K Y N S Q N Q S N N Q F V L Y R I T E A T K T V G S D T F Y S F K
Y F N N N T Q H S S L F M L N E V K R A Q R Q V V A G L N F R I T
K L N A E N N A T F Y F K I D N V K K A R V Q V V A G K K Y F I D

Y E I K E G D C P V Q S G K - T W Q D C E Y K D A A K A A T G E C
Y S I V Q T N C S K E N F L F L T P D C K S L W N G D - - T G E C
F V A R E T T C S K E S N E E L T E S C E T K K L G Q - - S L D C

T A T V G K R S S T K F S V A T Q T C Q I T P A E G P 116
T D N A Y I D I Q L R I A S F S Q N C D I Y P G K D F 238
N A E V Y V V P W E K K I Y P T V N C Q P L G M I S L 360
```

(b)

```
Hum kin       164 V K R A Q R Q V V A G L N F R I T Y S I V Q T N C S K 190
Hum kin       286 V K K A R V Q V V A G K K Y F I D F V A R E T T C S K 312
Bov kin       163 V K R A Q R Q V V S G W N Y E V N Y S I A Q T N C S K 189
Bov kin       285 V K K A T V Q V V A G L K Y S I V F I A R E T T C S K 311
Rat kin       164 V K S A D R Q V V A G M N Y Q I I Y S I V Q T N C S K 190
Rat kin       286 V K K A T S Q V V A G T K Y V I E F I A R E T K C S K 312
Rat TI-kin    163 V K S A H S Q V V A G M N Y K I I Y S I V Q T N C S K 189
Rat TI-kin    285 V K K A T S Q V V A G V I Y V I E F I A R E T N C S K 311
Rat TII-kin   163 V K S A H S Q V V A G M N Y K I I Y S I V Q T N C S K 189
Rat TII-kin   285 V K K A T S Q V V A G T K Y V I E F I A R E T N C S K 311
Hum cys        49 V V R A R K Q I V A C V N Y F L D V E L G R T T C T K 75
Hum ste        40 A V Q Y K T Q V V A G T N Y Y I K V R A G D N K Y M H 66
Hum liv cys    40 A V S F K S Q V V A G T N Y F I K V H V G D E D F V H 66
Bov col CPI    42 V V R A R K Q V V S G M N Y F L D V E L G R T T C T K 68
Rat liv TPI    40 A I S F R R Q V V A G T N F F I K V D V G E E K C V H 66
Rat epi TPI    46 V V E Y K S Q V V A G Q I L F M K V D V G N G R F L H 72
```

Fig. 12. (a) Internally homologous domains in HMM-kininogen. Only the positions at which amino acids are the same in all three domains are outlined. Pairwise comparison of the three domains reveals the following degrees of amino-acid identity: domains 1 and 2, 20%; domains 1 and 3, 19%; domains 2 and 3, 31%. (b) Homology among cysteine protease inhibitors. The amino-acid sequences are: human kininogen (hum kin (58–60)), bovine kininogen (bov kin (74, 75)), rat kininogen (rat kin (77)), rat TI-kininogen (rat TI-kin (76)), rat TII-kininogen (rat TII-kin (76)), human cystatin (hum cys (1904)), human stefin (hum ste (1905)), human liver cystatin B (hum liv cys (1906)), bovine colostrum cysteine protease inhibitor (bov col CPI (1907)), rat liver thiol protease inhibitor (rat liv TPI (1908)), rat epidermal thiol protease inhibitor (rat epi TPI (1909)). Only those positions at which the residues are identical in at least 12 cases are outlined.

kininogen are such inhibitors (66–68). Next comes the bradykinin sequence R-P-P-G-F-S-P-F-R, and finally the light chain. Two regions in the light chain can also be distinguished: the first is unusually rich in histidine, while the second, towards the C terminus, has no particular distinguishing structural characteristics but provides the sites of attachment for eight of the nine carbohydrate moieties that are O-bound to the light chain.

The disulphide-bond pattern has not been determined experimentally for human HMM-kininogen, but it has been for human LMM-kininogen (69), which gives a basis for the assignment of disulphide bonds in HMM-kininogen. The heavy chains in LMM-kininogen and HMM-kininogen are identical, and since it is these that contain 17 of the 18 cysteine residues, the inference from LMM-kininogen to HMM-kininogen can be regarded as reliable. It emerges that the cysteine residue closest to the N terminus, Cys10, is part of the disulphide bridge that holds the two kininogen chains together after cleavage by kallikrein. In LMM-kininogen, this bridge is Cys10–Cys389, while in HMM-kininogen it is Cys10–Cys596. The 16 other cysteines are linked in pairs, counting from the N terminus (69), giving the bridges Cys65–Cys76, Cys90–Cys109, Cys124–Cys127, Cys188–Cys200, Cys211–Cys230, Cys246–Cys249, Cys310–Cys322 and Cys333–Cys352. The structure of the HMM-kininogen heavy chain thus includes eight covalent loops.

For the sake of completeness, it should be mentioned that complete amino-acid sequences are also known for kininogen from the cow and the rat. The bovine kininogen sequences have been determined by analysis both of protein (70–73) and of cDNA (74, 75), and it has emerged that there are two forms each of bovine HMM-kininogen and LMM-kininogen, unsurprisingly called I and II. The differences between HMM-kininogens I and II are precisely the same as those between LMM-kininogens I and II, in accordance with the fact that the I forms are transcribed from one gene and the II forms from another. It is not yet clear whether these two forms are alleles, implying the existence of only one kininogen gene, or whether there are two highly similar but independent kininogen genes. In humans, present results point to a single gene, but in this case there is no variation in the kininogen proteins (61).

Kininogen sequences from rats have been deduced from cDNA sequences and from genomic sequencing (76, 77). In addition to LMM-kininogen and HMM-kininogen, rats have an additional type of kininogen molecule called T-kininogen; there are in fact two T-kininogens, each with its own gene. These two T-kininogen genes are regulated quite independently of the gene for the other two kininogens (78). It has furthermore been shown that T-kininogen is identical with the 'major acute phase α_1–protein' from rats (79, 80). Thus there are three kininogen genes in rats, one of which corresponds to the human kininogen gene while the other two encode their respective T-kininogens. Interestingly, the placing of the intron–exon junctions in one of the rat T-kininogen genes is identical to the placing of these junctions in the human kininogen gene (81).

2.2. Mechanism of contact activation

We have now presented the *dramatis personae* in the process of contact activation; the stage upon which they appear is a negatively charged surface, and this is where the action takes place during almost the entire coagulation process. At this point a general remark should be made, and it will be made repeatedly during this discussion: blood coagulation is a surface phenomenon. The vast majority of its constituent steps take place on surfaces rather than in the liquid phase. This may seem strange, because almost all the participant proteins are plasma proteins, but, as we shall see, it is in fact perfectly logical.

The site of the reactions is a negatively charged surface. Let us forget for a moment that coagulation of blood takes place in the organism and just consider activating surfaces *in vitro*. For example, one can use glass (82–84), kaolin (85), ellagic acid (86), sulphatides (87), dextran sulphate (88) or negatively charged phospholipids such as phosphatidyl serine or phosphatidyl inositol phosphate (89). We shall return later to surfaces *in vivo*, although it will already be clear that the allusion to phospholipid surfaces *in vitro* is not without relevance to these.

Since the exact mechanism of contact activation is not known, we start by stating its result: the formation of factor XIa. How does this take place?

Before going into this, let us summarise the system with which we are dealing. We have: (i) a negatively charged surface, (ii) factor XII, (iii) plasma prekallikrein, (iv) factor XI and (v) HMM-kininogen. We still need to know how these interact.

2.2.1. *The interaction of HMM-kininogen with plasma prekallikrein and with factor XI*

Both plasma prekallikrein and factor XI form stoichiometric complexes with HMM-kininogen in plasma (90, 91). Plasma prekallikrein and HMM-kininogen form a 1:1 complex, while factor XI and HMM-kininogen form a complex in the molar ratio 1:2 (92).

The binding of plasma prekallikrein and factor XI takes place via the light chain in HMM-kininogen (47, 93, 94). The isolated light chain of HMM-kininogen binds to plasma prekallikrein with the same 1:1 stoichiometry (94–96) and the same affinity (97) as does intact HMM-kininogen.

The binding site for plasma prekallikrein on the the light chain of HMM-kininogen can be localised to a 40–residue segment close to the C terminus (amino acids 185–224 in the light chain, corresponding to 556–595 in intact

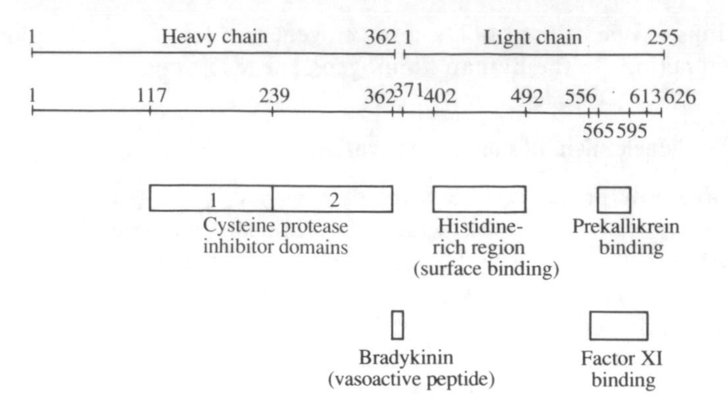

Fig. 13. A schematic representation of HMM-kininogen as a multifunctional protein.

HMM-kininogen) (98). The region of greatest importance for binding has since been narrowed down to 31 residues (amino acids 194–224 in the light chain of HMM-kininogen, corresponding to 565–595 in the intact molecule) (99).

The binding site for factor XI in HMM-kininogen has also been investigated (99), and it was found that the minimal requirement for a binding of factor XI to HMM-kininogen that resembled the binding of factor XI to the light chain (99) or to intact HMM-kininogen (93) was a 58-residue stretch in HMM-kininogen's light chain. This stretch (residues 556–613) overlaps with the peptide needed for the binding of plasma prekallikrein to HMM-kininogen, as noted above. The overlap of the binding sites is in accordance with the observation (93) that plasma prekallikrein and factor XI compete with one another in binding to HMM-kininogen.

The coagulation activity of HMM-kininogen resides in its light chain (100). Apart from the binding sites for plasma prekallikrein and factor XI, the light chain also contains the region responsible for the binding of these complexes to anionic (negatively charged) activating surfaces. This is the histidine-rich region mentioned above (101).

We are now in a position to make a sketch of the functional domains of HMM-kininogen, as shown in Figure 13.

To complete the picture of the binding between HMM-kininogen and the two ligands, it should be mentioned that both plasma prekallikrein (102) and factor XI (103) attach themselves to HMM-kininogen by way of their four-domain N-terminal structure. It is thus clear why factor XI binds two molecules of HMM-kininogen, as factor XI contains two identical polypeptide chains. The constants of these binding equilibria have been measured under various conditions (92, 93, 97–99). A dissociation constant of approximately 15 nM for the interaction of both plasma prekallikrein and plasma kallikrein with HMM-kininogen has been measured (97), while the corresponding constant for the interaction between HMM-kininogen and factors XI and XIa is approximately 32 nM (92); these values are confirmed by the general observation that it requires a stronger eluent to remove plasma prekallikrein from a column of immobilised HMM-kininogen than it does to remove factor XI. If these dissociation constants are compared with the concentration of these proteins in plasma, it becomes clear that most of the plasma prekallikrein and factor XI circulate bound to HMM-kininogen.

2.2.2. *Contact activation*

We have now come far enough to begin to see the actual molecular processes of contact activation. The starting-point for this is the negatively charged surface, to which the participating proteins can bind as soon as they encounter it.

Factor XII binds directly to the surface by its N-terminal region (104). A closer localisation of the binding site has been carried out by studies employing a monoclonal antibody that blocked the surface binding (105, 106).

The binding site was first localised to the amino-acid residues between positions 134 and 153. This assignment was based on the identification of the plasma kallikrein peptides of factor XII that were recognised by the immobilised antibody (105). However, studies employing factor XII peptides made by recombinant methods have since shown that the surface-binding region is more likely to be between residues 1 and 28, since this region binds the same monoclonal antibody much more strongly (106). The reason that this was not found in the first experiments may be that plasma kallikrein cleaves factor XII in the middle of the surface-binding region found later, thereby destroying this epitope.

As described above, plasma prekallikrein and factor XI bind to the surface via the light chain in HMM-kininogen, which contains binding sites for these two zymogens and also contains the histidine-rich region that mediates binding to the negatively charged surface.

All the necessary factors are thus bound in each others' vicinity – a plausible prerequisite for further reaction.

The principle of activation is simple, even if details are still unclear. Factor XII is activated to give factor XIIa, which activates plasma prekallikrein to give plasma kallikrein, and this in turn activates more factor XII to give yet more factor XIIa. The factor XIIa also activates factor XI to give factor XIa, and it is the factor XIa that brings about the next step of coagulation by the intrinsic pathway.

Thus the whole process is triggered off by the appearance of the first factor XIIa. Where does this come from?

One attractive hypothesis suggests that the binding of the inactive zymogen factor XII to the negatively charged membrane surface causes a conformational change that gives factor XII a certain activity with respect to the substrates plasma prekallikrein and factor XI, which are also present on the surface. This is supported by the observation that factor XII and factor XIIa are equally effective in activating plasma prekallikrein *in vitro*, while factor XIIa is two to four times faster than factor XII in activating factor XI (107). The activation of factor XI by factor XII is thus greatly accelerated when plasma prekallikrein and kallikrein are present, converting the factor XII into the more active factor XIIa. However, these experiments employed concentrations of plasma prekallikrein and factor XI that were 60 times higher than in normal plasma.

A second hypothesis is the 'autoactivation' hypothesis (108), according to which the initial activation is the cleavage of a small quantity of factor XII by a trace amount of factor XIIa of unknown origin. This leads to the cleavage of plasma prekallikrein to give plasma kallikrein, the activation of more factor XII, and so forth. In this connection it is noteworthy that plasma kallikrein can also activate itself *in vitro* under certain conditions (109).

An important fact in connection with contact activation is that the binding of factor XII to a negatively charged surface increases dramatically the rate of activation, presumably owing to conformational effects at the surface, possibly including the exposure of the bond to be cleaved. It has been reported that surface-bound factor XII is 500 times more sensitive to plasma kallikrein than factor XII free in solution (110), and in the author's opinion this is a key observation.

The suggestion that a conformational change of factor XII may play an important part in the mechanism of contact activation has also received support from studies with a monoclonal antibody that binds much more strongly to factor XIIa than to factor XII. This antibody can initiate contact activation in plasma, and this may be because it induces the activatable conformation of factor XII by binding to it (111).

Contact activation leads within seconds to a powerful activation of factor XII to factor XIIa and of plasma prekallikrein to give plasma kallikrein. The rapid activation of factor XII seen in normal plasma fails to occur in plasma that lacks plasma prekallikrein or HMM-kininogen (112). As mentioned earlier, these plasma types have abnormally long coagulation times *in vitro*. Factor XII becomes bound to the activating surface with the same speed in each case, which indicates that it is not enough just to bind factor XII to a negatively charged surface in order to attain a rate of activation on a par with that *in vivo*. Furthermore, surface-bound factor XII has no significant enzymic activity with respect to small peptide substrates (113).

Di-*iso*propyl fluorophosphate (DFP) inhibits serine proteases irreversibly by reaction with their active site. It also reacts slowly with the zymogen plasma prekallikrein, so that when this is cleaved by factor XIIa the product is an inhibited plasma kallikrein that can no longer convert factor XII to

XIIa. Surface-bound factor XII is not in a position to cleave DFP-treated plasma prekallikrein (114). On the other hand, factor XIIa is perfectly able to cleave DFP-treated plasma prekallikrein.

We can conclude that factor XIIa is very much more active than factor XII, and the necessary conditions for rapid contact activation thus seem to be this fact plus the heightened sensitivity of factor XII towards activation when it is bound to a surface.

As soon as factor XIIa is formed, it can activate plasma prekallikrein to give plasma kallikrein and thus reinforce the incipient activation of factor XII. After a short time, the quantity of factor XIIa formed has become sufficient to ensure the rapid activation of factor XI to XIa, which leads to the next step in the process of coagulation.

We have still not explained what it is that initiates the entire reaction chain. There are two current hypotheses about how this may happen.

The first proposes that there are normally trace quantities of factor XIIa circulating in plasma. On attachment to a negatively charged surface, these become better able to activate surface-bound plasma prekallikrein than they are in the liquid phase. This may be a result of some conformational change of α-factor XIIa at the surface, favouring the splitting of prekallikrein, or it could be due simply to the higher concentrations of factor XIIa and its substrate at the charged surface, which cause a higher frequency of encounters between enzyme and substrate. A third possibility could also be that the binding of factor XIIa to the surface protects it from its physiological inhibitor, $C\bar{1}$-inhibitor, a molecule that will be described in sections 8.1.2 and 15.1.1.

The second hypothesis proposes that surface-induced effects facilitate the activation of factor XII to give factor XIIa. Three possibilities immediately come to mind: (i) an unidentified proteinase, (ii) traces of plasma kallikrein and (iii) traces of factor XIIa (autoactivation). The choice between these possibilities is still a matter of personal taste.

Once plasma kallikrein is formed, it can, in addition to cleaving factor XII, also cleave HMM-kininogen, releasing bradykinin. This makes the HMM-kininogen two-chained, and it has been shown that this cloven form has a greater affinity for negatively charged surfaces than has single-chained HMM-kininogen (115). This suggests that single-chained HMM-kininogen is a pro-cofactor, requiring proteolytic cleavage in order to attain maximal activity in coagulation.

The processes of contact activation are hard to treat kinetically, as they influence each other to a large degree, but some effort has been made to derive kinetic parameters (109, 116, 117). Most studies have been purely qualitative. Both qualitatively and quantitatively, it can be shown that negatively charged surfaces in conjunction with HMM-kininogen accelerate dramatically the mutual activation of factor XII and plasma prekallikrein and the activation of factor XI by factor XIIa.

At this point it is appropriate to warn the reader of the dangers of

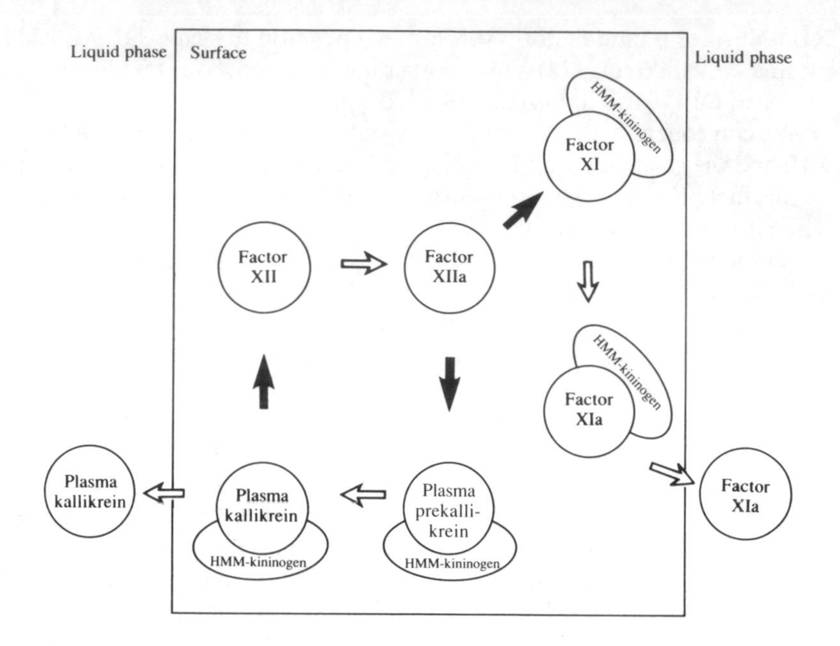

Fig. 14. A schematic representation of the contact activation of blood coagulation. →, enzymic activity; ⇒, consequence of enzymic activity.

uncritical interpretation. Studies of contact activation have employed many kinds of non-physiological conditions, especially with regard to the artificial contact surface, and, while some workers have used plasma systems, others have used purified coagulation factors.

To summarise this section (Fig. 14): in the presence of a negatively charged surface and HMM-kininogen, factor XII and plasma prekallikrein activate each other. This leads to the activation of factor XI to give factor XIa, a process in which HMM-kininogen and the charged surface also play a part. The rôle of HMM-kininogen as a co-factor in mediating directly the binding of plasma prekallikrein and factor XI to the activating surface appears to be enhanced by the proteolytic excision of the bradykinin peptide from HMM-kininogen by plasma kallikrein.

It should finally be mentioned that we in fact do not know how contact activation is initiated. Most models suggested so far presuppose the presence of one or another active enzyme in trace quantities. Furthermore, the significance of contact activation for coagulation *in vivo* is also unknown, since individuals deficient in factor XII, plasma prekallikrein or HMM-kininogen show no tendency towards haemophilia. It therefore seems probable that the true physiological rôle of contact activation is in connection with some other process, such as inflammation.

3. Activation of factor IX

The next step in the intrinsic pathway – if this complex process can be thought of as proceeding in 'steps' – is the activation of factor IX to give factor IXa. This activation is catalysed by the factor XIa that we have seen emerging as the end-product of contact activation. The reaction is usually regarded as a liquid-phase phenomenon; however, since factor XIa is largely bound to the activating negatively charged surface, a considerable part (if not all) of the activation of factor IX to IXa will occur at the same surface. Before discussing the mechanism of activation, let us briefly examine the properties of factor IX. Deficiency of factor IX is termed haemophilia B or Christmas disease (the allusion being personal rather than seasonal) (118, 119). Haemophilia B has the same symptoms as classical haemophilia (haemophilia A).

3.1. Factor IX

Factor IX, reviewed in reference (120), circulates in the bloodstream as a single-chained plasma glycoprotein. It is the zymogen for the active serine proteinase factor IXa, which participates in the intermediate phase of blood coagulation's intrinsic pathway.

The concentration of factor IX in normal human plasma lies between 3 μg/ml and 5 μg/ml (≈80 nM) (121, 122). Its molecular mass is estimated by SDS-polyacrylamide gel electrophoresis to be 55–57 kDa (122, 123), with a carbohydrate content of about 17% (123). Factor IX is the first of several vitamin K-dependent proteins that we shall meet in the coagulation system.

Protein-sequencing methods have been used to determine the amino-acid sequence of bovine factor IX (124) and part of the sequence of the human factor (123); the latter was completed by cDNA-sequencing (125–128) (Appendix, Fig. A5). The cDNA sequence includes a signal/propeptide of 39, 41 or 46 residues (the term 'signal/propeptide' is explained and discussed in Chapter 20). The uncertainty arises because three Met codons are found in close proximity. The complete sequence for canine

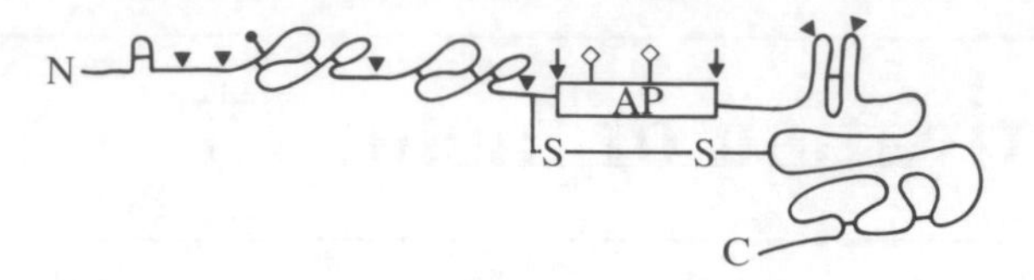

Fig. 15. The domain structure of factor IX. → shows the positions of cleavage by factor XIa, ▼ the positions of the introns, ◇ the potential attachment sites for N-bound carbohydrate and ● the O-bound carbohydrate (see text). (AP, activation peptide.)

factor IX has been determined by cDNA-sequencing (129); in this case, the signal peptide, highly homologous to the human one, has unambiguously 39 residues.

The gene structure of human factor IX is known, not only from the determination of intron/exon boundaries and transcription signals (127) but also from the sequencing of the entire gene (130). The factor IX gene is approximately 34 kbp long and has 8 exons. The exons vary in size from 25 bp to 1935 bp and the introns vary from 188 bp to 9473 bp, comprising more than 90% of the length of the gene (127, 130). The intron positions in the factor IX gene are set in relation to the amino-acid sequence in Figure 15. The gene is located in the long arm of the X chromosome, in the region designated Xq27.1 (131–136).

The molecular mass of factor IX, deduced from its amino-acid sequence, is 47.0 kDa, giving just under 57 kDa if the 17% carbohydrate is included, in agreement with the physically determined value. The carbohydrate is N-bound and is presumed to be attached to Asn157 and Asn167, the only Asn residues found in the consensus sequence for N-glycosylation (N-X-S/T, see Chapter 19). The glycosylation pattern has not yet been verified experimentally, but strong circumstantial evidence is provided by the fact that the activation peptide, which contains both the Asn residues mentioned, is strongly glycosylated. Bovine factor IX contains an unusual carbohydrate structure, attached at Ser53 and consisting of serine-bound glucose to which two xylose molecules are bound (137). Recently, it has been shown that human factor IX bears at the same position (Ser53) a glucose with one xylose molecule attached (138).

The structure of factor IX, which consists of 415 amino-acid residues (Fig. 15), displays a number of characteristics common to several other components of the coagulation system. Starting from the N terminus, we encounter first the 'Gla domain'. Bovine factor IX contains 12 γ-carboxyglutamic acid (Gla) residues among its first 40 amino acids (124); these are glutamic acid residues that have been carboxylated after translation. The number in human factor IX is not known accurately, but it is assumed to be similar. This post-translational carboxylation of glutamic acid residues is dependent upon vitamin K, and is described in more detail in Chapter 21. After the Gla domain, there follow two growth-factor domains possessing homology with EGF and then the peptide released upon activation (activation peptide). At the C terminus we find the serine-proteinase domain.

Although the pattern of disulphide bonds has not been demonstrated experimentally, these are assumed to be placed in the same way as in homologous domains. So the following disulphides are expected to exist: in the Gla domain, Cys18–Cys23; in the first growth-factor domain, Cys51–Cys62, Cys56–Cys71 and Cys73–Cys82; in the second growth-factor domain, Cys88–Cys99, Cys95–Cys109 and Cys111–Cys124; and in the serine-proteinase domain, Cys132–Cys289, Cys206–Cys222, Cys336–Cys350 and Cys361–Cys389.

Residue 64 in the first growth-factor domain is modified in a unusual way: it is a β-hydroxy aspartic acid residue (β-OH-Asp) (139). The same modification is found at precisely the same place in bovine factor IX (140). The β-hydroxylation of Asp64 in human factor IX is only partial (approximately 30%), while the corresponding residue in the bovine factor is nearly all converted to the β-hydroxy derivative (140, 141).

The activation of factor IX to give factor IXa is catalysed by factor XIa and takes place in two steps (123, 142). First, the Arg145–Ala146 bond is cleaved, which converts factor IX to a two-chained activation intermediate without catalytic activity of its own, sometimes referred to as factor IXα. Thereafter, the bond between Arg180 and Val181 is cleaved, with the formation of factor IXa and the activation peptide. The latter is an 11-kDa peptide with 35 residues. It contains a high proportion of carbohydrate and both the presumed glycosylation sites of factor IX. In addition, position 3 of the activation peptide, corresponding to position 148 of factor IX, is dimorphic, occupied by either alanine or threonine (128).

Factor IXa is thus a two-chained, non-glycosylated serine proteinase with a molecular mass of about 43 kDa; its light chain (about 18 kDa by electrophoresis) is derived from the N-terminal region of the precursor and its catalytically active heavy chain (about 28 kDa) from the C terminus. It is believed that the two chains are held together by a disulphide bond between Cys132 and Cys289 of the zymogen. The catalytic groups in factor IXa are believed to be His41, Asp89 and Ser185 of the heavy chain, corresponding to His221, Asp269 and Ser365 in factor IX.

3.2. Haemophilia B

As already stated, a deficiency of functionally active factor IX gives rise to haemophilia B (for further details see, for example, the reviews in references 143–145). Many of these types of haemophilia have been characterised at a molecular level (Table 3).

Table 3. *Characterised defects in factor IX*
A. Point mutations

Location	Mutation	Consequence	Haemophilia	References
5' not transcribed	T→A	regulation	severe→mild	(146)
Exon 1	A→G	regulation	severe→mild	(147, 148)
Exon 2	G→A	Arg(–4)→Gln	severe	(149–153)
Exon 2	C→T	Arg(–4)→Trp	severe	(149)
Exon 2	(G→T/C)	Arg(–1)→Ser	severe	(154)
Exon 2	A→C	Gla7→Asp	severe	(155)
Exon 2	C→T	Gln11→stop	severe	(155)
Exon 2	G→A	Gla27→Lys	severe	(156)
Exon 2	G→A	Arg29→Gln	mild	(148)
Exon 2	C→T	Arg29→stop	severe	(149)
Exon 2	A→C	Gla33→Asp	moderate	(148)
Exon 4	A→G	Asp47→Gly	moderate	(157)
Exon 4	A→C	Gln50→Pro	severe	(158)
Exon 4	C→G	Pro55→Ala	mild	(149, 159)
Exon 4	G→A	Gly60→Ser	mild	(148, 156, 160)
Exon 4	A→G	Asp64→Gly	moderate	(149)
Exon 5	(G→C)	Gly114→Ala	severe	(155)
Exon 5	A→T	Asn120→Tyr	severe	(149)
Exon 6	G→A	Arg145→His	mild	(148, 161–163)
Exon 6	(C→T)	Arg145→Cys	moderate	(164, 165)
Exon 6	(C→T)	Gln173→stop	severe	(148)
Exon 6	G→A	Arg180→Gln	severe	(166)
Exon 6	(C→T)	Arg180→Trp	severe	(167, 168)
Exon 6	unknown	Arg180→Glu	severe	(167)
Exon 6	(G→T)	Val181→Phe	severe	(167)
Exon 6	(G→T)	Val182→Phe	severe	(169)
Exon 6	G→A	Trp194→stop	severe	(149)
Exon 7	T→G	Cys222→Trp	severe	(148)
Exon 7	G→A	Ala233→Thr	mild	(148)
Exon 8	G→A	Arg248→Gln	moderate	(156)
Exon 8	C→T	Arg248→stop	severe	(149)
Exon 8	C→T	Arg252→stop	severe	(170, 171)
Exon 8	A→G	Asn260→Ser	mild	(148)
Exon 8	(G→C)	Ala291→Pro	severe	(155)
Exon 8	C→T	Thr296→Met	severe	(148)
Exon 8	G→A	Gly311→Arg	moderate	(148)
Exon 8	G→A	Arg333→Gln	severe	(149, 172)
Exon 8	T→C	Cys336→Arg	severe	(149)
Exon 8	C→T	Arg338→stop	severe	(173, 174)
Exon 8	(G→T)	Gly363→Val	unknown	(159)
Exon 8	C→A	Pro368→Thr	unknown	(167)

Table 3. (*cont.*)

Location	Mutation	Consequence	Haemophilia	References
Exon 8	C→T	Arg390→Val	severe	(175)
Exon 8	(C→T)	Arg390→Val	moderate	(176)
Exon 8	G→A	Gly396→Arg	severe	(177)
Exon 8	T→C	Ile397→Thr	moderate	(148, 177, 178)
Exon 8	T→C	Ile397→Thr	severe	(148, 179)
Exon 8	T→C	Trp407→Arg	severe	(148)
Exon 8	A→T	Lys411→stop	severe	(177)
Intron 5	G→T	splicing site loses function	severe	(180)

In cases where a mutation is only described at the level of protein, the possible nucleotide substitution is given in parentheses.

B. Deletions

Exons deleted	Size of deletion	Inhibiting antibody	Haemophilia	Reference
Part of 1	1 bp	no	severe→mild	(147)
1–7 and part of 8	>41 kbp	yes	severe	(181)
5–6	10 kbp	no	severe	(182)
4	2.8 kbp	no	severe	(183)
all	>114 kbp	yes	severe	(184)
5 and 7–8	5 kbp + 9–29 kbp	yes	severe	(184)
all	>34 kbp	yes	severe	(185)
1–3	9 kbp	yes	severe	(173)
1	11–35 kbp	unknown	severe	(173)
7	1.5 kbp	no	severe	(173)
4–5	8 kbp	no	severe	(173)
all	>60 kbp	no	severe	(173)
all	>34 kbp	yes	severe	(186)
part of 5	1 bp	no	severe	(187)
all	>34 kbp	yes	severe	(188)
6–8	23 kbp	yes	severe	(189)
all	>36 kbp	no	severe	(190)
part of 2	3 bp	no	severe	(149)
part of 2	1 bp	yes	severe	(149)
part of 8	8 bp	yes	severe	(149)
part of 8	2 bp	no	severe	(149)
part of 6	13 bp	unknown	severe	(148)
part of intron 4	4 bp	unknown	mild	(148)

Haemophilia B patients can be divided into three groups according to how their defect in coagulation activity is related to their deficiency in factor IX. A large number of haemophilia B patients in fact have normal levels of factor IX antigen, and these are designated as CRM^+ (cross-reacting material positive). The second group has a reduced level (10–50% of normal) of factor IX antigen and is called CRM^r (cross-reacting material reduced). Patients in the third group either show strongly depressed levels of factor IX antigen or else lack it altogether, so they are termed CRM^- (cross-reacting material negative). CRM^- patients are more likely to develop antibodies against the factor IX that is administered to them in their treatment than CRM^+ patients are. It is generally assumed that CRM^+ patients bear mutations that interfere with normal factor IX function, while CRM^- patients have defects in which greater or lesser portions of the factor IX gene are deleted or affected in a similar drastic manner. Patients with less than 1–2% of normal factor IXa activity are severely affected, but patients with more than 5% of normal factor IXa activity are only mildly affected. Between these two extremes the affliction with haemophilia B is said to be moderate.

Characterised defects associated with haemophilia B with a normal or a reduced level of factor IX antigen in the bloodstream (CRM^+ and CRM^r) include amino-acid substitutions in the signal/propeptide and substitutions on the factor IX polypeptide chain.

Observed mutations that affect the signal/propeptide include a replacement of Arg(–1) by serine (154) and in five independent cases a replacement of Arg(–4) by glutamine (149–153). There is also a documented replacement of Arg(–4) by tryptophan (149). All of these substitutions cause the factor IX to circulate with 18 'extra' amino acids at its N terminus, owing to incomplete processing. As will be described in more detail in Chapter 20, signal/propeptides are made up of a signal peptide important in secretion and a propeptide part whose rôle is to indicate to the appropriate enzymes where the glutamic acid residues for γ-carboxylation are sited, i.e., the Gla domain. In normal synthesis, the entire signal/propeptide is removed in two steps, but in the two mutants described the propeptide part is not removed, and this is enough to give defective factor IX molecules. The defects result in a lack of the ability to bind surfaces, either because the molecules are not correctly γ-carboxylated or because the propeptide interferes with the binding.

Known amino-acid substitutions in factor IX polypeptide chains are shown in Figure 16 and listed in Table 3. It should be remembered that the defective molecules have not been investigated exhaustively for mutations in all cases. They could thus in principle contain more than one point mutation, but this is usually regarded as improbable.

In many of the characterised haemophilia B defects in which the level of factor IX antigen is greatly reduced (CRM^-), there are greater or lesser

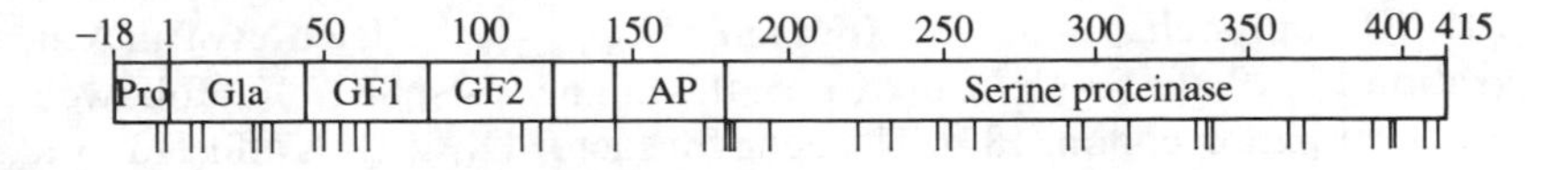

Fig. 16. Localisation of haemophilia B variants in relation to the amino-acid sequence of pro-factor IX. The exact positions of the amino-acid substitutions are given in Table 3. Positions at which point mutations have been detected are indicated by vertical bars.

deletions in the factor IX gene and also point mutations that introduce stop codons (see Table 3), but there are also further examples. A point mutation has been characterised in the GT donor splicing site at the 3′ side of the sixth exon in the factor IX gene, where G at position 21,165 is replaced by T (180), and an insertion of 6 kbp in the fourth intron is thought to be responsible for the moderate haemophilia $B_{\text{El Salvador}}$ (191). There is also an unusual haemophilia B_{Leyden}, and other cases of haemophilia B with the same phenotype, in which the level of factor IX is low from birth until puberty, whereafter both the level and the activity of factor IX rise concomitantly until at most 50–60% of the normal levels are reached (192, 193). The explanation for this could lie in the fact that the male sex hormone testosterone plays a regulatory rôle, an explanation supported by the observations that female carriers of the gene for haemophilia B_{Leyden} have constant levels of factor IX, and that treatment with anabolic steroids raised the factor IX level in an 8-year-old patient (194, 195). Three mutations have been characterised in connection with the haemophilia B_{Leyden} phenotype. One is a point mutation T→A at position (−20) relative to the start of transcription (147); the second is a point mutation A→G at position 13 (exon 1, upstream of start codon); and the third is a deletion of the same A in position 13 of exon 1 (147).

The deletion mutants referred to above range from a single nucleotide deletion to the deletion of the entire factor IX gene. Some of the partial deletions have been mapped accurately. In the gene for factor IX_{Seattle}, about 10 kbp are deleted, including exons 5 and 6, so that factor IX_{Seattle} is expressed lacking amino-acid residues 85 to 195. An abnormal 36-kDa molecule that cross-reacts with antibodies against factor IX has been found in affected patients (196). The precise structure of this 36-kDa protein is not known. Deletion of exons 5 and 6 is expected to give rise to a shift in reading frame and a termination codon only six codons after the deletion, since exons 4 and 7 are not in phase with one another. The gene for factor $IX_{\text{Strasbourg}}$ has a deletion that removes exon 4 and therewith residues 47–84, corresponding to the first growth-factor-like domain in factor IX (183). In this case, deletion will not give a shift in reading frame, and one can therefore picture a factor $IX_{\text{Strasbourg}}$ molecule as lacking one growth-factor domain only (a defect still sufficient to cause haemophilia). In the gene for factor IX_{Chicago} there are two deletions, removing exon 5 and exons 7 and 8

(184). The only defect in the gene for factor $IX_{Seattle2}$ is the deletion of an A at position 17699, the first base in exon 5: this changes Asp85 to Val followed by a termination codon (187). The gene for factor $IX_{London1}$ is affected by a deletion of 23 kbp, starting 704 bp upstream from the fifth exon, which means that the rest of the factor IX gene is absent (189). The genes for factor $IX_{London11}$, factor $IX_{Malmö1}$ and factor $IX_{London12}$ all have small deletions (2, 8 and 16 bp respectively). These lead to reading-frame shifts and the introduction of stop codons (149). This is also the case for a 13–bp deletion described in reference (148).

There is no doubt that more and more variants of factor IX will be characterised in time to come, and especially the point mutations will contribute to a better understanding of the biology of factor IX.

Finally, there is a characterised mutation that causes severe haemophilia B in dogs. This defect is the mutation of G→A in the codon for position 379, changing Gly379 to Glu (197).

3.3. Activation of factor IX by factor XIa

The activation of factor IX to give factor IXa requires the enzyme factor XIa and calcium ions as the only co-factors. Their concentration in plasma lies between 2.2 mM and 2.5 mM. There is no requirement for other proteins or negatively charged surfaces such as anionic phospholipids (103, 198, 199). We shall examine the function of calcium in some detail.

In the presence of EDTA, factor XIa and the isolated light chain of factor XIa activate factor IX at equal rates. However, in the presence of calcium ions, factor XIa is 2000 times more active than the light chain. At the same time, factor XIa and the isolated light chain of factor XIa have the same Ca^{2+}-independent activity in splitting an artificial oligopeptide substrate. It follows from this that the heavy chain of factor XIa plays an important part in determining the calcium-dependence of the activation of factor IX (103). The function of the heavy chain in factor XIa thus appears to be the mediation of a Ca^{2+}-dependent interaction between factor XIa and factor IX. This is supported by the observation that monoclonal antibodies against the heavy chain in factor XIa inhibit the factor XIa-catalysed activation of factor IX (200, 201). The contribution of the heavy chain of factor XIa thus probably lies in making the factor XIa specific for the substrate factor IX in the presence of calcium. There are therefore two functional domains in the heavy chain of factor XIa, since this chain mediates the binding of factor XI/XIa to HMM-kininogen and that of factor XIa to factor IX. With the help of monoclonal antibodies, it has been shown that these two binding sites do not overlap (202). It is believed that the effect of calcium upon the binding of factor IX to factor XIa is due to the binding of ions to factor IX and not to factor XIa, and it probably involves the phospholipid binding site of factor IX (203).

The binding of calcium ions to factor IX/IXa appeared at one time to differ between the bovine and human factors. Human factor IX was reported to have 16 Ca^{2+}-binding sites with a K_d value of approximately 0.6 mM (198, 204), while bovine factor IX possessed two high-affinity Ca^{2+}-binding sites with K_d approximately 0.1 mM and 10–11 weaker sites with K_d approximately 1.3 mM (205). The two high-affinity Ca^{2+}-binding sites in bovine factor IX were later found to be Gla-independent; this was shown by prior enzymic removal of the Gla domain (206, 207). With this factor IX lacking the Gla domain, the value of K_d was found to be approximately 85 μM. Similar measurements were carried out for human factor IX from which the Gla domain had been removed, and it was found that the human factor IX also has two high-affinity sites for Ca^{2+} with K_d approximately 52 μM (208).

The high-affinity Ca^{2+}-binding sites are important for the conformation of factor IX/IXa, and possibly also for the activatability of factor IX and/or for the activity of factor IXa. As mentioned above, the high-affinity Ca^{2+} sites are independent of the Gla domain, and the presence of a β-OH-Asp residue at position 64 (in the first growth-factor domain) has given rise to the idea that this modified amino-acid side-chain is related to the binding of calcium ions. Support for this idea came from the fact that other vitamin K-dependent coagulation factors show a somewhat similar correlation between the presence of Gla-independent Ca^{2+}-binding sites and the presence of β-OH-Asp in growth-factor domains (see also section 21.2). However, there is also evidence against this. For factor IX, it should be remembered that both the human and the bovine molecules have two high-affinity Ca^{2+}-binding sites, even though only about one-third of the human factor molecules contain the modified residue (141, 208). Furthermore, it seems unreasonable that a single β-OH-Asp residue should be responsible for two high-affinity sites. An isolated growth-factor domain produced by peptide synthesis, identical (except for the β-OH group) to the first growth-factor domain in human factor IX, has been shown to bind calcium ions (209).

Mutagenesis studies on human factor IX (210) have also provided information on the first growth-factor domain, in that the replacement of amino-acid residues containing carboxylic acid groups (Asp47, Asp49, Asp64 and Glu78) by non-carboxylic residues invariably reduces dramatically the coagulation activity. It was further shown that the activity affected was not the activation of factor IX by factor XIa, but the activation of factor X by factor IXa.

Following the discovery that β-hydroxylation of Asp side-chains is blocked by inhibitors of 2–oxoglutarate-dependent dioxygenases (211–213), it has been possible to produce recombinant factor IX that contains a non-hydroxylated Asp residue in position 64. This non-hydroxylated factor IX behaves like authentic factor IX in a one-stage clotting assay (212), so it is not yet clear what rôle β-OH-Asp plays in the coagulation process.

For the sake of completeness, we mention an observation that factor IX binds Fe^{3+} *in vitro* (214); however, this binding has no known function and the rôle of β-OH-Asp is still hypothetical.

The low-affinity binding sites for Ca^{2+} involve the Gla domain (residues 1–40). It is the Gla residues that, in conjunction with the calcium ions, are responsible for the binding of factor IX/factor IXa to negatively charged phospholipid surfaces. Furthermore, the Gla residues are essential for the activity of factor IXa in coagulation. This will be discussed in section 4.2.

The conclusions of this section are thus: the activation of factor IX to give factor IXa is catalysed by factor XIa and requires calcium ions, but it does not require a negatively charged surface such as an anionic phospholipid (the important rôle of which will be seen in the next few sections) or a protein co-factor.

The influence of phospholipids upon the activation of factor IX has been debated but not settled, because results from different laboratorics are in conflict. Some report a stimulation by phospholipids (215, 216), others no effect (103) and yet others an inhibition (217, 218). These differences are presumably due to differences in experimental technique or conditions such as phospholipid concentration.

It should be added that although there is no kinetic requirement for a surface in the activation of factor IX, it is highly probable that the activation *in vivo* does actually occur at a surface, since the catalyst for the reaction, factor XIa bound to its co-factor HMM-kininogen, is located at a surface. It is always important to bear in mind that even though an enzymic process may occur optimally under certain specifiable minimal conditions, the situation *in vitro* is not necessarily a true replica of that *in vivo*.

4.

Activation of factor X in the intrinsic pathway

4.1. Components of the Xase complex

The next step along the path of blood coagulation is the activation of factor X to give factor Xa. This is effected by factor IXa, the formation of which was discussed in the previous chapter. In addition to factor IXa, the activation of factor X requires the co-factors factor VIIIa, calcium and anionic phospholipid; without these it cannot take place at the necessary speed. The lack of functional factor VIII is the cause of the classical haemophilia (haemophilia A). First, we shall examine the components of the activation process.

4.1.1. *Phospholipid*

The phospholipid surface will play a prominent rôle in coming chapters, so a brief introduction to phospholipids is given here.

Phospholipids are components of membranes. For example, about one-half of the mass of the plasma membranes of mammals is lipid, while the rest is made up of membrane proteins. The lipids in plasma membranes consist mainly of phospholipids, cholesterol and, in smaller amounts, glycolipids (gangliosides).

Figure 17 shows the most important classes of phospholipid: phosphatidyl choline, phosphatidyl ethanolamine, sphingomyelin, phosphatidyl serine and phosphatidyl inositol.

The four phosphatidate molecules are derived from glycerol, while sphingomyelin is derived from sphingosine. The fatty-acid chains in phospholipids normally contain an even number of carbon atoms between 14 and 24; the numbers encountered most frequently are 16 and 18.

Phospholipid composition varies from membrane to membrane. Membranes contain two layers of lipid molecules, and it is worth noting that these are very nearly always asymmetric, that is, the inside and the outside differ (219). Figure 18 shows an example – the blood platelet membrane,

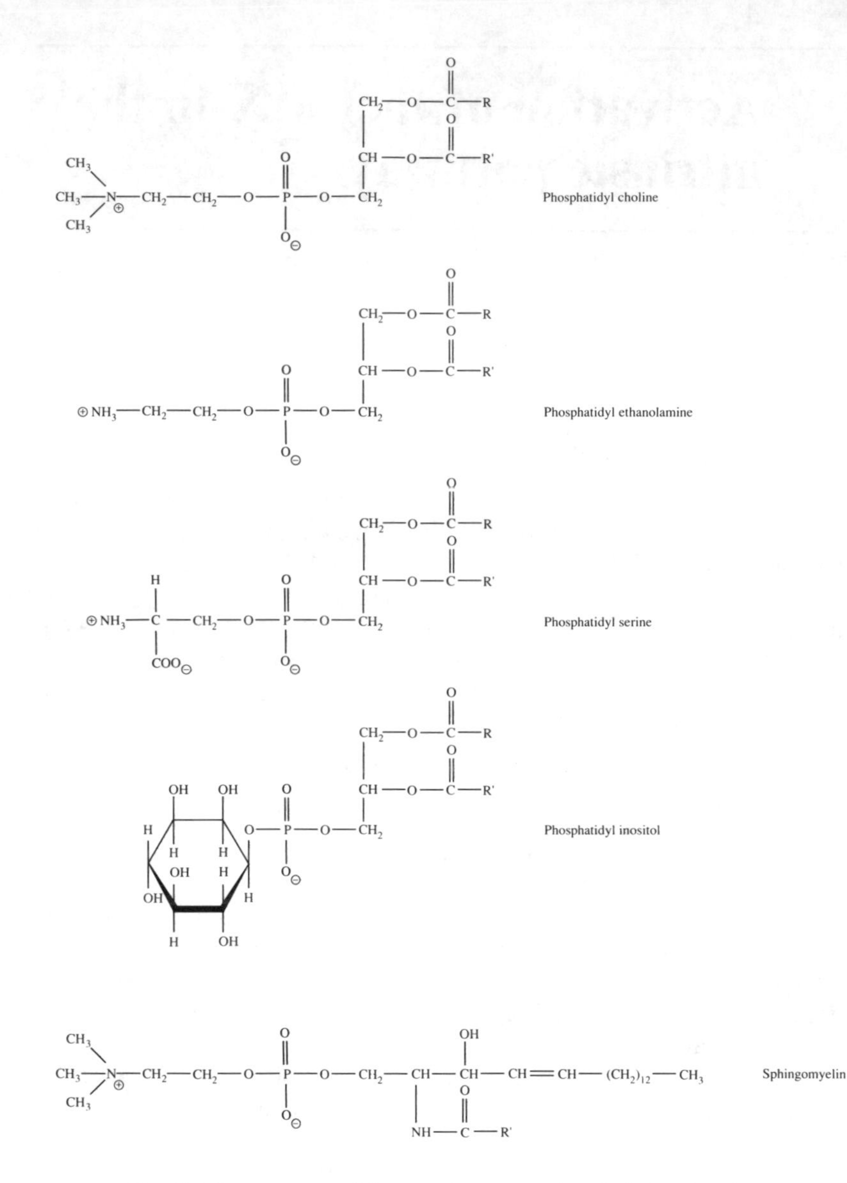

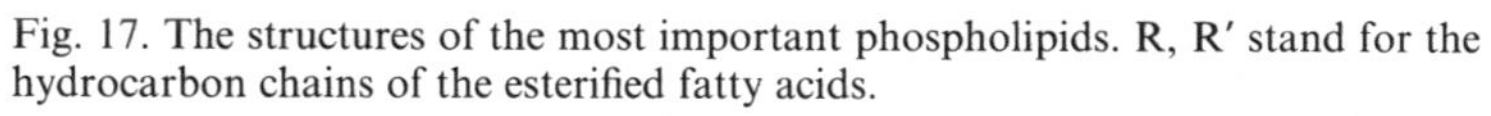

Fig. 17. The structures of the most important phospholipids. R, R′ stand for the hydrocarbon chains of the esterified fatty acids.

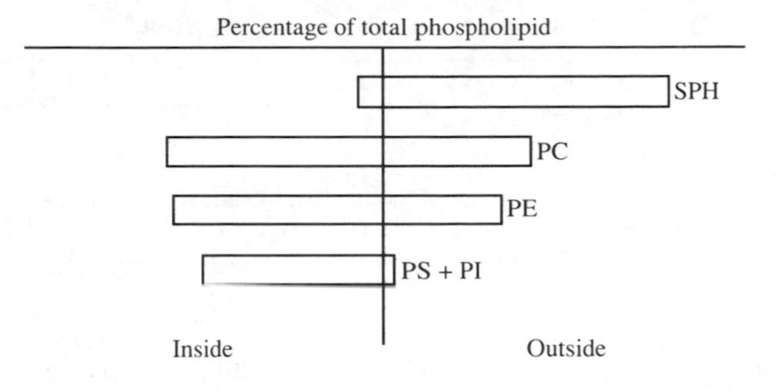

Fig. 18. Suggested distribution of the various phospholipids between the inner and outer layers of the porcine blood platelet membrane. (SPH, sphingomyelin; PC, phosphatidyl choline; PE, phosphatidyl ethanolamine; PS, phosphatidyl serine; PI, phosphatidyl inositol.) (Based on reference 220.)

which will later become important for other reasons (Chapter 22). There is a clear difference between the lipids on the inner side and those on the outer side (220). Those on the outer side tend to be electrostatically neutral, such as sphingomyelin, while those on the side directed in towards the cytoplasm include most of the membrane's acidic (negatively charged) phospholipids, such as phosphatidyl serine and phosphatidyl inositol.

4.1.2. *Factor VIII*

For reviews of this factor, see, for example, references (221–224) and (225).

The suggestion that the abnormal coagulation behaviour in haemophilia A was caused by the lack of a functional plasma factor was first made in 1911 (226). Since then many attempts have been made to purify and characterise factor VIII from plasma. Factor VIII circulates in the form of a complex with another protein, von Willebrand factor, which plays an important part in haemostasis, a topic that we shall return to in Chapter 24. The von Willebrand factor probably protects the factor VIII from premature degradation, to which factor VIII is relatively sensitive (228). The characterisation of factor VIII has been made more difficult by the fact that von Willebrand factor is a very large multimer and occurs in a large molar excess over factor VIII, so that in the isolation of [factor VIII.von Willebrand factor] complex a large quantity of von Willebrand factor is co-isolated (227).

For this reason, it is customary to use the terms factor VIIIC or factor VIIIC:Ag ('C' and 'C:Ag' for 'clotting antigen') for the potentially coagulation-stimulating part of the complex [factor VIII.von Willebrand factor], while factor VIIIR:Ag ('R' and 'R:Ag' for 'related antigen') is used for the von Willebrand factor alone.

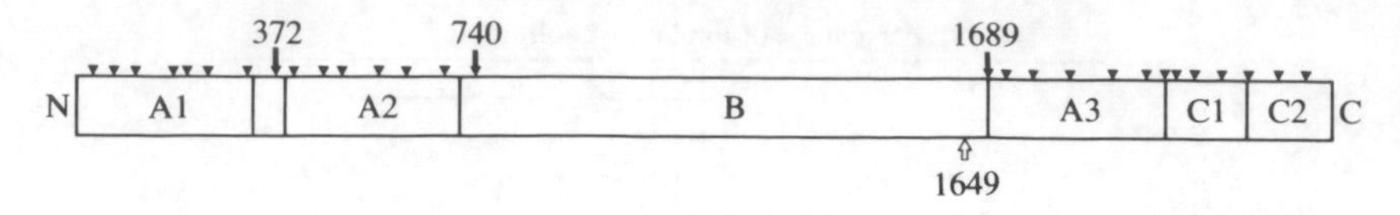

Fig. 19. The domain structure of factor VIII. → shows the positions of cleavage by thrombin, ⇒ the position of the bond cleaved when two-chained factor VIII is formed and ▼ the positions of the introns. A1, A2, A3, B, C1 and C2 designate the domains in factor VIII.

Not until 1980 did it prove possible to isolate pure bovine factor VIII (229), and since then porcine (230, 231) and human (232, 233) factors VIII have also been purified. The concentration of factor VIII in human plasma is relatively low – 100–200 ng/ml, or about 0.5 nM (233) – and its molecular mass has been estimated to be at least 250 kDa, so that the analysis of factor VIII by the methods of protein chemistry has been extremely difficult. Analyses of the carbohydrate content of factor VIII have not been published, but factor VIII is known to be a glycoprotein (234).

The small stretches of amino acids that could be sequenced allowed the cloning both of cDNA and of genomic DNA for factor VIII (235–239). From this, the amino-acid sequence of factor VIII (Appendix, Fig. A6) and the structure of its gene were determined. The cDNA sequence included a signal peptide of 19 amino-acid residues. The only differences between the published amino-acid sequences are in position 56, which is either Asp (235–237, 239) or Val (238), and in position 1241, which is either Asp (235–237, 239) or Glu (238).

The gene for factor VIII covers 186 kbp and possesses 26 exons. The exons vary from 69 bp to 3106 bp in length, and the introns vary from 207 bp to 32.4 kbp. The exon sequences amount to a mere 9 kbp of the gene's 186 kbp. The gene for factor VIII is located on the X chromosome; it is adjacent to the site denoted Xq27.3, which in turn is adjacent to the gene for factor IX (240).

The molecular mass, as calculated from the 2332–residue-long sequence, is 264.8 kDa. Analysis of the amino-acid sequence shows that factor VIII consists of three kinds of domain, termed A, B and C. The structure of factor VIII can thus be described as A1–A2–B-A3–C1–C2 (Fig. 19). The degree of identity among the A domains taken pairwise is approximately 30%, with about 20% of the amino acids being conserved in all three (Fig. 20a). The two C domains show a degree of identity of 37% (Fig. 20b).

Factor VIII shows homology with two other plasma proteins – ceruloplasmin and factor V (237, 238, 241, 242). Ceruloplasmin is the most important copper-binding protein in plasma, and both its protein sequence (243) and its cDNA sequence (244, 245) have been determined. The gene for ceruloplasmin is located at position q21–24 on chromosome 3 (244, 246). Roughly speaking, ceruloplasmin consists of three A domains. The structure of factor V will be described later (section 5.1.1) in connection with the activation of prothrombin.

Ceruloplasmin contains six copper ions bound in three different ways. Two of them are bound in a manner referred to as type 1, on the basis of sequence homology with the type 1 copper-binding protein plastocyanin. This homology implies that ceruloplasmin's type 1 copper ions are co-ordinated to four amino-acid residues in the C-terminal region of two of the internally homologous A domains. These four amino-acid residues comprise two histidines, one cysteine and one methionine, and in ceruloplasmin these are found in the last two A domains. The positions of these amino acids are His637, Cys680, His685 and Met690 for the one co-ordinating unit, and His975, Cys1021, His1026 and Met1031 for the other. In factor VIII, domains A1 and A3 both contain these four residues at homologous positions, but it is not known at present whether factor VIII binds copper or whether these residues are involved in the metal ion binding that supports the structure of factor VIII. There is no evidence that factor VIII possesses any of the enzymic activity of ceruloplasmin. The location of the potentially copper-binding sites in factor VIII are His267, Cys310, His315 and Met320 in the A1 domain and His1954, Cys2000, His2005 and Met2010 in the A3 domain.

The disulphide-bond pattern of factor VIII has not yet been determined, and, as the disulphide pattern of ceruloplasmin is also partly unknown (247), no conclusive information can be adduced from homology.

Purified factor VIII is rarely obtained single-chained, but appears to occur in two fragments. One of these has a molecular mass of about 80 kDa and is frequently seen as a double band in SDS-polyacrylamide gel electrophoresis (248–252). Determination of the N-terminal amino-acid sequence has shown that this fragment is derived from the C-terminal region of factor VIII (248, 250). The other fragment is in reality a whole series of fragments with molecular masses between about 90 kDa and 210 kDa (248–252), all of which have the same N-terminal amino-acid sequence as that deduced from the cDNA of the single-chained factor VIII (248, 250). These fragments are therefore all derived from the N-terminal sequence of factor VIII, and are probably either trimmed or else degraded differently, either in plasma or under purification. The largest fragment (210 kDa) may correspond to the N-terminal fragment that arises in the splitting of the newly synthesized single-chained factor VIII between Arg1648 and Glu1649; this would also give rise to the 80-kDa fragment, as Glu1649 is the N-terminal residue of this fragment.

As mentioned above, factor VIII is rarely obtained in its single-stranded form; indeed, this has only been reported once (249), and even here most of the factor VIII molecules were multicatenate. This has suggested the idea that the splitting of factor VIII at the Arg1648–Glu1649 bond may be a natural part of the secretion process. Studies of the synthesis of

(a)

```
A1      1 A T R R Y Y L G A V E L S W D Y M Q S D L G E L P V D A R - F
A2    380 K T W V H Y I A A E E E D W D Y A P L V L A P - - - D D R S Y
A3   1694 K T R H Y F I A A V E R L W D Y G M S S S P H V L - R N R A Q

P P R V P K S F P F N T S V V Y K K T L F V E F T D H L F N I A K P R P
K S Q Y L N N G P Q R I G R K Y K K V R F M A Y T D E T F K T - - - R E
S G S V P Q - - - - - - - - - - F K K V V F Q E F T D G S F T Q P L Y R G

P W - - - M G L L G P T I Q A E V Y D T V V I T L K N M A S H P V S L H
A I Q H E S G I L G P L L Y G E V G D T L L I I F K N Q A S R P Y N I Y
E L N E H L G L L G P Y I R A E V E D N I M V T F R N Q A S R P Y S F Y

A V G V S Y W K A S E G A E Y D D Q T S Q R E K E D D K V F P G G S H T
P H G I T D V R P L Y S R R L P K G V K H L - - K D F P I L P G E I F K
S S L I S Y E E D Q R Q G A E P - - - - - - - - - R K N F V K P N E T K T

Y V W Q V L K E N G P M A S D P L C L T Y S Y L S H V D L V K D L N S G
Y K W T V T V E D G P T K S D P R C L T R Y Y S S F V N M E R D L A S G
Y F W K V Q H H M A P T K D E F D C K A W A Y F S D V D L E K D V H S G

L I G A L L V C R E G S L A K E K T Q - - T L H K F I L L F A V F D E G
L I G P L L I C Y K E S V D Q R G N Q I M S D K R N V I L F S V F D E N
L I G P L L V C H T N T L N P A H G R Q V T V Q E F A L F F T I F D E T

K S W H S E T K N S L M - - - - - - - - - Q D R D A A S A R A W P K M H T
R S W Y L - T E N I Q R F L P N P A G V Q L E D P E F Q - A S N I M H S
K S W Y F - T E N M E R N C R A P C N I Q M E D P T F K - E N Y R F H A

V N G Y V N R S L P G L I G C H R K S V Y W H V I G M G T T P E V H S I
I N G Y V F D S L Q - L S V C L H E V A Y W Y I L S I G A Q T D F L S V
I N G Y I M D T L P G L V M A Q D Q R I R W Y L L S M G S N E N I H S I

F L E G H T F L V R N H - - - R Q A S L E I S P I T F L T A Q T L L M D
F F S G Y T F K H K M V - - - Y E D T L T L F P F S G E T V F M S M E N
H F S G H V F T V R K K E E Y K M A L Y N L Y P G V F E T V E M L P S K

L G Q F L L F C H I S S H Q H D G M E A Y V K V D S C 329
P G L W I L G C H N S D F R N R G M T A L L K V S S C 711
A G I W R V E C L I G E H L H A G M S T L F L V Y S N 2019
```

(b)

```
C1 2020 K C Q T P L G M A S G H I R D F Q I T A S G Q Y G Q - - - - W A
C2 2173 S C S M P L G M E S K A I S D A Q I T A S S Y F T N M F A T W S

P K L A R L H Y S G S I N A W S T K E - - P F S W I K V D L L A P M I I
P S K A R L H L Q G R S N A W R P Q V N N P K E W L Q V D F Q K T M K V

H G I K T Q G A R Q K F S S L Y I S Q F I I M Y S L D G K K W Q T Y R G
T G V T T Q G V K S L L T S M Y V K E F L I S S S Q D G H Q W T L F F Q

N S T G T L M V F F G N V D S S G I K H N I F N P P I I A R Y I R L H P
N - - G K V K V F Q G N Q D S F T P V V N S L D P P L L T R Y L R I H P

T H Y S I R S T L R M E L M G C D L N 2172
Q S W V H Q I A L R M E V L G C E A Q D L Y 2332
```

recombinant factor VIII in mammalian cell culture reinforce this view (253), since the newly synthesized factor VIII is single-chained in the endoplasmic reticulum, while the double-chained form begins to appear in the Golgi apparatus. It is not known what enzyme carries out this cleavage.

As mentioned above, factor VIII is a glycoprotein, and the factor VIII polypeptide contains 25 potential sites for N-glycosylation. Nineteen of them are located in the B domain, and most of these are utilised (253). It has been found that the reason for the appearance of the 80-kDa chain as a double band in SDS-polyacrylamide gel electrophoresis is partial glycosylation of Asn1810 or Asn2118 (254). In addition, factor VIII is O-glycosylated in both the 210-kDa chain and the 80-kDa chain (253), and sulphated at tyrosine residues in regions characterised by unusually large numbers of acidic amino-acid residues (255). These regions are located respectively between the A1 and A2 regions, after the A2 region and before the A3 region; their function will be discussed below. They include six tyrosine residues at positions 346, 718, 719, 723, 1664 and 1680, and it has been shown that Tyr1664 and Tyr1680 are both sulphated (256).

The mutual cohesion of the two factor VIII chains is dependent upon metal ions, presumably Ca^{2+} (230, 248, 250, 251, 257).

Factor VIII only attains maximal activity in coagulation if it is activated by thrombin or a similar proteinase (258). Knowledge of the complete amino-acid sequence has allowed the determination of the cleavage sites. Purified, thrombin-activated factor VIII appears to consist of three fragments of molecular masses 50 kDa, 43 kDa and 73 kDa, since the appearance of these three fragments during digestion is concomitant with the appearance of factor VIII activity (248). The chains of activated factor VIII are probably held together in the same metal-ion-dependent manner as the chains of factor VIII itself (259, 260).

The activation of factor VIII by thrombin appears to be correlated with the proteolytic cleavage of the bond Arg740–Ser741, followed by Arg372–Ser373 and Arg1689–Ser1690 (238, 248). *In vitro*, the first step is the cleavage of Arg740–Ser741 and formation of the N-terminal 90-kDa subfragment from the 210-kDa fragment (unless exposure to proteases has already made this or other cuts). The other two cleavages follow, cutting the 90 kDa again into an N-terminal 50-kDa and a C-terminal 43-kDa subfragment, and producing a C-terminal 73-kDa fragment.

It thus appears that activated factor VIII is a trimer, the activity of which is conditional upon the cleavage of the three Arg–Ser bonds mentioned.

Fig. 20. (a) Internally homologous A domains in factor VIII. Only the positions at which amino acids are the same in all three domains are outlined. The degrees of identity from pairwise comparisons are shown in Table 5. (b) Internally homologous C domains in factor VIII. Only the positions at which amino acids are the same in both domains are outlined. The degree of identity is included in Table 5.

However, this is not strictly true. Gel-filtration studies of the thrombin-activated factor VIII have shown that the coagulation activity is associated with a dimer of the 73-kDa and 50-kDa fragments which is eluted well separated from the 43-kDa fragment (260). The molecular mass of the functional unit is therefore around 125 kDa, which corresponds well with results obtained by an electron-scattering technique (261).

Recombinant-DNA techniques have further made it possible to replace the critical arginine residues by isoleucines, which gave the molecule specific resistance to cleavage by thrombin. The modification Arg740→Ile did not affect the ability to be activated by thrombin, but this ability was lost following either of the modifications Arg372→Ile and Arg1169→Ile (262). It is also interesting that the modification of Arg1648 to Ile failed to affect the ability of factor VIII to be activated by thrombin. We must thus conclude that the emergence of factor VIII activity is correlated with the cleavage of bonds Arg372–Glu373 and Arg1689–Glu1690, while the cleavage of bonds Arg1648–Ser1649 and Arg740–Ser741, although preceding the other two cleavages *in vivo*, are not related to the emergence of activity.

It should be mentioned that factor Xa is in a position to activate factor VIII and then inactivate it again (248). The cleavages by which factor Xa activates factor VIII are the same as those made by thrombin, and the subsequent inactivation occurs by means of the cleavage of both the 50-kDa and the 73-kDa fragments. The positions of the inactivating cleavage are Arg1721–Ala1722 in the 73-kDa fragment and probably Arg336–Met337 in the 50-kDa fragment.

The activated form of factor VIII with maximal coagulation activity is called factor VIIIa. It possesses a domain structure comprising two A domains (A1 and A3) and two C domains (C1 and C2); the 50-kDa fragment contains the A1 domain and the 73-kDa contains the A3 and the two C domains. Thus the A2 and B domains are released in the activation of factor VIII. Perhaps domain B can be regarded as a very large activation peptide.

The factor VIII molecule has been dissected by genetic engineering methods, and the results have pointed uniformly to the conclusion that the B domain is largely superfluous (263–267). The various factor VIII-like constructions used have all lacked large parts of the B domain. Examples of this include the following:

(i) 880 amino acids, from Ile759 to Gln1638, were deleted (263).
(ii) 90-kDa and 80-kDa fragments were synthesized separately and allowed to associate, giving a factor VIII molecule with the B domain virtually absent (264).
(iii) 766 amino acids, from residues 797 to 1562, were deleted (265).
(iv) The chains of factor VIII were synthesized independently as deletion mutants (266). For the heavy chain, the first 1293 residues plus a 46–

residue nonsense C-terminal extension were employed, and for the light chain the first 115 amino-acid residues plus residues 1455 to 2332, the result of a deletion between two *BclI* restriction sites.

(v) 896 residues, Pro771 to Asp1666, were deleted (267). This is the only mutant so far in which the cleavage site Arg1648–Glu1649 has been removed.

The results with these factor VIII constructs enable at least three conclusions to be drawn. First of all, the B domain is not needed for the synthesis of functional factor VIII, since the fragment chains formed active factor VIII when they had been prepared independently. Secondly, both the 90-kDa and the 80-kDa fragments are needed in functional factor VIII. Thirdly, the B domain is not necessary for the interaction with von Willebrand factor, as the 'synthetic' factor VIII also bound to immobilised von Willebrand factor.

So what is the function of the B domain? This cannot be answered with certainty, but one possibility is that it participates in the protection of factor VIII from proteolysis. Another has been revealed by studies on the synthesis of recombinant factor VIII in mammalian cell culture (253, 268): it was found that a large fraction of the primary factor VIII translation product was retained in the endoplasmic reticulum, binding to the protein BiP (immunoglobulin heavy-chain-binding protein), a native component of the endoplasmic reticulum. Since the first deletion mutant mentioned above (263), lacking a large part of the B domain, is secreted much more effectively than native factor VIII, it can be conceived that incomplete glycosylation of the B domain may result in secretion-incompetent factor VIII molecules, held back in the endoplasmic reticulum by BiP.

The interaction between factor VIII and von Willebrand factor is of especial importance, as almost all factor VIII molecules *in vivo* are complexed with von Willebrand factor. This binding has been found to take place via the 80-kDa fragment derived from the C terminus (269–271). It probably involves elements of the amino-acid sequence between Val1670 and Arg1689, since monoclonal antibodies against this epitope inhibit the interaction between factor VIII and von Willebrand factor (270, 271).

It is interesting to note the effect of von Willebrand factor upon the activation by thrombin of factor VIII. It will be remembered that the thrombin-catalysed activation of purified factor VIII resulted in a bimolecular non-covalent complex with chains of 50 kDa and 73 kDa derived from the 90-kDa chain and the 80-kDa chain respectively (260). The activation by thrombin of factor VIII in complex with von Willebrand factor results in the liberation of precisely these 50-kDa and 73-kDa fragments of factor VIII from the complex, while the 43-kDa fragment, which represents the remainder of the 90-kDa chain, remains bound to the von Willebrand factor (272). This also underlines the fact that the functional factor VIIIa in its physiological state is a heterodimer of the

50-kDa chain and the 73-kDa chain. This binding of the 43-kDa fragment to von Willebrand factor after the activation of factor VIII by thrombin is somewhat surprising, in view of the fact that isolated 90-kDa fragments or their heavier precursors have never been shown to bind directly to von Willebrand factor. An indirect indication of a possible interaction between the heavy chain of factor VIII and von Willebrand factor has come from reconstitution experiments in which factor VIII was re-formed from isolated subunits in the presence of Ca^{2+} or Mn^{2+} (257). The reconstitution proceeded up to five times more rapidly in the presence of von Willebrand factor than in its absence. This effect is, however, not necessarily related to binding of factor VIII's heavy chain to von Willebrand factor, since large quantities of von Willebrand factor did not inhibit reconstitution. A possible explanation could be that the binding of the light factor VIII chain to von Willebrand factor could lead to a conformational change, resulting in a faster reassociation. At the same time, it must be mentioned that secreted factor VIII made in mammalian cell culture is stabilised substantially if von Willebrand factor is included in the growth medium (253).

Interesting results have accrued from the study of porcine factor VIII, as in recent years it has become possible to obtain this factor in a high degree of purity (230, 231). Like human factor VIII, the porcine factor circulates as a dimer of various fragments derived from the N and C termini. An invariant C-terminal 76-kDa fragment associates with a variable N-terminal fragment of molecular mass 166 kDa, 130 kDa or 82 kDa (230, 231, 238). Short stretches of amino-acid sequence obtained from fragments of purified porcine factor VIII (241, 242) allowed the cloning of some of its genomic DNA, and by using this the human cDNA could also be isolated and cloned (238). The published genomic sequence for porcine factor VIII (263) covers principally the B domain, but also overlaps with the C terminus of the A2 and the N terminus of the A3 domains. If the deduced human and porcine amino-acid sequences are compared, two observations about the B domain can be made. First, about 25% of the B domain is deleted in the porcine factor, in comparison with the human; and second, the degree of identity between the two amino-acid sequences is much lower in the B domain than in other domains. The porcine factor VIII can be activated both by factor Xa and by thrombin (273); the latter is the more efficient. In contrast to human factor VIII, the products of activation by factor Xa are different from those of activation by thrombin (273): thrombin-activated porcine factor VIII behaves like a heterotrimer of chains with molecular masses 44 kDa, 35 kDa, and 69 kDa (274). This seems superficially to be different from thrombin-activated human factor VIII, but although thrombin-activated factor VIII has a coagulating activity as a heterodimer of 50-kDa and 73-kDa fragments, as stated above, it is possible that the trimeric structure is stabler. Porcine factor VIII interacts via its light chain with porcine von Willebrand factor (275).

4.1.3. *Haemophilia A*

We have already mentioned that a lack of factor VIII causes haemophilia A (reviewed, for example, in references 276–279), and, as in the case of haemophilia B (deficiency in functionally active factor IX), some of the underlying molecular defects have been mapped. The task of identification is complicated by the considerable size of the factor VIII gene (186 kbp) and by the very small quantities in which mRNA for factor VIII is produced.

It has been found that haemophilia A patients have widely varying levels of functional factor VIII, and, as for haemophilia B, they are classified into the categories CRM^+ and CRM^-, according to the result of a cross-reactivity test with antibodies against factor VIII antigen (280, 281). The degree of haemophilia A depends upon the fraction of the normal factor VIII activity that is reached in the patient, with $<1\%$ being regarded as severe and $>5\%$ as mild haemophilia A. Intermediate cases are termed moderate. In some instances, haemophilia A patients being treated with preparations of factor VIII develop antibodies against it.

The defects in factor VIII that have been characterised at a molecular level are shown in Table 4. They can in the main be classified as (i) greater or smaller deletions, or (ii) point mutations giving rise to the loss of Taq I restriction sites.

The reason that the loss of Taq I restriction sites reveals many point mutations is that the recognition sequence of Taq I contains the dinucleotide CpG. CpG is believed to be a 'hot spot' for mutagenesis (305, 306). Cytosine in CpG is the most frequently used methylation site in human DNA, and 5–methyl cytosine deaminates spontaneously to give thymine. Taq I restriction sites therefore offer the possibility both of C→T and of G→A mutations in the coding strand (depending upon which strand the C→T transition took place in). In the factor VIII gene, C→T mutations have also been demonstrated in CpG dinucleotides that do not correspond to Taq I restriction sites but which affect physiologically important cleavage sites in the factor VIII molecule. Furthermore, two haemophilia A patients with non-viral insertions of length 3.8 kbp and 2.3 kbp in exon 14 (307) have been described. As was the case for defective factor IX molecules, the location of one point mutation does not necessarily mean that the mutation found is the only important one, or indeed the one causing haemophilia A; however, this is generally assumed to be the case.

4.1.4. *Factor X*

Factor X occurs in plasma as a two-chained glycoprotein. It is the zymogen for the active serine proteinase factor Xa. In normal human plasma the concentration lies around 10 μg/ml ($\approx$170 nM). The molecular mass of factor X is 59 kDa, distributed between a light chain (17 kDa) and a heavy chain (42 kDa). The carbohydrate part makes up approximately 15% of the

Table 4. *Characterised defects in factor VIII*
A. Point mutations defined by *Taq*I restriction sites

Location	Mutation	Consequence	Haemophilia	Inhibiting antibody	Reference
Exon 7	A→G	Glu272→Gly	moderate	no	(282)
Exon 18	C→T	Arg1941→stop	severe	yes	(283, 284)
Exon 18	C→T	Arg1941→stop	severe	no	(285)
Exon 22	C→T	Arg2116→stop	severe	no	(284, 285)
Exon 22	G→C	Arg2116→Pro	severe	unknown	(286)
Exon 23	C→T	Arg2147→stop	severe	unknown	(286)
Exon 23	C→T	Arg2147→stop	severe	yes	(287)
Exon 24	C→T	Arg2209→stop	severe	yes	(284, 287, 288)
Exon 24	C→T	Arg2209→stop	severe	no	(287)
Exon 24	C→T	Arg2209→stop	severe	unknown	(289)
Exon 24	G→A	Arg2209→Gln	severe	no	(287, 290)
Exon 24	G→A	Arg2209→Gln	severe	unknown	(289)
Exon 26	C→T	Arg2307→stop	severe	no	(284, 288)
Exon 26	C→T	Arg2307→stop	severe	yes	(284)
Exon 26	G→A	Arg2307→Gln	mild	no	(291)
Exon 26	G→T	Arg2307→Leu	mild	no	(292)
Intron 2	C→T	unknown	severe	no	(288)
Intron 4	G→A	cryptic splicing site?	mild	no	(293)
Intron 25	A→G	unknown	severe	no	(288)

B. Other point mutations characterised

Location	Mutation	Consequence	Haemophilia	Inhibiting antibody	Reference
Exon 8	C→T	Arg336→stop	severe	no	(294)
Exon 8	C→T	Arg372→Cys	moderate	no	(295)
Exon 8	C→T	Arg372→His	mild	no	(296)
Exon 14	A→T	Tyr1680→Phe	mild	no	(297)
Exon 14	C→T	Gln1686→stop	severe	unknown	(297)
Exon 14	C→T	Arg1689→Cys	severe	no	(294)
Exon 14	C→T	Arg1689→Cys	moderate	no	(297, 298)

C. Deletions

Exons deleted	Size of deletion	Inhibiting antibody	Haemophilia	References
23–25	39 kbp	yes	severe	(288)
26	21.9 kbp	no	severe	(288)
11–19 (7/11–22)	60 (>80) kbp	yes	severe	(276, 283)
6	7 kbp	no	severe	(299)
part of 14	2.5 kbp	no	severe	(299)
24–25	> 7 kbp	no	severe	(299)
23–25	>16 kbp	no	severe	(299)
22	5.5 kbp	no	moderate	(299)
15–18	13 kbp	yes	severe	(300)
all	>210 kbp	no	severe	(276, 301)
23–26	unknown	yes	severe	(302)
1–22	>127 bp	yes	severe	(303)
26	>2 kbp	no	severe	(304)
1–5	>35 kbp	no	severe	(284)
part of 3	1.7–2 kbp	no	severe	(284)
7–9	15–20 kbp	yes	severe	(284)
part of 14	2–2.5 kbp	no	severe	(284)
14	12–16 kbp	yes	severe	(284)
26	>1.8 kbp	no	severe	(284)
part of 8	2 bp	unknown	severe	(297)

molecule's mass, and it is bound exclusively to the heavy chain (308). Factor X is the second vitamin K-dependent protein that we have met in the blood coagulation system. Factor X is also known as Stuart factor or Stuart-Prower factor, after the first documented cases of factor X deficiency (309, 310).

The amino-acid sequences of the light (311, 312) and the heavy (313) chains of bovine factor X and also of the light chain of human factor X (314) have been determined at the protein level. Furthermore, the complete amino-acid sequences of human (Appendix, Fig. A7) and bovine factor X have also been deduced from cDNA-sequencing (315–319), which in addition revealed a 40-residue signal/propeptide (317, 318). The gene structure of human factor X is also nearly completely known, in that all the exon–intron boundaries have been documented with the single exception of exon 1, which was absent from the genomic clones used (317). The factor X gene is approximately 27 kbp long and consists of eight exons 25–612 bp in length separated by introns that are estimated to be 950 bp to 7.4 kbp long (317, 320). The positions of the introns in relation to the amino-acid sequence of factor X are shown in Figure 21. The gene for factor X is located at position q34 of chromosome 13 (321, 322).

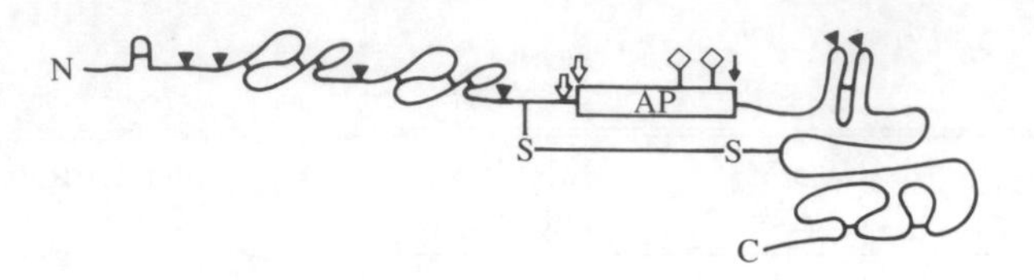

Fig. 21. The domain structure of factor X. → shows the position of cleavage by factor IXa, ⇒ the positions of the bonds that are cleaved when two-chained factor X is formed, ▼ the positions of the introns, and ◇ the potential attachment sites for N-bound carbohydrate. (AP, activation peptide.)

The molecular mass of the amino-acid sequence deduced for human factor X is 50.0 kDa, giving 58.8 kDa after inclusion of the 15% carbohydrate. The deduced structure has 448 amino-acid residues in a single chain (Appendix, Fig. A7) that reveals strong structural similarity with factor IX. However, factor X circulates in the bloodstream as a heterodimer (Fig. 21); this is formed by the excision of three residues (Arg140-Lys141-Arg142) during or after translation. The resulting complex, with a light chain from the N terminus and a heavy chain from the C terminus of the primary transcript is the form that is isolated from plasma.

Examining the structure of the light chain, proceeding as usual from the N terminus, we find first a Gla domain containing 11 Gla residues in the first 39 positions, corresponding to a modification of all the Glu residues in this region (314). In bovine factor X there are 12 Gla residues in this region (323). Thereafter come two growth-factor domains with homology with EGF, and a short peptide that makes up the C-terminal part of the light chain. The N-terminal region of the heavy chain is an activation peptide and the rest of it the serine proteinase domain. In the activation peptide we encounter another difference between human and bovine factor X. The bovine factor occurs in two forms, called X1 and X2, differing in that the tyrosine residue at position 18 in the activation peptide of factor X2 is sulphated (324). Human factor X occurs in one form only, but it is not yet known whether this is sulphated.

Residue 63 in the light chain of both human and bovine factors X is β-hydroxy aspartic acid; it is located in the first growth-factor domain (140, 314). The disulphide bonds in bovine factor X have been detected experimentally, with the exception of the last two bonds in the second growth-factor domain (325), and they are in agreement with what would be expected on the basis of homology. The following disulphide bonds in human factor X can be inferred from this: in the Gla domain Cys17–Cys22; in the first growth-factor domain Cys50–Cys61, Cys55–Cys70 and Cys72–Cys81; in the second growth-factor domain Cys89–Cys100, Cys96–Cys109 and Cys111–Cys124; between the chains Cys132 (light chain) to Cys160 (heavy chain); and in the serine-proteinase domain Cys59–Cys64, Cys79–Cys95, Cys208–Cys222 and Cys233–Cys261.

The activation of factor X to give factor Xa, catalysed by factor IXa, is the result of cleavage of the bond Arg52–Ile53 in the heavy chain, which releases an 11-kDa glycosylated activation peptide. In fact, all the carbohydrate in factor X is located in this activation peptide, on Asn39 and Asn49. Factor Xa is thus a two-chained serine proteinase consisting of a light chain (17 kDa) disulphide-bonded to a heavy chain (31 kDa) that accommodates the catalytic site. This is believed to involve His42, Asp88 and Ser185 in the heavy chain of factor Xa, corresponding to His94, Asp140 and Ser237 in the heavy chain of factor X.

The genetically determined deficiency of functional factor X is comparatively rare, and shows considerable heterogeneity (326, 327). The genetic basis for factor X deficiency has been determined in two cases. In one of these, Arg366 was changed to Cys because of a C→T mutation (328); in the other, exons 7 and 8 were found to be deleted (329).

4.2. Activation of factor X

Since negatively charged phospholipid is known to be a co-factor in the activation of factor X, the first prerequisite for this activation is clearly a negatively charged surface. Taking this as given, we now examine the enzyme complex that carries out the activation.

The enzyme itself is factor IXa, and, as described earlier (section 3.1), it consists of a serine proteinase with two chains connected by a disulphide bond. The heavy chain (28 kDa) is derived from the zymogen's C terminus and contains the catalytically active serine residue, while the light chain comes from the N terminus of the zymogen and contains the 12 Gla residues and the partly hydroxylated Asp residue.

The Gla residues are involved in the calcium-dependent binding of factor IXa (and factor IX) to the negatively charged phospholipids at the surface where activation takes place. The side-chain of γ-carboxy-glutamic acid can be converted (with a mixture of formaldehyde and morpholine at slightly acidic pH) to γ-methylene-glutamic acid. The conversion of only one of the γ-carboxy-glutamic acid residues is sufficient to destroy completely the factor's activity in coagulation (330), and this modification also results in a dramatic reduction in its ability to bind to phospholipid vesicles (331). In spite of the ambiguity frequently associated with the interpretation of data from chemically modified proteins, it is hard to escape the significance of this result. Likewise, factor IXa from which the Gla domain has been removed proteolytically shows a very low activity in coagulation ($<0.5\%$), even though its enzymic activity with respect to artificial small oligopeptide substrates is unaffected (206).

Thus the binding of factor IXa to phospholipids depends critically upon the Gla residues at the N terminus of the light chain.

The next protein involved, as a non-enzymic co-factor, is factor VIIIa, which, as stated in the previous section, consists of two fragments (50 kDa and 73 kDa) whose mutual cohesion is dependent upon calcium ions.

Which parts of factor VIIIa are involved in what interactions is still a matter of uncertainty. The following is based upon indirect evidence and a comparison with factor V/Va.

Interactions between factor IXa and factor VIIIa probably involve the heavy chain of factor IXa, since a monoclonal antibody against an epitope in this chain inhibits the ability of factor VIIIa to stimulate the activation of factor X (332). This monoclonal antibody fails to interfere with the activation of factor IX to factor IXa, irrespective of whether the activation is catalysed by factor XIa and calcium or by factor VIIa, tissue factor and calcium (see section 6.2); neither does the antibody prevent the inhibition of factor IXa by antithrombin III (see section 8.1.3). In contrast, an excess of the [factor VIIIa.von Willebrand factor] complex does inhibit the binding of the antibody to factor IXa and *vice versa*, so that these binding modes are competitive (332). Similarly, it appears that the light chain in factor IXa can interact with factor VIIIa, as the mutation of the first growth-factor domain in factor IX (mentioned in section 3.3) impaired factor VIIIa's co-factor function (210). All these mutagenesis experiments were aimed at the first growth-factor domain, and they would lead one to believe that it is this domain that is involved in the binding of factor IX to factor VIIIa. However, recent mutagenesis experiments suggest a different conclusion. Chimeric factor IX molecules were produced by exchanging the first growth-factor domain or both growth-factor domains of factor IX with the corresponding growth-factor domains from factor X (333). Both of these chimeric factor IX molecules were activated normally by factor XIa, and both appeared to be correctly carboxylated and also approximately 60% hydroxylated at the appropriate Asp residue (333). The activated form of the chimeric molecule in which the first growth-factor domain came from factor X showed nearly normal coagulation activity; however, the molecule in which both growth-factor domains came from factor X had greatly reduced activity in coagulation (333). This indicates that it may be the second growth-factor domain in factor IXa that is important for the interaction between this factor and factor VIIIa.

Which region or regions in factor VIIIa bind factor IXa is not known, but the following evidence enables us to make a qualified guess.

Factor VIIIa is inactivated proteolytically by protein Ca (this reaction is described in section 8.2.4). Protein Ca cleaves the bond between Arg336 and Met337 in the 50-kDa fragment of factor VIIIa (334). This cleavage releases a 36-residue oligopeptide that contains 13 Glu or Asp residues but only three Lys or Arg residues, and therefore bears a large negative charge. Since this cleavage inactivates factor VIIIa, the peptide released must have some functional importance (248). It is also known that factor IXa bound to factor VIIIa protects the factor VIIIa from inactivation by protein Ca (335, 336), and that the epitopes for two antibodies that inhibit factor VIIIa are located between Met337 and Arg372 (337, 338). These observations make it plausible that the region 337–372 is important for the binding of factor IXa to factor VIIIa. Furthermore, it has been shown that recombinant factor VIII in which the acidic region between positions 336 and 372 have been deleted does not possess coagulation activity after cleavage by thrombin (339). This is analogous to the inhibition of factor Va by protein Ca, which also prevents the binding of factor Xa and of prothrombin to factor Va. It would be consistent with this analogy if the oligopeptide were involved in the protein–protein interaction. However, this argument should not be taken as implying that no further parts of factor VIIIa are involved in the binding of factor IXa.

Factor VIII and factor VIIIa both bind to phospholipid, independently of calcium ions. It appears that factor VIIIa binds much more strongly to phospholipid than factor VIII does, but this may be because factor VIII is always found in complex with von Willebrand factor, which could, by steric hindrance, impede binding to the phospholipid (340, 341). The use of factor VIII, both of natural origin and prepared from recombinant DNA, has shown that it is the C-terminally derived 80-kDa fragment that mediates the binding of factor VIII to negatively charged phospholipid (342, 343). Therefore, it can be expected that the binding of factor VIIIa occurs by way of the 73-kDa fragment, which contains the A3, C1 and C2 domains. A more precise localisation of the phospholipid-binding site has been carried out with the help of synthetic peptides (344) and antibodies (345); it shows that the C2 domain is essential for this function. This agrees well with the observation that the two C domains are homologous to the first 150 amino-acid residues of a carbohydrate-binding lectin called discoidin I (degree of identity approximately 20%). Discoidin I is a tetrameric galactose-binding lectin required for cellular adhesion in the slime mould *Dictyostelium discoideum*, and it also binds to negatively charged phospholipids (346).

We thus have a picture of the enzyme complex ('Xase') that activates factor X to give factor Xa. Its empirical formula is [factorIXa.factorVIIIa. Ca^{2+}_{n}.phospholipid]: an activated factor VIII molecule binds both to phospholipid and to factor IXa, which also binds in a Ca^{2+}-dependent mode to the phospholipid by way of its light chain and the Gla residues contained therein. This is indicated schematically in Figure 22.

The Xase enzyme complex catalyses the conversion of factor X to factor Xa. Where does the factor X come from? There are two possibilities: from the phospholipid surface, for which the factor X zymogen has a certain

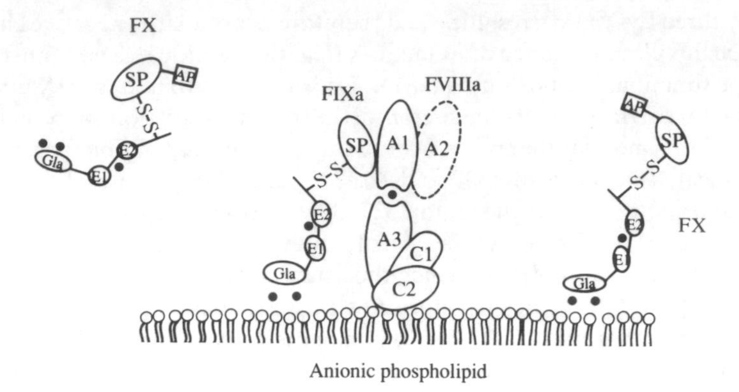

Fig. 22. A schematic representation of the Xase enzyme complex. F, factor; •, Ca^{2+}. Gla, E1, E2 and SP represent Gla domain, first growth-factor domain, second growth-factor domain and serine-proteinase domain in factors IXa and X. AP represents the activation peptide in factor X. A1, A2, A3, C1 and C2 are the domains in factor VIIIa. The A2 domain is shown stippled, as its presence is uncertain. It is not known whether the Xase enzyme complex accepts factor X from the fluid phase or from the surface, or which parts of Xase interact with factor X, but studies with synthetic peptides indicate that the amino-acid sequence between positions 125 and 161 in the heavy chain factor X are important for this.

binding affinity, or from solution. Little attention has been paid to this question, and different studies of the kinetics of the conversion of factor X to Xa have not given consistent results.

As regards the bovine system (347–350), it was originally believed that phospholipid lowered the Michaelis constant K_m for factor X, possibly to a level corresponding to the concentration in plasma (347), while factor VIIIa raised considerably the value of V_{max} for the conversion of factor X to factor Xa without affecting K_m to anything like the same extent (347, 348). The increase in V_{max} can be attributed to at least two main influences: increased catalytic efficiency of factor IXa when it is bound to factor VIIIa, and increased surface affinity of factor IXa in the presence of factor VIIIa (349). Recently, additional relevant results have been obtained from a continuous-assay system (350). Here it was found that K_m for factor X depended only upon the presence of calcium ions, and was not affected by phospholipid or by factor VIIIa. The observed value of K_m lay in the range of physiological concentrations of factor X. The presence of phospholipid raised the value of k_{cat} (defined as V_{max}/K_m) for the conversion of factor X, and factor VIIIa acting together with phospholipid raised it even more.

For the human system (351–354), studies have been published in which it was possible to calculate kinetic constants (351, 352), in addition to other studies in which the kinetic constants in the presence of phospholipid and factor VIIIa were not calculated (353, 354). Taken together, these results indicate that phospholipid lowers K_m while factor VIIIa both lowers K_m still further and raises V_{max}. As in the bovine system, human factor VIIIa increases the surface affinity of factor IXa (355).

Since it is always difficult to compare kinetic data, the preferred substrate of the Xase complex (surface- or solution-phase factor X) has not been identified with certainty. However, the burden of evidence at present does not indicate that the surface-phase factor X is preferred. This is because the artificial surfaces used for this have a much higher surface charge density than natural membrane surfaces, and this can distort the concentration ratios by enhancing binding to an unnatural degree.

The situation may be summarised by saying that there is an active complex that is based upon the ability of factor VIIIa to act as a strongly binding receptor for factor IXa when it resides on coagulation-active surfaces (see, for example, reference 355) in the presence of calcium. Xase is the receptor for factor X; the factor X is probably in solution but also possibly pre-bound to the charged surface. The actual catalyst is factor IXa. The reaction product is factor Xa, formed from factor X by the cleavage of the bond Arg52–Ile53 in the heavy chain of factor X and the concomitant release of the activation peptide.

The binding of factor X to Xase has been investigated with synthetic peptides that covered 65% of the factor X sequence, in order to locate the region of factor X where the interaction with Xase takes place (356). Three of these synthetic peptides, covering positions 125 to 141, 142 to 161 and 275 to 289 in the heavy chain of factor X, inhibited its activation in a system of purified coagulation factors. It is therefore believed that the peptide regions mentioned are important for the recognition of factor X by the Xase complex. The same synthetic peptides also inhibited the activation of factor X by tissue factor/factor VIIa in the extrinsic pathway.

5.

Activation of prothrombin

5.1. Components of the prothrombinase complex

The main catalytic function of factor Xa, once formed, is the activation of prothrombin to give thrombin. The reaction takes place in the presence of calcium ions, a negatively charged phospholipid and factor Va as co-factor. This is a further example of a process localised at a surface, and it is mechanistically analogous to the activation of factor X to give factor Xa. The active enzyme complex is often referred to as prothrombinase, and it consists of factor Xa, factor Va, Ca^{2+} and phospholipid. It is thus analogous to Xase, which contains factors IXa and VIIIa alongside Ca^{2+} and phospholipid, as described in the foregoing chapter.

5.1.1. *Factor V*

Factor V (also called pro-accelerin; reviewed in references 225 and 357) was discovered in the middle of the 1940s (358). It circulates as a single-chained glycoprotein both in bovine (359, 360) and in human plasma (361–363). Sedimentation-equilibrium analysis has led to an estimated molecular mass of 330 kDa for the bovine factor (360, 364), which agrees well with estimates for both bovine and human factor V based on denaturing gel electrophoresis. The carbohydrate content of human factor V has been found to be 13% (361), and its concentration in normal human plasma is about 7 μg/ml ($\approx$20 nM) (365).

The amino-acid sequence of short stretches of bovine (241, 242) and human factor V have been determined, and these partial sequences have provided the basis for cDNA cloning and thus the determination of the sequences of the entire human factor V molecules (366–368). The corresponding gene structure of factor V is not known. The gene is located on chromosome 1 at position q21–25 (369, 370).

The amino-acid sequence deduced from the cDNA (366–368) for human pro-factor V (Appendix, Fig. A8) contains 2224 amino-acid residues, of

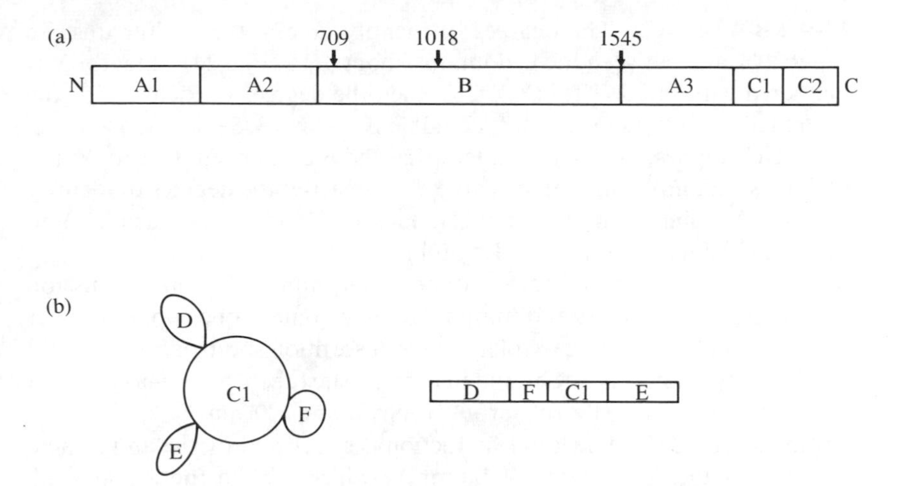

Fig. 23. (a) The domain structure of factor V. → shows the positions of cleavage by thrombin. A1, A2, A3, B, C1, and C2 designate the domains in factor V. (b) A schematic representation of bovine factor V based on electron micrographs (375). D, F, C1 and E stand for factor V fragments defined as in reference 379.

which 28 make up a signal peptide; factor V thus consists of 2196 residues. The only difference between the two derived sequences is at position 1257, which can be occupied by Leu (366) or by Ile (368). If the calculated molecular mass of 249 kDa is compared with the electrophoretically estimated molecular mass of 330 kDa, it appears that the carbohydrate content must be substantially greater than the measured value of 13%. The amino-acid sequence of factor V includes 37 potential N-glycosylation sites, and the factor contains both N-bound and O-bound carbohydrate (361). The complete amino-acid sequence for bovine factor V has also been inferred from the cDNA sequence (371).

The structure of factor V (Fig. 23a) resembles that of factor VIII. Starting at the N terminus, we find first of all two mutually homologous domains (A domains); these are also homologous to the A domains in factor VIII and to the three internally homologous domains in ceruloplasmin. There follows a large fragment without homology with any known protein sequence, called the B domain, then a third A domain and finally, at the C terminus, two other internally homologous domains (C domains), homologous to the C domains in factor VIII.

Both the human and the bovine factor V molecules have been studied by electron microscopy (372–375). They appear to consist of a relatively large central domain surrounded by three smaller domains (372, 375) as shown in Figure 23b.

The structure of factor V can thus, like that of factor VIII, be denoted

A1-A2-B-A3-C1-C2. The degrees of identity among the A domains are about 30% and between the C domains about 37% (Fig. 24). If factor V is compared with factor VIII (Fig. 25), then the degrees of identity for the different domains are: A1, 36%; A2, 44%; B, 14%; A3, 39%; C1, 44%; C2, 42%. This implies, as already stated, that the B domains in factors V and VIII are structurally unrelated. Table 5 summarises the degrees of identity between A domains in factor VIII, factor V and ceruloplasmin, and between C domains in factor VIII and factor V.

It has been shown by both atomic absorption and atomic emission spectroscopy that bovine and human factor V contain one copper ion per factor molecule, while ultra-violet/visible absorption spectroscopy showed that the copper ion was not bound in a type 1 site (absorption maximum at 610 nm) or a type 3 site (absorption maximum 320 nm) as defined for ceruloplasmin (376). The latter deduction is supported by the amino-acid sequence of the A domains of factor V, which, unlike the A1 and A3 domains of factor VIII, contain none of the amino-acid constellations believed to be necessary for the co-ordination geometry of type 1 copper sites.

As in factor VIII, most of the potential N-glycosylation sites of factor V are localised in the B domain (25 out of 37). In factor V, this domain also contains many repetitions of variations on a nine-residue theme with the consensus sequence D-L-S-Q-T-T/N-L-S-P and two consecutive 17-residue stretches identical at 15 positions. The C domains in both factor V and factor VIII are identical in 20% of their residues with discoidin I, which is a tetrameric lectin from the slime mould *Dictyostelium discoidum* and binds both galactose and negatively charged membranes (367).

It is not known whether factor V contains tyrosine-O-sulphate, but there are two regions containing many acidic side-chains in which potentially sulphatable tyrosines are present. These two regions lie between Asp659 and Tyr698, and between Glu1507 and Asp1525, that is, they flank the B domain and separate it from the A2 and A3 domains respectively.

In order to take part in coagulation, factor V (like factor VIII) has to be activated by thrombin (359, 377–379), and the resulting factor Va molecule consists of two chains whose non-covalent association is calcium-dependent. Bovine factor Va possesses a 94-kDa heavy chain – also called fragment D – derived from the N terminus of factor V, and a 74-kDa light chain – fragment E – from the C terminus (379). Human factor Va has the same structure, with a 105-kDa heavy chain (fragment D) from the N terminus and a light chain from the C terminus, termed fragment F1-F2,

Fig. 24. (a) Internally homologous A domains in factor V. Only the positions at which amino acids are the same in all three domains are outlined. The pairwise degrees of identity are shown in Table 5. (b) Internally homologous C domains in factor V. Only the positions at which amino acids are the same in both domains are outlined. The degree of identity is included in Table 5.

(a)

```
A1      1 A Q L R Q F Y V A A Q G I S W S Y R P E P T N S S L N L S V T
A2    319 M K R W E Y F I A A E E V I W D Y A P V I P A - - - - - - - - N
A3   1549 G N R R N Y Y I A A E E I S W D Y S E F V - - - - - - - - - - -

S F K K I V - - - - - - - - Y R E Y E P Y F K K E K P Q S T I S - - - - -
M D K K Y R S Q H L D N F S N Q I G K H Y K K V M Y T Q Y E D E S F T K
- - - Q R E T D I E D S D D I P E D T T Y K K V V F R K Y L D S T F T K

- - - - - - - - - - - G L L G P T L Y A E V G D I I K V H F K N K A D K
H T V N P N M K E D - G I L G P I I R A Q V R D T L K I V F K N M A S R
R D P R G E Y E E H L G I L G P I I R A E V D D V I Q V R F K N L A S R

P L S I H P Q G I R Y S K L S E G A S Y L D H T F P A E K M D D A V A P
P Y S I Y P H G V T F S P Y E D E V N S S F T S G R N N T M I R A V Q P
P Y S L H A H G L S Y E K S S E G K T Y E D D S P E W F K E D N A V Q P

G R E Y T Y E W S I S E D S G P T H D D P P C L T H I Y Y S H E N L I E
G E T Y T Y K W N I L E F D E P T E N D A Q C L T R P Y Y S D V D I M R
N S S Y T Y V W H A T E R S G P E S P G S A C R A W A Y Y S A V N P E K

D F N S G L I G P L L I C K K G T L T E G G T Q K T F D K Q I V L L F A
D I A S G L I G L L L I C K S R S L D R R G I Q R A A D I E Q Q A V F A
D I H S G L I G P L L I C Q K G I L H K D S N M P V D M R E F V L L F M

V F D E S K S W S Q - - - - - - - - - - - - - - - - - - - - - - S S S L
V F D E N K S W Y L E D N I N K F C E N P D E V K R D D P K F Y E S N I
T F D E K K S W Y Y E K K S R S S W R L T S S E M K K S H E F H A I N G

M Y T V N G Y V N G T M P D I T V C A H D H I S W - - - H L L G M S S G
M S T I N G Y V P E S I T T L G F C F D D T V Q W - - - H F C S V G T Q
M - - - - - - - - - - I Y S L P G L K M Y E Q E W V R L H L L N I G G S

P E L F S I H F N G Q V L E - - - Q N H H K V S A I T L V S A T S T T A
N E I L T I H F T G H S F I - - - Y G K R H E D T L T L F P M R G E S V
Q D I H V V H F H G Q T L L E N G N K Q H Q L G V W P L L P G S F K T L

N M T V G P E G K W I I S S L T P K H L Q A G M Q A Y I D I K N C 301
T V T M D N V G T W M L T S M N S S P R S K K L R L K F R D V K C 656
E M K A S K P G W W L L N T E V G E N Q R A G M Q T P F L I M D R 1877
```

(b)

```
C1 1878 D C R M P M G L S T G I I S D S Q I K A S E F L - - - - - G Y W
C2 2037 G C S T P L G M E N G K I E N K Q I T A S S F K K S W W G D Y W

E P R L A R L N N G G S Y N A W S V E K L A A E F A S K P W I Q V D M Q
E P F R A R L N A Q G R V N A W Q - - - - A K A N N N K Q W L E I D L L

K E V I I T G I Q T Q G A K H Y L K S C Y T T E F Y V A Y S S N Q I N W
K I K K I T A I I T Q G C K S L S S E M Y V K S Y T I H Y S E Q G V E W

Q I F K G N S T R N V M Y F N G N S D A S T I K E N Q F D P P I V A R Y
K P Y R L K S S M V D K I F E G N T N T K G H V K N F F N P P I I S R F

I R I S P T R A Y N R P T L R L E L Q G C E V N 2036
I R V I P K T W N Q S I T L R L E L F G C D I Y 2196
```

(a)

```
Factor VIII 1 A T R R - Y Y L G A V E L S W D Y M Q S D L G E L P V
Factor V    1 A Q L R Q F Y V A A Q G I S W S Y R P E - - - - - P T

D A R F P P R V P K S F P F N T S V V Y K K T L F V E F T D H L F N I
N S S L N L S V T S - - - - - - - - - - F K K I V Y R E Y - E P Y F K K

A K P R P P W M G L L G P T I Q A E V Y D T V V I T L K N M A S H P V
E K P Q S T I S G L L G P T L Y A E V G D I I K V H F K N K A D K P L

S L H A V G V S Y W K A S E G A E Y D D Q T S Q R E K E D D K V F P G
S I H P Q G I R Y S K L S E G A S Y L D H T F P A E K M D D A V A P G

G S H T Y V W Q V L K E N G P M A S D P L C L T Y S Y L S H V D L V K
R E Y T Y E W S I S E D S G P T H D D P P C L T H I Y Y S H E N L I E

D L N S G L I G A L L V C R E G S L A K E K T Q - T L H K F I - L L F
D F N S G L I G P L L I C K K G T L T E G G T Q K T F D K Q I V L L F

A V F D E G K S W H S E T K N S L M Q D R D A A S A R A W P K M H T V
A V F D E S K S W - - - - - - - - - - - - - - - - S Q S S S L M Y T V

N G Y V N R S L P G L I G C H R K S V Y W H V I G M G T T P E V H S I
N G Y V N G T M P D I T V C A H D H I S W H L L G M S S G P E L F S I

F L E G H T F L V R N H R Q A S L E I S P I T F L T A Q T L L M D L G
H F N G Q V L E Q N H H K V S A I T L V S A T S T T A N M T V G P E G

Q F L L F C H I S S H Q H D G M E A Y V K V D S C P E E P Q L R M K N
K W I I S S L T P K H L Q A G M Q A Y I D I K N C P K - - - - - - - -

N E E A E D Y D D D L T D S E M D V V R F D D D N S P S F I Q I R S V
- - - - - - - - - - - - - - - - - - - - - - - - K T R N L K K I T R E

A K K H P K T W V H Y I A A E E E D W D Y A P L V L A P D D R S Y K S
Q R R H M K R W E Y F I A A E E V I W D Y A P V I P A N M D K K Y R S

Q Y L N N G P Q R I G R K Y K K V R F M A Y T D E T F K T R E A I Q H
Q H L D N F S N Q I G K H Y K K V M Y T Q Y E D E S F T K H T V N - P

- - - E S G I L G P L L Y G E V G D T L L I I F K N Q A S R P Y N I Y
N M K E D G I L G P I I R A Q V R D T L K I V F K N M A S R P Y S I Y

P H G I T D V R P L Y S R R L P K G V K H L - - K D F P I L P G E I F
P H G V T F S P Y E D E V N S S F T S G R N N T M I R A V Q P G E T Y

K Y K W T V T V E D G P T K S D P R C L T R Y Y S S F V N M E R D L A
T Y K W N I L E F D E P T E N D A Q C L T R P Y Y S D V D I M R D I A

S G L I G P L L I C Y K E S V D Q R G N Q I M S D K R N V I L F S V F
S G L I G L L L I C K S R S L D R R G I Q R A A D I E Q Q A V F A V F

D E N R S W Y L T E N I Q R F L P N P A G V Q L E D P E F Q A S N I M
D E N K S W Y L E D N I N K F C E N P D E V K R D D P K F Y E S N I M

H S I N G Y V F D S L Q - L S V C L H E V A Y W Y I L S I G A Q T D F
S T I N G Y V P E S I T T L G F C F D D T V Q W H F C S V G T Q N E I

L S V F F S G Y T F K H K M V Y E D T L T L F P F S G E T V F M S M E
L T I H F T G H S F I Y G K R H E D T L T L F P M R G E S V T V T M D

N P G L W I L G C H N S D F R N R G M T A L L K V S S C 711
N V G T W M L T S M N S S P R S K K L R L K F R D V K C 656
```

(b)

```
Factor VIII  1689 R S F Q K K T R H Y F I A A V E R L W D Y G - M S
Factor V     1545 R S N N G N R R N Y Y I A A E E I S W D Y S E F V

S S P H V L R N R A Q S G S V P Q F K K V V F Q E F T D G S F T Q P L
Q R E T D I E D S D D I P E D T T Y K K V V F R K Y L D S T F T K R D

Y R G E L N E H L G L L G P Y I R A E V E D N I M V T F R N Q A S R P
P R G E Y E E H L G I L G P I I R A E V D D V I Q V R F K N L A S R P

Y S F Y S S L I S Y E - - - - - - - - E D Q R Q G A E P R K N F V K P
Y S L H A H G L S Y E K S S E G K T Y E D D S P E W F K E D N A V Q P

N E T K T Y F W K V Q H H M A P T K D E F D C K A W A Y F S D V D L E
N S S Y T Y V W H A T E R S G P E S P G S A C R A W A Y Y S A V N P E

K D V H S G L I G P L L V C H T N T L N P A H G R Q V T V Q E F A L F
K D I H S G L I G P L L I C Q K G I L H K D S N M P V D M R E F V L L

F T I F D E T K S W Y F T E N M E R N C R A P C N I Q M E D P T F K E
F M T F D E K K S W Y Y E K - - - - - - - K S R S S W R L T S S E M K K

N Y R F H A I N G Y I M D T L P G L V M A Q D Q R I R W Y L L S M G S
S H E F H A I N G M I Y - S L P G L K M Y E Q E W V R L H L L N I G G

N E N I H S I H F S G H V F T V R K K E E Y K M A L Y N L Y P G V F E
S Q D I H V V H F H G Q T L L E N G N K Q H Q L G V W P L L P G S F K

T V E M L P S K A G I W R V E C L I G E H L H A G M S T L F L V Y S N
T L E M K A S K P G W W L L N T E V G E N Q R A G M Q T P F L I M D R

K C Q T P L G M A S G H I R D F Q I T A S G Q Y G Q W A P K L A R L H
D C R M P M G L S T G I I S D S Q I K A S E F L G Y W E P R L A R L N

Y S G S I N A W S T K E - - - P F S - - - W I K V D L L A P M I I H G
N G G S Y N A W S V E K L A A E F A S K P W I Q V D M Q K E V T T T G

I K T Q G A R Q K F S S L Y I S Q F I I M Y S L D G K K W Q T Y R G N
I Q T Q G A K H Y L K S C Y T T E F Y V A Y S S N Q I N W Q I F K G N

S T G T L M V F F G N V D S S G I K H N I F N P P I I A R Y I R L H P
S T R N V M Y F N G N S D A S T I K E N Q F D P P I V A R Y I R I S P

T H Y S I R S T L R M E L M G C D L N S C S M P L G M E S K A I S D A
T R A Y N R P T L R L E L Q G C E V N G C S T P L G M E N G K I E N K

Q I T A S S Y F T N M F A T - W S P S K A R L H L Q G R S N A W R P Q
Q I T A S S F K K S W W G D Y W E P F R A R L N A Q G R V N A W Q A K

V N N P K E W L Q V D F Q K T M K V T G V T T Q G V K S L L T S M Y V
A N N N K Q W L E I D L L K I K K I T A I I T Q G C K S L S S E M Y V

K E F L I S S S Q D G H Q W T L F F Q N G K V K V - - F Q G N Q D S F
K S Y T I H Y S E Q G V E W K P Y R L K S S M V D K I F E G N T N T K

T P V V N S L D P P L L T R Y L R I H P Q S W V H Q I A L R M E V L G
G H V K N F F N P P I I S R F I R V I P K T W N Q S I T L R L E L F G

C E A Q D L Y 2332
C D - - - I Y 2196
```

Fig. 25. (a) Comparison of the amino-acid sequences of A1 and A2 domains of factors V and VIII. Only the positions at which amino acids are the same in both chains are outlined. The degree of identity is included in Table 5. (b) Comparison of the amino-acid sequences of A3, C1 and C2 domains of factors V and VIII. Only the positions at which amino acids are the same in both chains are outlined. The degrees of identity are shown in Table 5.

Table 5. *Comparison of the domains in factor VIII, factor V and ceruloplasmin (numerical values taken from reference 366). The A domains are numbered 1, 2 and 3, while the C domains are numbered 1 and 2. V, VIII and CP designate factor V, factor VIII and ceruloplasmin respectively.*

A. Percentage of identical residues among A domains

	VA1	VA2	VA3	VIIIA1	VIIIA2	VIIIA3	CPA1	CPA2	CPA3
VA1	100	31	32	36	30	27	31	32	21
VA2	31	100	29	29	44	31	35	34	35
VA3	32	29	100	33	31	39	32	34	39
VIIIA1	36	29	33	100	30	32	34	32	33
VIIIA2	30	44	31	30	100	33	38	35	36
VIIIA3	27	31	39	32	33	100	28	34	36
CPA1	31	35	32	34	38	28	100	37	39
CPA2	32	34	34	32	35	34	37	100	40
CPA3	21	35	39	33	36	36	39	40	100

B. Percentage of identical residues among C domains

	VC1	VC2	VIIIC1	VIIIC2
VC1	100	37	44	34
VC2	37	100	41	42
VIIIC1	44	41	100	38
VIIIC2	34	42	38	100

that appears in SDS-polyacrylamide gel electrophoresis as a double band with molecular masses 71 kDa and 74 kDa (378). Comparison of the amino-acid sequence of factor V with the N-terminal sequences of activation fragments from human and bovine factor V has shown that thrombin cleaves the bonds at Arg709–Ser710, Arg1018–Thr1019 and Arg1545–Ser1546 in human factor V. There is at present conflict between results on the order of appearance of the various fragments. However, there is agreement that, for human factor V, a low factor Va activity arises after the cleavage of the first two of the bonds mentioned, and that this results in the formation of the 105-kDa N-terminal fragment and a large C-terminal fragment. The cleavage of the latter at Arg1545–Ser1546 leads to a higher factor Va activity and to the 71–74-kDa fragment. The difference between the two light chains presumably lies in the C terminus or the mass of the attached carbohydrate.

This means that the heavy chain of factor Va consists largely of two A domains (A1 and A2), while the light chain of factor Va consists of an A domain (A3) and the two C domains, while the B domain is split off in two parts during the activation. This resembles the situation for factor VIII.

Bovine factor Xa is also able to activate bovine factor V to give factor Va, but the splitting pattern has been shown to be a little different from that of activation by thrombin (380). Thrombin-activated factor Va can further be split by factor Xa without loss of activity (381). Its heavy chain is cleaved at the bond that corresponds to Arg348–Ser349 in human factor V, so that there arise two fragments with molecular masses 56 kDa and 45 kDa, while the light chain is cleaved at the bond corresponding to Arg1765–Leu1766 in human factor V, so that this is converted to two chains with molecular masses of 30 kDa and 48 kDa (382).

Genetically caused deficiency of factor V is termed parahaemophilia. This is a rare disease (358, 383, 384); only about 150 cases of it have been reported (326), and none of these has been characterised at the molecular level.

5.1.2. *Prothrombin*

Like many of the other proteins that we have encountered, prothrombin, or factor II, is the zymogen of a serine proteinase that takes part in blood coagulation. Prothrombin is a single-chained glycoprotein with a molecular mass of 72 kDa as estimated by SDS-polyacrylamide gel electrophoresis; about 8% is carbohydrate (308). Its concentration in normal human plasma lies around 80 μg/ml ($\approx$ 1.1 μM). Prothrombin is also a vitamin K-dependent protein, the third that we meet in the coagulation system.

The amino-acid sequences of bovine and human prothrombin have been determined from the proteins (35, 385, 386) and from the respective cDNA sequences (387, 388). The cDNA sequence for bovine prothrombin includes a 43-amino-acid signal/propeptide (387), while the human cDNA includes only a 36-residue portion of this (388). The genes for human and bovine prothrombin have been characterised. The human prothrombin gene is located on chromosome 11 at position p11–q12 (246). It has been sequenced completely and possesses 14 exons; in all, it covers 20.2 kbp of genomic DNA (389). The exons are between 25 bp and 315 bp in length, the introns between 84 bp and 9447 bp. The gene sequence also includes the complete 43-residue signal/propeptide. The bovine gene has been characterised by the determination of the intron–exon junctions and analysis of their 5′ and 3′ flanking sequences (390, 391). This gene covers about 15.4 kbp of genomic DNA distributed over 14 exons and 13 introns that are located at positions corresponding to those in the human prothrombin gene. The sizes of the exons in the bovine gene are the same as those in the human gene, while the introns vary in size from 75 bp to about 7 kbp. Figure 26 shows the positions of the introns in the prothrombin gene in relation to the amino-acid sequence of prothrombin.

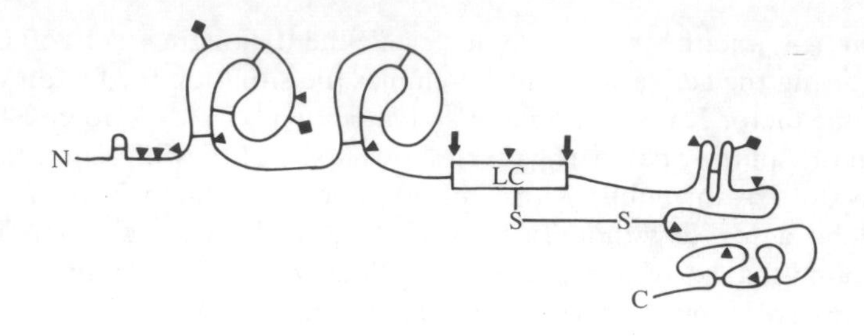

Fig. 26. The domain structure of prothrombin. → shows the positions of cleavage by factor Xa, ▼ the positions of the introns and ◆ the N-bound carbohydrate. (LC, light chain of thrombin.)

The molecular mass of the amino-acid sequence deduced from the cDNA is 65.7 kDa, giving some 71.5 kDa after inclusion of the 8% carbohydrate. Human prothrombin consists of 579 amino acids in a single chain (Appendix, Fig. A9). If the sequence derived from the cDNA is compared with published results from protein-sequencing, then some small differences are found (388), especially the absence from the cDNA-derived sequence of Glu266 and Glu267 found in the protein sequence (385). The overall structure of prothrombin is shown in Figure 26. The N terminus of prothrombin contains a Gla domain with 10 Gla residues among the first 32 amino acids of the chain. Thereafter, there appear two internally homologous domains called kringles (Fig. 27); we have seen one of these before, in factor XII, but they were first found in prothrombin (35). Finally, at the C terminus, we find the region that after activation of prothrombin by factor Xa becomes the active serine proteinase.

The nomenclature of the prothrombin and prothrombin activation intermediates that we shall use is shown in Figure 28. Prothrombin fragment 1.2 contains the Gla domain and the two kringles distributed in such a way that prothrombin fragment 1 includes the Gla domain and kringle 1, and prothrombin fragment 2 contains kringle 2. The rest of the prothrombin, that is to say the part that is not fragment 1.2, is termed prethrombin 2.

The disulphide-bond pattern in prothrombin has been determined for the bovine protein (35). It corresponds to the following disulphide bonds in human prothrombin: in the Gla domain, Cys17–Cys22; in the stretch between this and the first kringle, Cys47–Cys60; in the first kringle domain, Cys65–Cys143, Cys86–Cys126 and Cys114–Cys138; in the second kringle domain, Cys170–Cys248, Cys191–Cys231 and Cys219–Cys243; and in prethrombin 2, Cys293–Cys439, Cys348–Cys364, Cys493–Cys507 and Cys521–Cys551. The carbohydrate is bound to Asn78, Asn100 and Asn373. In contrast to the other vitamin K-dependent proteins that we have seen, neither human nor bovine prothrombin contains β-OH-Asp (141).

```
 65 C A E G L G T N Y R G H V N I T R S G I E C Q L W R S R Y P H
170 C V P D R G Q Q Y Q G R L A V T T H G L P C L A W A S A Q A K

K P E I N S T T H P G A D L Q E N F C R N P D S S N T G P W C Y T
A L S K H Q D F N S A V Q L V E N F C R N P D G D E E G V W C Y V

T D P T V R R Q E C S I P V C 143
A G K P G D F G Y C D L N Y C 248
```

Fig. 27. Internally homologous kringle domains in prothrombin. Only the positions at which amino acids are the same in both domains are outlined. The degree of identity is 30%.

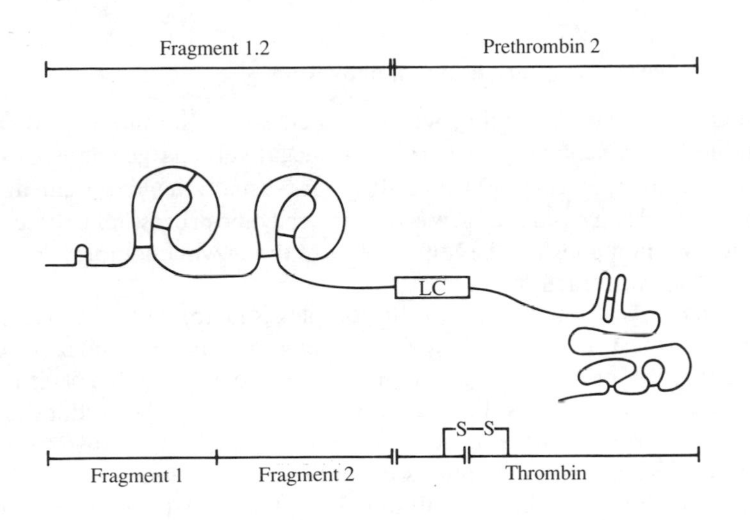

Fig. 28. Nomenclature for prothrombin fragments. (LC, light chain of thrombin.)

The activation of human prothrombin by factor Xa to give thrombin occurs with the proteolytic cleavage of two bonds: Arg271–Thr272 and Arg320–Ile321. A large activation peptide (33.5 kDa) and the catalytically active two-chained thrombin are formed. Thrombin consists of a light chain (6 kDa) disulphide-bonded to a heavy chain from the C terminus of prothrombin (32 kDa) that contains the catalytic site. This site is believed to involve His43, Asp99 and Ser205 in the heavy chain of thrombin, corresponding to His363, Asp419 and Ser525 in prothrombin.

Prothrombin fragment 1 has been crystallised and the three-dimensional structures of the kringle domain (392) and the Gla domain (393) have been solved with a resolution of 2.8 Å (0.28 nm). An interesting feature of the structure of the Gla domain in the presence of calcium ions is that six out of the ten Gla residues are located at the same edge of the Gla domain, and thus of fragment 1. In addition, the crystal structure of thrombin in complex with the thrombin inhibitor D-Phe-Pro-Arg-chloromethyl ketone has been solved to a resolution of 1.9 Å (0.19 nm) (394).

Although genetically caused deficiencies of prothrombin activity are rare, with fewer than 20 reported cases (see, for example, reference 395), five prothrombin variants have been characterised at the molecular level. In prothrombin$_{\text{Barcelona}}$ and prothrombin$_{\text{Madrid}}$ a single cleavage site is affected, with the replacement of Arg271 by Cys (163, 396). The other three variants contain amino-acid replacements in the proteinase domain. In prothrombin$_{\text{Tokushima}}$, Arg418 is replaced by Trp (397), while prothrombin$_{\text{Quick}}$ is the unusual instance of a person having two different thrombin molecules, both defective. Prothrombin$_{\text{Quick I}}$ contains the substitution Arg382→Cys and prothrombin$_{\text{Quick II}}$ the substitution Gly558→Val (398, 399).

5.2. Mechanism of prothrombin activation

As mentioned at the beginning of this chapter, the prothrombinase complex contains factor Xa, factor Va, Ca^{2+} and a negatively charged phospholipid (reviewed in references 400 and 401). As usual, the requirement for phospholipid as a co-factor means that the enzymic process takes place at a surface. We now consider the components of the enzymic complex and their interactions with each other.

The enzymically active part of the complex is factor Xa, which is a two-chained serine proteinase built up from a heavy and a light chain connected by a disulphide bridge. The light chain (17 kDa) comes from the N terminus of factor X and contains the Gla domain with its 11 Gla residues and a single β-OH-Asp residue; the heavy chain (31 kDa) is derived from the C terminus of factor X and contains the entire catalytic unit.

The Gla domain at the N terminus ensures that both factor X and factor Xa can bind to acidic phospholipids, with the help of calcium ions. If the Gla domain is removed from the light chain of bovine factor Xa, then both the ability to bind phospholipids and the activity in coagulation are abolished (residual activity <0.1%), even though the enzyme's activity with respect to small oligopeptide substrates remains unaltered (402, 403). Factor Xa's calcium-dependent binding to phospholipid is thus completely dependent upon the Gla residues. Bovine factor X/Xa contains a Gla-independent high-affinity binding site for calcium that is of importance for the molecule's conformation and thus possibly for its function (404). This binding site is probably located in the first growth-factor domain, as the isolated domain has been shown to bind calcium (405).

The non-enzymic co-factor in the prothrombinase complex is factor Va, which, as described in section 5.1.1, consists of two fragments from factor V, a heavy chain from the N terminus (105 kDa for human, 94 kDa for bovine factor Va) and a light chain from the C terminus (71–74 kDa for human, 74 kDa for bovine factor Va); these are held together in a manner dependent upon calcium.

We now examine the interactions between the various components of prothrombinase. Nearly all of the relevant studies have been performed with bovine coagulation factors, but it is not expected that there will be substantial differences between the bovine and the human systems. In short, factors Xa and Va form a 1:1 stoichiometric complex on the negatively charged lipid surface in the presence of calcium ions. It is not known what part of factor Xa binds to factor Va, but it is known that the absence of the Gla domain affects factor Xa's affinity for factor Va (402).

The regions in factor Va that are related to the binding of factor Xa are located in both chains. This is indicated by various lines of evidence. Studies with monoclonal antibodies against epitopes in the 105-kDa and 71–74-kDa chains from human factor Va have shown that factor Xa competes with these antibodies for binding to factor Va, which in turn indicates that both of the factor Va chains participate in interactions with factor Xa (406).

Studies have also been carried out with monoclonal antibodies against epitopes in the 94-kDa or the 74-kDa chains from bovine factor Va. None of these interfered with the binding to the phospholipid. Furthermore, the antibody against the 74-kDa chain blocked the binding of factor Xa to the phospholipid-bound factor Va, which indicates that the 74-kDa chain takes part in the binding of factor Xa (407). Three lines of evidence support the belief that both chains of bovine factor Va play an important part in the binding of factor Xa. (i) Factor Va binds to immobilised factor Xa, while neither of the isolated factor Va chains does so, suggesting that both chains are necessary for an optimal interaction between factor Va and factor Xa (408). (ii) The inactivation of factor Va by protein Ca is accompanied by the loss of the ability to bind to immobilised factor Xa (408), and this inactivation is correlated with the proteolytic cleavage of the heavy chain of both human and bovine factor Va (409, 410). (iii) The 74-kDa chain fails to bind factor Xa to membrane surfaces, even though intact factor Va does so (411).

The binding of bovine factors V and Va to phospholipid does not require exogenous calcium, but it is dependent upon negatively charged phospholipids (412–414). It is the light chain (74 kDa) of bovine factor Va that can bind to membranes (411, 414, 415), while the heavy chain (94 kDa) is located some distance away from the membrane surface (416). Even though the two C domains in the light chain of factor Va are homologous to discoidin I (see section 5.1.1), which itself has some affinity for negatively charged phospholipid, it appears rather to be the A3 domain that mediates the surface binding of factor Va (417). The argument for this is that the N-terminal product of the cleavage of the light chain of factor Va by factor Xa contains most of the A3 domain and none of the C domains, and precisely this product can displace surface-bound factor Va. The same study also indicated that factor Va's binding site on the phospholipid is defined by a particular number of anionic phospholipid molecules (417). The character of the interaction between factor V/Va and negatively charged phospholipid is at present unknown. Some results speak for an interaction that is primarily electrostatic (411, 413, 414), while others support a primarily hydrophobic interaction (412, 415). This disagreement, along with discrepancies in the measured dissociation constants (from 10^{-7} to 10^{-11} M), may be due to differences in experimental technique. Lipophilic light-sensitive reagents become attached to the light chain of factor Va (418, 419), which is taken as a confirmation that the interaction is basically hydrophobic, although the efficiency of labelling of factor Va was considerably lower than for integral membrane proteins (418).

The formation of the bovine prothrombinase complex from factor Xa, factor Va, negatively charged phospholipid and Ca^{2+} depends upon the initial formation of protein–phospholipid complexes (420). This means that factor Xa must be thought of as binding to a phospholipid surface with the help of calcium, and factor Va also binds to this before the two proteins meet and form the prothrombinase complex. This complex formation between factor Xa and factor Va appears to affect significantly the environment around the active site of factor Xa (421, 422).

The prothrombinase complex is thus assembled at the surface and is ready to activate prothrombin (Fig. 29). Where does the prothrombin come from?

We again have two possibilities, as we had in the case of the Xase complex (section 4.2). The substrate of prothrombinase can come either from the solution or from the phospholipid surface, for which prothrombin possesses some affinity. The phospholipid-binding domain in prothrombin is localised in fragment 1 (423, 424), and, as expected, the binding of prothrombin by anionic phospholipid is Gla-dependent. The absence of only three of the Gla residues abolishes the binding of phospholipid by prothrombin (425). The isolated Gla domain (residues 1–45) binds to negatively charged phospholipid in the presence of calcium ions, but with an association constant ten times greater than that of fragment 1 (426, 427). Prothrombin binds to prothrombinase by way of the heavy chain in factor Va (408, 428), but the interaction between these two is independent of calcium and phospholipid, and in prothrombin it can be attributed to fragment 2 (429, 430).

Before discussing the choice of substrate, we shall examine the activation of prothrombin to thrombin. Since two peptide bonds are broken, this must clearly proceed in at least two steps.

We look first at the activation of bovine prothrombin by bovine prothrombinase (Fig. 30). The first bond to be cleaved is Arg323–Ile324, leading to the formation of the two-chained meizothrombin. After this, the bond Arg274–Thr275 is broken, forming thrombin and fragment 1.2 (431, 432). In the absence of purified factor Va, these two bonds are cleaved in the reverse order (429, 433–435); this observation naturally led to the hypothesis that the formation of thrombin from prothrombin takes place

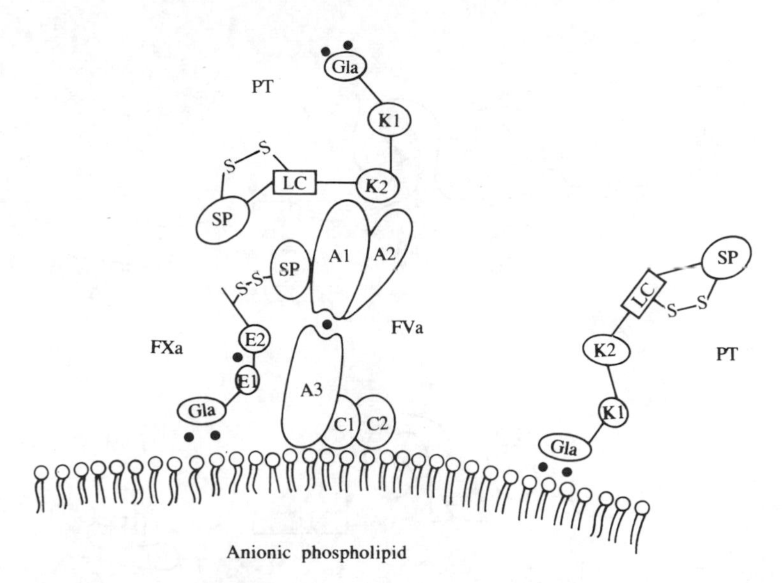

Fig. 29. A schematic representation of the prothrombinase enzyme complex. F, factor; PT, prothrombin; •, Ca^{2+}. Gla and SP represent Gla and serine-proteinase domains in factor Xa and prothrombin. E1 and E2 represent the first and second growth-factor domains in factor Xa. K1, K2 and LC represent the first and second kringle domains and thrombin's light chain in prothrombin. A1, A2, A3, C1 and C2 are the domains in factor Va. It is not known whether the prothrombinase enzyme complex accepts prothrombin from the fluid phase or from the surface.

by way of a non-covalent complex between fragment 1.2 and prethrombin 2 (436). Studies on the activation of prothrombin are complicated by the fact that both thrombin and meizothrombin cleave the bond (Arg156–Thr157) between fragments 1 and 2 of prothrombin.

The analogous activation of human prothrombin to thrombin by purified human prothrombinase is more complicated still, as human prothrombin can be cleaved at two points by thrombin and meizothrombin: at the bonds Arg155–Ser156 and Arg284–Ser285. As we saw for bovine prothrombin, the activation catalysed by prothrombinase leads first to meizothrombin by cleavage of the bond Arg320–Ile321. Subsequent cleavage of Arg271–Thr272 gives fragment 1.2 and thrombin (437). In the absence of factor Va, the initial cleavage takes place at Arg271–Thr272 (437). In contrast to activation in systems with purified components, the activation of human prothrombin in human plasma leads to the formation of fragment 1.2.3 as a principal product, since the bond Arg284–Ser285 is cleaved before Arg271–Thr272 (438). However, the significance of this is not known.

Kinetic investigations of the bovine prothrombinase complex (439, 440) show that factor Xa alone is able to activate prothrombin, even though this

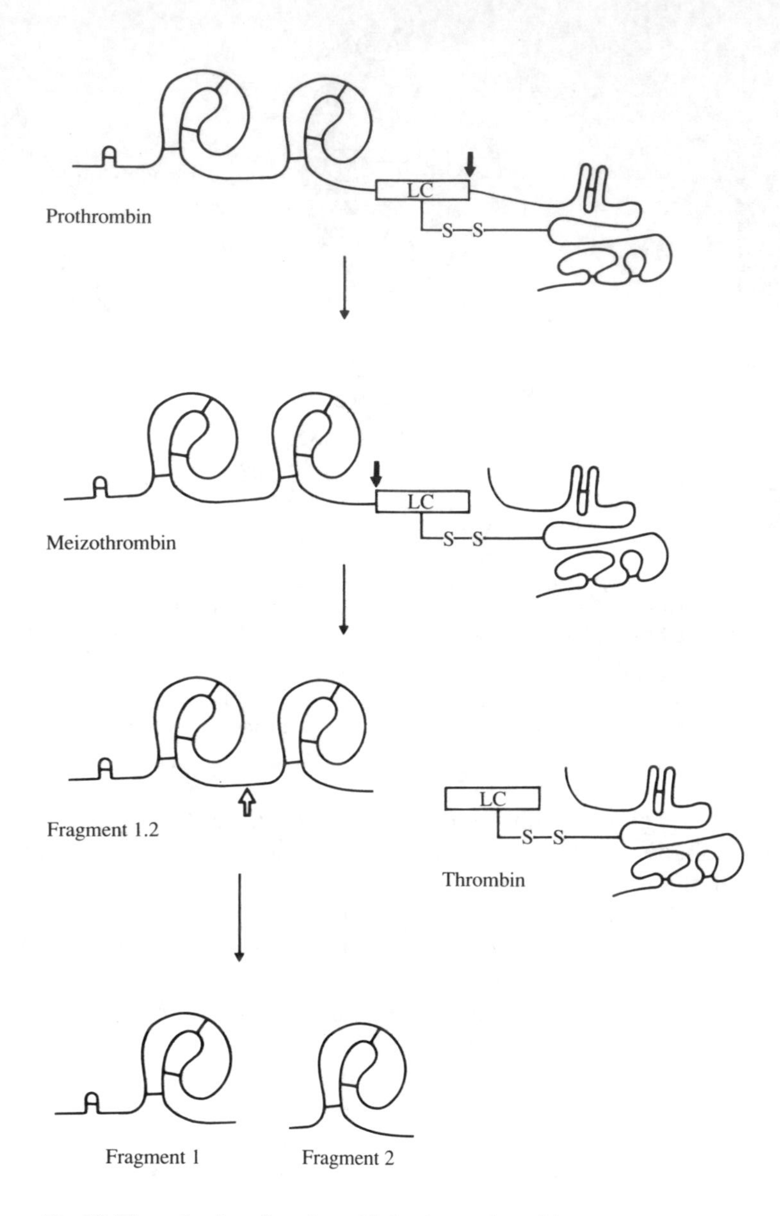

Fig. 30. The activation of prothrombin by the prothrombinase enzyme complex. → shows the sites of cleavage by factor Xa and ⇒ the sites of cleavage by thrombin and meizothrombin. (LC, light chain of thrombin.)

only happens very slowly. The addition of calcium has little effect on this, but the simultaneous presence of calcium and of negatively charged phospholipid depresses K_m to a value much lower than the concentration of prothrombin in plasma. If the prothrombinase complex is now made complete by the further addition of factor Va, then V_{max} rises dramatically, so that the complete prothrombinase complex activates prothrombin about 300 000 times faster than does factor Xa alone. The simplest – while not necessarily correct – interpretation of this in terms of the individual components of prothrombinase is that anionic phospholipid lowers K_m for prothrombin, while factor Va raises the value of V_{max} for the reaction.

There is no doubt that factor Va raises the value of V_{max} for the activation of prothrombin in at least three ways. First of all, factor Va strengthens the binding of factor Xa to the coagulation-active surface (421, 441, 442). Secondly, factor Va promotes specifically the cleavage of the bond Arg323–Ile324, at the expense of the cleavage of the bond Arg274–Thr275 (443), which raises k_{cat}. Thirdly, factor Va promotes the formation of a complex between factor Xa and prothrombin ([enzyme.substrate] complex). This last point touches upon the issue of the choice of substrate, and this calls for more detailed explanation.

The situation considered here is completely analogous to that which we discussed for the activation of factor X with the Xase complex (section 4.2). In that instance, kinetic results indicated that it is the fluid-phase factor X that is used as a substrate, rather than surface-bound factor X (350). For prothrombinase, there are basically two current conceptions of how the enzyme may function. According to one model, all three components have affinity for negatively charged phospholipid, so that the surface concentrations of all three are increased, thus facilitating the formation of the [enzyme.substrate] complex. The rate of the reaction is thus determined by the local concentration of prothrombinase and prothrombin on the phospholipid surface (439, 440, 444). The second model proposes that the presence of a anionic phospholipid membrane could give a better binding of substrate to the enzyme complex because of more favourable free energies caused by protein–protein and protein–membrane interactions around the active site (445).

A consequence of the first model is that the apparent K_m value for prothrombin is determined by the prothrombin concentration at the phospholipid surface, whereas the second model predicts that K_m should be a function of the concentration of free prothrombin. It has been shown that the binding of prothrombin and of factor X to artificial phospholipid surfaces depends upon the proportion of negatively charged lipids (such as phosphatidyl serine) present. If the phosphatidyl serine content in artificial phospholipid vesicles is reduced below 30%, then the binding of prothrombin and factor X decreases in proportion to the phosphatidyl serine content (446). Investigation of the way in which the proportion of phosphatidyl serine in artificial lipid vesicles affects K_m for the factor Xa-catalysed activation of prothrombin in the presence and absence of factor Va can thus reveal whether it is bulk or surface-bound prothrombin that determines the kinetics of the reaction. Experiments of this kind show that, in the absence of factor Va, K_m rises as a consequence of the reduced phosphatidyl serine content in phospholipid vesicles, while, in the presence of factor Va, no significant change in K_m (or in V_{max}) occurs, even at very low phosphatidyl serine levels, such as 2% (447, 448).

It thus appears that there is no correlation between the observed K_m values for the activation of prothrombin and the amount of prothrombin bound to the phospholipid. The K_m value seems rather to reflect a heightened affinity between the prothrombinase complex and free prothrombin.

There is as yet no broad agreement on which model is correct, but, as mentioned in the description of the Xase system (section 4.2), it seems that the model in which the substrate comes directly from the fluid phase is to be preferred. The justification for this is that natural coagulation membranes do not contain nearly as much negatively charged phospholipid as do the artificial lipids generally used in the experimental systems referred to here. The use of a high proportion of, for example, phosphatidyl serine in these experiments gives a substrate binding to the vesicles that scarcely has its counterpart *in vivo*, and which ultimately is the cause of the entire debate surrounding the choice of substrate.

Finally, it is appropriate to re-emphasize that the prothrombinase complex binds prothrombin, after which factor Xa cleaves Arg320–Ile321 (in human prothrombin), resulting in meizothrombin. This meizothrombin is then cleaved at the bond Arg271–Thr272 (or, alternatively, Arg284–Thr285) with the formation of fragment 1.2 and thrombin (or, alternatively, fragment 1.2.3 and thrombin with 13 amino acids fewer in the N terminus of the light chain). The thrombin formed is released from both fragment 1.2 and the prothrombinase enzyme complex, and it ends up as the first active enzyme in free solution (bulk phase), where it encounters substrates such as fibrinogen and factor XIII.

6.

Activations in the extrinsic pathway

6.1. Components of the extrinsic pathway

We have already seen how factor X is activated in the intrinsic pathway with the help of the enzyme complex Xase, which consists of factor IXa, factor VIIIa, calcium ions and negatively charged phospholipid. The present chapter is in the main a description of how the same activation takes place in the extrinsic pathway with the help of factor VIIa and tissue factor (reviewed in reference 449). Intrinsic and extrinsic pathways converge upon a fixed point – the activation of factor X.

6.1.1. *Factor VII*

Factor VII (earlier called pro-convertin) is the zymogen for an active serine proteinase in plasma that participates exclusively in the reactions of blood coagulation known collectively as the extrinsic pathway. Factor VII is a single-chained, vitamin K-dependent glycoprotein with a molecular mass of about 50 kDa (450, 451), of which some 10% is carbohydrate. The concentration of factor VII in normal human plasma is low, approximately 0.5 μg/ml ($\approx$ 10 nM) (452).

The amino-acid sequence of human factor VII was first deduced from the cDNA sequence (453) (Appendix, Fig. A10), and, since then, peptide sequences covering the whole of factor VII have been published (454). Bovine factor VII has also been sequenced completely at the protein level (455). The cDNA sequence for human factor VII includes a signal/propeptide of either 60 or 38 amino-acid residues (see below). The gene for human factor VII has also been sequenced completely. It covers some 12.8 kbp of genomic DNA (456). The gene has nine exons, from 25 bp to 1.6 kbp in length and separated by introns with lengths between 68 bp and 2574 bp. The unsettled issue of the length of the signal/propeptide was resolved when it was found that the second exon in the factor VII gene (exon 1b) is frequently removed by alternative splicing in the splicing process that leads

Fig. 31. The domain structure of factor VII. → shows the position of cleavage by factor Xa, ▼ the positions of the introns, ◆ the N-bound carbohydrate and ● the O-bound carbohydrate (see text).

to mature mRNA. If this exon is removed, then a signal/propeptide of 38 residues results, while if this exon is retained, the signal/propeptide is 60 residues long. Figure 31 shows the positions of the introns with respect to the amino-acid sequence. The chromosomal location of the gene for factor VII is position 13q34 (457–459), as revealed by chromosomal abnormalities that resulted in a deficiency of factor VII. It may be noted that the gene for factor X is located at the same position.

The molecular mass of human factor VII calculated from the sequence is 45.5 kDa, which with the inclusion of the 10% carbohydrate gives approximately 50 kDa. Human factor VII contains 406 amino-acid residues (Fig. 31), and it shows considerable similarity to factors IX and X. From the N terminus, factor VII contains a Gla domain with 10 Gla residues among the first 35 amino acids (454), after which there follow two growth-factor domains homologous to EGF. The C-terminal region of the factor is the potential serine-proteinase domain.

Bovine factor VII contains an additional Gla residue in the Gla domain, so that the total number of Gla residues is 11. It also contains a partially (20–30%) hydroxylated Asp residue at position 63, in the first growth-factor domain (139, 455). There is no β-OH-Asp at this position in human factor VII, in spite of the fact that the position is occupied by aspartic acid (454).

Human factor VII has carbohydrate chains N-bound at Asn145 and Asn322 (454), while bovine factor VII not only contains N-bound carbohydrate at Asn145 and Asn204 but also, like bovine factor IX, contains a trisaccharide with the formula ((xylose$_2$)glucose) that in factor VII is bound to Ser52 (137). In human factor VII, Ser52 is either modified by attachment of the same trisaccharide or by attachment of a xylose-glucose disaccharide as in the case of human factor IX (138).

The disulphide-bond pattern of factor VII has not been determined by experiment, but on the basis of homology with, for example, factor X it can be assumed to be as follows: in the Gla domain, Cys17–Cys22; in the first growth-factor domain, Cys50–Cys61, Cys55–Cys70 and Cys72–Cys81; in the second growth-factor domain, Cys91–Cys102, Cys98–Cys112 and Cys114–Cys127; and in the serine-proteinase domain, Cys135–Cys262, Cys159–Cys164, Cys178–Cys194, Cys310–Cys329 and Cys340–Cys368.

The activation of factor VII to give factor VIIa is catalysed by factor Xa and possibly by factor IXa. It is accompanied by the cleavage of the bond Arg152–Ile153 (453, 460), so factor VIIa is a two-chained serine proteinase with a light chain derived from the N terminus (152 amino acids, 20 kDa) disulphide-bonded to a heavy chain from the C terminus (254 amino acids, 30 kDa) that contains the catalytic groups. It is believed on the basis of homology considerations that a disulphide bond in factor VII connects cysteines 135 and 262 and holds the two chains together after activation, and likewise that the catalytic groups of factor VIIa are His41, Asp90 and Ser192 in the heavy chain, corresponding to His193, Asp242 and Ser344 in native factor VII.

An interesting feature of both bovine and human factors VII is their ability to react with DFP (450, 461), a reaction of which normally only active serine proteases (including activated serine proteinases) are capable. Single-chained bovine factor VII has a certain esterase activity with artificial substrates (462), and this has led to the suggestion that factor VII may have coagulation activity before it is activated (463). We shall return later to this point.

Factor VII deficiency was first described in 1951 (464), and since then about 150 cases have been documented (326).

6.1.2. *Tissue factor*

Tissue factor – also termed thromboplastin or factor III – is a lipid-dependent glycoprotein, found localised in many different tissue types (465). Under normal circumstances, both endothelial cells and circulating cells are regarded as inactive in the initiation of coagulation by way of the tissue factor system, while cells outside the vascular system can contain a large quantity of exposed tissue factor. Tissue factor consists of a protein component (the apoprotein) and lipid. Neither of these two components in purified form possesses coagulation activity, but the combination of these restores this activity (466, 467). The interactions between the phospholipid and the protein parts are purely hydrophobic and independent of calcium ions (468).

Tissue factor apoprotein was originally reported to have a molecular mass between 270 kDa and 330 kDa (469), on the basis of ultracentrifugation and non-denaturing electrophoresis. It was later shown that purified bovine (470) and human (471–474) tissue factor apoproteins are single-chained glycoproteins with a molecular mass, estimated by SDS-polyacrylamide gel electrophoresis, in the range 43–53 kDa.

The complete amino-acid sequence of tissue factor apoprotein has been deduced by cDNA-sequencing (475–478) (Appendix, Fig. A11) and a large part of this has been confirmed by protein-sequencing (477). The cDNA sequence includes a signal peptide of length 32 or 34 residues (see below). The complete DNA sequence of the gene for tissue factor apoprotein is

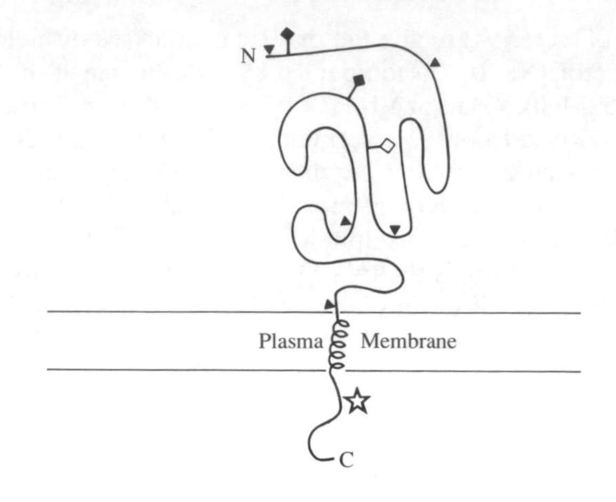

Fig. 32. The domain structure of tissue factor. ▼ shows the positions of the introns, ◆ the N-bound carbohydrate, ◇ the potential attachment sites for N-bound carbohydrate and ☆ the position of the esterified Cys245.

known (479). It covers 12.4 kbp of genomic DNA at position p21–22 on chromosome 1 (476, 480, 481), with six exons. The sizes of the exons vary from 112 bp to 1278 bp, while the introns vary between 607 bp and 4090 bp. The intron positions in relation to the amino-acid sequences are shown in Figure 32.

Tissue factor apoprotein consists of 263 amino-acid residues if the signal peptide is taken to be 32 residues long and 261 residues if the signal peptide is taken as being 34 residues long. Purified tissue factor apoprotein is heterogeneous at its N terminus; in sequencing experiments, two amino-acid sequences related by a phase shift of two residues are observed. This could be because the signal peptide can be cleaved off at two different points. The molecular mass of tissue factor apoprotein is thus calculated to be 29,593 or 29,449 Da respectively, both of which values are on the low side, even after inclusion of the estimated carbohydrate fraction.

Tissue factor apoprotein contains four potential N-glycosylation sites at Asn11, Asn124, Asn137 and Asn261. Studies of the protein sequence show that at least Asn11 and Asn137 are used (477). Asn261 can scarcely be used, as this residue is intracellular. It is not known whether tissue factor apoprotein contains O-bound carbohydrate, but total deglycosylation by chemical methods and N-deglycosylation by enzymes give molecules with almost the same estimated molecular mass (477), suggesting that it does not.

The domain structure of tissue factor apoprotein (Fig. 32) comprises an N-terminal extracellular domain (residues 1–219), a presumed transmembrane peptide (residues 220–242) and a C-terminal cytoplasmic peptide (residues 243–263). No homology has been detected between the amino-acid sequence of tissue factor apoprotein and that of any other protein.

An interesting detail of the amino-acid sequence of tissue factor apoprotein is the appearance three times of the tripeptide sequence W-K-S. This triplet is rare in itself, and its significance here is unknown.

The disulphide-bond pattern in tissue factor apoprotein has been determined, and consists of the S–S bonds Cys49–Cys57 and Cys186–Cys209 (482). This leaves a single unpaired cysteine with a different function: it lies in the cytoplasmic domain and, in the cell, it is linked via a thio-ester group to palmitic acid or stearic acid (482). This provides a possible explanation both for the small quantity of tissue factor apoprotein dimer that is often seen after purification of the factor and also for the heterodimeric form of tissue factor apoprotein that has been described (475, 483, 484), and which has been found to be a dimer of tissue factor apoprotein and haemoglobin α-chain (485). Any deacylation of Cys245 during purification will provide an excellent opportunity for this free cysteine to form a new disulphide bond with another tissue factor apoprotein molecule or with a different protein that contains free cysteine.

6.2. Tissue-factor-dependent activation processes

Since tissue factor is a membrane protein, all reactions in which it acts as a co-factor must clearly proceed at surfaces (reviewed in, for example, reference 486).

A molecule of tissue factor can bind one molecule of factor VII or VIIa. Complex formation between factor VII/VIIa and tissue factor is dependent upon calcium ions (468), and the binding of these two components to each other is therefore probably due to Gla residues in factor VII/VIIa. The presence of acidic phospholipids in tissue factor is not necessary for the binding of factor VII/VIIa to tissue factor. Purified bovine tissue factor apoprotein that has been reconstituted in phospholipid vesicles made up only of phosphatidyl choline binds both bovine factors VII and VIIa in a stoichiometric ratio of 1:1 and with dissociation constants of 13 nM and 4.5 nM respectively (487). The presence of acid phospholipids increases the strength of the binding, but not nearly as steeply as might be expected. The increase is greater for the binding of factor VII. This provides a further example of a binding to membranes that is dependent upon Gla and calcium ions, and yet is not dependent upon the negative charge of the phospholipid.

In fact, little more is known about the complex that consists of factor VII/VIIa, tissue factor and Ca^{2+}. In particular, it is not known precisely which parts of the protein chains interact with each other, except that it is the extracellular domain of tissue factor apoprotein that binds to factor VII/VIIa.

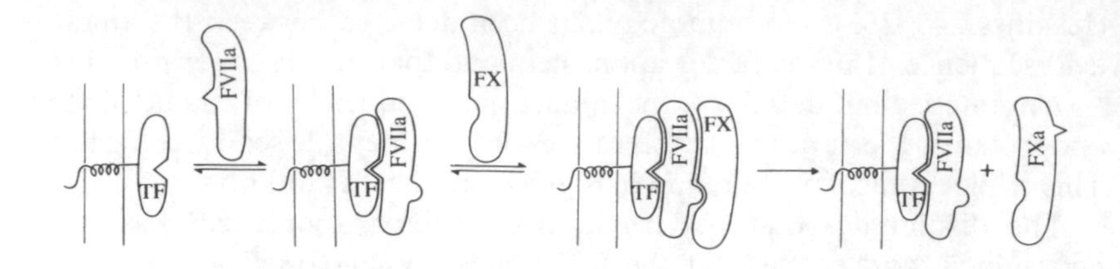

Fig. 33. 'Ordered-addition, essential-activation mechanism' for the activation of blood coagulation by tissue factor. (TF, tissue factor; FVIIa, factor VIIa; FX, factor X; Factor Xa, activated Factor X.)

Kinetic studies carried out in systems of bovine components have shown that reconstituted tissue factor is needed in order to confer enzymic activity upon factor VIIa (488, 489), and also that the presence of negatively charged phospholipids lowers K_m for factor X without changing k_{cat} for the activation of factor X to give factor Xa (489). The same investigation (489) has furthermore made it appear probable that the complex of tissue factor and factor VIIa employs free factor X, and not surface-bound factor X, as a substrate.

The mechanism of tissue factor is best described as an 'ordered-addition, essential-activation mechanism' (490) (Fig. 33). In this model, factor VIIa is bound to tissue factor and, in doing so, it undergoes a conformational change that gives it affinity for factor X. The binding of the factor X substrate causes factor VIIa to bind more strongly to tissue factor. Finally, factor VIIa catalyses the cleavage of factor X, which is thus activated to give factor Xa. It may be noted that factor VIIa cleaves the same bond in factor X as factor IXa does (491).

In addition to activating factor X to give factor Xa, factor VIIa, with the co-factors tissue factor and calcium, can also activate factor IX to give factor IXa (492). This process proceeds with the breakage of the same bonds that factor XIa cleaves, and in the same order (142). Mention is made above (section 4.3) of two chimeric factor IX molecules, of which one or both growth-factor domains were replaced by, respectively, one or two of the growth-factor domains of factor X (333). The activation of these two chimeric factor IX molecules by the [tissue factor.factor VIIa] complex leads to different results. The chimeric molecule with one replaced growth-factor domain gives a normal activated product, whereas the chimeric molecule with both growth-factor domains replaced is less effective in coagulation (333). Again, it appears that the second growth-factor domain in factor IX is important.

As mentioned earlier, it has been suggested that bovine factor VII bound to tissue factor has a weak enzymic activity, about 1% of that of factor VIIa (463), although there does not appear to be a comparable zymogen activity in human factor VII. There are several reasons that make this hypothesis unlikely to be true for human factor VII. For bovine factor VII it was argued that the treatment of the zymogen with DFP leads to the incorporation of DFP with pseudo-first-order kinetics, while the zymogen's activity in coagulation decays according to a different kinetic law (463). In contrast, the incorporation of DFP into bovine factor VIIa and the resulting reduction in coagulation activity follow the same pseudo-first-order kinetics. The coagulation activity of the zymogen preparation was therefore explained as a combination of decay of factor VII and small quantities of contaminating factor VIIa in the preparation; the coagulation activity of factor VII was calculated to be a little less than 1% of that of factor VIIa (463). Corresponding measurements of the decay of coagulation activity for human factors VII and VIIa reveal pseudo first-order decay concomitant with the incorporation of DFP (450), so that the same argument need not be used in the case of human factor VII/VIIa.

There are also several observations in purified systems that speak against the hypothesis that human factor VII possesses coagulation activity (493–495).

If the abilities of factors VII and VIIa to activate factor IX in the presence of tissue factor and calcium ions are compared, then it is true that the rates are equal; however, factor VII requires a lag period of 20 minutes that factor VIIa does not. When factor IX in these experiments was replaced with a particular variant of factor IX called factor IX_{BmLE}, which in its activated state is unable to activate factor VII, then no activation of factor IX took place; however, factor VIIa activated factor IX_{BmLE} with the same initial rate as that with which it activated normal factor IX (493).

The activation of factor X in the presence of tissue factor and calcium takes place at the same rate when the activating factor is VII as when it is VIIa, but with a lag phase of about 1 min for factor VII (493, 494). The following important experiment makes use of the fact that a mixture of antithrombin III and heparin inhibits factor Xa but not factor VII or VIIa. If factor X is pre-incubated with antithrombin III and heparin in order to inactivate possible traces of contaminating factor Xa, then factor X is still activated just as well by factor VIIa, but factor VII no longer shows any activity (494).

Studies with certain peptidyl chloromethyl ketones that function as highly effective inhibitors of factor VIIa and factor Xa, respectively, have likewise shown that factor VII is unable to activate factor X directly, even in the presence of tissue factor and calcium ions (495).

There is thus no firm evidence that human factor VII in complex with tissue factor and calcium ions possesses even a small enzymic activity, in contrast to what has been reported for the corresponding bovine system. It will be interesting to see the results of comparable experiments with purified bovine factors, and especially of experiments in which the activation of factor VII to give factor VIIa is blocked.

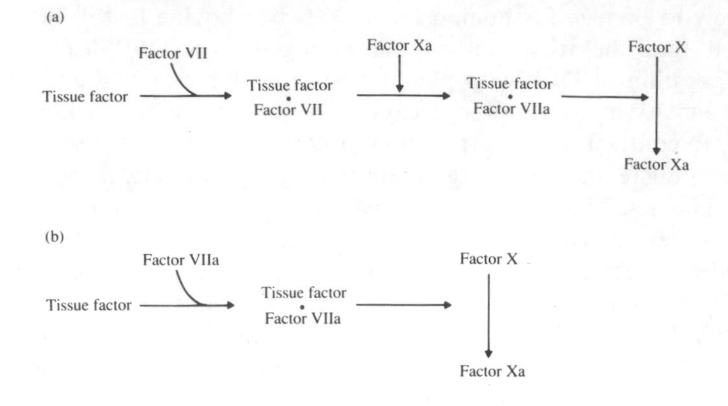

Fig. 34. Possible initiation mechanism of the extrinsic pathway of blood coagulation. (a) Initiation occurs via the cleavage of tissue-factor-bound factor VII by factor Xa. (b) Initiation occurs via the cleavage of factor X by tissue-factor-bound factor VIIa. (See text for a discussion of the origins of the factors VIIa and Xa.)

The results that have been obtained with the human system indicate strongly that the activation of factor VII bound to tissue factor is of major importance for the initiation of coagulation by tissue factor (494), as is also more or less the case for the bovine system. The difference between the two systems is that in the human system the activation of factor VII bound to tissue factor may well be the initial activation, while in the bovine system, if it is true that factor VII has coagulation activity, it is a 'mere' amplification process, and the initial activation is the activation of factors IX and X, catalysed by factor VII complexed with tissue factor.

How are we now to picture the initiation of blood coagulation in the extrinsic pathway (Fig. 34)?

The first step in this process is the exposure of tissue factor to plasma, opening the possibility for it to bind factor VII/VIIa. Since factor VIIa, in contrast to other activated coagulation factors, is relatively stable in plasma, with a half-life for decay of 2.5 hours (496), it is easy to imagine that a small quantity of factor VIIa becomes bound to the exposed tissue factor. There are two possibilities for the next step; these do not necessarily exclude one another. The first is that a trace quantity of factor VIIa becomes bound to tissue factor. This gives an enzyme complex that can activate factor X to give factor Xa, and this in turn can activate tissue-factor-bound factor VII, giving more factor VIIa and thus reinforcing the process. The second possibility is that the tissue-factor-bound factor VII is activated by traces of factor Xa from plasma, so that, in the same way, tissue-factor-bound factor VIIa is formed. These trace quantities of factor Xa could possibly lead to trace quantities of factor VIIa without tissue factor, that could take part in the first possible pathway suggested. However, one must of course ask whether it is physiologically reasonable to postulate the existence of trace quantities of free factor Xa. The answer must be that this is at present not known; however, small quantities both of prothrombin activation fragments (497) and of factor X activation peptide (498) have been found in normal plasma, which could indicate a certain, low factor Xa activity. The concentration of factor X activation peptide in plasma from patients deficient in factor VII is significantly lower than that in normal plasma. This indicates that the formation of factor X activation peptide depends upon the extrinsic pathway (498). Furthermore, it has been reported that normal plasma contains factor VIIa activity, in contrast to plasma from haemophilic patients (499), which does not contain this activity. This indicates that the presence of factor VIIa activity is dependent upon the intrinsic pathway.

Both of the mechanisms suggested lead to the formation of factor VIIa bound to tissue factor, which upon activation of factor X to give factor Xa both reinforces its own formation and brings the coagulation process one step further with the help of the prothrombinase complex.

There can be no doubt that factor Xa is by far the most important activator of both bovine and human factor VII *in vivo*. Co-factors needed for this activation are calcium ions and negatively charged phospholipid, but not factor Va (450, 451, 461, 500). In the bovine system, the presence of tissue factor apoprotein together with negatively charged phospholipids is particularly effective in accelerating the activation of factor VII by factor Xa (501). Human factor IXa, with the same co-factors, is also able to activate human factor VII, but human factor Xa is about 20 times more effective (502).

It will by now be clear to the reader that in all probability the initiation of blood coagulation by the extrinsic pathway is dependent upon the presence of trace quantities of factor Xa or factor VIIa. As we saw earlier, a similar situation is found for the initiation of contact activation (section 2.2.2).

These observations raise an interesting question. Is the coagulation system really constantly activated but normally kept under such tight control that no undesirable effects occur, or is it a stable system of zymogens that are activated by external stimuli? No definitive answer to this question can be given, but the author's personal bias is towards the former possibility.

7. Formation and stabilisation of fibrin

7.1. Proteins involved in fibrin formation and stabilisation

The final step in blood coagulation is the formation of a fibrin clot and its covalent stabilisation. This is the subject of this chapter – the use of thrombin to catalyse the conversion of fibrinogen to fibrin and the activation of factor XIII to give factor XIIIa. Factor XIIIa, once formed, stabilises the fibrin clot by covalent cross-linking. We shall therefore start by making the acquaintance of two new proteins – fibrinogen and factor XIII.

7.1.1. *Fibrinogen*

Fibrinogen (or factor I), the precursor of fibrin, is reviewed in references (503) and (504). It circulates in normal human plasma at a concentration between 2 mg/ml and 4.5 mg/ml ($\approx$6–13 μM), which is much higher than the other coagulation proteins that we have met so far. Fibrinogen is built up of six polypeptide chains linked by disulphide bonds: two Aα chains, two Bβ chains and two γ chains. The composition can appropriately be denoted $(A\alpha B\beta\gamma)_2$, since the fibrinogen molecule consists of two identical halves and can be thought of as a dimer of the AαB$\beta\gamma$ monomer. The total molecular mass of human fibrinogen is estimated to be 340 kDa, a figure that includes about 5% carbohydrate attached to the Bβ chain and the γ chain (505, 506).

The structures of all three chains in human fibrinogen have been determined at both protein and cDNA levels. The determination of the amino-acid sequence of the Aα chain, completed in 1979 (507, 508), was based on a number of previously published partial sequences (509–516). The sequence of the Bβ chain was also assembled (517, 518) from earlier published part-sequences (519–523). The sequence of the γ chain is published almost fully in reference (524) and can be put together from results in references (525) to (533). The sequences are collated in reference (534). The amino-acid sequences have since been derived from the cDNA sequences, for the Aα chain (535, 536), for the Bβ chain (537) and for the γ

chain (538). Finally, analyses of the gene structures of the Bβ (537) and γ (539–541) chains have been published. These genes contain 8 and 10 exons respectively.

We begin by examining the fibrinogen chains and their genes separately.

The Aα chain is the largest of the three chains in fibrinogen, with an estimated molecular mass of 63.5 kDa (found by ultracentrifugation (542)) or 73 kDa (SDS-polyacrylamide gel electrophoresis). The Aα chain consists of 610 amino-acid residues, as determined by protein-sequencing, corresponding to a molecular mass of 66.1 kDa. The chain contains no bound carbohydrate (534), although it has two potential N-glycosylation sites. The amino-acid sequence deduced from the cDNA sequence is 15 residues longer than the sequence found by analysis of the protein. This is due to a C-terminal extension and would give a molecular mass of 67.6 kDa. The function of these 15 C-terminal residues is unclear. Antibodies against an synthetic oligopeptide corresponding to residues 615–625 (the last ten residues of the extension) do not react with fibrinogen in plasma, although antibodies against residues 600–610 react strongly. This means that the extension is probably not present in plasma fibrinogen (543). The reason for this is likely to be a limited proteolytic trimming of the Aα chain, either in connection with the assembly of the fibrinogen molecule from the separate chains or in connection with its secretion into plasma. There is also a possibility that this processing takes place in the plasma itself. The bond cleaved is apparently Val610–Arg611, but this is unlikely to be the primary cutting site; it seems more probable that the first cleavage is between Arg611 and Gly612, after which the exposed C-terminal arginine residue is removed by a carboxypeptidase.

The sequence deduced from cDNA (Appendix, Fig. A12) is in broad agreement with the sequences determined at the protein level (507, 508); small discrepancies between these have been largely resolved by a third partial amino-acid sequence determined with purified foetal Aα chain (544). Three positions in the Aα chain may prove to be dimorphic: position 47 (Ser/Thr), position 296 (Thr/Ala) and position 312 (Thr/Ala). All of these forms have been found at the protein level (544, 545).

The signal peptide for the Aα chain is 19 or 16 residues long, and it is not yet clear whether there are two potential initiation codons. The amino-acid sequence of the Aα chain is to some extent homologous to the sequences of the other two fibrinogen chains. Within the Aα chain there are eight consecutive stretches of 13 residues displaying clear homology with each other (Fig. 35). This region covers positions 270 to 372.

The Aα chain in fibrinogen contains two phosphorylated serine side-chains. These are Ser3 and Ser 365 (546, 547). The fibrinogen molecule gradually loses its phosphate groups, so the degree of phosphorylation, believed to be 100% for newly synthesized fibrinogen, decreases slowly with time (548, 549).

It should be mentioned that, in addition, the entire gene for the Aα chain

270	P	S	S	A	G	S	W	N	S	G	S	S	G	282
283	P	G	S	T	G	N	R	N	P	G	S	S	G	295
296	T	G	G	T	A	T	W	K	P	G	S	S	G	308
309	P	G	S	A	G	S	W	N	S	G	S	S	G	321
322	T	G	S	T	G	N	Q	N	P	G	S	P	R	334
335	P	G	S	T	G	T	W	N	P	G	S	S	E	347
348	R	G	S	A	G	H	W	T	S	E	S	S	V	360
361	S	G	S	T	G	Q	W	H	S	E	-	S	G	372

Fig. 35. Internal homology in the Aα chain of fibrinogen. Only the positions at which amino acids are the same in at least five stretches are outlined.

of rat fibrinogen has been sequenced. This gene has five exons (551).

Although details have not been published, it is known that the human fibrinogen gene for the Aα chain has five exons (550), as does the corresponding rat gene (551). In addition, the complete amino-acid sequence for the bovine Aα chain has been derived from the protein (552), while the complete amino-acid sequence for lamprey Aα chain has been found by a combining partial peptide, cDNA and gene sequences (553).

The Bβ chain is the intermediate-length chain in fibrinogen, with an estimated molecular mass of 56 kDa (determined by ultracentrifugation (542)) or 60 kDa (SDS-polyacrylamide gel electrophoresis (554)). The Bβ chain consists of 461 amino-acid residues (Appendix, Fig. A13) with a calculated molecular mass of 52.3 kDa. This gives a molecular mass of approximately 55 kDa after taking into account the carbohydrate chain attached at Asn364. The N terminus of the Bβ chain is blocked because of cyclisation of the N-terminal Gln residue to pyroglutamic acid. There are three possible dimorphic positions in the Bβ chain: position 162 can apparently be occupied by either proline (517, 518, 537) or alanine (537), position 296 is reported to contain asparagine or arginine (517, 518, 537), and position 448 may be occupied by aspartic acid or lysine (535). Included in the cDNA sequence is a partial signal peptide sequence of 22 amino-acid residues, but sequencing of genomic DNA has shown that the signal peptide can be up to 30 residues long. However, it may contain only 27 or 16 residues on account of potential initiation codons at the appropriate positions (537). The gene for the Bβ chain has been characterised by the determination of the intron–exon junctions and by heteroduplex analysis. It covers about 10 kbp of genomic DNA and has eight exons with sizes ranging from 117 bp to 665 bp. The Bβ chain and the γ chain show more homology with each other than either of these does with the Aα chain (534, 555).

The complete amino-acid sequences of the bovine Bβ chain (556–558)

γ 401 H L G G A K Q A G D V 411
γ' 401 H L G G A K Q V R P E H P A E T G Y D S L Y P E D D L 427
γ55 401 H L G G A K Q V R P E H P A E T G Y D S L Y P 423

Fig. 36. Genetic variants of the γ chain of fibrinogen. The γ' and γ55 variants of the γ chain arise by alternative splicing of the primary transcript. How the four C-terminal residues in γ' are removed during the formation of γ55 is not known.

and the lamprey Bβ chain (559, 560) have been determined by a combination of protein-chemical evidence and cDNA-sequencing.

The γ chain is the shortest of the chains in fibrinogen, with an estimated molecular mass of 47 kDa (ultracentrifugation (542)) or 53 kDa (SDS-polyacrylamide gel electrophoresis (554)). It consists of 411 amino-acid residues (Appendix, Fig. A14) with a calculated molecular mass of 46.4 kDa, which with the inclusion of the carbohydrate chain on Asn52 becomes about 49 kDa. The sequence contains a single possible dimorphism at position 88, which may contain lysine (533, 538) or isoleucine (541). In the cDNA sequences a 26-residue signal peptide was found. The gene for the γ chain covers 10.5 kbp of genomic DNA and has ten exons (541). The lengths of the exons lie between 45 bp and 278 bp, while the the introns vary from 96 bp to 1638 bp.

There occurs a genetic variant of the γ chain, termed γ'. It has a somewhat greater molecular mass than the usual γ chain. Protein-sequencing studies have shown that the differences between the γ and the γ' chains occur at their C termini, where the amino-acid sequences diverge (Fig. 36). The two chains are identical up to Gln407, after which the γ chain contains only four further residues (-A-G-D-V), while the γ' chain continues for 20 residues (-V-R-P-E-H-P-A-E-T-G-Y-D-S-L-Y-P-E-D-D-L) (561). Studies of the γ-chain gene have shown that the γ' chain arises when an alternative polyadenylation site in the ninth and last intron is used (539, 540). This means that the tenth exon is never reached, so that the intron sequence is not spliced out. The translation of what would normally be an intron continues for 20 codons; the 21st codon signals termination, resulting in the C terminus described. Exon 10 thus codes only for the final four amino acids in the normal γ chain and the non-translated 3′ mRNA sequence.

There is one further γ-chain variant, referred to as γ55. It is identical to γ' except that the final four residues from the C terminus are absent (Fig. 36) (562). It is not known how this variant arises. The frequencies of occurrence of the three forms of γ chain are 93.5% (γ), 5% (γ') and 1.5% (γ55) (563).

The amino-acid sequence for bovine γ chain has been determined at the protein level (552), while the complete amino-acid sequences of both rat (564) and lamprey (565) γ chains have been determined by cDNA-

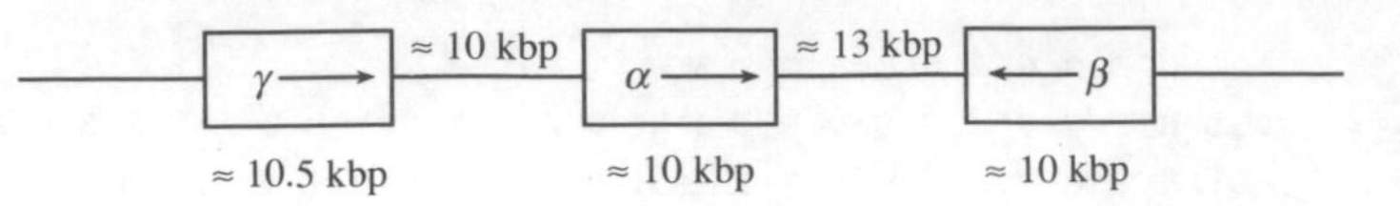

Fig. 37. The organisation of the fibrinogen genes. The arrows give the direction of transcription.

sequencing. The complete nucleotide sequence of the rat gene for the γ chain has also been determined (566). Just as for the human γ chain, there are two variants of the rat γ chain that arise by aberrant processing of the primary transcript (564, 566). The most frequent γ chain makes up 90–95% of the total, and, like its human counterpart, it is formed by use of all of the ten exons of the rat gene. The γ' variant arises through defective excision of intron 9 and the use of the polyadenylation site in exon 10 such that the mRNA for the γ' chain contains 513 extra nucleotides between exons 9 and 10. The translation of this mRNA comes to a halt after 12 residues, when a termination codon is encountered.

Human fibrinogen also contains tyrosine-O-sulphate (567), but it is not known at which residue. In other species, this modification has been found in fibrinopeptide B, but human fibrinopeptide B (residues 1 to 14 in the Bβ chain) contains no tyrosine at all.

The genes for all three fibrinogen chains lie within a 50-kbp region on chromosome 4, at band q26–32 on the long arm (568–570). The Aα gene is in the middle, flanked by the γ and Bβ genes. The Aα and γ genes are transcribed in the same direction on the same DNA strand, while the Bβ gene is on the other strand and is therefore transcribed in the opposite direction (Fig. 37). The close proximity of all three chains may be related to the fact that the synthesis of mRNA for the three chains is regulated in a co-ordinated way, and all three chains are produced when fibrinogen synthesis is stimulated artificially (571). The advantage of the co-ordination, when the three chains are always required in equal amounts, is obvious.

Once the three chains have been synthesised, they must be assembled correctly, and it appears that the following path is taken. There is an intracellular pool of Aα and γ chains in the endoplasmic reticulum, and these combine independently with Bβ chains while the latter are still being synthesised on the polysomes. Thus the first products of the synthesis are AαBβ and B$\beta\gamma$ structures, and these combine with γ and Bβ chains respectively to give the fibrinogen half-molecule AαB$\beta\gamma$, which finally dimerises to give the complete fibrinogen molecule $(\mathrm{A}\alpha\mathrm{B}\beta\gamma)_2$ (572). It should be noted here, first, that the entire structure is strongly bound together by disulphide bonds, and, secondly, that it is the Bβ chain around which the entire structure collects, which indicates a possible 'core' rôle for this chain. These results could indicate that the chains are synthesised at different rates, with the synthesis of Bβ being the slowest. It is suggestive

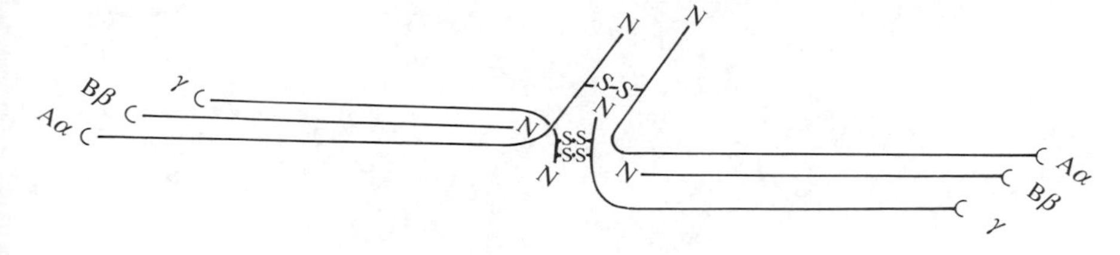

Fig. 38. Disulphide bonds that link the two fibrinogen half-molecules. Note that the two disulphide bonds connecting the γ chains make the two halves of the molecule antiparallel.

that the Bβ chain is the one transcribed in a direction opposite to that of the other two.

The strong disulphide bonding referred to involves as many as 29 S–S bridges, all of which have been located. Three of these hold the half-molecules (AαBβγ) associated, and they are arranged such that the two Cys28 residues in the Aα chains are connected; a similar bond is made by the Cys8 residue of one γ chain with the Cys9 residue of the other, and *vice versa* (Fig. 38) (519, 573). So it is the Aα and the γ chains that hold the dimer together. Note that the two disulphide bonds that connect the two γ chains lead to antiparallel half-molecules of fibrinogen.

Within each half-molecule, the three chains are linked by four disulphide bonds in their N-terminal regions. This means that the central portion of the fibrinogen molecule contains 11 disulphide bonds in all. The part of the fibrinogen molecule containing these bonds is therefore called N-DSK (N-terminal disulphide knot). Within each half-molecule the four interchain disulphide bridges are: AαCys36–BβCys65, AαCys45–γCys23, AαCys49–BβCys76 and BβCys80–γCys19 (519) (Fig. 39). The last two disulphide bridges can also be arranged differently: AαCys49–BβCys80 and BβCys76–γCys19 (545). It is worth mentioning that the second, third and fourth disulphide bonds mentioned make up what is called a disulphide ring, and that all the six cysteine residues involved in it reside in the sequence C-P-S/T-G/T-C (Fig. 39).

The disulphide ring is followed by a region of about 110 amino acids in all three chains that is believed to have a coiled-coil structure, after which comes a second disulphide ring (Fig. 40). The sequences between the Cys residues involved in the second disulphide ring are not so well conserved as those in the first, but the spacing of the cysteines is still the same in all chains: -C-X-X-X-C-. The structure of the second ring is presumably analogous to that of the first. The disulphide bridges are: AαCys161–γCys139, AαCys165–BβCys193 and BβCys197–γCys135 (574).

The pattern of the six cysteine residues in disulphide rings is thus formally: the first (closer to the N terminus) Cys in the Aα chain is bound to the second Cys in the γ chain, and the first Cys in the γ chain is bound to the

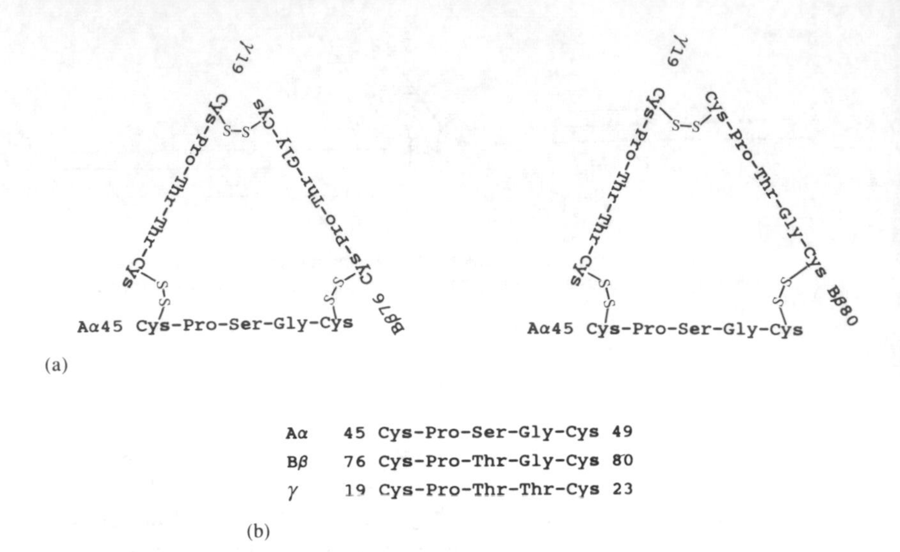

Fig. 39. (a) The two alternative disulphide-bond patterns in the first disulphide ring of fibrinogen. (b) The amino-acid sequences around the Cys residues that participate in the disulphide ring.

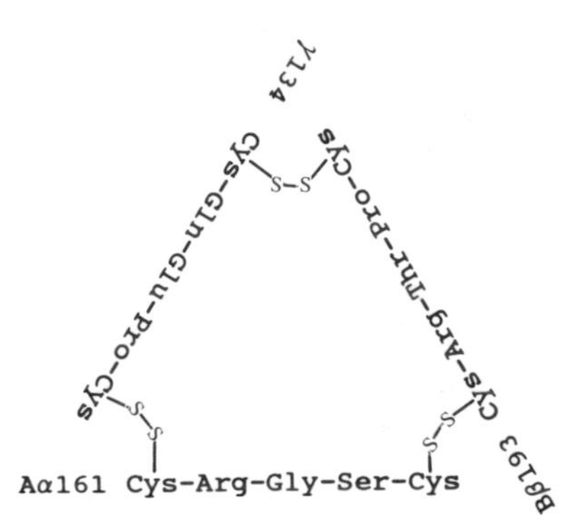

Fig. 40. The disulphide-bond pattern in the second disulphide ring of fibrinogen.

second Cys in the Bβ chain; finally, the first Cys in the Bβ chain is bound to the second Cys in the Aα chain, closing the ring.

In addition to these intercatenary disulphide bonds, there are six intracatenary disulphide bonds within each half-molecule: AαCys442–AαCys472, BβCys201–BβCys286, BβCys211–BβCys240, BβCys394–BβCys407, γCys153–γCys182 and γCys326–γCys339 (574, 575). Figure 41 shows a schematic model for the way in which fibrinogen is built up.

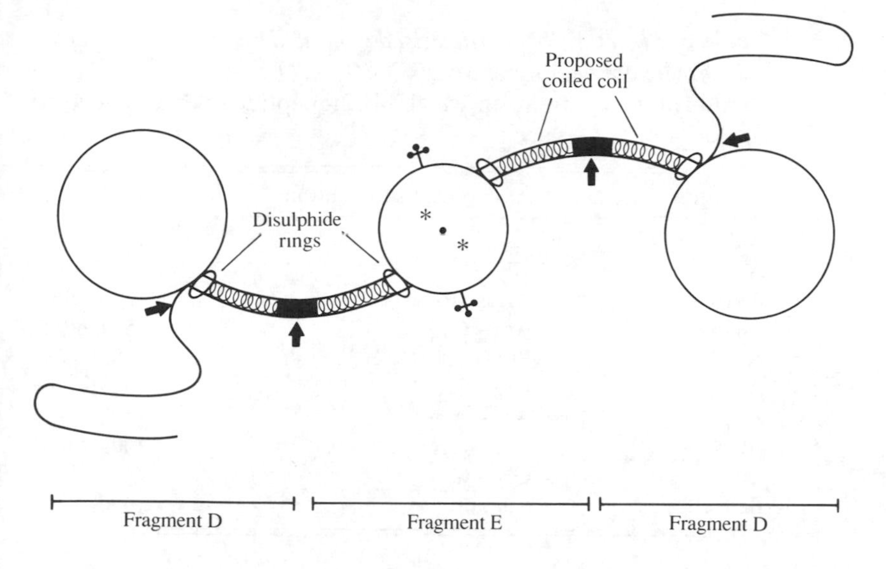

Fig. 41. A schematic model for the fibrinogen molecule. • represents the hypothetical point through which the rotation axis of the molecule passes. → shows positions that are especially sensitive to cleavage by plasmin. * shows fibrinopeptide A and † shows fibrinopeptide B. In this model, the two fibrinopeptides A point up out of the paper.

Human and bovine fibrinogen each contain three high-affinity binding sites for Ca^{2+} (576, 577). In human fibrinogen, two of these sites have been localised to the γ chain between Gln311 and Met336 (one binding site in each γ chain) (578, 579), while the third binding site is believed to be in N-DSK (580).

Electron-microscopic studies (581) of human and bovine fibrinogen indicate a trinodular structure (Fig. 41), in which the central domain corresponds to fragment E (582, 583) and N-DSK (584), while each of the two outer domains corresponds to one of the two fragments D (582, 583, 585, 586). Fragments D and E are degradation products of fibrinogen that arise when it is digested by plasmin.

Fibrinogen is converted to fibrin by thrombin, which cleaves it at the bond Arg16–Gly17 in the Aα chains and at the bond Arg14–Gly15 in the Bβ chains. This releases two molecules each of the fibrinopeptides A and B, and the fibrinogen $(A\alpha B\beta\gamma)_2$ is converted to fibrin $(\alpha\beta\gamma)_2$, which polymerises spontaneously to give a 'soft' clot.

Various genetic defects are known to be associated with the fibrinogen molecule (e.g. reference 587). Those for which the genetic basis has been defined are summarised in Table 6.

Table 6. *Structurally characterised genetic defects in fibrinogen (see, for example, references 503, 504, 587)*
A. Fibrinogen variants in which fibrinopeptide release is delayed or does not take place

Location	Amino-acid substitution	References
Aα7	Asp→Asn	(588)
Aα12	Gly→Val	(589)
Aα16	Arg→His	(590–596)
Aα16	Arg→Cys	(597–601)
Aα18	Pro→Leu	(602)
Aα19	Arg→Ser	(603)
Aα19	Arg→Asn	(598)
Bβ14	Arg→Cys	(604, 605)
Bβ15	Gly→Cys	(602)
Bβ9–72	deletion	(606)

B. Fibrinogen variants with normal fibrinopeptide release but poor polymerisation

Location	Amino-acid substitution	References
Bβ335	Ala→Thr	(607)
γ275	Arg→His	(608)
γ275	Arg→Cys	(609, 610)
γ308	Asn→Lys	(611)
γ310	Met→Thr	(612)
γ319–320	deletion	(613)
γ329	Gln→Arg	(614)
γ330	Asp→Val	(615)
γ330	Asp→Tyr	(616)
γ375	Arg→Gly	(602)

7.1.2. *Factor XIII*

Factor XIII is the zymogen for the enzyme that stabilises the fibrin clot by catalysing the formation of covalent cross-links in it. Factor XIII in plasma consists of four chains held together by non-covalent forces. The chains are two identical pairs, so factor XIII is an a_2b_2 tetramer with a total molecular mass estimated to be 320 kDa (617). Factor XIII is a glycoprotein with its carbohydrate attached to the b-chain. Its concentration in normal human plasma has been found to be 21 μg/ml ($\approx$70 nM) (618). The total concentrations of a- and b-chains in plasma are about 140 nM and 270 nM. Thus, while virtually all the a-chains in plasma are in the tetramer, there is an excess of free, monomeric b-chains at a concentration of about 130 nM.

There is also an intracellular factor XIII, for example in blood platelets, placenta, macrophages and megacaryocytes, occurring in the form a_2. (It is true that small quantities of b-chains are found in bovine platelets (623), presumably in complex with the a-chain, but these are not believed to be significant.)

The amino-acid sequence of the human a-chain has been determined both by direct protein-sequencing (619) and by cDNA-sequencing (620, 621), while the complete sequence of the human b-chain has been determined by cDNA-sequencing only (622).

The cDNA sequence of the a-chain codes for methionine in position −1, and the polypeptide sequence deduced for the cDNA preceding this does not resemble any known signal peptide. In the vast majority of cases so far investigated, secreted polypeptides have a signal peptide, but this does not seem to be the case for the a-chain of factor XIII. Why is this? Both the extracellular and the intracellular forms of the a-chain have N-acetyl serine at the N terminus. A conceivable way of making a protein for both extra- and intracellular use would be to synthesise the former with a normal signal peptide and the latter without signal peptide, for example by splicing the primary transcript in different ways in different tissues. The analysis of genomic DNA for factor XIII a-chain has not cleared up this point (624). Since all evidence at present indicates that the extracellular and intracellular chains are identical, the mechanism of its secretion must remain an open question – as, indeed, is the site of its synthesis (e.g. reference 629).

The a-chain of factor XIII contains 731 amino-acid residues (Appendix, Fig. A15) with a calculated molecular mass of 80.5 kDa. It is claimed that the a-chain contains 1.5% carbohydrate, but none of the six potential N-glycosylation sites is used (619). The gene for the a-chain covers more than 160 kbp and has 15 exons (624). The sizes of the exons vary from 89 bp to 1688 bp; those of the introns have not been published. The gene is located on chromosome 6 (625–627), band p24–25 (628).

The a-chain contains nine free cysteine residues and no disulphide bonds. Its amino-acid sequence shows a general homology with transglutaminase from guinea-pig liver (35%) (630), but except for two very short stretches no other homology with other known proteins has been found. These stretches are amino acids 468–479, homologous to Ca^{2+}-binding stretches in other proteins. In the stretch around the cysteine residues of the active site (Cys314), there is a certain homology to the amino-acid sequences close to the active sites in thiol proteases. If the sequences of the a-chain, as determined by sequencing of the protein and of cDNA, are compared with one another, then five positions are found with possible dimorphism. These are positions 77 (Arg/Gly), 78 (Arg/Lys), 88 (Phe/Leu), 650 (Ile/Val) and 651 (Gln/Glu).

The cDNA sequence determined for the b-chain includes the last 19 residues of the signal peptide. The b-chain is built up of 641 residues (Appendix, Fig. A16) with a combined molecular mass of 70.0 kDa, which on inclusion of the experimentally determined 8.5% carbohydrate gives about 76.5 kDa. The b-chain consists of 10 internally homologous stretches of some 60 amino acids each (Fig. 42). Each stretch contains four Cys residues and two presumed disulphide bridges. The gene for factor XIII b-chain lies on chromosome 1 at position q31–32.1 (631), and although the complete genomic structure has not yet been published, the mapping of some of the intron–exon junctions has shown that several of the stretches referred to are coded by one exon each. 60-amino-acid domains of this kind are found in many other proteins, and this type of domain is termed SCR (short consensus repeat). The other known proteins showing this kind of structure are human β_2-glycoprotein I (632), human complement factor B (633, 634), human and murine complement C4b-binding protein (635, 636), human complement component C2 (637), human and murine complement factor H (638, 639), human complement C3b-receptor (CR1) (640, 641), murine CRy (homologous to human CR1) (642), human and murine complement C3d-receptor (CR2) (643–645), the human complement components C1r (646–648) and C1s (649, 650), human decay-accelerating factor (DAF) (651, 652) and human membrane co-factor protein (MCP) (653) from the complement system, human complement factor I (654), human complement components C6 (655, 656) and C7 (657), human interleukin-2 receptor (658) and a 35-kDa protein from vaccinia virus (659). In addition, factor XIII b-chain contains the tripeptide R-G-D, which in several proteins is a part of the recognition structure for their respective receptors, although the function of this sequence in the b-chain of factor XIII is unknown.

The investigation of factor XIII and the individual chains of factor XIII by electron microscopy has shown that isolated a-chains are globular (6 nm by 9 nm), while b-chains are thin, flexible rods (2–3 nm by 30 nm) (660). One could thus imagine plasma factor XIII as a dimer of two globular a-chains with two b-chain monomers coiled around it.

Thrombin activates factor XIII to give factor XIIIa by cleaving the bond Arg37–Gly38 in both a-chains.

7.2. Fibrin formation

The formation of fibrin sounds simple in principle, as the only chemical reaction that occurs is the thrombin-catalysed cleavage of the Arg16–Gly17 peptide bond in the Aα chains and the Arg14–Gly15 in the Bβ chains; this triggers off spontaneous polymerisation and the precipitation of the

(a)

```
  5 C G F P H V E N G R I A Q Y Y Y T F K S F Y F P M S I D K K L S F
 71 C T K P D L S N G Y I S D V K L L Y K - - - - - - - I Q E N M H Y
133 C L A P E L Y N G N Y S T T Q K T F K - - - - - - - V K D K V Q Y

F C L A G Y T T E S G R Q E E Q T T C T T E G W S P E P R C 67
G C A S G Y K T T G G K D E E V V Q C L S D G W S S Q P T C 126
E C A T G Y Y T A G G K K T E E V E C L T Y G W S L T P K C 188
```

(b)

```
193 C S S L R L I E N G Y F H P - V K Q T Y E E G D V V Q F F C H E N
254 C P P P P L P I N S K I Q T - H S T T Y R H G E I V H I E C E L N
316 C E E P P F I E N G A A N L - H S K I Y Y N G D K V T Y A C K S G
376 C K H P P V V M N G A V A D G I L A S Y A T G S S V E Y R C N E Y
504 C T S P P L I K H G V I I S S T V D T Y E N G S S V E Y R C F D H

Y Y L S G S D L I Q C Y N F G W Y P E S P V C 247
F E I H G S A E I R C E D G K W T - E P P K C 307
Y L L H G S N E I T C N R G K W T - L P P E C 369
Y L L R G S K I S R C E Q G K W S - S P P V C 430
H F L E G S R E A Y C L D G M W T - T P P L C 558
```

(c)

```
434 C T V N V D Y M N R N N I E M K W K Y E G K - - V L H G D L I D F
562 C T L S F T E M E K N N L L L K W D F D N R P H I L H G E Y I E F

V C K Q G Y D L S P L T P L - S E L S V Q C N R G E V K Y P L C 495
I C R G D T Y P A E L Y I T G S I L R M Q C D R G Q L K Y P R C 626
```

Fig. 42. Internal homology between SCR domains in the factor XIII-b chain. The 10 SCR domains are classified into three groups on the basis of their degrees of identity. The sequences are shown only between the first and last Cys residues in each domain. Group 1 is made of domains 1 to 3, group 2 of domains 4 to 7 plus 9 and group 3 of domains 8 and 10.

(a) Group 1. Only the positions at which amino acids are the same in all three domains are outlined. Pairwise comparison of the three domains reveals degrees of amino-acid identity between 36% and 43%. (b) Group 2. Only the positions at which amino acids are the same in at least four domains are outlined. Pairwise comparison of the five domains reveals degrees of amino-acid identity between 31% and 43%. (c) Group 3. Only the positions at which amino acids are the same in both domains are outlined. Their degree of amino-acid identity is 37%.

insoluble fibrin polymer as a clot. However, the process is in fact far from simple, as we shall now see.

To picture the polymerisation of fibrin, we first require a brief consideration of symmetry. Without this it is easy to get the wrong idea. Since the two γ chains in fibrinogen are held together in an antiparallel manner, the half-molecules must also be antiparallel to one another. However, this does not mean that equivalent positions in the two half-

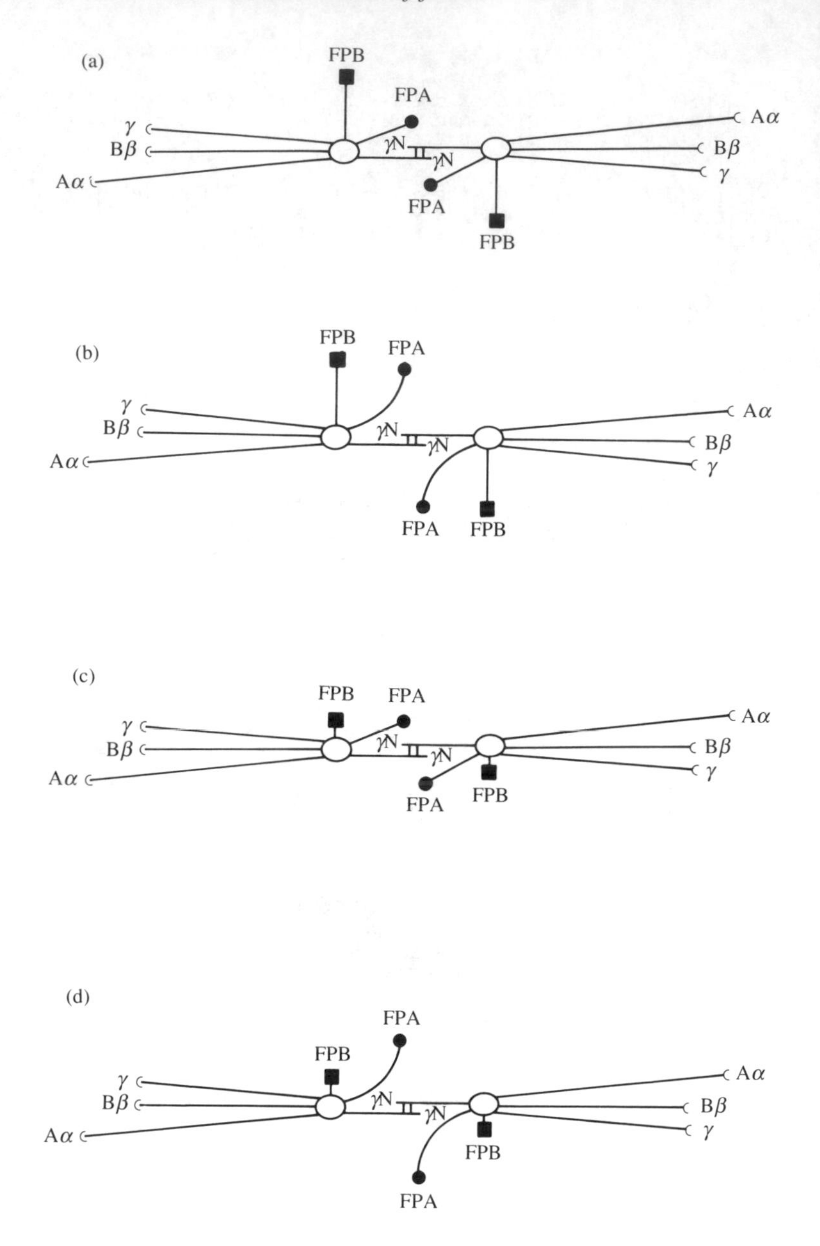

Fig. 43. Schematic models of fibrinogen, emphasising the placing of the fibrinopeptides. Common to all models is the fact that, on account of the antiparallel configuration of the half-molecules, the complex has rotation symmetry about an axis passing through a point between the two disulphide bonds that connect the two γ chains. (FPA and FPB, the fibrinopeptides; γN, the N terminus of the γ chain.)

molecules are necessarily placed on opposite sides of the molecule: they merely have opposite orientations. Thus, the fibrin molecule can be drawn in various ways, without any of these being in conflict with the antiparallel placing of the chains, as shown schematically in Figure 43. (Here the reader should be cautioned of the dangers inherent in the representation of complex three-dimensional systems by two-dimensional diagrams.) It should be noted that the fibrinopeptides can perfectly well be placed parallel to each other on the same side of the fibrinogen molecule, in spite of the antiparallel nature of the rest of the molecule; this is possible because the antiparallel orientation of the half-molecules in the Aα and Bβ chains arises after covalent linking to the antiparallel γ chains.

Thrombin, which is formed from prothrombin with the help of the prothrombinase complex at a coagulation-active surface, does not stay bound to the surface but is released into the plasma. Here, sooner or later, it collides with a fibrinogen molecule. Fibrinogen possesses two binding sites for thrombin; the principal one of these is associated with the Arg16–Gly17 bond in the Aα chain, the first bond in fibrinogen to be cleaved by thrombin. The results of investigations of synthetic fibrinogen peptides (661–667), set in relation to the phenotypes of genetically defective fibrinogen molecules in which position 7 or 12 is altered (588, 589), show that the stretch between Asp7 and Arg16 makes up an essential element in the primary interaction between thrombin and fibrinogen. The secondary binding site for thrombin in fibrinogen has also been partially localised (668, 669), as thrombin is bound not only by fragment E from both fibrin and fibrinogen but also by *des*-A,B-N-DSK (this is N-DSK – see section 7.1.1 – from which fibrinopeptides A and B have been split off). The secondary binding site for thrombin is thus narrowed down to a fragment consisting of α (Gly17–Met51), β (Val55–Met118) and γ (Tyr1–Lys53) (the numbering corresponds to the fibrinogen sequence). Attempts to narrow this down still further have led to conflicting results. In one study it was observed that free α (Gly17–Lys78) bound to immobilised thrombin, which none of the other chains in fragment E did (668), while another investigation showed that all three fragment E chains, isolated and immobilised, bound free thrombin (669). It is interesting in this connection that the N-terminal cyanogen bromide fragment of the Bβ chain of bovine fibrinogen binds rather poorly

(a) The FPBs and γNs lie in the plane of the paper. The N-terminal regions of the two Aα chains are parallel to one another and point upwards out of the paper. These chains, which include the FPAs, are thus close to each other and on the same side of the monomer. The two FPB chains lie on opposite sides of the monomer and point away from it. (b) The same model as in (a), but including a kink in the N-terminal regions of the two Aα chains, so that the two chains of FPA are above the plane of the paper but point in opposite directions. (c) As model (a), but with the FPBs protruding down into the paper. The two FPAs thus lie close together on one side of the complex and the two FPBs on the other. (d) As model (c) but with Aα kinked as in model (b).

to bovine thrombin, unlike the corresponding fragment from Aα chain (670).

When fibrinogen is cleaved by thrombin, the first step is a cleavage of the bond Arg16–Gly17 in the Aα chains, with the release of fibrinopeptide A and the formation of fibrin I. The kinetic constants for this reaction have been determined: the Michaelis constant is 7 μM and the k_{cat} value is 85 s^{-1} (671).

Since the fibrinogen molecule contains two fibrinopeptide A parts, it is logical to ask whether the two Arg16–Gly17 bonds are cleaved independently, or whether the kinetics of cleavage are different if one of these has been cleaved already. Before discussing this, we introduce some relevant information concerning the fate of fibrin I after thrombin-catalysed cleavage of the Arg16–Gly17 bond in the Aα chain of fibrinogen. The cleavage exposes a previously hidden polymerisation site (called A) in N-DSK, which can bind specifically to a polymerisation site (called a) in the C-terminal region (672). This polymerisation proceeds, end to end, in two staggered rows, where each molecule overlaps with the halves of two successive molecules in the other row (Fig. 44). This produces linear protofibrils with a cross-sectional area twice that of a single fibrin I molecule, a model supported by studies employing both light-scattering (673) and electron-microscopic (674) techniques.

Two answers to the kinetic question raised above can now be discussed (Fig. 45). If the cleavage of the two Arg16–Gly17 bonds in the two fibrinogen monomers proceeds in two independent steps, and we adopt the fibrinogen model shown in Figure 45a, then there will arise a temporary pool of molecules that inhibit polymerisation, since most of the fibrin I molecules will be monofunctional instead of bifunctional. End-to-end polymerisation requires that both polymerisation sites in N-DSK are accessible. The kinetics of polymerisation of fibrin I fail to show any such inhibition. One must therefore suppose either that the cleavage of the second Arg16–Gly17 bond is much faster than that of the first (675, 676), or that the fibrinogen model in Figure 45a is wrong. If instead we accept the model in Figure 45b, then we can well assume that the cleavage of the two Arg16–Gly17 bonds proceeds independently without inhibition of polymerisation (677). Which of these models corresponds more closely to reality is not known, but the model in Figure 45a and the explanation based upon the different cleavage rates of the two Arg16–Gly17 bonds is generally favoured. However, it should be noted that Figure 45b in many ways gives a simpler explanation of this and other observed polymerisation phenomena, such as branching of the protofibril and lateral polymerisation without release of fibrinopeptide B (see below). In addition, further polymerisation models could be proposed, for example on the basis of other models shown in Figure 44.

The two polymerisation sites (A and a) have been roughly located. For

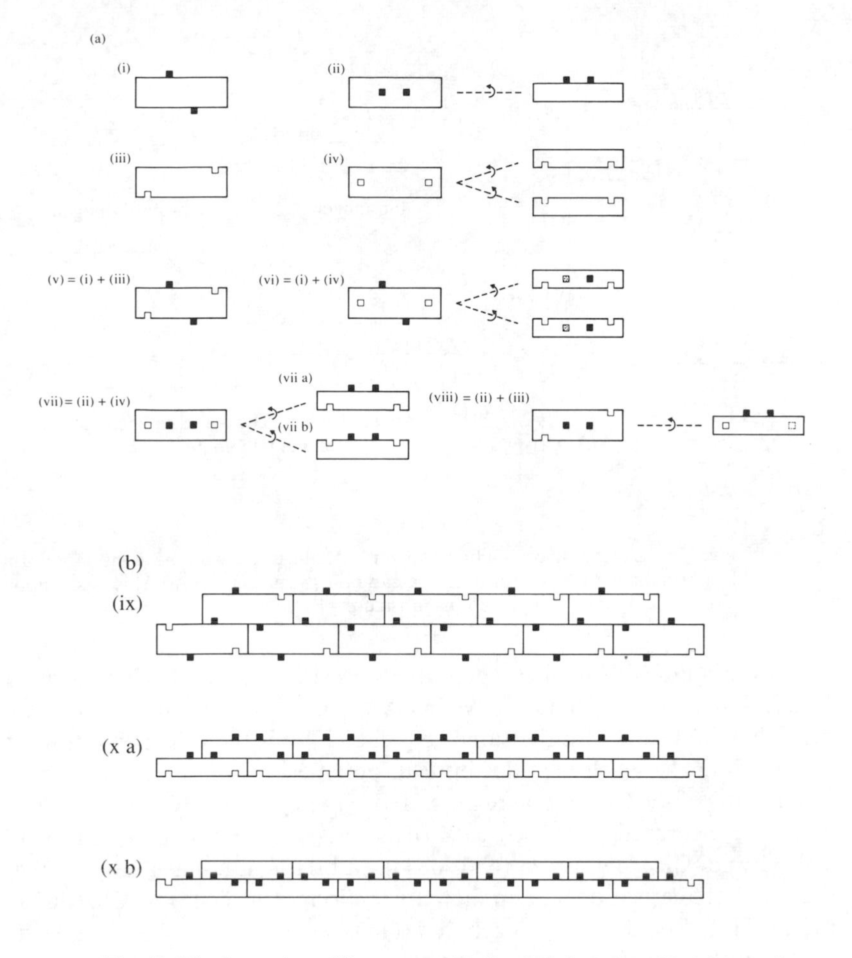

Fig. 44. The formation of linear protofibrils of fibrin I. (Modified from reference 677.)

(a) Schematic models of fibrin I. These models are based upon Figures 43a (ii) and 43b (i). In Figure 43b, the two chains of fibrinopeptide A point in opposite directions, so that the exposed association sites 'A' in the fibrin I model (i) also point in opposite directions. In Figure 43a, the chains of fibrinopeptide A point is the same direction, so that the exposed association sites 'A' in the fibrin I model (ii) also do this. In model (ii), the molecule is also shown rotated 90° about its long axis. The location of the association site 'a' relative to fibrinopeptide A is not known, but two possibilities are shown in models (iii) and (iv), in which the association sites 'A' have been omitted for clarity. In model (iv), the association sites 'a' are located on the same side of the paper, either below or above, as shown in the two representations of the molecule rotated by 90° about its long axis. Models (v)–(viii) show the various combinatorial possibilities. (b) The formation of linear protofibrils of fibrin I based upon the models in Figure 44a. Model (ix) shows polymerisation of the fibrin I model (v), and (x) shows polymerisation of the fibrin I model (vii).

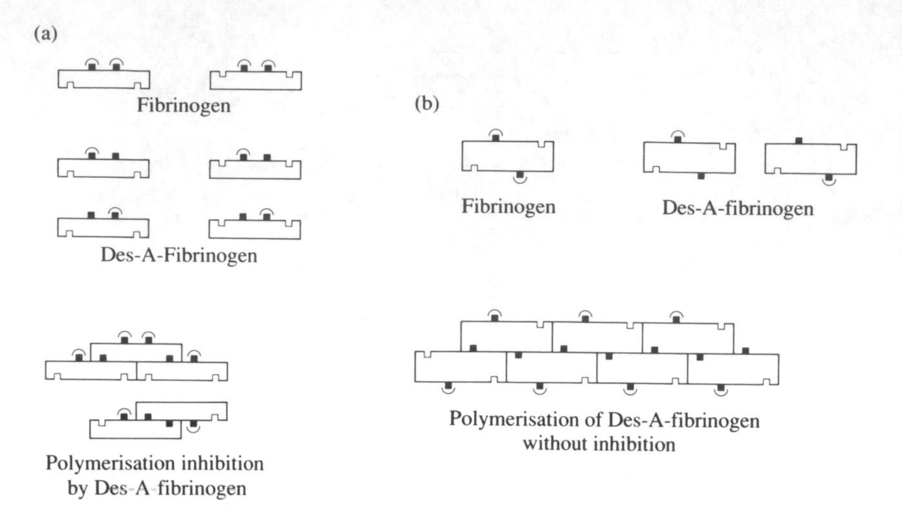

Fig. 45. The rate of release of fibrinopeptide A, illustrated with the models for fibrin I. Fibrinopeptide A is indicated as a cap on association site A. (a) is based upon Figure 44(vii) and (b) is based upon Figure 44(v). See text.

N-DSK, studies with synthetic peptides have shown that the newly formed N terminus of the α chain (Gly-Pro-Arg-) is a part of site A, because peptides that begin with the same sequence inhibit the polymerisation of fibrin (678, 679). Such peptides bind to polymerisation site a, which is in fragment D of the C-terminal region. In this region, the polymerisation site was once incorrectly localised to short stretches in the γ chain (Thr374–Val411 and Thr374–Glu396 (680, 681)). Subsequent studies have shown that the polymerisation sites in fact cover a longer stretch: Lys356/Ala357 to Val411 (578). It cannot be excluded that stretches on the N-terminal side of this in the γ chain are also of significance (682). In addition, nearly all characterised polymerisation-defective fibrinogen variants with substitutions in the γ chain have these substitutions lying between Arg275 and Asp330. In the same way, it cannot be excluded that the α and/or β chains contribute to site a.

Once the protofibrils have been formed, the bond between Arg14 and Gly15 in the β chain is cleaved and fibrinopeptide B is released, with the formation of fibrin II. This is a somewhat over-simplified representation: fibrinopeptide B begins to be split off as soon as fibrinogen is exposed to thrombin, but the rate with which this happens is very low at the beginning, so that virtually none of the fibrinopeptide B is released until after most of the fibrinopeptide A has been released. The removal of fibrinopeptide B results in the formation of a further association site in N-DSK (called B) and this can bind to a complementary site (called b) (683). The site b is thought to come into being through end-to-end association of protofibrils

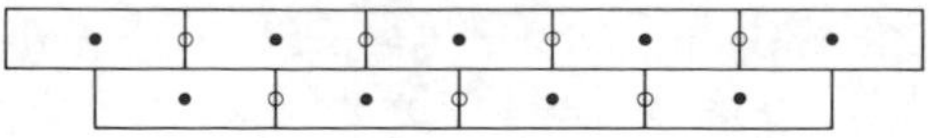

Fig. 46. A suggestion for the geometry of the lateral polymerisation of fibrin. • represents association site (B), which points upwards from the paper, and ○ represents association site (b), which points downwards into the paper. Each rectangle represents a fibrin II molecule. Each of the two chains in the protofibril can bind to other protofibrils, giving rise to lateral polymerisation.

(684). The binding of B to b brings about the lateral growth of the fibrin clot (684). The characterisation of the association sites B and b has not yet proceeded very far. Synthetic peptides, beginning with the amino-acid sequence Gly-His-Arg- and corresponding to the newly emerged N terminus in the β chain, do not inhibit the polymerisation of fibrin; they do, on the other hand, bind both to fibrinogen and to fragment D (678, 679). This binding occurs at a site in fragment D that is not the same as that where the above-mentioned peptides containing Gly-Pro-Arg- bind (679). It has also been shown that the newly emerged N terminus in the β chain – in particular the His residue – plays a part in the end-to-end association of fibrin, but not necessarily in the lateral association of fibrin protofibrils (685). The mechanism of the latter process is still unknown.

It is worth mentioning in this connection that the models used to describe the lateral growth of fibrin polymer normally presuppose two B and two b sites in each fibrin monomer. This is an unnecessary assumption, and at present the only evidence supporting it is the fact that two fibrinopeptide B molecules are removed. However, it is possible to polymerise fibrin I laterally without removing fibrinopeptide B (677, 686), although the process is faster when the latter is absent, and one can imagine a lateral polymerisation based upon a single B association site in N-DSK and a single b association site arising from end-to-end association in the protofibrils. It should be noted that, in any case, B and b must be located on opposite sides of the protofibril in order to achieve lateral polymerisation.

Figure 46 gives a schematic representation of the possible course of events in the polymerisation of fibrin.

We have thus followed the formation of a fibrin clot from spontaneously polymerising fibrin monomers. The next step is the covalent stabilisation of the fibrin clot.

7.3. Fibrin stabilisation

The components required for the stabilisation of fibrin, beyond the fibrin polymer itself, are calcium ions and factor XIIIa. The latter is also referred to as fibrin-stabilising factor. It stabilises the fibrin clot by introducing covalent cross-links. We shall now see how this takes place.

Like the conversion of fibrinogen to fibrin, the proteolytic activation of factor XIII to give factor XIIIa is catalysed by thrombin. Thrombin cleaves both a-chains in factor XIII ($[a_2b_2]$) between Arg37 and Gly38 (687, 688). This results in the release of two 37-residue activation peptides and the formation of factor XIII′ ($[a'_2b_2]$). This cleavage does not require the presence of calcium ions, but these are needed for the formation of the enzymically active factor XIIIa (689). Factor XIII′ ($[a'_2b_2]$) dissociates in the presence of calcium ions to give one $[a'_2]$ dimer and two b monomers. At the same time, the cysteine residue at the active site is exposed (690–694). Thus factor XIIIa has the structure $[a^*_2]$ and is a thiol enzyme.

The activation of factor XIII *in vitro* requires Ca^{2+} concentrations above 10 mM, but the presence of physiological amounts of fibrinogen reduces the required calcium concentration to that encountered in plasma (695). This effect involves the central region in the Aα chain in fibrinogen (696). Factor XIII circulates as a complex with fibrinogen (697), and the activation of factor XIII by thrombin to give factor XIII′ is more rapid in the presence of fibrinogen (698, 699). The process is accelerated especially effectively by non-cross-linked fibrins I and II, which are the physiological substrates for factor XIIIa (700–704). Polymerised fibrins I and II are, as discussed earlier (section 7.2), nothing other than polymerised *des*-A-fibrinogen and polymerised *des*-A,B-fibrinogen – that is, polymerised forms of fibrinogen from which the fibrinopeptide A or, respectively, both fibrinopeptides have been split off. Their ability to promote the activation of factor XIII is lost when catalytically competent factor XIIIa is formed. This implies that, when fibrin is cross-linked in a reaction catalysed by factor XIIIa, the fibrin loses its activity as an enhancer of the activation of factor XIII by thrombin. This may be of physiological significance in ensuring that the formation of factor XIIIa ceases once its substrates have become cross-linked (702). Figure 47 sums up present knowledge of the activation of factor XIII.

The binding site for factor XIII on fibrinogen is located on the a-chain (697), between Gly38 and Lys513 (705); the corresponding dissociation constant is about 10^{-8} M (697). In fibrinogen, factor XIII binds to both the Aα and the Bβ chains (706).

We shall now examine the events that follow the formation of catalytically active factor XIIIa.

Factor XIIIa is a transglutaminase. The cross-linking reaction that it catalyses takes places between glutamine side-chains of one polypeptide and lysines of the other. The reaction involves nucleophilic attack by the ϵ-nitrogen atom of lysine (termed the *donor* residue) and the expulsion of amide nitrogen as NH_3 from glutamine (the *acceptor* residue) (Fig. 48).

The primary function of factor XIIIa is the cross-linking of the spontaneously polymerised fibrin monomers, such that the clot formed becomes stabilised covalently by isopeptide bonds (707–709). The 'soft clot'

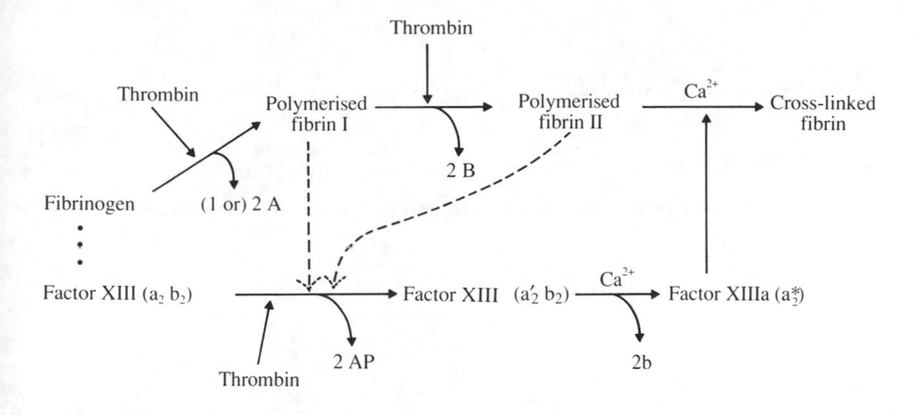

Fig. 47. The activation of factor XIII. ●●● indicates the fact that factor XIII circulates in complex with fibrinogen. Dashed arrows indicate that polymerised fibrins I and II stimulate the activation of factor XIII by thrombin. (A, fibrinopeptide A; B, fibrinopeptide B; AP, factor XIII activation peptide.)

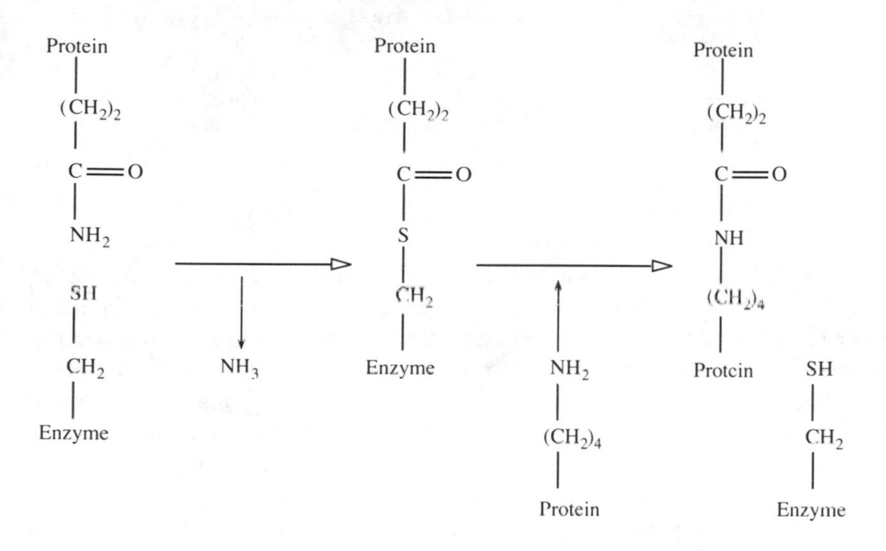

Fig. 48. A suggested enzymic mechanism for the formation of γ-glutamyl-ϵ-lysine cross-links catalysed by factor XIIIa.

is converted to a 'hard clot'. The cross-linking reactions result in the formation of intermolecular γ-chain dimers (710) and α-chain polymers (554, 711). The two γ chains that cross-react to give a dimer are thus present in two different fibrin molecules, just as the α chains in α-chain polymers come from different fibrin molecules.

The two γ chains are cross-linked by two isopeptide bonds towards the C

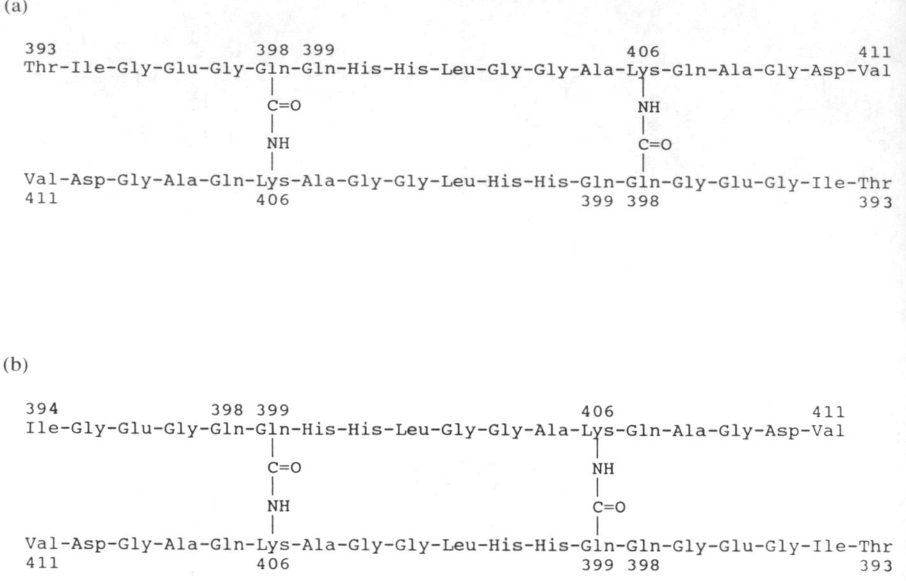

Fig. 49. Isopeptide cross-links in the C-terminal region of two cross-linked fibrin γ chains. (a) As shown in reference (713). (b) As shown in reference (714).

termini of the chains (712, 713) (Fig. 49). The two Lys406 residues are connected either to the two Gln398 residues (713) or the two Gln399 residues (714), and the chains therefore lie antiparallel to one another. In the α chains, the physiologically functional glutamines have been found to be Gln328 and Gln366 (numbers from the Aα chain), while the number and location of the lysines involved have not been determined. A strong candidate is Lys508 (716), and it is thought that other lysines may be found in the region of sequence between residues 518 and 581, which include five lysines (715). It should be remembered that the two Gln residues in a particular α chain must be cross-linked to Lys residues in α chains from two other fibrin monomers, if polymerisation of the α chains is to take place.

As described above, the dimerisation of γ chains is an important part of the stabilisation process, and it results in the formation of dimers by cross-linking of two fragments D. The degradation by plasmin of cross-linked fibrin leads as expected to the formation of fragment D dimers, but in addition, unexpectedly, to trimers and tetramers of fragment D (717). There are two conceivable mechanisms for the formation of these structures. One is a cyclical cross-reaction of the reactive residues present, such as Gln399(A)–Lys406(B), Gln399(B)–Lys406(C) and Gln399(C)–Lys406(A). The other is a linear cross-reaction of the form Gln399(A)–Lys406(B) and Gln399(B)–Lys406(C) without the final ring-closing cross-link (717).

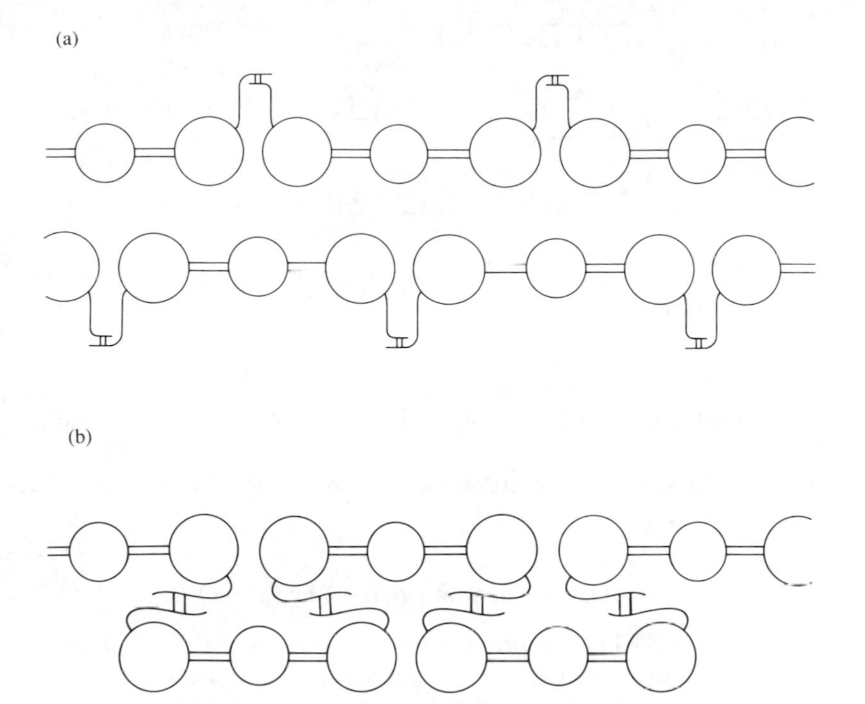

Fig. 50. Two models for the cross-linking of fibrin γ chains. The models show the fibrin protofibril with fibrin as a trinodular structure. (a) Model in which the cross-linking of the γ chain occurs in the same chain in the protofibril. (b) Model in which the cross-linking of the γ chain takes place between the chains in the protofibril.

These trimeric and tetrameric D fragments must come from branching-points in protofibrils or in laterally polymerised fibrin, and their existence also has implications for the manner in which fragment D dimers are formed. There are two possibilities for forming fragment D dimers by interconnection of γ chains (Fig. 50). One is the serial connection of consecutive γ chains, in which the protofibrils therefore consist of two parallel bound chains of indefinite length, not connected by covalent bonds (718). The other possibility is a structure in which each D fragment is cross-linked to a D fragment in the other half-protofibril strand (719, 720). If mechanical strength is a primary criterion, then the second possibility seems more likely, and this would also provide the simplest explanation for the occurrence of trimers and tetramers of fragment D (717).

β chains of fibrin do not take part in cross-linking. This could be because they are more centrally placed and less accessible; this would correspond well with the suggested 'core' function mentioned previously (section 7.1.1).

The cross-linking of chains in fibrin, and especially the cross-linking of α chains to α-chain polymers, results in a five-fold increase in the mechanical strength of the clot (721). This explains, at least partly, why genetic

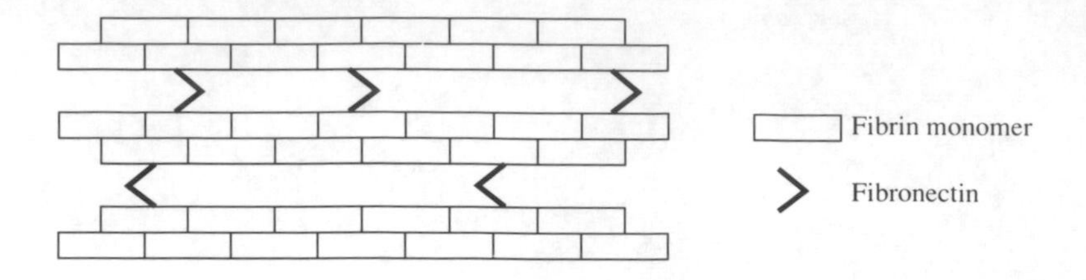

Fig. 51. The cross-linking of the α chains of fibrin via fibronectin. (Redrawn from reference 731.)

deficiencies in factor XIII and other molecular disturbances in the stabilisation of fibrin lead to poor wound healing and to secondary bleeding (722).

Further cross-linking reactions catalysed by factor XIIIa are discussed in the next section.

7.4. Other cross-linking reactions catalysed by factor XIIIa

Although factor XIIIa has a primary rôle as a stabiliser of fibrin, there are other cross-linking reactions that it catalyses. These are of interest in connection with haemostasis, with tissue repair and with fibrinolysis.

When factor XIIIa cross-links fibrin monomers, there occurs simultaneously a cross-linking of α_2-plasmin inhibitor to the α chains of fibrin (723–726). α_2-plasmin inhibitor is the most important inhibitor of the fibrin-degrading enzyme plasmin (see section 9.4.2). These cross-links are between Gln2 of α_2-plasmin inhibitor (726) and probably Lys287 in the α chain of fibrin (727). The evidence for the involvement of Lys287 is circumstantial: it is based upon the finding that α_2-plasmin inhibitor can be cross-linked to the corresponding Lys303 in the Aα chain of fibrinogen. This incorporation does not interfere with the lysine residues that are used for cross-linking in fibrin.

Plasma fibronectin can also be cross-linked to Lys residues in fibrin α chains (728, 729), probably via Gln3 (730). The lysine residues involved have not been identified, but they are known not to be the same as those that cross-link to α_2-plasmin inhibitor. Up to one molecule of plasma fibronectin per fibrin monomer can be incorporated *in vitro*, but the degree of incorporation can never be so high *in vivo*, since the concentration of fibronectin in plasma is an order of magnitude lower than that of fibrinogen (731). As plasma fibronectin consists of two chains with a cross-linking position on each, it is conceivable that plasma fibronectin takes part in a sort of alternative cross-linking of the fibrinogens' α chains (Fig. 51). This may be reflected in the observation that fibrin fibres become larger and denser when they form in the presence of plasma fibronectin (732). Since plasma fibronectin can also cross-link to collagen with the help of factor XIIIa (733, 734), it is possible that the factor XIIIa-mediated cross-linking of fibronectin plays a part in the process of tissue repair.

In connection with the reactions of haemostasis, it is also interesting that both von Willebrand factor (735) and thrombospondin (736, 737) can be cross-linked to fibrin by factor XIIIa, and that von Willebrand factor also cross-links to collagen (738).

Several other proteins have been shown to contain glutamine residues that take part in isopeptide-bond formation catalysed by factor XIIIa, but the significance of this is not known. These proteins include factor V (739, 740), vitronectin (741), vinculin (742) and histidine-rich glycoprotein (743).

Further processes connected with haemostasis are described in Chapter 24.

8. Regulation of blood coagulation

Once a clot has been formed, the processes of coagulation are complete. However, there is another topic that we havc not yet covered, and it is one of central importance: the regulation of the various coagulation processes. The reader will recollect that we have up to now described the formation of seven active serine proteinases (factors VIIa, IXa, Xa, XIa and XIIa, plasma kallikrein and thrombin) and four acceleratory co-factors (factors Va and VIIIa, HMM-kininogen and tissue factor), and all of these come into action in the steps leading up to the conversion of fibrinogen to fibrin. A system of this kind cannot just be allowed to free-wheel. The success of the coagulation process is due to the finely tuned modulation and regulation of all of the partial proteolytic digestions that occur. Too little or too much activity would be equally damaging for the organism. Regulation is a central issue in blood coagulation.

For normal organisms, there is no lack of potential for coagulation as expressed in the activity of the proteinases. The principle of regulation of coagulation is thus restriction of the lifetimes of the active components. This is achieved with the help of inhibitors. We shall be concerned with two principal methods of inhibiting enzymic reactions *in vivo*. One is the formation of a (usually) covalent complex between a proteinase and an inhibitor, and the other is the inactivation of a co-factor by proteolytic digestion.

8.1. Inhibition by the formation of a [proteinase.inhibitor] complex

An inhibitor in the sense to be used here is a protein that is made in the organism and that is able to depress enzymic activity. All the relevant proteinases in the coagulation system are serine proteinases, so the following description will be restricted to the inhibition of proteinases of this sort (for a review, see, for example, reference 744).

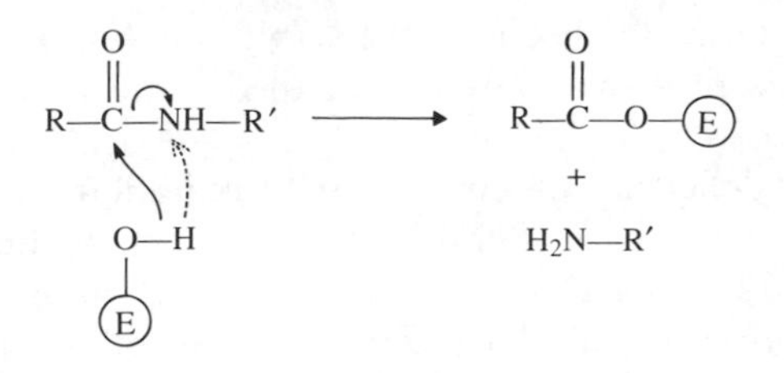

Fig. 52. A proposed mechanism for the inhibition of serine proteinases by serpins. (E, serine proteinase.)

8.1.1. *Complex formation*

The inhibition of a serine proteinase by complex formation with a serine-proteinase inhibitor is thought usually to take place by the splitting of a peptide bond in the inhibitor. This leads to a stable [proteinase.inhibitor] complex and often to the release of an activation peptide from the inhibitor. The [proteinase.inhibitor] complex is stable under denaturing conditions, but it can be made to dissociate with the help of a nucleophile and/or extreme pH values, suggesting strongly the existence of a covalent link between the enzyme and its inhibitor.

Serine-proteinase inhibitors function first of all as competitive inhibitors by binding to the active site of the serine proteinase via a region of their own sequence that resembles the sequence of the usual substrate. This can be achieved with great speed and specificity, as the coagulation proteinases are themselves highly specific.

After this, a bond in the amino-acid sequence of the proteinase inhibitor is broken. This is believed to take place in such a way that the serine residue at the active site is bound covalently, and therefore stably, to the carbonyl carbon atom from the peptide bond that is cleaved. The serine proteinase and its inhibitor are therefore presumed to be connected by an ester bond (Fig. 52).

The formation of the [proteinase.inhibitor] complex is thus a perfectly normal proteolytic cleavage, except in that the acyl-enzyme intermediate is not hydrolysed. The inactivation or inhibition of serine proteinases is thus virtually irreversible.

8.1.2. *What inhibits what?*

The principal inhibition of the serine proteinases of the coagulation system is carried out by only three serine-proteinase inhibitors. These are the $C\overline{1}$-inhibitor, antithrombin III and, with a less important rôle, the α_1-proteinase inhibitor. All three inhibitors belong to a special protein family of homologous proteins, termed collectively the serpin family (for **ser**ine-**p**roteinase **in**hibitor). This protein family is described in more detail in Chapter 26.

The genetically caused deficiency of any of these three inhibitors results in serious illness, but only in the case of antithrombin III can the illness be traced unambiguously to resulting abnormalities in coagulation. Both $C\bar{1}$-inhibitor and α_1-proteinase inhibitor have especially important parts in other processes that are more likely to be related to the symptoms observed.

The amino-acid sequences of all the three inhibitors mentioned have been determined, and they can be found in the following sources: for α_1-proteinase inhibitor, references (745) and (746); for antithrombin III, references (747) and (748); for $C\bar{1}$-inhibitor, reference (749).

$C\bar{1}$-inhibitor appears to be the most important inhibitor of factor XIIa (750) and plasma kallikrein (751), although the latter can also be inhibited by the general proteinase inhibitor α_2-macroglobulin (751).

Factor XIa appears to be inhibited largely by α_1-proteinase inhibitor (752) and, to a lesser extent, by antithrombin III (753) and $C\bar{1}$-inhibitor (754). $C\bar{1}$-inhibitor and α_1-proteinase inhibitor are described in more detail in sections 15.1.1 and 26.1

Factor IXa and factor Xa can react *in vitro* both with α_1-proteinase inhibitor and with antithrombin III, but these reactions are unlikely to be of physiological importance. There is reason to believe that the inactivation of factors IXa and Xa proceeds by a fundamentally different mechanism, namely, by proteolytic degradation of the co-factors that they use (see below, section 8.2.4).

The most important inhibitor of thrombin is antithrombin III, and its inhibitory effect is enhanced considerably by the presence of heparin (755, 756).

The only serine proteinase that we have failed to mention in this context is factor VIIa. The physiological regulation of this factor is more complex. It is not dependent upon antithrombin III (757, 758), since this fails to inhibit factor VIIa; instead, it depends upon the presence of factor Xa, tissue factor and an extrinisic pathway inhibitor (see section 8.1.4).

8.1.3. *Antithrombin III*

Antithrombin III is a single-chained glycoprotein with a molecular mass estimated to lie between 54 kDa and 68 kDa (756, 759, 760), of which 9% is reportedly carbohydrate (759). Its concentration in normal human plasma is 150 μg/ml ($\approx$2.5 μM) (761).

The amino-acid sequence of human antithrombin III has been determined almost completely at the protein level (747) and has been deduced both partially (763) and wholly (748, 762) by cDNA-sequencing (Appendix, Fig. A17). The cDNA sequence includes a signal peptide 32 amino acids in length. The gene for antithrombin III covers approximately 19 kbp; it contains six exons (764) and is located on chromosome 1 at position q23–25 (765, 766).

Human antithrombin III has 432 amino-acid residues, and the calculated molecular mass is 49 kDa. Carbohydrate is attached to Asn96, Asn135, Asn155 and Asn192, and the protein's three disulphide bonds are Cys8–Cys128, Cys21–Cys90 and Cys247–Cys430 (747).

Antithrombin III has two functional sites. One is the reactive site that acts in the inhibition of thrombin, and the other is the binding site for heparin, the co-factor that influences the rate with which thrombin is inhibited.

The reactive site – the bond that thrombin cleaves – is the peptide bond Arg393–Ser394 (767, 768). The cleavage by thrombin produces a 39-amino-acid peptide derived from the C terminus. Because of the disulphide bond between Cys247 and Cys430, the activation peptide is not released from the proteinase-inhibitor complex.

Heparin is a sulphated glucosaminoglycan (769, 770). The region that binds antithrombin III has been identified as a pentasaccharide sequence. Various sulphate groups in this pentasaccharide sequence have been found to be essential for the binding of antithrombin III.

The exact structure of the heparin-binding site in antithrombin III is not known, and the mechanism of action of heparin is not fully understood either. It is probable that the [heparin.antithrombin III] complex is held together by electrostatic interactions between the negatively charged groups in the heparin (e.g. sulphate) and Lys/Arg residues in antithrombin III. Naturally occurring antithrombin III mutants with lowered heparin affinity are altered at Pro41 (771) or at Arg47 (772–776). Chemical modifications have implicated Trp49 (777), Lys107 (778), Lys114 (779) and Lys125 (778–780) in the binding of heparin. It is also interesting to note that a naturally occurring antithrombin III variant, in which the carbohydrate attachment site at Asn135 is not used, has an enhanced affinity for heparin (781), perhaps because the carbohydrate side-chain usually present causes steric hindrance. Conversely, it has been shown that the genetic variant Ile7→Asn, which contains an additional carbohydrate attachment site, has a diminished affinity for heparin (782).

Many genetic variants of antithrombin III have been characterised, and Table 7 provides a summary of these.

8.1.4. *Extrinsic pathway inhibition*

As described in the section on activation in the extrinsic pathway (Chapter 6), this is put into operation by the exposure of tissue factor to plasma. We saw that the formation (or the presence) of factor Xa was of major importance for the rate of activation, since factor Xa is the enzyme that best

Table 7. *Characterised variants of antithrombin III*

Name	Amino-acid substitution	References
Rouen III	Ile7→Asn	(782)
Basel	Pro41→Leu	(771)
Clichy	Pro41→Leu	(783)
Clichy 2	Pro41→Leu	(783)
Franconville	Pro41→Leu	(783)
Alger	Arg47→Cys	(772)
Tours	Arg47→Cys	(773)
Toyama	Arg47→Cys	(774)
Paris	Arg47→Cys	(783)
Paris 2	Arg47→Cys	(783)
Rouen	Arg47→His	(775)
Rouen II	Arg47→Ser	(776)
Hamilton	Ala382→Thr	(784)
Hvidovre	Ala384→Pro	(785)
Esbjerg	Ala384→Pro	(785)
Charleville	Ala384→Pro	(783)
Cambridge	Ala384→Pro	(786)
Glasgow	Arg393→His	(787, 788)
Sheffield	Arg393→His	(789)
Chicago	Arg393→His	(790)
Avranches	Arg393→His	(783)
Northwick Park	Arg393→Cys	(787, 788)
Milano	Arg393→Cys	(791)
Pescara	Arg393→Pro	(792)
Denver	Ser394→Leu	(793)
Aalborg	Ser394→Leu	(785)
Tønder	Ser394→Leu	(785)
Oslo	Ala404→Thr	(794)
Utah	Pro407→Leu	(795)

activates tissue-factor-bound factor VII. The regulation of activation by the extrinsic pathway calls for more than just the inhibition of factor Xa or VIIa – the exposure of tissue factor must be regulated. The way in which the extrinsic pathway is regulated or inhibited is therefore different from the other mechanisms of regulation in coagulation, and it makes use of an inhibitor that is variously termed extrinsic pathway inhibitor (EPI) or lipoprotein-associated coagulation inhibitor (LACI) (reviewed in reference 796).

The story of EPI/LACI goes back to the beginning of this century, but present ideas about it date from 1957 (797), when it was shown that serum contains a substance (anti-convertin) that inhibits not tissue factor itself,

```
 19 L K L M H S F C A F K A D D G P C K A I M K R F F F N I F T R
 90 Q Q E K P D F C F L E E D P G I C R G Y I T R Y F Y N N Q T K
182 E F H G P S W C L T P A D R G L C R A N E N R F Y Y N S V I G

Q C E E F I Y G G C E G N Q N R F E S L E E C K K M C T R D N A N
Q C E R F K Y G G C L G N M N N F E T L E E C K N I C E D G P N G
K C R P F K Y S G C G G N E N N F T S K Q E C L R A C K K G F I Q

R I I K T T L 89
F Q V D N Y G 160
R I S K G G L 252
```

Fig. 53. Internally homologous domains in EPI/LACI. Only the positions at which amino acids are the same in all three domains are outlined. Pairwise comparison of the three domains reveals the following degrees of amino-acid identity: domains 1 and 2, 39%; domains 1 and 3, 38%; domains 2 and 3, 34%.

but rather the calcium-dependent complex between tissue factor and factor VIIa (then called convertin). Not until twenty-five years later was it discovered that a plasma protein that inhibited the [factor VIIa.tissue factor] complex required the presence of factor X in order to function (798). This mechanism has since been confirmed (799–802), and it has also been shown that it is the [factor VIIa.tissue factor] complex, and not tissue factor alone, that is inhibited. In this way, anti-convertin was rediscovered as EPI and LACI.

EPI/LACI purified from the medium used to grow a heptoma cell-line has a physically determined molecular mass of 38 kDa (803), while plasma appears to contain two forms with molecular masses of approximately 40 kDa and 32 kDa (804). Its concentration in plasma is believed to be 100 ng/ml ($\approx$2–3 nM) (803) and the name LACI is due to the fact that about one-half of the protein is found in the lipoprotein fraction of plasma (798, 805–807). This physical association arises partly because EPI/LACI occurs in a form that is disulphide-bound to apolipoprotein A-II (808). The amino-acid sequence of EPI/LACI, including a 28-residue signal peptide, has been deduced from the sequence of the cDNA (809) (Appendix, Fig. A18).

EPI/LACI consists of 276 amino-acid residues and contains three potential glycosylation sites at Asn117, Asn167 and Asn228. The calculated molecular mass is nearly 32 kDa. It is therefore probable that the two forms found in human plasma are glycosylated and non-glycosylated, respectively, but as yet there is no direct evidence for this. The amino-acid sequence can be divided up into a strongly negatively charged domain at the N terminus, three internally homologous domains that also are homologous to the Kunitz-type inhibitor family, and a strongly positively charged domain at the C terminus. The internal degrees of identity vary from 34% to 39% (Fig. 53).

Functional studies, both on authentic and on recombinant EPI/LACI, indicate a two-step mechanism for the inhibition of the extrinsic pathway (804, 805, 810).

EPI/LACI can bind and inactivate factor Xa without the help of calcium ions, but the binding can only take place if the unmodified active site of factor Xa is present. The inactivation is prevented by chemical modification of the Arg residues in EPI/LACI (804). Recombinant EPI/LACI, in which Arg107 – at the presumed active site in the second Kunitz-inhibitor-like domain of EPI/LACI – is mutated to an isoleucine, fails to inhibit factor Xa. In contrast, recombinant EPI/LACI molecules, in which Lys36 and Arg199 – corresponding to the presumed active sites in the first and third Kunitz-inhibitor-like domains – are mutated respectively to isoleucine and leucine, do inhibit factor Xa (810). It is therefore assumed that the second Kunitz-inhibitor domain binds to and inhibits factor Xa as the first step in the inhibition of the extrinsic pathway. The second step in the inhibition is thus the Ca^{2+}-dependent binding of the [EPI/LACI.factor Xa] complex to the [factor VIIa.tissue factor] complex. Of the recombinant EPI/LACI that has been modified by site-directed mutagenesis at Lys36, Arg107 and Arg199, EPI/LACI(Ile36) and EPI/LACI(Ile107) are unable to inhibit [factor VIIa.tissue factor], while EPI/LACI(Leu199) is just as effective in inhibiting the extrinsic pathway as is authentic EPI/LACI (810). This indicates that the first Kunitz-inhibitor domain binds factor VIIa when it is complexed to tissue factor, and that factor Xa is bound to the second Kunitz-inhibitor domain. Figure 54 shows schematically the mechanism of the inhibition of the extrinsic pathway by EPI/LACI.

8.2. Inhibition by proteolytic inactivation

The inhibition reactions that involve proteolytic cleavage and subsequent inactivation appear to make up the control mechanism that regulates the activities of factors IXa and Xa. This does not mean that factors IXa and Xa are inactivated by an additional protein cleavage reaction; the proteolytic attack is directed towards the accelerating co-factors Va and VIIIa. The proteolytic inactivation of these co-factors is sufficient to block the activities of the enzymes in question.

In order to understand the mechanisms of these inactivations, we must make the acquaintance of three new proteins – protein C, protein S and thrombomodulin (see, for example, references 811 and 812).

8.2.1. *Protein C*

Protein C (reviewed in reference 813) is the fifth of the vitamin K-dependent proteins that we encounter in connection with blood coagulation. It is a glycoprotein that circulates in the plasma as the zymogen for the serine proteinase protein Ca (also termed activated protein C, or APC). The molecular mass of human protein C has been estimated by SDS-

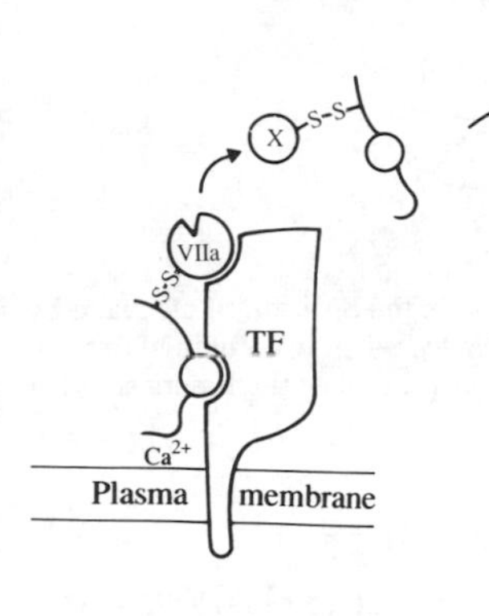

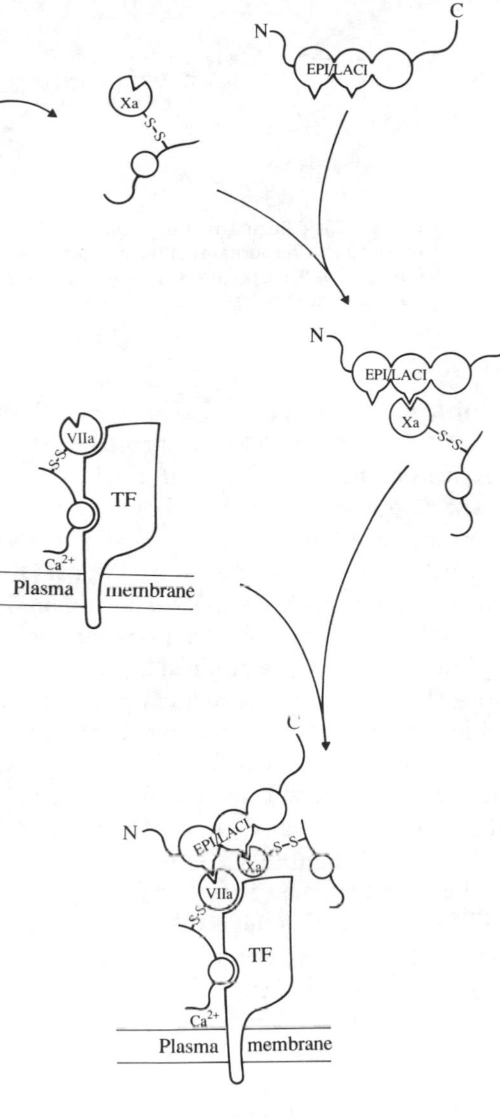

Fig. 54. A schematic model for the mechanism of action of extrinsic pathway inhibitor EPI/LACI. (TF, tissue factor; VIIa, factor VIIa; X, factor X; Xa, factor Xa.)

polyacrylamide gel electrophoresis to be 62 kDa, of which 23% is carbohydrate (814). In human plasma, protein C is found in two forms: some 25% of the protein C is one-chained, and the remaining 75% is cloven and thus two-chained (815). The two-chained protein C possesses a 21-kDa light chain disulphide-bonded to a 41-kDa heavy chain (814). The concentration of protein C in normal human plasma lies around 3–4 μg/ml ($\approx$60 nM) (816).

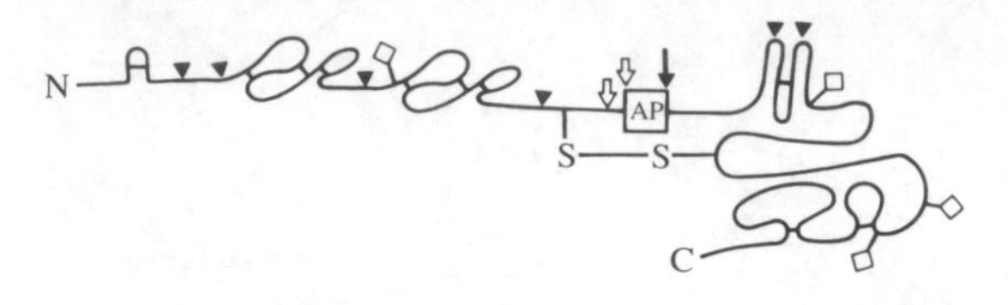

Fig. 55. The domain structure of protein C. → shows the position of cleavage by thrombin, ⇒ the positions of the bonds that are cleaved when two-chained protein C is formed, ▼ the positions of the introns, and ◇ the potential attachment sites for N-bound carbohydrate (see text). (AP, activation peptide.)

Protein C was first identified in bovine plasma in 1976 (817). It was later shown to be identical to the protein autoprothrombin IIa (818), first described in the early 1960s (819).

The amino-acid sequence of bovine protein C has been determined both by protein-sequencing (820, 821) and by cDNA-sequencing (822). Human protein C has been sequenced at the level of DNA only, both as cDNA (823) and as genomic DNA (824, 825). The human signal/propeptide, revealed by the cDNA-sequencing, is 42 amino-acid residues long. The gene structure of protein C has been determined from the exon–intron junctions (825) and, with the exception of the first exon and intron, by the sequencing of the whole gene (824). The gene for protein C covers 11.2 kbp of genomic DNA and possesses nine exons (825). The exons vary in size from 25 bp to 885 bp, and the introns from 92 bp to 2668 bp (824). The placing of the introns in the amino-acid sequence is indicated in Figure 55. The gene for protein C is localised on chromosome 2 at position q14–21 (826–829).

The molecular mass of protein C deduced from the amino-acid sequence is 47.5 kDa when account is taken of the γ-carboxylation of the glutamine residues, the β-hydroxylation of an Asp residue, and the removal of the Lys-Arg dipeptide from between the light and heavy chains when they are formed. The two-chained structure of protein C thus resembles that of factor X (section 4.1.4). If the 23% of carbohydrate is added in, the expected molecular mass of protein C is 61.6 kDa, in agreement with the experimental estimate (above).

Single-chained protein C contains 419 amino-acid residues (Appendix, Fig. A19). Bicatenate protein C is formed from its unicatenate precursor by the excision of the dipeptide Lys156–Arg157, such that the N-terminally derived light chain consists of 155 amino-acid residues, while the C-terminally derived heavy chain consists of 262 residues. Protein C displays structural homology with factors VII, IX and X, and Table 8 sums up these four proteins' various degrees of kinship. The overall structure of protein C is shown in Figure 55. Starting from the N terminus of protein C, we find first the Gla domain, that is thought to contain 9 Gla residues among its

Table 8. *Percentage of identical residues between factor VII (FVII), factor IX (FIX), factor X (FX) and protein C (PC)*

	FVII	FIX	FX	PC
FVII	100	41	41	39
FIX	41	100	47	36
FX	41	47	100	37
PC	39	36	37	100

first 29 amino-acid residues. These Gla residues have not been found experimentally, but are placed on the assumption of complete modification of all the Glu residues in the region (the next does not come until position 55). For comparison, the first 35 amino-acid residues of bovine protein C include 11 experimentally detected Gla residues (820). The Gla domain is followed by two growth-factor domains. In the bovine protein C, the first growth-factor domain contains a β-OH-Asp residue in position 71 (830), and, since amino-acid analysis shows the presence of one β-OH-Asp in human protein C and its cDNA sequence codes for Asp at this position, the human protein is also taken to have β-OH-Asp here. After the growth-factor domain, the light chain terminates; the final amino acids include a Cys residue believed to provide the connection with the heavy chain. The C-terminal region of the heavy chain is the potential serine-proteinase domain. The N terminus of the heavy chain is the activation peptide, later removed.

The disulphide-bond pattern of protein C has not been demonstrated experimentally, but can be inferred from homology with other proteins. On this basis, the following disulphide bridges can be expected in the two chains of protein C: in the Gla domain, Cys17–Cys22; in the first growth-factor domain, Cys50–Cys69, Cys63–Cys78 and Cys80–Cys89; in the second growth-factor domain, Cys98–Cys109, Cys105 Cys118 and Cys120–Cys133; between the two chains, Cys141 (light chain)–Cys120 (heavy chain); and in the serine-proteinase domain, Cys39–Cys55, Cys174–Cys188 and Cys199–Cys227. There are also two additional Cys residues in the first growth-factor domain, in positions 59 and 64, that may form a further disulphide bond in this domain.

The activation of protein C to give protein Ca is carried out by thrombin, which cleaves the bond between Arg12 and Leu13 in the heavy chain, releasing a 12-amino-acid activation peptide. This is remarkable not only because an Arg–Leu bond is cleaved, but also because the N-terminal residue of the heavy chain of activated protein Ca thus becomes leucine. No other serine proteinase has Leu at the N terminus of the chain that is proteolytically active; the others have Ile or Val instead. The reason for regarding this as important is that the N-terminal amino group, which is released in the activation of serine-proteinase zymogens, participates in a salt-bridge that is needed for the activity of the proteinase.

Protein Ca thus consists of an N-terminally derived light chain (21 kDa) disulphide-bonded to a C-terminally derived heavy chain (40 kDa) with serine-proteinase activity. The catalytic apparatus is believed to involve His42, Asp88 and Ser191 in the heavy chain, corresponding to His54/His211, Asp100/Asp257 and Ser203/Ser360 in bicatenate and unicatenate protein C, respectively.

It may also be mentioned that bovine protein C has an unusual attachment site for carbohydrate, as the Asn residue in the sequence N-E-C (amino acids 170–172 of the heavy chain) is glycosylated. It could be imagined that the human protein C is glycosylated in the homologous position in the same amino-acid sequence (N-E-C; residues 172–174 in the heavy chain). In addition, human protein C contains three potential N-glycosylation sites: at Asn97 in the light chain and at Asn91 and Asn156 in the heavy chain.

Genetically caused deficiency of functional protein C is pathologically serious. In heterozygotes the risk of thrombosis is heightened (831, 832), while homozygotic cases are afflicted from birth by a potentially fatal thrombotic disease that requires immediate treatment (833–835). The genetic causes of deficiency in protein C described hitherto are threefold:

1. the mutation of the Arg306 codon to a termination codon (836),
2. the mutation of Trp402 to Cys (836) and
3. the mutation of Arg169 to Trp (837).

(The positions given are for single-chained protein C.) Since the first of these mutations naturally gives rise to an incomplete molecule, and the second changes a normally invariant tryptophan residue, they are assumed to be the causes of the respective diseases, even though the sequences of the mutant molecules have not been checked in their entirety. The third mutation alters the arginine residue at which thrombin cleaves normal protein C, and, since it is the only mutation that has been found by the analysis of all coding sequences in the protein C gene of the afflicted patient, there is no doubt that it is the mutation responsible for the deficiency. (In this case, the patient lacked one entire protein C allele and was therefore affected by two different defects.)

8.2.2. *Thrombomodulin and the activation of protein C*

As mentioned in section 8.2.1, protein C is activated by thrombin to give protein Ca, but this occurs at a significant rate only in the presence of an endothelial-cell glycoprotein co-factor called thrombomodulin (838–840), reviewed, for example, in reference (841). Thrombomodulin in fact functions as a high-affinity receptor for thrombin on the endothelial surface ($K_d \approx 0.5$ nM) (839), since, as we have seen (section 5.2), thrombin is released into the plasma as a consequence of activation by the prothrombinase complex. The binding to thrombomodulin changes the specificity of thrombin from the cleavage of fibrinogen and the activation of factors V, VIII and XIII to the activation of protein C (842–844). In addition, thrombin also loses its ability to activate blood platelets (845). The explanation for the unexpected ability of thrombin to cleave an Arg–Leu bond presumably lies in the binding to thrombomodulin. Electron spin resonance studies have shown that the complex formation between human thrombin and rabbit thrombomodulin leads to conformational changes at the active site of the thrombin (846), even though other studies have not revealed large conformational changes (842, 843).

Thrombomodulin has been purified from rabbit lung (840), bovine lung (843) and human placenta (847), and it should be noted that most investigations with thrombomodulin up to now have been carried out in heterologous systems.

Partial amino-acid sequences have been deduced from cDNA for both bovine (848) and murine thrombomodulin (849), and the complete sequence of human thrombomodulin has been deduced in the same way (850–852) (Appendix, Fig. A20). Included in the cDNA sequence is a signal peptide of 18 residues. The gene for thrombomodulin is located on chromosome 20 (851). It covers 3.7 kbp and, surprisingly, it contains no introns (852, 853).

The amino-acid sequence of human thrombomodulin is made up of 557 amino-acid residues with a calculated molecular mass of 58.7 kDa, to which carbohydrate should be added. Treatment of thrombomodulin with neuramidase and O-glycanase reduces its molecular mass by about 10 kDa, while treatment with neuramidase and N-glycanase reduces it by about 8 kDa, so thrombomodulin apparently contains both O-bound and N-bound carbohydrate (850). There are five potential N-glycosylation sites in thrombomodulin, at Asn29, Asn97, Asn98, Asn364 and Asn391.

Thrombomodulin appears to be made up of five regions (Fig. 56). The first defined region (residues 1–226) shows no homology with other known proteins, and could thus well make up several hitherto undefined domains. The second region (residues 227–462) consists of six growth-factor domains, while the third (residues 463–497) contains several serine and threonine residues as possible sites for O-glycosylation. The last two regions are a putative transmembrane peptide (residues 498–521) and the cytoplasmic region (residues 522–557). Thrombomodulin from a mouse endothelial cell line contains phospho-O-serine in this cytoplasmic region (849). The disulphide-bonding pattern in thrombomodulin has not been determined experimentally and can only be projected, by homology, for the growth-factor regions.

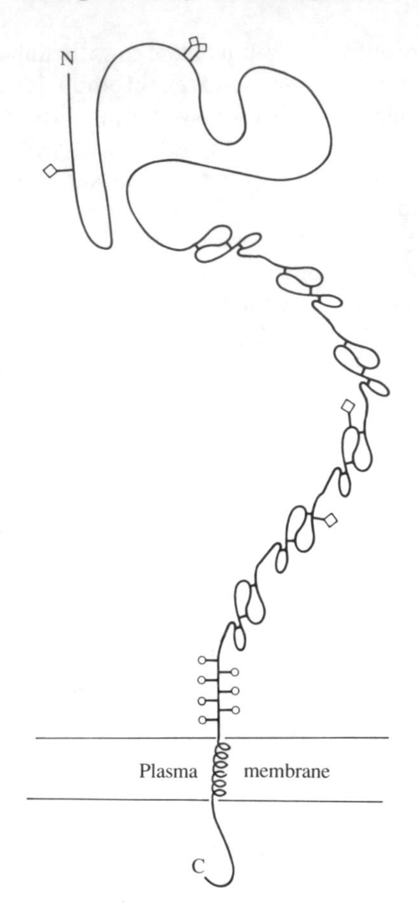

Fig. 56. The domain structure of thrombomodulin. ◇ shows potential attachment sites for N-bound carbohydrate and ○ the region containing potential attachment sites for O-bound carbohydrate.

The higher-order structure of thrombomodulin is reminiscent of the LDL receptor (LDL = low-density lipoprotein), even though there is no general sequence homology between these two. Thrombomodulin takes part in the endocytosis of thrombin. The physiological significance of this endocytotic process is unclear, but the regulation of the process is complex, since the binding of thrombin stimulates it (854) and protein C inhibits it (855).

Thrombomodulin has at least three functions in the coagulation system (more precisely, perhaps, the anti-coagulation system), and these will now be examined. They are:

a) the binding of thrombin, resulting in modulation of substrate specificity;
b) the inhibition of thrombin, in which the thrombomodulin acts as a co-factor;

c) the activation of protein C, in which the thrombomodulin likewise acts as a co-factor.

The binding of thrombin to thrombomodulin alters the substrate specificity of the thrombin and thus opposes coagulation. As stated above, the binding effectively destroys the specificity for catalysing the formation of fibrin, the activation of factors V, VIII and XIII and the activation of platelets, and a new specificity arises: the activation of protein C. The manner in which thrombomodulin alters the specificity of thrombin is not yet completely clear, and before discussing current ideas about this we shall mention another, related function. The binding of human thrombin to rabbit thrombomodulin stimulates the inactivation of thrombin by human antithrombin III by accelerating it some four- to eightfold (856, 857), an effect that resembles earlier observations of the inactivation of thrombin by an endothelial cell-surface antithrombin III co-factor (858). In contrast, it has been shown in bovine systems that the binding of thrombin to thrombomodulin either had no effect upon the inhibition of antithrombin III (859) or, indeed, afforded protection against inhibition (843). The two functions mentioned of rabbit thrombomodulin must be separate from its third function, the activation of protein C, because it is possible to isolate two forms of rabbit thrombomodulin that are equally good at promoting the activation of protein C but of which only one form both prevents the cleavage of fibrinogen and at the same time promotes the inactivation by antithrombin III of bound thrombin (860). The two forms of rabbit thrombomodulin have different molecular masses and charges, because one of them has an anionic domain; this probably consists of a sulphated galactosaminoglycan and not a heparin-related polysaccharide structure (861).

On the basis of studies with monoclonal antibodies against rabbit thrombomodulin and heparin-neutralising proteins, an integrated model of the mechanism of action of thrombomodulin has been proposed (Fig. 57) (861). In this model, it is assumed that the anionic domain in thrombomodulin interacts with bound thrombin, and thus prevents the thrombin from cleaving substrates such as fibrinogen, by making it inaccessible; at the same time the thrombomodulin promotes the inhibition of antithrombin III without affecting the activation of protein C. This model is also compatible with results for bovine thrombomodulin that have led to the proposal of a model based upon overlapping binding domains (843) that basically is the same model as that for rabbit thrombomodulin.

The change of the specificity of thrombin at the same time as its binding to thrombomodulin is not merely an ingenious refinement – it is essential in order to prevent the enzyme complex from being completely covered with fibrin. Indeed, it has been shown that, in the cleavage of fibrinogen, thrombomodulin is a competitive inhibitor of thrombin (856, 862).

The activation of protein C requires the presence of calcium ions but not

(a)

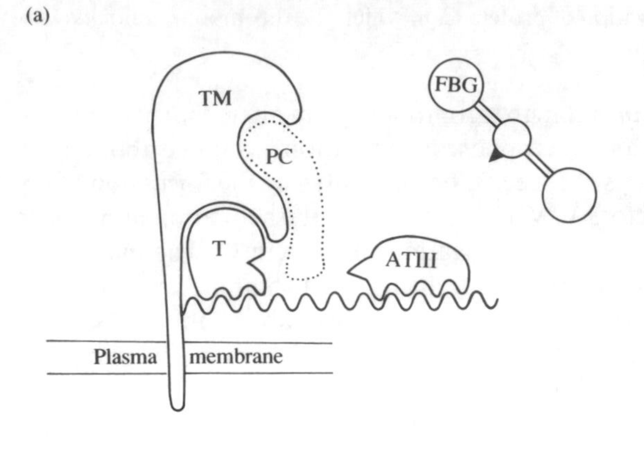

(b)

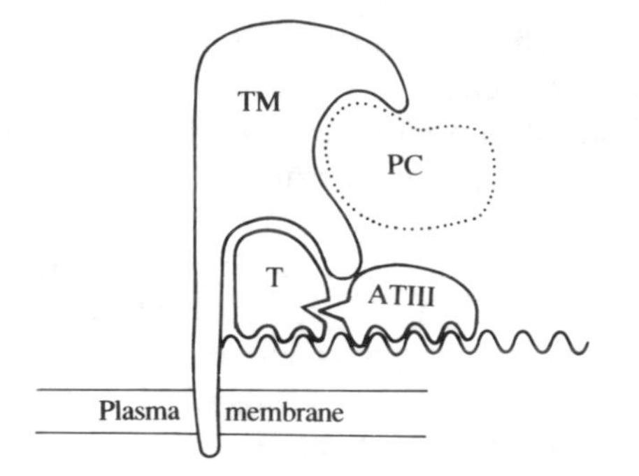

Fig. 57. A suggested mechanism for the action of rabbit thrombomodulin. (T, thrombin; TM, thrombomodulin; PC, protein C; ATIII, antithrombin III; FBG, fibrinogen.) The wavy line symbolises the anionic domain in the form of a sulphated galactosaminoglycan. The binding of thrombin to thrombomodulin prevents the cleavage of fibrinogen but promotes the activation of protein C (a) and the inhibition of thrombin by antithrombin III (b).

of negatively charged phospholipid (840), even though the rate of activation is raised a little in the presence of anionic lipid (863). Bovine protein C has been shown to contain a strong Gla-independent binding site for Ca^{2+} (404, 864), and the binding of Ca^{2+} to this introduces a conformational change in protein C such that activation by the [thrombin.thrombomodulin] complex can take place (865). Both in bovine and in human protein C, the strong binding site lies in the region of the light chain containing the growth-factor domains (866, 867). This has been determined in bovine protein C by direct measurement of calcium binding to a proteolytically prepared fragment consisting of amino-acid residues 43–

143 from the light chain disulphide-bonded to residues 108–131 from the heavy chain (867) (the growth-factor domains cover residues 45 to 135 in the light chain). In human protein C, the Ca^{2+}-binding site has been defined with the help of a conformation-specific monoclonal antibody that recognises, in a calcium-dependent manner, an epitope in the growth-factor domains (866). Furthermore, mutagenesis studies of recombinant protein C, prepared in mammal cell culture, indicate that the presence of Asp at position 71 in the first growth-factor domain is of decisive importance for the binding of calcium (868). If Asp71 is replaced by Glu, then the biological activity of the mutant protein C is only one-tenth of that of native protein C, irrespective of whether this is isolated from plasma or by cloning and expression of cDNA. The conformation-specific antibody mentioned above fails to recognise the mutated protein C molecule, which therefore does not possess the Ca^{2+}-dependent epitope. It is, however, not known whether this amino-acid substitution depresses the biological activity by steric hindrance or by the inability of glutamic acid to be hydroxylated.

In this way, the binding of Ca^{2+} to a domain in the light chain of protein C can affect the conformation of protein C such that activation by thrombin/prothrombin can take place. The binding of Ca^{2+} affects the conformation of the entire molecule (404) including, specifically, the conformation around the activation peptide, so that a conformation-specific monoclonal antibody against an epitope that includes the activation peptide recognises this only in the presence of calcium ions (869).

For rabbit and human thrombomodulin, both the binding of thrombin to thrombomodulin and the thrombomodulin's activity as a co-factor when the thrombin activates protein C can be traced to the six growth-factor domains (870–872). A small fragment of rabbit thrombomodulin, probably consisting of growth-factor domains 5 and 6 and beginning at a position just before growth-factor domain 5, corresponding to Phe407 in human thrombomodulin, was able to bind to thrombin without possessing co-factor activity in the activation of protein C (870). However, it is intriguing that this short fragment does inhibit the thrombin-catalysed cleavage of fibrin (871). This brings into question the correctness of some of the assumptions of the above model. A larger fragment from rabbit thrombomodulin, covering the last part of growth-factor domain 2 and the whole of growth-factor domains 3–6, and corresponding to residues 310–486 in human thrombomodulin, functions like thrombomodulin in that it is able both to bind thrombin and to act as co-factor in the activation of protein C (871). The introduction of deletion mutants into recombinant human thrombomodulin has led to analogous conclusions (872).

We thus have a somewhat remarkable, and certainly unusual, state of affairs. The receptor (thrombomodulin) uses its growth-factor domains for two purposes: the binding of thrombin and the generation of co-factor activity in the activation of protein C. It should also be remembered that human thrombomodulin undergoes calcium-dependent conformational rearrangement, as detected by calcium-dependent, conformation-specific monoclonal antibodies (873), perhaps as a consequence of the presence of β-OH-Asp groups. In support of this, β-OH-Asp has been detected by amino-acid analysis (874).

To summarise: thrombin binds to thrombomodulin. Thereafter, either it is inhibited by antithrombin III or it activates protein C to give protein Ca by cleavage of the Arg12–Leu13 bond in the heavy chain.

The subsequent behaviour of protein Ca brings us into contact with protein S.

8.2.3. *Protein S*

Protein S functions, as we shall see in more detail, as a co-factor in the processes catalysed by protein Ca. It is not a zymogen. Protein S is a glycoprotein with a single polypeptide chain, an estimated molecular mass of 69 kDa, a carbohydrate content of about 8% (859, 875) and a concentration in normal human plasma of around 10 μg/ml free protein S ($\approx$150 nM) (876). Protein S circulates in plasma in two forms. One form is free protein S and the other is a 1:1 stoichiometric complex with another protein, complement C4b-binding protein, which we shall meet later; as its name implies, it is a part of the complement system (877, 878). About 40% of the protein S present is taken to be in the free form, so that the total concentration of protein S in plasma is some 25 μg/ml ($\approx$350 nM). Protein S is the sixth vitamin K-dependent protein that we encounter.

Ninety-five per cent of the primary structure of bovine protein S has been determined by protein-sequencing (879), and cDNA-sequencing has revealed the full sequence including a signal/propeptide 41 residues in length (879). Human protein S, on the other hand, has been sequenced almost exclusively on the basis of cDNA (880–882) (Appendix, Fig. A21). The three cDNA sequences differ from each other at five positions: –31 (Pro/Leu), –16 (Leu/Phe), 180 (Leu/Pro), 222 (His/Tyr) and 304 (Tyr/Asp). The two deviations at positions –16 and 222 will be discussed briefly in Chapter 21. Yet another dimorphism has been found at position 460, in which a Ser is replaced by Pro. This results in the loss of a carbohydrate attachment site (883). The human cDNA sequence contains in addition a 41-residue signal/propeptide (881, 882). The gene for protein S has not yet been characterised fully, but it is known to cover more than 45 kbp genomic DNA (884) and it is estimated to contain at least 11 introns (827). The gene is located on chromosome 3 (827, 885, 886), and there is a report that two protein S genes may be present (886), of which one is a pseudogene (887).

The molecular mass of the amino-acid sequence deduced from the cDNA is 70.7 kDa, which together with about 8% carbohydrate gives approximately 76 kDa for the complete protein.

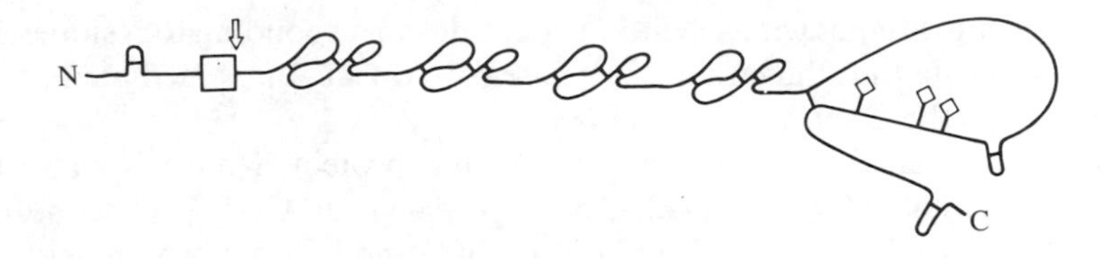

Fig. 58. The domain structure of protein S. ⇒ shows the position presumed to be sensitive to cleavage by thrombin and ◇ the potential attachment sites for N-bound carbohydrate.

The structure of human protein S is shown in Figure 58. The polypeptide chain contains 635 amino-acid residues with three potential glycosylation sites according to the consensus sequence N-X-S/T; these are at Asn458, Asn468 and Asn489. Protein S shows a certain degree of structural similarity with other vitamin K-dependent proteins. In common with these, protein S possesses a Gla domain at the N terminus; there are 11 glutamic acid residues among the first 36 (the next Glu does not come until position 87), and all 11 are presumably modified. This would agree exactly with bovine protein S, for which the positions of the Gla residues have been determined experimentally (888). The Gla domain is followed by a stretch of some 35 residues showing no homology with other vitamin K-dependent proteins. This short stretch is especially sensitive to cleavage by thrombin, and it is followed by no fewer than four growth-factor domains, unlike the other vitamin K-dependent proteins we have met, which contain only two.

In bovine protein S, the first growth-factor domain is furnished with a β-OH-Asp group, as seen earlier, at position 95 (879). The surprising aspect of bovine protein S is that the next three growth-factor domains contain at their homologous positions – 136, 178 and 217 – not the expected β-hydroxy aspartic acid residue but β-hydroxy asparagine (β-OH-Asn) instead (889). Human protein S appears to contain a β-OH-Asn residue in each of the second and third growth-factor domains only, and no modified residues in the fourth (880). The presence of β-OH-Asn in proteins is a relatively recent discovery; this modification is also found in complement components C1r and C1s (see sections 12.1.2, 12.1.3 and 21.2).

The remaining 391 amino acids – residues 245–635 – possess no homology with any of the other vitamin K-dependent proteins, and their function is not clear. This part of protein S is, however, homologous to a protein of completely different function. This homology was first detected by comparing protein S with the androgen-binding protein from rats (890). Thereafter, rat androgen-binding protein was found to be 68% homologous with human sex hormone-binding globulin (891, 892), and these two proteins are probably equivalent in the two species. Thus protein S is homologous to sex hormone-binding globulin (892, 893), and this is indeed surprising. Studies with a synthetic peptide corresponding to residues 605–614 indicate that the binding of protein S to C4b-BP is mediated at least partly by the C-terminal region (894).

Individual disulphide bonds in bovine protein S have been verified experimentally: Cys17–Cys22, Cys408–Cys434 and Cys597–Cys624. If this pattern is extrapolated to human protein S and homology with epidermal growth factor is used to place the disulphide bonds of the growth-factor domains, then only four unpaired cysteine residues remain, and these can also be placed with reasonable plausibility. The disulphide bonding pattern of human protein S thus adduced is: in the Gla domain, Cys17–Cys22; in the thrombin-sensitive domain, Cys47–Cys72; in the first growth-factor domain, Cys80–Cys93, Cys85–Cys102 and Cys104–Cys113; in the second growth-factor domain, Cys120–Cys134, Cys130–Cys143 and Cys145–Cys158; in the third, Cys164–Cys176, Cys171–Cys185 and Cys187–Cys200; in the fourth, Cys206–Cys215, Cys211–Cys224 and Cys226–Cys241; and in the C-terminal region, Cys247–Cys527, Cys408–Cys434 and Cys598–Cys625. The bonds not based upon homology are 47–72 and 247–527.

Genetically caused deficiency of protein S gives a heightened risk of thrombosis (895–897), but so far only one case has been described at the genetic level; in this instance a partial gene deletion was found (898).

As stated, protein S does not require proteolytic cleavage before it can function optimally as a co-factor for protein Ca. We therefore proceed to examine the processes in which proteins Ca and S participate.

8.2.4. *Proteolytic inactivation of factors Va and VIIIa*

The targets of the proteolytic activity of protein Ca are factors Va and VIIIa – the two major co-factors in blood coagulation.

Protein Ca abolishes the coagulation activity of these two co-factors by restricted proteolytic cleavage; these processes require the presence of both calcium ions and phospholipid. It is to be noted here that factor V and factor VIII are greatly inferior substrates for protein Ca in comparison with factors Va and VIIIa (229, 248, 408, 410, 899, 900).

Like the other vitamin K-dependent proteins, both protein C and protein Ca can bind to negatively charged phospholipids in the presence of calcium ions, and this is dependent upon the presence of the Gla residues close to the N terminus. The Gla domain in protein Ca has been shown to be essential for the inhibition of coagulation (865).

It has in fact been found that protein Ca alone, together with calcium ions and negatively charged phospholipid, is not a very good inactivator of factors Va or VIIIa. This may reasonably be attributed to the fact that, of all the vitamin K-dependent proteins, protein C and factor VII have by far the lowest affinity for anionic lipid membranes (901). As the affinity for anionic phospholipid surfaces is low, collisions between protein Ca and the membrane-bound co-factors Va and VIIIa are inevitably both infrequent and short-lived. The rate of inactivation of factor Va has been found to be proportional to the extent of binding of protein Ca to the phospholipid.

This problem is solved by protein S, which acts as a co-factor for protein Ca (902). Protein S has no enzymic activity in itself, but it acts as a co-factor by enhancing the binding of protein Ca to phospholipid. The complex between proteins Ca and S has a stoichiometry of 1:1 (903). Studies with the conformation-specific monoclonal antibody mentioned earlier (section 8.2.2), that displays a Ca^{2+}-dependent recognition of an epitope in the growth-factor domain of protein C, indicate that one growth-factor domain, or both, takes part in the interaction with protein S. This is because the antibody inhibits the ability of protein Ca to prevent coagulation in normal plasma (868). This hypothesis is further supported by the observation that a fragment of the light chain of protein C, that contains the Gla domain and both growth-factor domains with their disulphide bonds intact, functions as a competitive inhibitor of protein Ca activity in normal plasma (868).

As mentioned above, protein S does not require proteolytic cleavage before it can display anticoagulatory activity. In fact, the opposite situation is observed: the cleavage of protein S, at least by thrombin, results in its inactivation. This is partly because the affinity of the protein S for negatively charged phospholipid is reduced, and partly because the complex with protein Ca is labilised and the two proteins part company (876, 904, 905).

Bovine protein S is cleaved by thrombin at two positions, Arg52–Ala53 and Arg70–Ser71. An 18-residue peptide is released (888). It seems paradoxical that thrombin appears to have a double rôle in activating protein C and at the same time in inactivating protein S, which is a co-factor necessary for the functioning of protein Ca. It can of course be imagined that this is a complex mechanism that contributes to the regulation of the activity of protein Ca, but the cleavage of protein S by thrombin is in all probability not physiologically relevant, since it has been shown that thrombin in the presence of physiological concentrations of calcium fails to cleave protein S and that thrombomodulin inhibits this cleavage (906). In the absence of calcium ions, protein S is very sensitive to even small concentrations of thrombin, but the Ca^{2+} concentration in plasma (2.5 mM) is enough to abolish this sensitivity completely. It is true that purified protein S is found partly in the cleaved form, but this should rather be attributed to the use of calcium-free buffers during purification than to the presence of proteolytically cleaved protein S in plasma.

Protein S contains two or three strong binding sites for Ca^{2+}, and these do not involve Gla residues (907). The modification of the Gla residues in protein S to γ-methylene-Glu residues, carried out in the absence of calcium ions, results in the loss of anticoagulatory activity by reduction of activity as co-factor for protein Ca (908). This appears in turn to be due to loss of the ability to bind lipids, which thus must be important for the interaction with protein Ca. In the presence of calcium ions, some protection is afforded to the Gla residues, and the modification of these does not necessarily eliminate co-factor activity. There may thus be different functional classes of Gla residues (908).

It seems probable that even though protein Ca in itself has some affinity for binding to negatively charged phospholipid, the presence of protein S is required in order to strengthen this interaction. In agreement with this, studies with modified protein S have indicated that the binding of lipid by protein S is a prerequisite for the exposure of the binding site for protein Ca (908). The formation of the complex of protein S and protein Ca is thus dependent both upon calcium ions and phospholipid. The binding of proteolytically active protein Ca to surfaces with the help of protein S raises substantially the probability of collision between protein Ca and factors Va and VIIIa respectively. In this way, the anticoagulatory activity of protein Ca is increased.

We now examine the effect of the assemblage [protein Ca.protein S.Ca^{2+}.phospholipid] upon factors Va and VIIIa. As mentioned earlier, proteolytic cleavage takes place, resulting in the loss of the co-factors' activity. The protein Ca-catalysed cleavage of factors VIIIa and Va in the absence of protein S is inhibited by the binding of factor IXa and factor Xa to their respective co-factors (909, 910), but this effect is partly cancelled out by protein S (911, 912). Furthermore, it appears that it is the light chain in each co-factor that binds protein Ca (428, 913, 914).

The cleavage reactions catalysed by protein Ca that are correlated with the disappearance of co-factor activity occur in the fragments that originally arise from the N termini of factors VIII and V (Fig. 59).

In human factor VIII, the heavy chain is cleaved by protein Ca between Arg336 and Met337 (248, 334). Cleavage at other sites, as yet undefined, also takes place (248, 900), but the 336–337 cleavage is probably the one correlated with the loss of factor VIIIa's co-factor activity. As the direct cleavage of factor VIIIa has not yet been investigated, the importance of the 336–337 cleavage is deduced from the following circumstantial evidence. It has been observed that factor VIII is cleaved by protein Ca *in vitro* (248, 900). The first large fragment formed weighs 68 kDa and this is subsequently cleaved to give fragments of 45 kDa and 23 kDa. The 45-kDa fragment comes from the N terminus of factor VIII. Since it is normally factor VIIIa that is cleaved by protein Ca, the 45-kDa fragment must arise by cleavage within the sequence of the 50-kDa light chain of factor VIIIa, engendered when factor VIII is activated. Thus, the 68-kDa and 23-kDa fragments do not arise *in vivo*. In factor VIII/VIIIa the C-terminally derived chain is not cleaved by protein Ca.

In bovine factor Va, bovine protein Ca produces breaks both in the N-

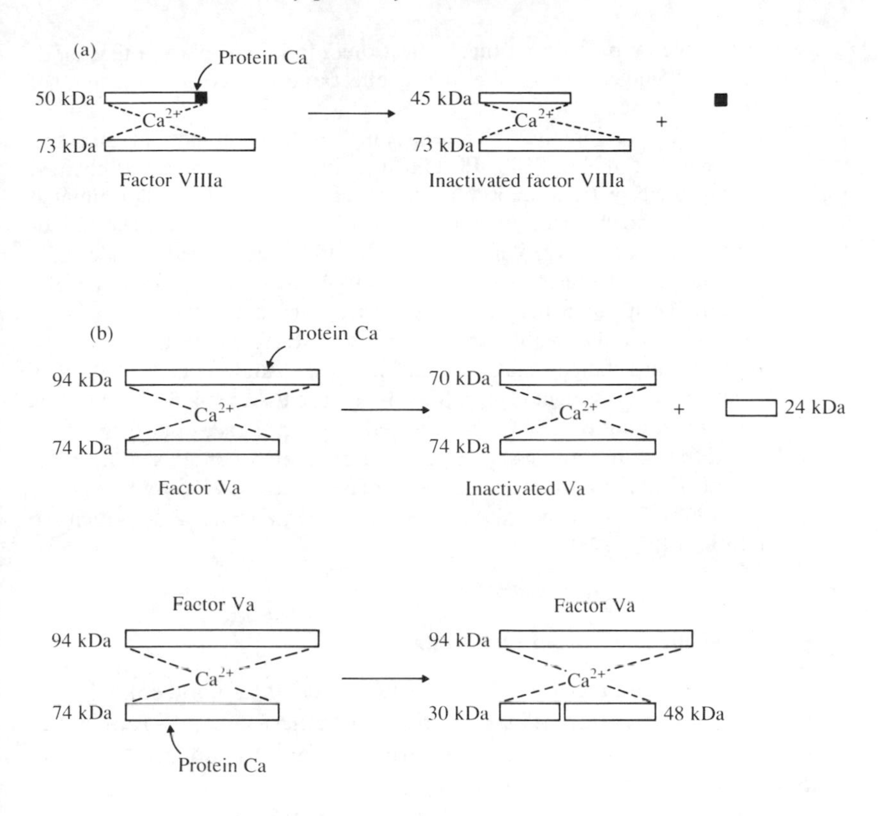

Fig. 59. The cleavage and inactivation of factors VIIIa and Va by protein Ca. (a) Protein Ca inactivates factor VIIIa by cleaving the bond Arg336–Met337 in the 50-kDa fragment of factor VIIIa, releasing a strongly anionic 36-residue peptide from the C terminus of the 50-kDa fragment. (b) Protein Ca inactivates factor Va by cleaving the bond Arg506–Gly507 in the 94-kDa fragment. The cleavage of the 74-kDa fragment by protein Ca does not inactivate factor Va.

terminally derived 94-kDa chain and in the C-terminally derived 74-kDa chain, but, as mentioned above, it is the cleavage of the 94-kDa chain that destroys the co-factor activity. The cleavage of importance in bovine factor Va occurs in a position corresponding to Arg506–Gly507 in human factor V (382). The cleavage by protein Ca of the 74-kDa chain has no significance for the activity of the co-factor. The cleavage occurs at the position corresponding to Arg1765–Leu1766 in human factor V (382), which is where factor Xa can also cleave the 74-kDa chain.

For the sake of completeness, we mention three further modulating factors.

1. Protein S bound to C4b-binding protein has no protein Ca co-factor activity (895, 915), which means that it is the concentration of free protein S that is physiologically relevant in this context.

2. It has been reported that there is a protein in bovine plasma that binds protein S and exercises a stimulating effect upon the co-factor activity of protein S (916).
3. A protein Ca inhibitor has been purified from human plasma and characterised (917, 918). Protein Ca inhibitor is a single-chained glycoprotein with a molecular mass of 57 kDa and an estimated concentration in normal human plasma of 5 μg/ml ($\approx$90 nM). The amino-acid sequence of this protein, along with a 19-residue signal peptide, have been deduced from the sequence of cDNA (919). The protein has 387 residues (Appendix, Fig. A22) and a calculated molecular mass of 43.8 kDa, with an unknown amount of carbohydrate, for which there are three potential attachment sites at Asn230, Asn243 and Asn319 (919). The reactive site of protein Ca inhibitor is the bond Arg354–Ser355, and its ability to inhibit protein Ca is increased by the presence of heparin (918). However, it has been shown that plasma also contains a heparin-independent inhibitor of protein Ca in plasma that may be physiologically important. This second inhibitor has been identified as an α_1-proteinase inhibitor (920–923).

8.3. Other mechanisms of inhibition

There is a further regulatory mechanism associated with blood coagulation that deserves brief mention. It involves the competitive neutralisation of the surface-bound negative charges that are essential for so many of the processes involved in blood coagulation.

The first protein shown to participate in this mechanism is β_2-glycoprotein I, which both inhibits contact activation (924) and also inhibits the action of the prothrombinase complex (925). The sequence of β_2-glycoprotein I is, as mentioned above (section 7.1.2), made up of five SCR domains (632). A corresponding inhibitory effect on contact activation has since been described for the b-chain of factor XIII (926), which is made up of 10 SCR domains, and for histine-rich glycoprotein (743), which contains a region with homology to the histidine-rich region in HMM-kininogen.

Other participants in this mechanism are various members of a protein family, for which the term 'lipocortin family' will be used here. Not all of its members have been shown to act as coagulation inhibitors. There are at least four independent reports of the isolation and functional characterisation of anticoagulant Ca^{2+}- and phospholipid-binding proteins (927–931): the determination of their amino-acid sequences by cDNA-sequencing (932–935) has since revealed that they were all the same protein. This protein has also been characterised and sequenced as endonexin II (936, 937). The protein(s) in question show(s) homology with proteins in the lipocortin family (938–941), as do the three most recently characterised anticoagulation proteins from placenta (942).

The major problem with these anticoagulant proteins in the lipocortin family is that they are all intracellular, so that it is hard to assess their physiological significance in the context of coagulation.

The first members of the lipocortin family to be discovered were lipocortin and calpactin. The members of this family all bind phospholipid and calcium, so it is not surprising that the family includes anticoagulant proteins.

Part II Fibrinolysis

9. Components of the fibrinolytic system

'Fibrinolysis' is the term applied to the set of processes that are necessary to re-dissolve the fibrin clot, the formation of which has been described in Part I. Thus, fibrinolysis covers not only the actual dissolution of the fibrin clot, but also the steps leading to its dissolution and their regulation.

Even though the formation of a stabilised fibrin clot is an essential part of the coagulation process, its removal is no less essential for the organism, since large pieces of fibrin can (and, as is well known, often do) cause serious problems for the circulatory system.

The fibrinolytic processes to be described here require acquaintance with some new proteins (Table 9): plasminogen, urokinase (u-PA), tissue plasminogen activator (t-PA), α_2-plasmin inhibitor (α_2-PI) and the inhibitors of the two plasminogen activators (PAIs). The properties of these proteins suffice to explain the regulatory mechanism for the dissolution of fibrin clots.

9.1. Plasminogen

Plasminogen is a single-chained glycoprotein that circulates in plasma as the zymogen of the fibrin-cleaving enzyme plasmin. Its molecular mass is estimated to be 92 kDa and its concentration in normal human plasma is around 200 μg/ml ($\approx$2 μM) (943). The proportion of carbohydrate in plasminogen is less than 2%.

Human plasminogen has been sequenced at the level of protein (944–946), and in one study partially (947), in another completely (948) at the level of cDNA (Appendix, Fig. A23). The signal peptide has 19 amino-acid residues. Unmodified plasminogen has a calculated molecular mass of 88.4 kDa (948), and inclusion of the carbohydrate gives the expected value of around 92 kDa. The gene for human plasminogen covers about 55 kbp and has 19 exons (947, 949). Its structure has been determined completely, but details have not been published. The gene resides on chromosome 6 at location q26–27 (950, 951). There is an unidentified gene homologous to the

Table 9. *The proteins of the fibrinolytic system*

Protein	Concentration in plasma	Physically-estimated molecular mass (kDa)	Number of amino-acid residues	Calculated molecular mass (kDa)	Gene size (kbp)	Chromosome (and position, if known)
Plasminogen	2 μM	92	791	88.4	55	6q26-27
Urokinase (u-PA)	40 pM	50–54	411	47	6.4	10q24–qter
Tissue plasminogen activator (t-PA)	70 pM	72	530	59	32.7	8
α_2-plasmin inhibitor	1 μM	67	452	50.5	16	18p11.1–q11.2
Plasminogen activator inhibitor 1 (PAI-1)	500 pM	52	379	42.8	12.2	7q21.3–22
Plasminogen activator inhibitor 2 (PAI-2)	–	47	415	46.6	16.5	18q21–23
Protease nexin	–	43	378/379	39.5	unknown	unknown

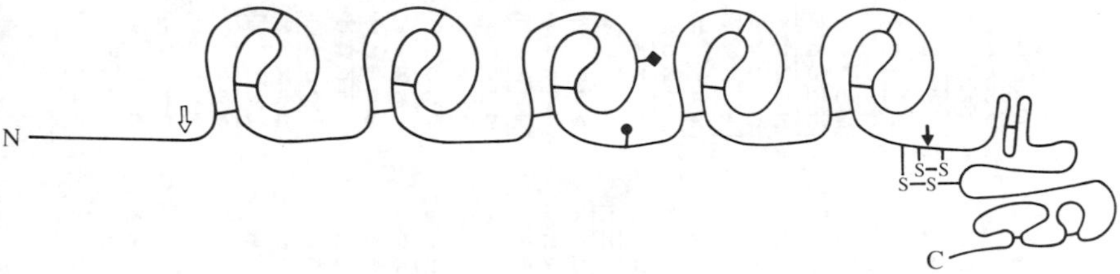

Fig. 60. The domain structure of plasminogen. → shows the position of cleavage by plasminogen activators, ⇒ the position of cleavage by plasmin, ◆ the N-bound carbohydrate and ● the O-bound carbohydrate.

plasminogen gene on chromosome 2, position q11–p11 (952). Bovine (953) and porcine (954, 955) plasminogens have been sequenced completely as proteins, and partial sequences are known for chicken (956), horse, sheep, goat and dog (957) plasminogens. The complete sequence of rhesus monkey plasminogen has been determined from cDNA (958).

The molecule of human plasminogen, as derived from the cDNA sequence, consists of 791 amino-acid residues (Appendix, Fig. A23), one more than found by protein sequencing. The additional residue is Ile65. Plasminogen can contain two carbohydrate chains, located at Asn289 and Thr346. However, Asn289 is only partially glycosylated, which leads to the isolation of human plasminogen in two forms, denoted I and II (959). Plasminogen I has carbohydrate attached at Asn289 and Thr346, while plasminogen II is glycosylated at Thr346 only (960–962).

As seen in Figure 60, plasminogen is built up (from the N terminus) of an 80-residue sequence followed by five domains that are homologous both to each other and to the two kringle domains in prothrombin. The homology relationships of kringle domains 1 to 5 are compared in Figure 61. At the C terminus of kringle 5 is a piece of connecting strand that leads to the molecule's potential serine-proteinase domain. The connecting strand also contains the peptide bond that is cleaved when plasminogen is activated to give plasmin.

The activation of plasminogen, by one of the activators u-PA or t-PA, results from cleavage of the bond Arg561–Val562. This cleavage results in a heavy chain derived from the N terminus, bound by two disulphide bridges to a light chain, from the C terminus, that contains the catalytically active serine residue. The disulphide pattern has been determined experimentally, with the exception of five S–S bonds, and these can be placed by inference from homologous domains (945). The disulphide bonds are as follows: in the N-terminal region, Cys30–Cys54 and Cys34–Cys42; in kringle domain 1, Cys84–Cys162 (inferred), Cys105–Cys145 and Cys133–Cys157; in kringle domain 2, Cys166–Cys243 (inferred), Cys187–Cys226 and Cys215–Cys238; in kringle domain 3, Cys256–Cys333 (inferred), Cys277–Cys316

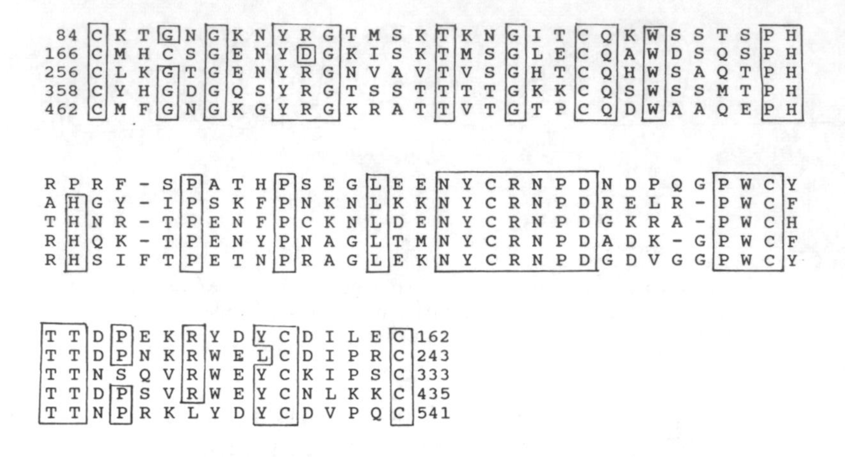

Fig. 61. Internally homologous kringle domains in plasminogen. Only the positions at which amino acids are the same in at least four domains are outlined and only the amino-acid sequence between the first and last Cys residues in each domain is shown. Pairwise comparison of the five kringle domains reveals the following degrees of amino-acid identity: kringle 1 and 2, 49%; kringle 1 and 3, 47%; kringle 1 and 4, 53%; kringle 1 and 5, 56%; kringle 2 and 3, 52%; kringle 2 and 4, 49%; kringle 2 and 5, 47%; kringle 3 and 4, 52%; kringle 3 and 5, 52%; kringle 4 and 5, 53%.

and Cys305–Cys326; in kringle domain 4, Cys358–Cys435, Cys379–Cys418 and Cys407–Cys430; in kringle domain 5, Cys462–Cys541, Cys483–Cys524 and Cys512–Cys536; between the second and third kringle domains, Cys169–Cys297; between the chains, Cys548–Cys666 and Cys558–Cys566; and in the serine-proteinase domain, Cys588–Cys604, Cys680–Cys747 (inferred), Cys710–Cys726 and Cys737–Cys765 (inferred). The catalytic apparatus in the light chain of plasmin is believed to be made up of His42, Asp85 and Ser180, corresponding to His603, Asp646 and Ser 741 in plasminogen.

Plasminogen and plasmin both possess binding sites for lysine and analogues of lysine such as ϵ-aminocaproic acid or tranexamic acid (Fig. 62). Common to most of the lysine analogues is a dipole with one elementary unit of charge at each end separated by 6.8 Å (0.68 nm). The lysine-binding sites are thought to be of importance for the fibrin-binding property of plasminogen and plasmin, since a correspondence between their lysine-binding and fibrin-binding sites has been found (963–965).

Native plasminogen possesses a high-affinity binding site for ϵ-aminocaproic acid ($K_d = 9$ μM), along with four or five much weaker sites ($K_d = 5$ mM) (966). Native plasminogen (also called Glu-plasminogen on account of its N-terminal Glu residue) is converted to Lys-plasminogen by the cleavage of peptide bond Lys77–Lys78 and the consequent removal of the 77 residues closest to the N terminus. This cleavage is catalysed by

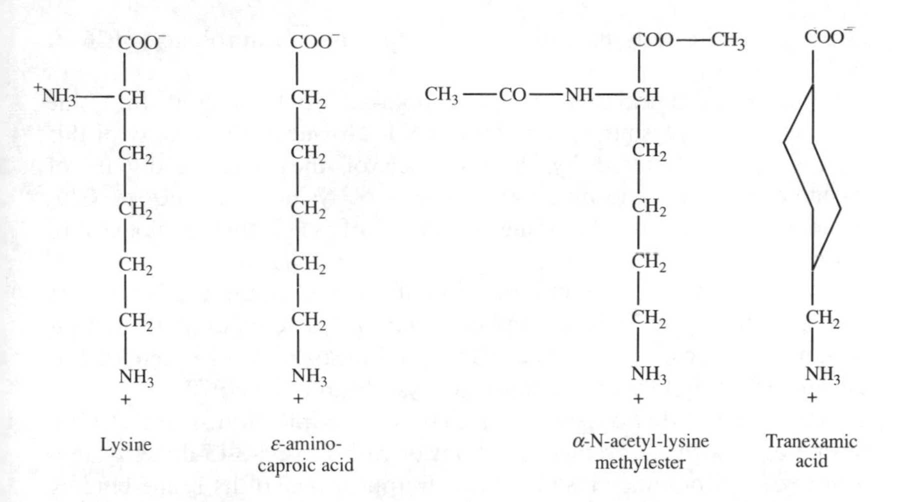

Fig. 62. Lysine and examples of lysine analogues that function as ligands for kringles.

plasmin itself, and it alters several of the properties of plasminogen. One such change is an alteration of the binding properties towards ϵ-aminocaproic acid (967), such that the strong site loses some of its affinity (K_d increases to 35 μM) while one of the weak binding sites becomes stronger (K_d decreases to 260 μM). The two or three other lysine-binding sites are not affected by the transition from Glu-plasminogen to Lys-plasminogen (967). Precisely the same general result is found for the binding of tranexamic acid (968).

Before examining the binding of plasminogen to fibrinogen and fibrin, its physiologically important property, we shall take a look at the positions of its lysine-binding sites, and see how much is known about their structure. A reasonable model for the placing of these sites has been derived from studies on intact native plasminogen and various proteolytic fragments thereof. We take the arguments one by one.

Intact, native plasminogen binds to lysine-Sepharose, and therefore must possess one or more lysine-binding sites. When plasminogen is digested by elastase, three main fragments result: 'K1 + 2 + 3', containing kringles 1, 2 and 3 and corresponding to Tyr80–Val338 or Tyr80–Val354; 'K4', containing kringle K4 (Val 355–Ala440); and 'mini-plasminogen', containing kringle 5 plus the remaining C-terminal sequence, which includes the serine-proteinase domain (944). K1 + 2 + 3 and K4 still bind to lysine-Sepharose, while mini-plasminogen has lost this ability; therefore, lysine-binding sites are not found on the latter fragment (944). K1 isolated from K1 + 2 + 3 has been shown to contain a high-affinity lysine-binding site (its K_d value for the binding of ϵ-aminocaproic acid is 17 μM), while K4 has a binding site for lysine with a K_d value for ε-aminocaproic acid of 36 μM (969).

It thus seems logical to infer that the high-affinity binding site for lysine in intact, native plasminogen resides in K1. However, the affinity of this binding site is affected by the presence of the N-terminal part of plasminogen, since the binding constants of lysine analogues to Glu-plasminogen are not the same as those of the same analogues to Lys-plasminogen.

Studies on the chemical modification of intact plasminogen have given support to this hypothesis, and indicate that the lysine-binding site in K4 is the same site whose binding affinity for ε-aminocaproic acid is increased by the conversion of Glu-plasminogen to Lys-plasminogen (970).

The basis for this conclusion is that the modification of intact Glu-plasminogen with 1,2-cyclohexanedione or with 1-ethyl-3-(3-dimethylaminopropyl)-carbodiimide results in the disappearance of its lysine-binding properties (971). The former reagent modifies the guanidino group in arginine residues, while the latter modifies carboxylic acid groups. The presence of ε-aminocaproic acid protects the Glu-plasminogen against these modifications, and it may be guessed that essential carboxylic acid and guanidino groups bind the ε-amino groups and the α-carboxylate groups of lysine (or its analogues).

The observation that the elastase-catalysed digestion of 1,2-cyclohexanedione-treated plasminogen results in fragments of which only K4 binds to lysine-Sepharose shows that the lysine-binding site in K4 is latent in plasminogen and that the lysine-binding site in K1+2+3 is the one that gives plasminogen its affinity for lysine (970). The modification of K4 with 1,2-cyclohexanedione likewise results in the loss of the lysine-binding properties of this fragment.

The structure of the lysine-binding sites has been investigated both by chemical modification and by proton magnetic resonance. Chemical modification of K4 has led to the suggestion that Asp56, Arg70 and Trp71 in K4, corresponding to Asp412, Arg426 and Trp427 in plasminogen, are of importance for the binding of lysine (972, 973). In these studies tryptophan was modified with dimethyl (2-hydroxy-5-nitrobenzyl)-sulphonium bromide. Proton NMR of K4 has indicated that Trp61, Phe63 and Trp71 in K4, corresponding to Trp417, Phe419 and Trp427 in plasminogen, make up a part of the lysine-binding site (974). As far as K1 is concerned, it appears that Arg34, corresponding to Arg117 in plasminogen, participates in the binding of lysine, and that both Arg32 and Arg34, corresponding to Arg115 and Arg117 in plasminogen, participate in the binding of fibrin (975).

At this point, the reader should be reminded that conclusions drawn from chemical modification studies must always be taken with a pinch of salt. Even if the modification of a particular amino-acid residue results in an

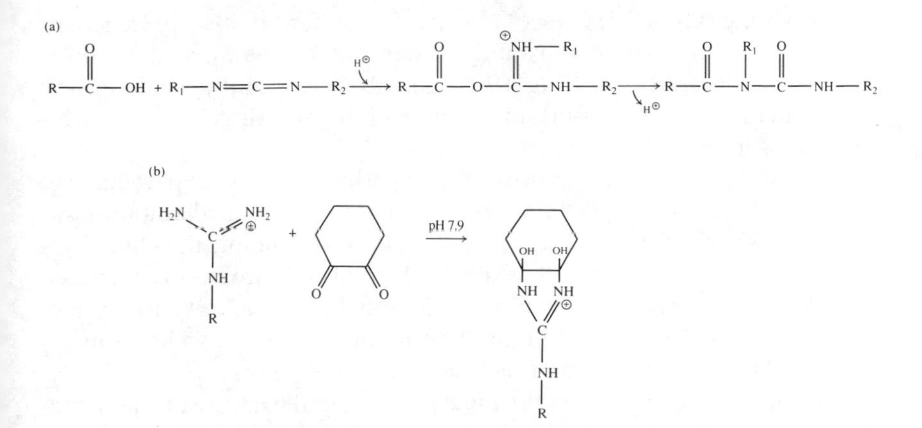

Fig. 63. Chemical modification of amino-acid residues used in studies of the lysine-binding sites in plasminogen. (a) The reaction between a carboxylic acid group and 1-ethyl-3-(3-dimethylaminopropyl)-carbodiimide. R represents protein bearing the carboxylic acid group; R1 represents the ethyl group; R2 the N,N-dimethyl aminopropyl group. (b) The reaction between a guanidino group and 1,2-cyclohexanedione. R represents the protein bearing the guanidino group.

alteration of the behaviour of the protein in question, this need not mean that the modified residue is essential for the property that is altered. This has several reasons. (1) The modification procedure can affect other things beside the amino-acid residue of interest. (2) Modification frequently involves the addition of a relatively bulky adduct to the functional group on a side-chain. The first reason scarcely requires discussion, but the second deserves a little more attention. Figure 63 shows the modification reactions used for plasminogen. It is in each case apparent that the structures introduced demand a great deal of space relative to the amino acids' side-chains. This can cause local crowding that displaces other side-chains, so that the steric effect is passed on over some distance. Thus, even when the modified residues are identified exactly, functions changed by modification need not be functions in which the modified amino acids play a direct or even an indirect part. It is thus not necessarily the case that Asp412, Arg426 and Trp427 are a part of the lysine-binding site of K4, or that Arg115 and Arg117 are a part of the fibrin-binding site of K1, even though it is well established that modification of these residues affects the lysine-binding properties of plasminogen.

As mentioned above, both plasminogen and plasmin bind to fibrinogen and fibrin, apparently in part by way of the lysine-binding sites just described. However, this does not account fully for the binding, as mini-plasminogen, which, as mentioned, does not bind to lysine-Sepharose, in fact binds better to fibrin than K1+2+3 or K4 fragments do (965). The binding of K1+2+3 is also to some extent dependent upon the presence of carbohydrate at Asn289 (see above), in as far as the presence of carbohydrate at Asn289 substantially weakens the binding of fibrin to K1+2+3 (965). The binding affinity of fibrin for mini-plasminogen is lower than that of Lys-plasminogen and higher than that of Glu-plasminogen (965, 976).

At first sight, it appears confusing, and perhaps somewhat paradoxical, that the affinity of Lys-plasminogen for ϵ-aminocaproic acid and tranexamic acid should be lower than that of Glu-plasminogen, while Lys-plasminogen has a much higher affinity for fibrin than Glu-plasminogen has. Equally surprisingly, mini-plasminogen has no affinity for lysine-Sepharose, while K1+2+3 and K4 both bind it well, yet it has a much higher affinity for fibrin than these two fragments do.

The probable explanation for these facts lies in the nature of the lysine-binding sites. These have been defined with the help of molecules containing a COO^- group and an NH_3^+ group separated by 6.8 Å (0.68 nm): lysine itself, and the analogues ϵ-aminocaproic and tranexamic acids. But there are no such molecules in intact fibrinogen or fibrin. In a polypeptide chain, lysine has only its ϵ-NH_3^+ group free, while the α-COO^- group is blocked. The only possible exception, a C-terminal lysine, does not exist in intact fibrinogen or fibrin. The binding of plasmin(ogen) to fibrin(ogen) does not, therefore, necessarily involve the lysine-binding sites. With the use of α-N-acetyl-lysine methyl ester, a weak lysine-binding site has been detected in mini-plasminogen – possibly in K5 – that could be responsible for the initial binding of plasminogen to fibrin(ogen) (977).

It is not known with certainty at which point plasminogen binds to fibrin/fibrinogen. However, studies of the binding of fibrin fragments to Lys-plasminogen have indicated that amino-acid residues Aα52–78 and Bβ119-120 are important for this interaction (978).

Apart from binding to fibrin(ogen), plasminogen also binds to its own specific inhibitor; this topic we shall return to later.

The genetically caused lack of – or defect in – plasminogen occurs only very rarely. In some cases, its molecular basis has been determined. Two different defects in plasminogen have been described. In one, Ala601 is replaced by Thr (949, 979, 980), and in the other Val355 is replaced by Phe (949).

9.1.1. *Apolipoprotein (a)*

At this point, it is appropriate to include a section on a protein with very remarkable properties. This protein is the apoprotein of lipoprotein (a), which is an LDL-like protein (review in reference 981). Lipoprotein (a) has been the object of much interest, since its concentration in human plasma is significantly correlated with the risk of heart disease (982–990).

Plasma concentrations vary in individuals from below the detection limit to 1 mg/ml (982). Apolipoprotein (a) – apo(a) – is a large glycoprotein with considerable heterogeneity in size; forms have been found with molecular masses ranging from 300 kDa to 700 kDa (see, for example, reference 991).

Partial amino-acid-sequencing (992, 993) and cDNA-sequencing (994) have revealed that apo(a) is homologous to plasminogen. The amino-acid sequence of apo(a) as deduced from the cDNA sequence contains 4529 residues plus a 19-residue signal peptide, so apo(a) is an unusually large molecule.

The structure of apo(a) is indeed unusual, as we now shall see. The first 16 residues in the signal peptides of apo(a) and plasminogen are identical, and after that the two sequences diverge. However, they become similar again after the point where, in plasminogen, kringle 4 begins. The first three residues of kringle 4 are identical to the final three residues in the signal peptide of apo(a). Thereafter, apo(a) contains 37 domains displaying homology with kringle 4, one domain homologous to kringle 5 and one homologous to the serine-proteinase domain of plasminogen.

Of the 37 domains of apo(a) homologous to kringle 4 of plasminogen, 24 are identical at the nucleotide level, while 4 are identical at the protein level and differ by only three nucleotides from the first 24. All these 28 kringle-4-like domains in apo(a) are thus identical polypeptides, and their degree of identity with kringle 4 is 67%. The remaining 9 kringle-4-like domains in apo(a) differ by between 11 and 71 nucleotides from the first 24 identical domains of apo(a), and one of them has a 24-bp deletion. The degree of identity of these 9 domains with the kringle 4 region of plasminogen lies between 62% and 75% (Fig. 64). The domain of apo(a) homologous to kringle 5 and the serine-proteinase domain have respectively 83% and 88% identical amino acids with respect to the corresponding domains in plasminogen (Fig. 64). Interestingly, the gene for apo(a) has been found to be tightly linked to the gene for plasminogen at location q26–27 on chromosome 6 (951, 995, 996).

Apo(a) is not cleaved by plasminogen activators, as its residue corresponding to Arg561 in plasminogen is a serine (Ser4308). Apo(a) has not been shown to possess actual or potential proteolytic activity over against oligopeptides. However, the fact that the presumed catalytic apparatus in the serine-proteinase domain is intact, involving His4350, Asp4393 and Ser4479, makes it reasonable to expect that there may be still some connection in which apo(a) functions as a proteinase. In addition, there is evidence that apo(a) can bind to fibronectin and cleave it (997). It is therefore surprising that the serine-proteinase domain in rhesus monkey apo(a) does not contain an intact catalytic apparatus, as was shown by sequencing of partial cDNA clones (958): the His and Ser residues are replaced with a Cys and an Asn respectively. Furthermore, the domain homologous to kringle 5 is missing in rhesus monkey apo(a), in which the

(a)

```
K1                  E Q S H V V Q D C Y H G D G Q S Y R G T Y S T T V
K2 - K29      A P T E Q R P G V Q E C Y H G N G Q S Y R G T Y S T T V
K30           A P T E Q R P G V Q E C Y H G N G Q S Y R G T Y S T T V
K31           A P T E Q R P G V Q E C Y H G N G Q S Y Q G T Y F I T V
K32           A L T E E T P G V Q D C Y Y H Y G Q S Y R G T Y S T T V
K33           A P T E Q S P G V Q D C Y H G D G Q S Y R G S F S T T V
K34           A P T E Q S P T V Q D C Y H G D G Q S Y R G S F S T T V
K35           A P T E N S T G V Q D C Y R G D G Q S Y R G T L S T T I
K36           A P P E K S P V V Q D C Y H G D G R S Y R G I S S T T V
K37           A P T E Q T P V V R Q C Y H G N G Q S Y R G T F S T T V
PLG K4        A P P E L T P V V Q D C Y H G D G Q S Y R G T S S T T T

T G R T C Q A W S S M T P H Q H N R T T E N Y P N A G L I M N Y C
T G R T C Q A W S S M T P H S H S R T P E Y Y P N A G L I M N Y C
T G R T C Q A W S S M T P H S H S R T P E Y Y P N A G L I M N Y C
T G R T C Q A W S S M T P H S H S R T P A Y Y P N A G L I K N Y C
T G R T C Q A W S S M T P H Q H S R T P E N Y P N A G L T R N Y C
T G R T C Q S W S S M T P H W H Q R T T E Y Y P N G G L T R N Y C
T G R T C Q S W S S M T P H W H Q R T T E Y Y P N G G L T R N Y C
T G R T C Q S W S S M T P H W H R R I P L Y Y P N A G L T R N Y C
T G R T C Q S W S S M I P H W H Q R T P E N Y P N A G L T E N Y C
T G R T C Q S W S S M T P H R H Q R T P E N Y P N D G L T M N Y C
T G K K C Q S W S S M T P H R H Q K T P E N Y P N A G L T M N Y C

R N P D A V A A P Y C Y T R D P G V R W E Y C N L T Q C S D A E G
R N P D A V A A P Y C Y T R D P G V R W E Y C N L T Q C S D A E G
R N P D P V A A P Y C Y T R D P S V R W E Y C N L T Q C S D A E G
R N P D P V A A P W C Y T T D P S V R W E Y C N L T R C S D A E W
R N P D A E I R P W C Y T M D P S V R W E Y C N L T Q C L V T E S
R N P D A E I S P W C Y T M D P N V R W E Y C N L T Q C P V T E S
R N P D A E I R P W C Y T M D P S V R W E Y C N L T Q C P V M E S
R N P D A E I R P W C Y T M D P S V R W E Y C N L T R C P V T E S
R N P D S G K Q P W C Y T T D P C V R W E Y C N L T Q C S E T E S
R N P D A D T G P W C F T M D P S I R W E Y C N L T R C S D T E G
R N P D A D K G P W C F T T D P S V R W E Y C N L K K C S G T E A

T A V A P P T V T P V P S L E A P S E Q   K1
T A V A P P T V T P V P S L E A P S E Q   K2 - K29
T A V A P P T I T P I P S L E A P S E Q   K30
T A F V P P N V I L A P S L E A F F E Q   K31
S V L A T L T V V P D P S T E A S S E E   K32
S V L A T S T A V - - - - - - - - S E Q   K33
T L L T T P T V V P V P S T E L P S E E   K34
S V L T T P T V A P V P S T E A P S E Q   K35
G V L E T P T V V P V P S M E A H S E A   K36
T V V A P P T V I Q V P S L G P P S E Q   K37
S V V A P P P V V L L P D V E T P S E E   PLG K4
```

Fig. 64. (a) Internally homologous kringle 4 domains of apolipoprotein (a). (K1, kringle domain 1, and so on.) The amino-acid sequence of plasminogen kringle 4 (PLG K4) is also shown. Only the positions at which amino acids are the same in all domains are outlined. (b) Homology between the kringle-5-like domain in apolipoprotein (a) (APO(a) K5) and plasminogen kringle 5 (PLG K5). Only identical amino acids are outlined. The degree of identity is 83%. (c) Homology between the serine-proteinase-like domain in apolipoprotein (a) (APO(a)) and the serine-proteinase domain of plasminogen (PLG). Only identical amino acids are outlined. The degree of identity is 88%.

(b)

```
PLG K5       462 C M F G N G K G Y R G K R A T T V T G T P C Q D W
APO(a) K5   4209 C M F G N G K G Y R G K K A T T V T G T P C Q E W

A A Q E P H R H S I F T P E T N P R A G L E K N Y C R N P D G D V
A A Q E P H R H S T F I P G T N K W A G L E K N Y C R N P D G D I

G G P W C Y T T N P R K L Y D Y C D V P Q C 541
N G P W C Y T M N P R K L F D Y C D T P L C 4288
```

(c)

```
PLG       542 A A P S F D C G K P Q V E P K K C P G R V V G G C V
APO(a)   4289 A S S S F D C G K P Q V E P K K C P G S I V G G C V

A H P H S W P W Q V S L R T R F G M H F C G G T L I S P E W V L T
A H P H S W P W Q V S L R T R F G K H F C G G T L I S P E W V L T

A A H C L E K S P R P S S Y K V I L G A H Q E V N L E P H V Q E I
A A H C L K K S S R P S S Y K V I L G A H Q E V N L E S H V Q E I

E V S R L F L E P T R K D I A L L K L S S P A V I T D K V I P A C
E V S R L F L E P T Q A D I A L L K L S R P A V I T D K V M P A C

L P S P N Y V V A D R T E C F I T G W G E T Q G T F G A G L L K E
L P S P D Y M V T A R T E C Y I T G W G E T Q G T F G T G L L K E

A Q L P V I E N K V C N R Y E F L N G R V Q S T E L C A G H L A G
A Q L L V I E N E V C N H Y K Y - - - - - - - - - T C A E H L A R

G T D S C Q G D S G G P L V C F E K D K Y I L Q G V T S W G L G C
G T D S C Q G D S G G P L V C F E K D K Y I L Q G V T S W G L G C

A R P N K P G V Y V R V S R F V T W I E G V M R N N 791
A R P N K P G V Y A R V S R F V T W I E G M M R N N 4529
```

kringle 4 domain passes directly over to the serine-proteinase domain. The general significance – or even presence – of proteolytic activity in apo(a) is therefore still uncertain.

The physiologicai function of lipoprotein (a), and thus of apo(a), is unknown, and the concentrations of these two proteins in plasma do not appear to depend upon dietary factors. The medical significance of lipoprotein (a) has already been mentioned – its correlation with the incidence of heart disease. The mechanism of this has not been discovered. However, it has been shown that lipoprotein (a) inhibits the binding of plasminogen and of plasmin both to fibrin and to a cellular plasminogen receptor (998–1001), a process that of course interferes with the fibrinolytic process and which may therefore have adverse consequences.

9.2. Urokinase

One of the two physiological plasminogen activators is urokinase (u-PA, reviewed in reference 1002). Urokinase was originally isolated as a two-chained enzyme from urine, but it has been shown that urokinase is synthesized as a single-chained, glycosylated zymogen, pro-urokinase (1003–1005). Its level in plasma is low, 1–2 ng/ml (≈40 pM), so that urokinase is rarely isolated from plasma, but rather from cell culture. The molecular masses of pro-urokinase and urokinase are about the same, and are estimated by SDS-polyacrylamide gel electrophoresis to be 50–54 kDa (1006–1008).

At the protein level, all of human urokinase (1009–1011) and 85% of human pro-urokinase (1012) have been sequenced. A partial (1013) and a complete (1014) sequence have been published from cDNA of human pro-urokinase, which is synthesized with a 20-residue signal peptide (Appendix, Fig. A24). Both porcine and murine pro-urokinases, including a 20-residue signal peptide, have been sequenced completely at the cDNA level (1015, 1016). For all three, the complete genomic sequence is known, including both introns and exons (1015, 1017, 1018). The human pro-urokinase gene covers 6387 bp of genomic DNA (1017) and has 11 exons. The lengths of the exons vary from 28 bp to 1106 bp and those of the introns from 146 bp to 989 bp. The gene for human pro-urokinase has been located at position q24–qter on chromosome 10 (1019, 1020).

Human pro-urokinase consists of 411 amino-acid residues (Fig. 65), giving a total molecular mass of nearly 47 kDa, which with the inclusion of a carbohydrate chain on Asn302 agrees with the molecular mass estimated by electrophoresis. Starting from the N terminus, pro-urokinase contains first a growth-factor domain, a kringle domain, a connecting strand and a C-terminal serine-proteinase domain. The disulphide pattern has not been determined experimentally, but on the basis of the known homology the following S–S bonds can be expected: in the growth-factor domain, Cys11–Cys19, Cys13–Cys31 and Cys33–Cys42; in the kringle domain, Cys50–Cys131, Cys71–Cys113 and Cys102–Cys126; between the chains, Cys148–Cys279; and in the serine-proteinase domain, Cys189–Cys205, Cys197–Cys268, Cys293–Cys362, Cys325–Cys341 and Cys352–Cys380.

The conversion of pro-urokinase to the two-chained urokinase occurs when the peptide bond between Lys158 and Ile159 is cleaved. Urokinase thus contains a light chain (20 kDa) from its precursor's N terminus disulphide-bonded to a C-terminally derived heavy chain (30 kDa); the latter contains the catalytic apparatus, which is believed to involve the residues His46, Asp97 and Ser198, corresponding to His204, Asp255 and Ser356 in pro-urokinase.

In amino-acid-sequencing studies on purified two-chained urokinase from urine, the C-terminal residue of the light chain has been identified as

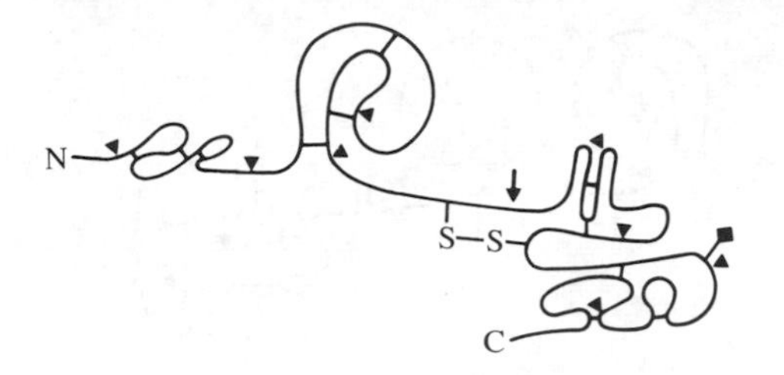

Fig. 65. The domain structure of pro-urokinase. → shows the position of cleavage by plasmin, ▼ the positions of the introns and ◆ the N-bound carbohydrate.

Phe157 (1011). It therefore appears that Lys158 is somehow removed within the organism after the cleavage of the bond Lys158–Ile159.

9.3. **Tissue plasminogen activator**

Tissue plasminogen activator (t-PA) is the second of the physiological plasminogen activators (reviewed in, for example, references 1021 and 1022). As the name suggests, this plasminogen activator was originally isolated from tissue. However, it is not found exclusively in tissue. Its concentration in plasma is low (5 ng/ml, or roughly 70 pM) (1023, 1024), as is the case for pro-urokinase. t-PA is thought to be synthesized mainly in endothelial cells, from which it is secreted into the plasma. It is synthesized as a single-chained glycoprotein with a molecular mass of approximately 72 kDa.

The structure of human t-PA has been elucidated by cDNA-sequencing (1025, 1026) (Appendix, Fig. A25), and it includes a 32-residue signal peptide (combined results from (1025) and (1027)). Nearly 95% of the deduced amino-acid sequence has been confirmed by protein-sequencing (1028). t-PA, when purified, appears with two different N-terminal amino-acid sequences, commencing respectively 32 and 35 residues after Met1 in the translation product. This is due to a dimorphism (S/G) 36 residues after Met1; the long form of t-PA, with a signal peptide of 32 residues, has Ser in this position while the short form has Gly (1029). The presence of Gly in this position creates an alternative cleavage site for the arginine-specific protease that removes the propeptide. Here, the numbering will be based upon the long t-PA molecule.

The gene for t-PA has been characterised, first by localisation of the intron–exon junctions (1030, 1031) and later by the sequencing of the entire gene (1032). The t-PA gene covers 32,720 bp of genomic DNA from the start of transcription to the polyadenylation site; it has 14 exons between 43 bp and 914 bp in length, and 13 introns between 111 bp and 14,257 bp. The t-PA gene is located on chromosome 8 (1020, 1033). The amino-acid

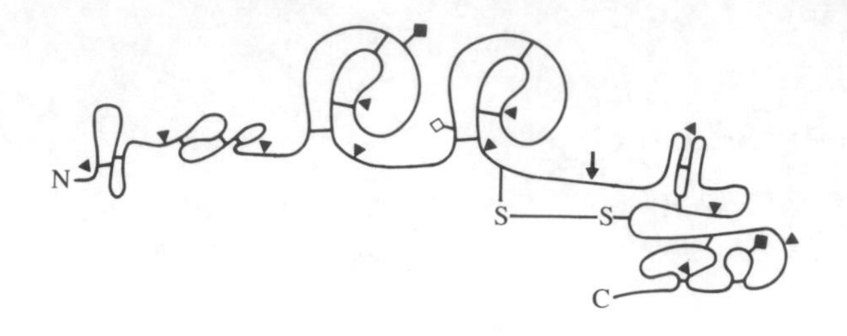

Fig. 66. The domain structure of t-PA. → shows the position of cleavage by plasmin, ▼ the positions of the introns, ◆ the N-bound carbohydrate and ◇ the potential attachment site for N-bound carbohydrate.

sequence of a plasminogen activator from the salivary gland of a vampire bat (*Desmodus rotundus*) is, on the basis of structural homology, to be regarded as a type of t-PA rather than u-PA (1034).

t-PA has 530 amino-acid residues, giving a combined molecular mass of about 59 kDa, to which should be added the carbohydrate chains bound at Asn120 and Asn451, and possibly also Asn187. There is a fourth site, at Asn221, that fulfils the criterion N-X-S/T, but this is not utilised, because the X in this case is proline (1028). Like many other of the proteins that we have encountered, t-PA is a multi-domain protein (Fig. 66). From the N terminus, its domains are: a finger domain (this is a domain with homology of type I as defined in fibronectin), a growth-factor domain, two kringle domains, a connecting strand and a serine-proteinase domain. The disulphide bonds can be inferred from the homologies, and the following pattern has been obtained: in the finger domain, Cys9–Cys39 and Cys37–Cys46; in the growth-factor domain, Cys54–Cys65, Cys59–Cys76 and Cys78–Cys87; in the first kringle domain, Cys95–Cys176, Cys116–Cys158 and Cys147–Cys171; in the second kringle domain, Cys183–Cys264, Cys204–Cys246 and Cys235–Cys259; between the chains, Cys267–Cys398; and in the serine-proteinase domain, Cys310–Cys326, Cys318–Cys387, Cys412–Cys487, Cys444–Cys460 and Cys477–Cys505. There remains an unpaired cysteine residue, Cys86 in the growth-factor domain.

The conversion of single-chained t-PA to two-chained t-PA occurs upon cleavage of the bond Arg278–Ile279 (Arg275–Ile276 in the original numbering, which however fails to take account of the fact that unicatenate t-PA has three amino-acid residues at the N terminus that are missing in bicatenate t-PA (1027)). This cleavage results in the formation of a heavy chain (36 kDa) from the N terminus and a light chain (32 kDa) from the C terminus of the precursor. The latter contains the presumed catalytic apparatus, comprising His47, Asp96 and Ser203, corresponding to His325, Asp374 and Ser481 in single-chained t-PA. The cleavage is catalysed by plasmin, both at Arg278–Ile279 and at Arg3–Ser4.

The vampire bat salivary gland plasminogen activator mentioned above is 85% identical to t-PA, but differs from it by lacking kringle domain 2. All the other domains of human t-PA are present. Another difference lies in the fact that salivary gland plasminogen activator is not cleaved by plasmin (1034).

9.4. Regulatory proteins of fibrinolysis

As was the case for coagulation, a key part of the mechanism of fibrinolysis is the regulation of the proteolytic enzymes involved by inhibitors. The serine proteinases formed are held in check by serine-proteinase inhibitors. Plasmin is far too potent an enzyme to be let off the rein, so it has its own, specific, fast-acting inhibitor: α_2-plasmin inhibitor (α_2-PI). Plasminogen activators, that, as we shall see, catalyse the formation of plasmin with high efficiency, also require strict regulation in order to prevent the breakdown of the physiological processes. For this purpose, there are several inhibitors of plasminogen activator (PAIs) that are also termed 'fast acting'.

9.4.1. *α_2-plasmin inhibitor*

α_2-plasmin inhibitor (α_2-PI) was first described by three groups, independently, in 1976 (1036–1038). It is reviewed in reference (1035). It is a single-chained glycoprotein (11–14% carbohydrate) with a physically determined molecular mass of 67 kDa, and it circulates in normal human plasma with a concentration of 70 μg/ml (≈ 1 μM).

Most of the sequence of α_2-PI has been determined by protein-sequencing (1039), and both partial and complete cDNA sequences have been reported (1040–1042) (Appendix, Fig. A26). In one case, the cDNA sequence also contains the complete sequence of a 39-residue signal peptide (1042). The genomic organisation of the structural gene for α_2-PI, which lies at the chromosomal position 18p11.1–q11.2 (1043), has been described; the gene contains 16 kbp of DNA and 10 exons (1044). The sizes of the exons vary from 17 bp to 1169 bp, and those of the introns from 105 bp to about 6 kbp.

α_2-PI has 452 amino-acid residues, corresponding to a molecular mass of 50.5 kDa, to which the 11–14% carbohydrate bound at Asn87, Asn256, Asn270 and Asn277 must be added (1039). However, the physical estimate of the molecular mass is considerably greater (see above), which has not yet been explained. α_2-PI further contains a tyrosine-O-sulphate residue at position 445 (1045). The disulphide-bond pattern in α_2-PI has been shown by experiment to involve the bonds Cys31–Cys113 and Cys64–Cys104 (1039). Like the other serine-proteinase inhibitors we have met, α_2-PI belongs to the serpin family, and the reactive site is at the bond Arg364–Met365 (1040, 1046).

The genetically caused deficiency of α_2-PI is only a rare occurrence, but it has been described in at least 10 families (1035), and in three of these cases the genetic basis has been found. In one of them (α_2-PI$_{\text{Enschede}}$) the level of α_2-PI antigen in the plasma is normal, but the functional level is only about 3% of normal. The reason for this is an inserted codon that specifies an extra alanine residue between positions 355 and 356 (1047), which changes α_2-PI from being a plasmin inhibitor to being a plasmin substrate (1047, 1048). In the second case, the three nucleotides that specify Glu137 have been deleted (1049). In the third case, α_2-PI cannot be detected in the plasma by either immunological or functional assays. Investigation of the genomic DNA from the patient concerned revealed the insertion of a single base pair between the codons specifying residues 440 and 441 in α_2-PI. This insertion causes a reading-frame error, which in turn results in the replacement of the normal 12 C-terminal residues with 178 completely different ones (1050).

9.4.2. *Plasminogen activator inhibitors*

Plasminogen activator inhibitors (PAIs), reviewed in reference (1051), are a class of relatively recently characterised molecules. They were found after a search motivated by the recognition that the rates of inhibition of plasminogen activators by the known serine-proteinase inhibitors were insufficient to account for the regulation of plasminogen activators *in vivo*.

In fact, there are probably four different PAIs: PAI-1 (endothelial PAI) (1052–1054), PAI-2 (placental PAI) (1055), PAI-3 (1056) and protease nexin (1003). PAI-1 denotes the PAI activities found in endothelial cells, normal plasma and serum and blood platelets, since these PAIs are probably identical. PAI-2 is the term used for the PAI activity found in placenta, monocytes and monocyte cell lines. PAI-3 has been found to be identical with protein Ca inhibitor (1057).

PAI-1, PAI-2 and protease nexin all inhibit both urokinase and t-PA (1053, 1055, 1058), while PAI-3 inhibits urokinase only. The different distribution patterns of PAIs throughout the organism make it seem likely that PAI-1 is the most important PAI in plasma under normal conditions, although the higher concentration of PAI-3 implies that it has a certain importance for the inhibition of urokinase.

The molecular mass of the glycoprotein PAI-1 is estimated by SDS-polyacrylamide gel electrophoresis to be about 52 kDa. This is in reasonable agreement with the molecular mass calculated on the basis of the cDNA sequence of human PAI-1, which is 42.8 kDa (1059–1062) (Appendix, Fig. A27) when the existence of three potential N-glycosylation sites is taken into account (Asn residues at positions 209, 265 and 329). The cDNA sequence includes that of a signal peptide of 23 residues (1060, 1061). The amino-acid sequences of rat PAI-1 (1063) and bovine PAI-1 (1064) have also been determined by cDNA-sequencing. The gene for PAI-1 has been characterised (1065–1067) and fully sequenced (1066). The PAI-1 gene covers 12.2 kbp of genomic DNA, with nine exons (1065). The sizes of the exons vary from 84 bp to 1764 bp (1066). Intron 7 contains a cryptic acceptor splicing site; when it is used, it results in an insertion of seven amino-acid residues, and the resulting molecule lacks PAI activity (1060, 1065). The structural gene for PAI-1 is located on chromosome 7 at position q21.3–22 (1068).

PAI-1 contains 379 amino-acid residues with homology with the other members of the serpin family, and the reactive site has been located at the Arg346–Met347 bond (1069). Surprisingly, PAI-1 contains no Cys residues. Purified PAI-1 from fibrosarcoma cells shows heterogeneity at the N terminus, believed to be due to removal of the signal peptide by signal peptidase at variously 21 or 23 residues after the commencement of translation (1069).

The concentration of PAI-1 in human plasma varies with the age of the individual, but generally lies around 25–30 ng/ml ($\approx$500 pM) (1070). In plasma, PAI-1 is also protected by complexation with a protein identified as vitronectin (1071–1074).

PAI-2 (reviewed in reference 1075) is a protein with a molecular mass of 47 kDa (1055). The placental form appears to be identical, both biochemically and immunologically, to the PAI in monocytes and monocytic cell lines (1076, 1077). PAI-2 reacts more rapidly with urokinase than with t-PA (1077), and is not normally found in plasma except during pregnancy, when its level in plasma rises to high values (1078). The structure of PAI-2 has been determined by cDNA-sequencing (1079–1082) (Appendix, Fig. A28), and the gene's structure has been documented (1083). The PAI-2 gene, which lies on chromosome 18 at position q21–23 (1080, 1083), covers 16.5 kbp and contains eight exons. The exons vary in size from 68 bp to 986 bp and the introns from 340 bp to 3.85 kbp. There is currently some debate about whether a part of the amino-acid sequence deduced from the cDNA sequence functions as a signal peptide or not, as the N terminus of purified PAI-2 is blocked, so that it is hard to decide where the mature PAI-2 actually starts in comparison with the cDNA sequence. The N-terminal residues of the mature protein could possibly function as a signal peptide, even though this would be an atypical signal peptide, and a possible cleavage point for the signal peptidase has been suggested at Ala22–Ser23 (1080, 1081). There are, however, three arguments against this. First of all, the presumed signal peptide sequence is atypical and is encoded by a part of a relatively large exon, while it is more usual for signal peptides to have their own exon. Secondly, PAI-2 is not normally secreted, so it has no need for a signal peptide. Thirdly, a monocytic cell line is known from which PAI-2 is secreted, and this PAI-2 has an N-terminal sequence that corresponds to the sequence found by cDNA-sequencing, starting with methionine (1084), so that a separate signal-peptide sequence obviously is not strictly needed.

The deduced amino-acid sequence of 415 residues corresponds to a molecular mass of 46.6 kDa, in agreement with the value estimated by electrophoresis. This implies that the three potential N-glycosylation sites at Asn75, Asn115 and Asn339 are not used. PAI-2 found in the plasma during pregnancy has a higher molecular mass ($\approx$60 kDa) (1085) than does cellular PAI-2, and this might be due to the use of the glycosylation sites. However, the actual mechanism of secretion remains unexplained.

PAI-2 is a member of the serpin family, and the reactive site has been found to be the bond Arg380–Thr381 (1086). A comparison of the four published cDNA sequences and the genomic sequence indicates the presence of three polymorphic positions in the amino-acid sequence of PAI-2: positions 120 (D/N), 404 (N/K) and 413 (S/C). Protein-chemical evidence indicates the presence of serine at position 413 (1086).

Protease nexin is a proteinase inhibitor with a considerably broader spectrum than PAI-1 and PAI-2, and in addition to urokinase and t-PA it also inhibits, for example, thrombin and plasmin. It is secreted *inter alia* by fibroblasts, but it is not found in plasma. Its molecular mass is 43 kDa, with a carbohydrate content of 6% (1058). The amino-acid sequence of protease nexin has been determined by cDNA-sequencing (1087) (Appendix, Fig. A29), and the molecule is synthesized with a signal peptide of 19 residues. There are two forms of protease nexin cDNA, differing by an insertion of three nucleotides that alter an arginine residue at position 310 to a Thr-Gly dipeptide (1087). The former has an amino-acid sequence (deduced from the cDNA) that agrees exactly with the deduced amino-acid sequence of human glia-derived nexin (1088, 1089). Protease nexin consists of 378 or 379 residues, corresponding to a molecular mass of 39.5 kDa. Since there are two potential N-glycosylation sites at Asn99 and Asn140, the total molecular mass appears to agree with that estimated by physical methods. As its similarity to the others leads one to expect, protease nexin belongs to the family of serpins, and the reactive site is taken to be the bond Arg345–Ser346.

10. Mechanisms in fibrinolysis

10.1. Fibrinolysis

The word 'fibrinolysis' naturally means the dissolution of fibrin, and in an organism this means the dissolution of the stabilised fibrin clot. It is reviewed in reference (1090). The principle of plasmin-catalysed fibrinolysis is the formation of relatively small, readily soluble peptide fragments by the proteolytic degradation of the fibrin clot. However, this is necessarily preceded by processes that culminate in the formation of plasmin, and these will now be examined.

10.2. Activation of u-PA and t-PA

The conversion of plasminogen to plasmin requires proteolytic cleavage by one of the plasminogen activators, i.e., by u-PA or t-PA. We examine first the activation of u-PA.

As discussed earlier, u-PA circulates in the form of a single-chained pro-urokinase. We might therefore expect that this form would have to be activated to give bicatenate urokinase, in order to activate plasminogen and produce plasmin. However, at present this is not generally accepted, and there is disagreement about whether pro-urokinase itself has an inherent, low activity (see below). The conversion of pro-urokinase to urokinase that occurs by the cleavage of the bond Lys158–Ile159 can *in vitro* be catalysed both by plasmin and by plasma kallikrein (1091). This means either that the formation of urokinase is catalysed by the enzyme that it activates, or else that it is catalysed by an enzyme produced in the course of contact activation.

It has proved difficult to reach a conclusion on whether pro-urokinase has an inherent, low enzymic activity, and the problem is made the more intractable by the fact that plasmin, once present, catalyses the conversion of pro-urokinase to urokinase. It was originally claimed that human pro-urokinase produced by recombinant techniques in *E. coli* could activate plasminogen directly (1092), but pro-urokinase from a human source (urine or cell culture) is considerably worse or indeed unable to carry out this activation (1008, 1093–1097). Modified recombinant pro-urokinase, in which Lys158 has been changed to Gly or Glu (1098), or to Val or Met (1099), is not cleaved by plasmin in the same way as native pro-urokinase, and these mutant molecules possess plasminogen-activating activity equal to only 0.5% of that of urokinase; this is taken to indicate that pro-urokinase really does have a certain inherent enzymic activity, even though this is very low. It must be concluded that this question is still unresolved.

It could be regarded as surprising that urokinase can have any physiological significance in connection with fibrinolysis, since the activity of urokinase in plasma could perhaps lead to the activation of plasminogen independently of other factors (systematic activation). This is not least because urokinase has a very low affinity for fibrin, unlike the other physiological plasminogen activator t-PA (see below). If urokinase is the activator, then the activation of plasminogen should have no direct connection with the fibrin clot. However, two relevant observations must be made here. First, it is quite possible that the activation of plasminogen by urokinase is systematic and independent of fibrin, as this activation plays only a marginal physiological rôle in fibrinolysis, while still being a principal part of the general mechanism of tissue degradation and repair (1100). Secondly, pro-urokinase could have a certain affinity for fibrin that is lost after cleavage and conversion to urokinase (1007).

To consider the second of these points first: there is disagreement in the literature on whether a high-affinity binding of pro-urokinase to fibrin can be demonstrated (1008, 1092, 1093) or not (1007, 1101, 1102). In any case, fibrin has a stimulatory effect upon the activation of plasminogen by pro-urokinase in plasma systems (1092, 1103); this activity is attributed to selective activation of fibrin-bound plasminogen (1104).

In connection with a possible rôle of urokinase over and above fibrinolysis, it should be noted that various cell types – for example, monocytes and fibroblasts – have a high-affinity urokinase receptor with K_d approximately 0.1 nM (1105). The receptor binds pro-urokinase and urokinase equally well (1106). The purpose of this is not endocytosis, but rather to place the plasminogen-activating function on the outer side of the cell. The urokinase receptor is a membrane protein with an estimated molecular mass of 50–60 kDa (1107, 1108). The region of pro-urokinase/urokinase that binds to the urokinase receptor has been narrowed down to amino-acid residues 13–33, which lie in the growth-factor domain (1109, 1110). Interestingly, it is the corresponding homologous sequence in EGF that binds to the EGF receptor (1111, 1112).

Unlike pro-urokinase, t-PA has long been known to possess a marked affinity for fibrin (1113). There is still discussion of whether the affinity of single-chained t-PA for fibrin is the same as that of double-chained t-PA

(1114–1117). For both forms of t-PA, this binding is stimulated considerably when the fibrin is partially degraded by plasmin (1116). The ability of t-PA to activate plasminogen is especially dependent upon the presence of fibrin, and possibly also upon whether the t-PA is in its one- or its two-chained state (1114, 1115, 1118–1120). There is general agreement that bicatenate t-PA is more active that unicatenate, both with respect to small synthetic oligopeptide substrates (1114, 1115) and to plasminogen in the absence of fibrin (1115, 1119). There are conflicting reports on the activation of plasminogen by t-PA in the presence of physiological quantities of fibrin: (a) both forms of t-PA activate plasminogen equally well (1114, 1115, 1120); or (b) the two-chained form is two to five times more active than the one-chained form (1118, 1119). Four of these studies employed recombinant t-PA, in which the plasmin cleavage site Arg278–Ile279 was mutated respectively to Glu–Ile (1115), to Gly–Ile (1119, 1120) or to Ser–Ile (1120). However, one cannot exclude the possibility that the mutant molecules, in which Arg278 had been replaced by Glu or Ser, were still cleaved, since no controls to this effect were undertaken. In contrast, the one study in which this control was successfully carried out, involving the mutant Arg278→Gly (1119), was the one that showed the difference in activity, which should perhaps therefore be regarded as significant.

The region in t-PA responsible for the interaction with fibrin has been localised at the N-terminally derived heavy chain (1121, 1122). The light chain, from the C terminus, consists, as mentioned in section 9.3, of the serine-proteinase domain, and as an isolated fragment it is able to activate plasminogen to give plasmin independently of the presence of fibrin (1121). In contrast, the heavy chain binds to lysine-Sepharose and to fibrin (1122). t-PA thus contains, like plasminogen, a lysine-binding site; this has been localised more precisely, at the second kringle domain (1122–1124). Further, studies with deletion mutants of t-PA have shown that the finger domain is also involved in the binding of fibrin, but without involving a lysine-binding site (1123–1127). A single study of deletion mutants has also indicated that the first kringle domain may also possess a site that binds lysine (1128).

t-PA also binds to endothelial cells in culture with reported values of the dissociation constant (K_d) between 18 nM and 240 nM (1129–1131). This binding can be inhibited with a monoclonal antibody against the finger domain and with a synthetic peptide corresponding to residues 7–17 of t-PA (1132).

The efforts to understand the biology of plasminogen activators are naturally motivated by the desire to develop pharmaceutica for improved thrombolytic therapy. This has produced a large number of recombinant t-PA and u-PA molecules, of which some have been described here. In addition to mutations in and deletions of domains in the plasminogen activators, hybrid plasminogen activators have been made by recombination of domains from the two PAs (1133–1137). Furthermore, hybrids of plasmin's heavy chain containing the fibrin-binding site and the catalytic chains of plasminogen activators have been constructed (1138, 1139). We shall not go into further depth with these recombinants, since, although of great potential importance, they do not give further insight into the basic biochemistry of normal fibrinolysis.

10.3. Activation of plasminogen

A recurrent detail in the description of blood coagulation was the idea of localisation on a surface, and this has not been emphasized in the discussion of fibrinolysis up to now. However, it has been present implicitly, since the phospholipid surface is simply replaced by a fibrin surface or matrix. Because of this, the discussion still involves strictly localised processes, and this is the key to understanding them.

The coagulation of blood requires a surface in order to raise the rates of certain enzymic reactions and the degrees of binding of certain proteins by others. In addition, attachment to a surface is needed both in order to keep the catalytic activities where they are needed and to prevent them from operating at other places where they could perhaps do damage. Precisely the same considerations apply to the processes and components of the fibrinolytic process.

The localising element in fibrinolysis is fibrin – appropriately enough, since this is also the substrate that is ultimately to be degraded.

Since plasminogen has affinity for fibrinogen and for fibrin, plasminogen will always be bound to every fibrin clot. How is it activated at the right time to give plasmin? There are two plasminogen activators circulating as single-chained proteins, and we start by considering them separately.

Single-chained t-PA possesses affinity for fibrin. It therefore binds to the fibrin clot when this is formed. The binding to fibrin gives it a certain proteolytic activity towards plasminogen and cleaves this at the Arg561–Val562 bond, converting it to plasmin. This plasmin is able both to cleave bonds in the fibrin clot and to convert the single-chained t-PA to the (probably) more active two-chained t-PA. It has been shown that the fibrinogen Aα-chain fragment 148–160 stimulates the t-PA-dependent activation of plasminogen, and that Lys157 of Aα plays a part in this process (1140–1142). Here it should be mentioned that fibrinogen does not enhance the ability of t-PA to activate plasminogen, and that a monoclonal antibody against a synthetic peptide corresponding to Aα-chain residues 148–160 recognises fibrin without recognising fibrinogen (1143). These facts imply that Aα-chain residues 148–160 of fibrinogen are not exposed until the fibrinogen becomes converted to fibrin.

The initial, partial degradation of the fibrin clot appears to enhance and strengthen the binding both of t-PA (1116) and of plasminogen (1144) to the fibrin, presumably by the exposure of new and stronger binding sites in the form of C-terminal lysine residues that stimulate the fibrinolytic effect.

The possible rôle of pro-urokinase in fibrinolysis is harder to evaluate. We have several possibilities, depending upon whether or not pro-urokinase can be regarded as having significant enzyme activity and/or binding affinity for fibrin.

If pro-urokinase has no enzymic activity, either in the presence or the absence of fibrin, then fibrinolysis by urokinase must wait for the activation of the pro-urokinase. As mentioned in section 9.2, the formation of this urokinase can be catalysed by plasmin or by plasma kallikrein. Since urokinase has no significant affinity for fibrin, there is nothing that retains the plasminogen activator activity in the region of the clot.

If pro-urokinase possesses enzymic activity, then there must be some mechanism that prevents the emergence of plasminogen activation in plasma. A theoretical possibility might be that pro-urokinase first becomes active in the presence of fibrin, but this would presuppose a marked affinity for fibrin, and this has not been demonstrated. It would also make the situation very similar to that of t-PA. Another possibility might be that the presence of fibrin counteracts an inhibition that otherwise puts a check on the enzymic activity of pro-urokinase (1092). A third possibility is that pro-urokinase is unable to cleave plasminogen until the conformation of the plasminogen has been altered by binding to partially degraded fibrin (1104).

So how is plasminogen activated in the course of normal fibrinolysis? At present, no definitive answer can be given. The following sketch illustrates a possible mechanism, based upon reference (1145), that is consonant with the facts that are available.

Fibrin-bound, single-chained t-PA activates fibrin-bound plasminogen to give plasmin by the cleavage of the Arg561–Val562 peptide bond. The plasmin thus formed cleaves some of the bonds of fibrin, exposing new binding sites for plasminogen, and it also cleaves t-PA at the Arg278–Ile279 bond, raising the activity of the t-PA. These newly exposed plasminogen-binding sites are made up of C-terminal lysine residues, but the plasminogen bound to them does not appear to be important for t-PA-dependent fibrinolysis (1145). However, it does appear that pro-urokinase has the possibility of activating this plasminogen and thus working co-operatively with t-PA (Fig. 67).

10.4. **Degradation of fibrin by plasmin**

We have reached the point of the production of plasmin, which is able to degrade the insoluble fibrin clot to the level of soluble fibrin fragments. How does this degradation occur?

A large amount of our knowledge of this is derived from the plasmin-

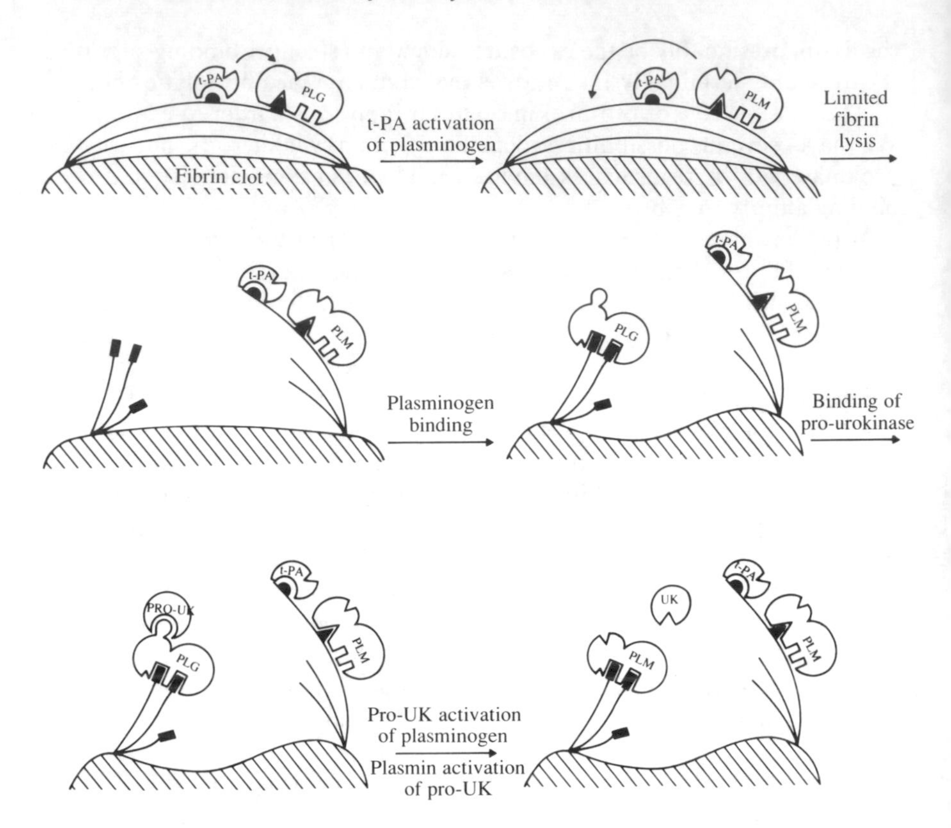

Fig. 67. A possible mechanism of plasminogen activation, based upon reference (1145). (PLG, plasminogen; PLM, plasmin; Pro-UK, pro-urokinase;UK, urokinase.)

catalysed cleavage of fibrinogen (Fig. 68), that results in the formation of fragments D and E as final products (see, for example, Figure 41 and reference 503). Fragments D and E are not of fixed size, but are primary fragments that have been exposed to a greater or a lesser amount of proteolytic trimming. The early form of fragment D consists of Aα105–197, Bβ134–461 and γ63–411, while the form that appears later begins at position 111 in the Aα chain and in position 86 or 89 in the γ chain. Cleavage can also occur in the C-terminal region in the γ chain. In fragment E, the early form comprises Aα1–78, Bβ54–122 and γ1–62, and this form is only modified slightly in the later stages of degradation. The first cleavage that plasmin carries out in fibrinogen is between Lys583 and Met584 in the Aα chain, after which the bonds Lys206–Met207 and Lys230–Ala231 in the same chain are rapidly broken. The result is the release of a 40-kDa fragment from the C terminus of the Aα chain. The remaining part of the fibrinogen molecule is called fragment X; this is cleaved in all three chains in

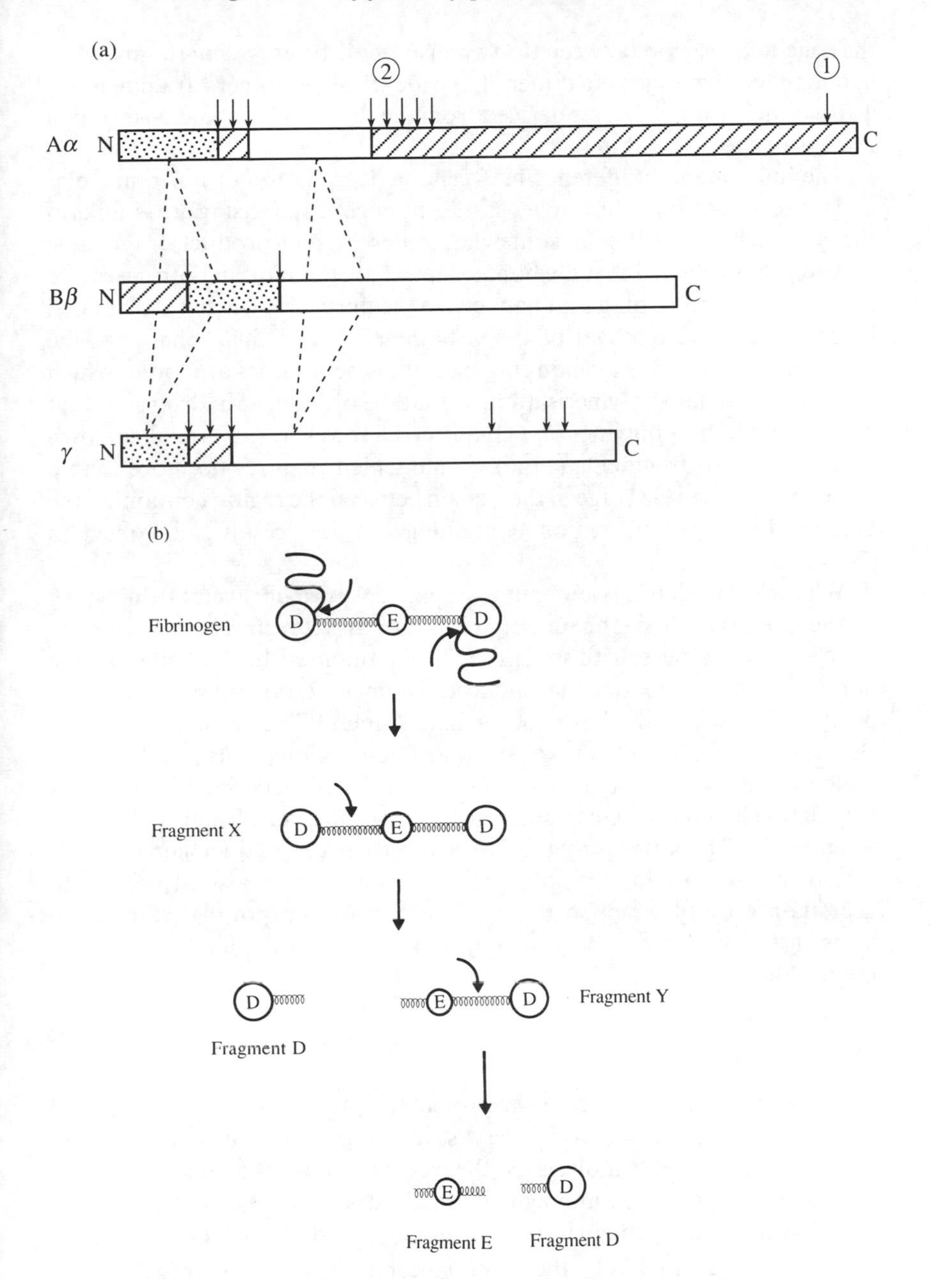

Fig. 68. (a) A schematic diagram of the plasmin-catalysed degradation of fibrinogen. Numbered arrows indicate the primary cleavage sites in the Aα chain of fibrinogen and dashed lines indicate the two disulphide rings. Fragment E is shown dotted and fragment D as hollow rectangles. The diagram shows the fibrinogen half-molecule. (b) A schematic view of the plasmin-catalysed degradation of fibrinogen. Arrows indicate plasminogen cleavage sites.

the coiled-coil region between the two disulphide rings, giving fragments D and E. Since fibrinogen is a dimer of two identical monomers, fragment E is also a dimer, with its halves derived from the disulphide-linked N-terminal regions of the original monomers.

The fundamental difference between the degradation of fibrin in a clot and the degradation of fibrinogen is the presence of stabilising cross-links in the clot, which results in somewhat different end-product fragments; however, it is believed that the bonds cleaved are the same in both cases (see Fig. 68). So the first thing that happens to the fibrin clot when it is degraded by plasmin is the removal of the C-terminal end of the α chains of the monomers. It is in this α-chain fragment that the residues are found which take part in α-chain polymerisation, so that the product of hydrolysis might be expected to be a polymer of α-chain ends. However, this is broken down rapidly to small fragments (1146), a step which requires no more than a plasmin-catalysed cleavage in the α chain between the region containing the Gln residues and the region containing the Lys residues involved in cross-linking.

What we have left is (viewed in a simplified way) a fibrin clot from which all the parts that give α-chain polymerisation have been removed. We are thus left, more or less, with strings of fibrin monomers held together by the non-covalent interactions that induced the fibrin to polymerise in the first place; in addition, we have the γ-chain dimers. The plasmin-catalysed cleavage of all three chains in the coiled-coil regions disrupts the entire structure into small fragments, such as are found experimentally (1147) (Fig. 69). The final fragments are in this case fragment E and a dimer of fragment D. The latter is due to cross-links between two γ chains.

Finally, it should be noted that fibronectin is also sensitive to degradation by plasmin, so that, although it is incorporated stably into fibrin clots, it can also be degraded by the same enzyme that degrades fibrin in the clot.

10.5. **Regulation of fibrinolysis**

So far, we have looked only at the proteolytic processes in fibrinolysis and ignored their regulation. As was the case for coagulation, the key element in the regulation of fibrinolysis is the restriction of the life-time of its proteolytically active components by inhibitors.

In fibrinolysis, two main inhibitors are employed: PAI-1 and α_2-PI. Both belong to the serpin family, the general mechanism of whose members was described in the section dealing with the regulation of coagulation (section 8.1.1).

PAI-1 inhibits the activators of plasminogen by means of a reactive site that consists primarily of the bond Arg346–Met347. (However, PAI-1 does not bind pro-urokinase (1003, 1148, 1149).) Purified PAI-1 exists mainly in

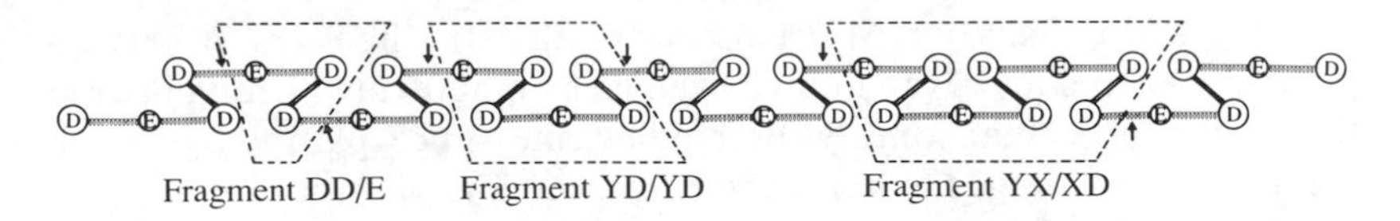

Fig. 69. The plasmin-catalysed dissolution of a fibrin clot. (Based upon references 1090 and 1147.)

its so-called latent form, which requires activation with, for example, a denaturing agent (1150) or, more interestingly, a negatively charged phospholipid (1151). In plasma, on the other hand, most of the PAI-1 is active (1152, 1153), probably as a result of complex formation with vitronectin. It has been shown that PAI-1 binds to fibrin (1154), which fits nicely with the fact that t-PA is activated in the presence of fibrin. The formation of a complex between urokinase and fibrin-bound PAI-1 releases the complex from fibrin, while complexes between t-PA and fibrin-bound PAI-1 remain bound to the fibrin (1155). With a view to therapeutic application, three mutant t-PA molecules resistant to inactivation by PAI-1 have been made by recombinant-DNA methods (1156). The mutations introduced were a deletion of residues 299 to 305 or point mutations of Arg307 to serine or glutamic acid, respectively. These mutant molecules, which are just as active towards plasminogen as is native t-PA, have presumably been altered in a region of importance for the interaction between t-PA and PAI-1.

Plasmin is inhibited by α_2-PI, and the reactive site has been found to lie at Arg364–Met365. However, α_2-PI has affinity for plasminogen, in a binding mediated by the lysine-binding site in the kringle domain 1 of plasminogen (964). The value of K_d for the interaction between Glu-plasminogen and α_2-PI is 4 μM, while K_d for the interaction between Lys-plasminogen and α_2-PI is about one-tenth as great, so the latter binding is ten times stronger (964). This interaction between the zymogen and the inhibitor of the active enzyme is, presumably, to a large degree responsible for the successful and efficient inhibition of plasmin by α_2-PI.

There are in fact two forms of α_2-PI in plasma, of which only one binds plasminogen (1157). The non-plasminogen-binding form of α_2-PI arises from the plasminogen-binding form in circulation (1158) by the removal of the 26 C-terminal amino-acid residues (1159). These 26 C-terminal amino-acid residues thus play a decisive part in the binding of plasminogen by α_2-PI, and it has further been shown that the C-terminal Lys452 is important in this interaction (1160). Lys436 is also important in this interaction (1161). The observation of two important lysine residues has led to the hypothesis that the interaction between α_2-PI and plasmin/plasminogen (and their other physiological ligands) takes place by way of two lysine-binding sites of plasmin(ogen) and correctly placed lysine residues in the ligand (1161). This means that the interaction between plasminogen and α_2-PI (and other ligands) cannot involve strong interactions with more than one lysine-binding site, since plasminogen only contains one strong site for lysine (section 9.1). It appears that the plasminogen-binding form of α_2-PI cross-links selectively to fibrin with the catalytic help of factor XIIIa (1162).

Kinetic studies of the inhibition of plasmin by α_2-PI have revealed that this is a two-stage reaction (1163, 1164) consisting of a very rapid, reversible complex formation followed by a slower, irreversible step. If the lysine-binding sites in plasmin are blocked or removed, then the overall rate of inhibition falls substantially, as is also the case if an artificial substrate is bound to the active site of the plasmin. The first step in the inhibition process is thus dependent upon the presence both of a free lysine-binding site and of a free active site in the plasmin.

Three further aspects of especial importance will now be discussed.

The first aspect emerges from the observation that the amount of α_2-PI cross-linked to the a chain of fibrin is roughly equal to the amount of Glu-plasminogen bound to the fibrin. This in fact means that the quantity of fibrin-bound α_2-PI is sufficient to inhibit all the plasmin that can be formed from the fibrin-bound plasminogen (724, 725). There are two immediate possibilities for overcoming this problem. (i) The conversion of Glu-plasmin(ogen) to Lys-plasmin(ogen) raises its affinity for fibrin and thus lowers the rate of inhibition of plasmin with α_2-PI, since this process depends upon a free lysine-binding site in plasmin. (ii) Plasminogen activators stimulate the binding of Glu-plasminogen to fibrin (1165–1167).

The second aspect to be mentioned is the protein 'histidine-rich glycoprotein' (HRG). HRG is a single-chained, multifunctional plasma glycoprotein with a known amino-acid sequence corresponding to a molecular mass of 67 kDa (1168). Its amino-acid sequence shows partial homology with regions in, *inter alia*, antithrombin III and HMM-kininogen. Its concentration in plasma is about 100 μg/ml ($\approx$ 1.5 μM) (1169, 1170). HRG binds to the lysine-binding site of plasminogen and in this way inhibits the binding of plasminogen to fibrinogen and fibrin (1169, 1170). It has been suggested that this may lead to a certain inhibition of the fibrinolytic process. The presence of lysine analogues inhibits the binding of plasmin(ogen) to fibrin (963), and the situation here could well be similar. In this connection, it is interesting to note that HRG also has a lysine residue at its C terminus.

The third aspect is related to the second. HRG in a 1:1 stoichiometric complex with plasminogen binds t-PA with high affinity (K_d= 15–30 nM) and raises the catalytic activity of plasminogen fortyfold (1171). Precisely the same observation has been made for the blood platelet protein thrombospondin, which forms a complex with plasminogen (1172) and at the same time binds t-PA with high affinity (K_d= 12–36 nM) and promotes

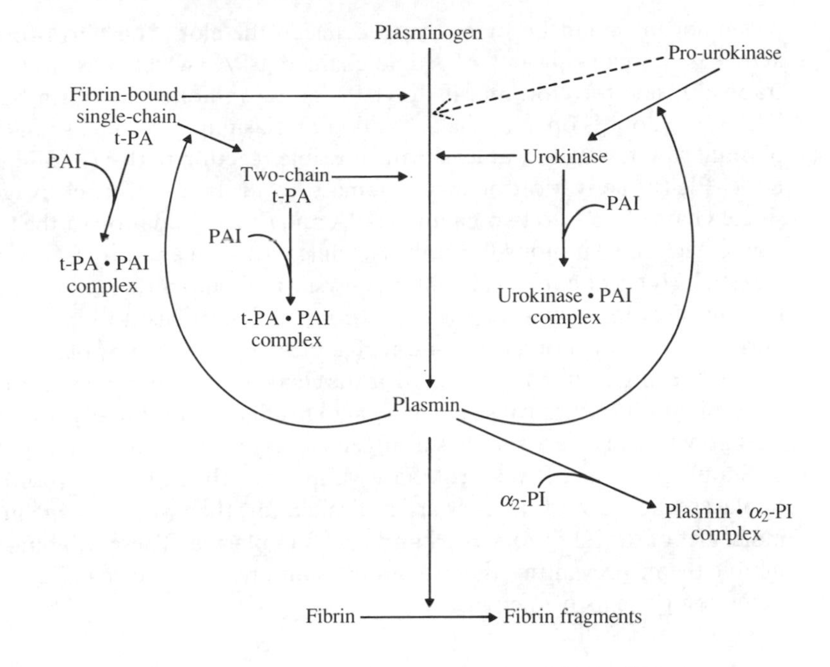

Fig. 70. The fibrinolytic system. (α_2-PI, α_2-plasmin inhibitor; PAI, plasminogen activator inhibitors.)

the activation of plasminogen (1171). The binding of plasminogen to thrombospondin proceeds most probably by way of kringle domain 5 in plasminogen (1173). It should be added that thrombospondin is released from blood platelets when these are exposed to thrombin – and that it is therefore present in the fibrin clot, and can in fact be cross-linked to it covalently by factor XIIIa (736, 737).

The presence of thrombospondin and HRG thus appears to make an important contribution to the activation of plasminogen to give plasmin. However, hasty conclusions should not be drawn, as there are still many interactions whose importance has not yet been elucidated. For example, thrombospondin and HRG can also form a complex with each other (1174), and HRG interacts with fibrinogen and fibrin (1175). Finally, HRG can act as a substrate for factor XIIIa (743), so that it may even be incorporated stably into the fibrin clot.

10.6. **A schematic version of fibrinolysis**

To conclude Part II, let us recapitulate by considering a brief sketch of the path of fibrinolysis (Fig. 70).

In the conversion of fibrinogen to fibrin, followed by polymerisation and formation of the clot, it may be assumed that a certain portion of the plasminogen bound becomes trapped inside the clot. The fibrin formed activates a small amount of single-chained t-PA, which has also been trapped inside the clot, and if this t-PA is not inhibited by fibrin-bound PAI-1, the door is open to the activation of plasminogen to plasmin. This plasmin can now follow at least four possible reaction paths: (i) inhibition by α_2-PI, (ii) activation of pro-urokinase to urokinase, (iii) cleavage of single-chained t-PA to two-chained t-PA and (iv) degradation of the fibrin clot. Reaction (i) inhibits fibrinolysis, while paths (ii) and (iii) reinforce it. Reaction (iv) is of course actual fibrinolysis, and one can imagine that the plasmin uses its lysine-binding sites to move from place to place on the fibrin, hydrolysing bonds at random. The cleavage of fibrin by plasmin will – by virtue of plasmin's specificity for the cleavage of peptide bonds on the C-terminal side of basic (especially lysine) residues – continually form new sites at which plasmin and plasminogen can bind to a C-terminal lysine.

Should plasmin, t-PA or urokinase escape from the clot and diffuse away to places where their functions are not called for, they will soon encounter molecules of the inhibitors α_2-PI and PAI-1 in plasma. These will bind and inhibit them, preventing them from causing any damage to tissue or to other components of the plasma.

Part III The Complement System

11.

Overview of the complement system

Up to now, we have concerned ourselves with the processes that restore an organism to its normal state after mechanical damage to its circulatory system. If, as a result of the mechanical damage, foreign bodies or organisms find their way into the bloodstream, coagulation alone does not provide sufficient help. In such situations, a different set of proteins can be called upon for assistance: the collective term for these proteins is the *complement system.*

The complement system is by no means the only one that plays an active part in combating foreign bodies in the bloodstream, but it is a part of the body's defences whose importance is frequently understressed.

In brief, the complement system is a set of proteins (Table 10) that, within the immune system, help to repel intruding micro-organisms.

The activation of the complement system can occur by two paths, 'classical activation' or 'alternative activation'. The two paths of activation are ultimately convergent, in that they both result in the formation of a cytolytic complex; this is called membrane attack complex (MAC), and its task is to destroy alien organisms by perforating their plasma or cell membrane. Not all organisms can be destroyed in this way, but the processes that lead to the formation of MAC cause at the same time the foreign body to be 'labelled', with the result that phagocytes can later recognise and remove it. In addition to labelling and puncturing foreign cells and organisms, the activation of the complement system also releases anaphylatoxins, which, *inter alia*, are attractants for these phagocytes.

The following treatment of the complement system falls into four main sections. These deal respectively with the classical and the alternative activation pathways, the cytolytic complex and the regulation of the complement system. In addition, there are two short sections that describe the anaphylatoxins and the complement receptors.

The complement system and particular parts of it have been considered in a number of recent review articles (1176–1205).

Table 10. *The proteins of the complement system*

Protein	Concentration in serum	Physically-estimated molecular mass (kDa)	Number of amino-acid residues	Calculated molecular mass (kDa)	Gene size (kbp)	Chromosome (and position, if known)
C1q $(ABC)_6$	150 nM	410	≈4050	460	–	–
A chain	–	–	224	24	unknown	1p
B chain	–	–	226	24	2.6	1p
C chain	–	–	≈225	unknown	unknown	unknown
C1r	400 nM	90	688	78.3	unknown	12p13
C1s	400 nM	85	673	74.8	unknown	12p13
C2	200 nM	100	732	81	18	6p21
C3	6 μM	185	1637	173.9	35–40	19p13.2–13.3
α chain	–	110	992	112.8	–	–
β chain	–	75	645	71.1	–	–
C4	1.8–2.5 μM	200	1695	186.7	22 + 16/22[a]	6p21
α chain	–	95	748	82.1	–	–
β chain	–	75	656	71.6	–	–
γ chain	–	30	291	33.0	–	–
C5	350 nM	190	≈1650	unknown	unknown	9q32–34
α chain	–	115	999	112.3	–	–
β chain	–	75	≈650	–	–	–
C6	600 nM	105–128	913	102.4	unknown	unknown
C7	600 nM	92–121	821	91.1	unknown	unknown
C8	350 nM	150	1251/1252	142.5	–	–
α chain	–	64	553	61.5	unknown	1
β chain	–	64	536/537	60.7	unknown	1
γ chain	–	22	182	20.3	unknown	1
C9	850 nM	71	538	60.8	> 80	5p13

Factor B	2 μM	90–93	739	83.0	6	6p21
Factor D	65 nM	23	218/222	24.0	unknown	unknown
C$\bar{1}$ inhibitor	2 μM	104	478	52.9	17	11p11.2–q13
Factor H	3 μM	150	1213	137.1	unknown	1q
C4b-binding protein (C4b-BP)	450 nM	570	7 × 549 + unknown	–	–	–
chain 1	–	70	549	61.5	unknown	1q
chain 2	–	45	unknown	unknown	unknown	unknown
Factor I	400 nM	50 + 38	317 + 244	35.4 + 27.6	unknown	4
Properdin	350 nM[b]	2–4 x 56	2–4 x 441[c]	47.9	unknown	Xp11.23–p21.1
S-protein	4–6 μM	83	459	52.4	3	unknown
Carboxypeptidase N	100 nM	300	1948	217.8	–	–
light chain	–	42–55	433	50.0	unknown	unknown
heavy chain	–	83–98	536	58.9	unknown	unknown
Decay-accelerating factor (DAF)	–	70	347	46.0	35	1q31–41
Membrane co-factor protein (MCP)	–	58/63	350	39.0	unknown	1q31–41
Complement receptor type 1A (CR1 A)	–	190	1998	219.0	130–160	1q32
Complement receptor type 2 (CR2)	–	145	1012/1067	110.0/117.0	unknown	1q32
Complement receptor type 3 (CR3)	–	260–265	1883/1884	207.5	–	–
α subunit	–	165–170	1136/1137	125.0	unknown	16p11–13.1
β subunit	–	95	47[illegible]	82.5	unknown	21q22.1–qter

[a] There are two coupled C4 genes–C4A and C4B–of which the former covers 22 kbp and the latter either 16 kbp or 22 kbp..

[b] The concentration stated is that of the properdin monomer.

[c] The number of residues stated refers to the murine properdin monomer.

12. Classical activation of the complement system

The classical pathway of complement activation employs the complement components C1, C2, C3 and C4.

12.1. Complement component C1

C1 is built up from three sub-components, called C1q, C1r and C1s, and which to begin with will be treated separately.

12.1.1. *C1q*

Human C1q has a concentration in normal human serum of about 70 μg/ml (≈150 nM) and is made up of no fewer than 18 polypeptide chains – six A chains, six B chains and six C chains. The amino-acid sequences of human C1q A chain and B chain have been determined fully at the protein level, while only a part of the C chain has been sequenced (1207–1211) (Appendix, Figures A30–A32). At the DNA level, a partial cDNA sequence and the complete sequences of the exons of the C1q B chain have been published (1212). It appears that the C1q B chain is synthesized with a 25-residue signal peptide and that the approximately 2.6-kbp-long gene only possesses one intron, of about 1.1 kbp (1212). The gene for the C1q B chain is found on the short arm of chromosome 1 (1213), as is also the case for the gene for C1q A chain (1214). The amino-acid sequence for the murine C1q B-chain has been deduced from the cDNA sequence (1215).

Human C1q A chain consists of 224 amino-acid residues and 12% carbohydrate, while the human C1q B chain consists of 226 residues with 8% carbohydrate. The precise size of the C1q C chain is not yet known, but it has about the same length as the other two and contains 4% carbohydrate. C1q thus contains some 8% carbohydrate on average (1216). The three C1q chains are homologous to one another, with a degree of identity of approximately 40% (Fig. 71). The molecular mass of C1q

```
A-chain   1 E D L C R A P D G K K G E A G R P G R R G R P G L K G
B-chain   1 Q L S C T G P P A I P G I P G I P G T P G P D G Q P G
C-chain   1 N T G C Y G I P G M P G L P G A P G K D G Y D G L P G

E Q G E P G A P G I R T G I Q - - - G L K G D Q G E P G P S G N P
T P G I K G E K G L P - G L A G D H G E F G E K G D P G I P G D P
P P G E P G I P A I K - G I R - - - G P P G Q K G E P G L P G H K

G K V G Y P G P S G P L G A R G I K G I K C T P G S P G N I K D Q
G K V G P K G P M G P K G G P G A P G A P G P K G E S G D Y K A T
G K D G P N G P P G M P G V P G P M G I P G E P G E E G R Y K Q K

P R P A F S A I R R - N P P M G G N V V I - F D T V I T N Q E E P
Q K I A F S A T R T I N V P L R R D Q T I R P D H V I T N M N N N
F Q S 92 C-chain

Y Q N H S G R F V C T V P G Y Y Y F T F Q V L S Q W E I N L S I V
Y E P R S G K F T C K V P G L Y Y F T Y H A S S R - - G N L C V N

S W S - R G Q V R R S L G F C D T T N K G L F Q V V S G G M V L Q
L M R G R E R A Q K V V T F C D Y A Y N - T F Q V T T G G M V L K

L Q Q G D Q V W V E K D P K K G H I Y Q G S E A D S V F S G F I L
L E Q G E N V F L Q A T D K N S L L G M E G - A N S I F S G F L L

- P C F S A 224 A-chain
F P D M E A 226 B-chain
```

Fig. 71. Homology between the A, B and C chains of C1q. Since the C-chain sequence is not fully known, comparison involving this chain is only partial. Only the positions at which amino acids are the same in all chains shown are outlined. Pairwise comparison of the chains reveals the following degrees of amino-acid identity: chains A and B, 40%; chains A and C, 44%; chains B and C, 48%.

derived from sedimentation equilibrium studies is 410 kDa (1217), but that calculated from the amino-acid sequences is 460 kDa.

All three C1q chains have the common property of containing a collagen-like region of about 80 residues located near the N terminus, while the rest of the molecule is globular. All three collagen-like regions contain a repeated tripeptide motif Gly-Xaa-Yaa, where Xaa is often proline and Yaa is often 5-hydroxy-lysine or 4-hydroxy-proline. In human C1q, the repeating tripeptide sequences begin at the 9th, 6th and 3rd amino-acid residues respectively of the A, B and C chains. There is, however, a region where this pattern is not adhered to perfectly, and this has consequences for the way in which the overall structure of C1q is built up (see below).

In the currently accepted model for C1q, the N-terminal parts of the C1q chains together form a collagen-like, triple-helix structure, in which each triple helix contains one A, one B and one C chain. In addition, the A, B and

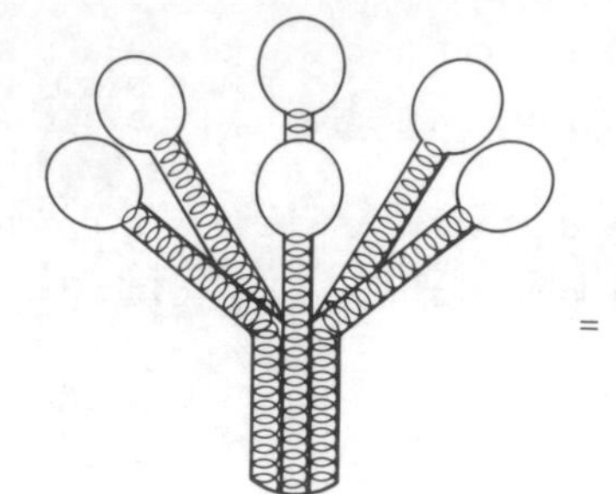

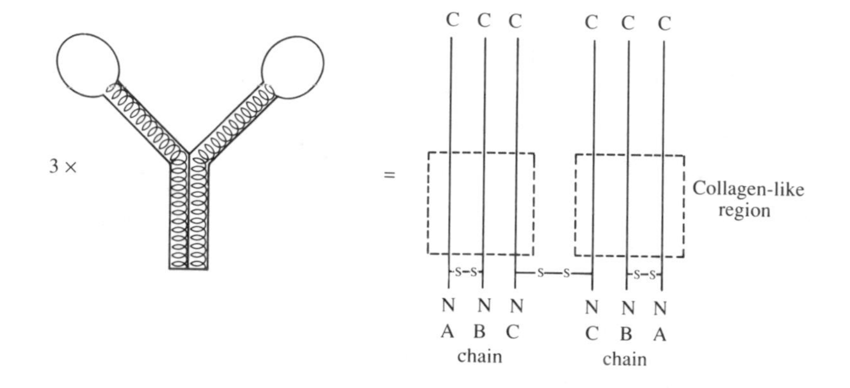

Fig. 72. A schematic model of C1q. (Based upon reference 1177.)

C chains of C1q are held together in dimers by disulphide bonds, giving six AB dimers and three CC dimers. In the complete C1q structure, these nine dimers are held together by non-covalent forces (Fig. 72). The disulphide bond that connects the A and B chains to one another links Cys4 in the A to Cys4 in the B chain, and the CC homodimers use Cys4 of both C chains (1209). Apart from these Cys residues, the C1q A chain contains Cys also at positions 131 and 168, while the B chain has Cys at positions 135, 154 and 171. Considerations of homology suggest that Cys154 in the C1q B chain is unpaired, while the other cysteine residues form intracatenary disulphide bridges.

The complete C1q structure resembles a bunch of six tulips (Fig. 72). This model is derived from electron-microscopic pictures (1218, 1219) and has since been confirmed by the results of neutron-scattering studies (1220). A more detailed model of the structure of the collagen-like part of C1q has also been proposed (1221).

The 'stalks' of the 'tulips' correspond to the collagen-like regions, and it should be observed that these have a kink in the middle, at a point correlated with the region where the primary structure shows a breakdown in the otherwise regular Gly-Xaa-Yaa pattern. In this region, the C1q A chain has an inserted amino acid (Thr38), the B chain has an additional tripeptide, and the C chain has an alanine residue where the repeating pattern would have demanded a glycine. All these three factors, especially the third, lead to the kink in the C1q structure, a point at which there is probably also an increased flexibility in the composite chain. It is particularly interesting to note that the intron found in the gene for the B chain of C1q is positioned such that it splits the codon for Gly36, which is located in the kink.

As already mentioned, the collagen-like regions of all C1q chains include the modified amino acids 4-hydroxy-proline and 5-hydroxy-lysine. In the A chain there are eight 4-hydroxy-proline and five 5-hydroxy-lysine residues, while the B chain has eleven 4-hydroxy-prolines and six 5-hydroxy-lysines and the C chain has thirteen 4-hydroxy-prolines and three 5-hydroxy-lysines. Of the fourteen 5-hydroxy-lysines of C1q, it appears that eleven are modified by the attachment of carbohydrate. The carbohydrate attached to 5-hydroxy-lysine is always the disaccharide group glucosyl-galactosyl, and C1q is the only known example of a plasma protein that carries this modification. The same applies for 4-hydroxy-proline, with the exception of the single 4-hydroxy-proline residue found in HMM-kininogen. The glycosylated 5-hydroxy-lysine residues are believed to be residues 11, 26, 45, 75 and 78 in the C1q A chain, residues 32, 35, 71 and 83 in the B chain and residues 47 and 56 in the C chain. N-bound carbohydrate has been found at Asn124 of the A chain of C1q.

Deficiency of C1q is rare (see reference 1195 for a general review of hereditary abnormalities in the complement system). In only one such case has the molecular basis for this abnormality been elucidated; it was found to be a mutation of Arg150 to give a termination codon in the B chain (1222).

12.1.2. *C1r*

C1r is another of the protein components of C1. It invariably appears as a non-covalently linked dimer, $[C1r]_2$. Logically, it should perhaps be the dimer that is called C1r, since this is the functional form and also the form that is isolated; however, it has become conventional to use 'C1r' for the monomer.

$[C1r]_2$ is the zymogen for a homodimeric serine proteinase. It is a two-chained glycoprotein molecule containing 9.4% carbohydrate (1223). Its two identical monomers have a molecular mass of about 90 kDa each. The total molecular mass of $[C1r]_2$, estimated by sedimentation analysis, is 170 kDa (1224). The concentration of C1r in normal human serum is around 35 μg/ml ($\approx$200 nM for the dimer) (1225). The amino-acid sequence of human

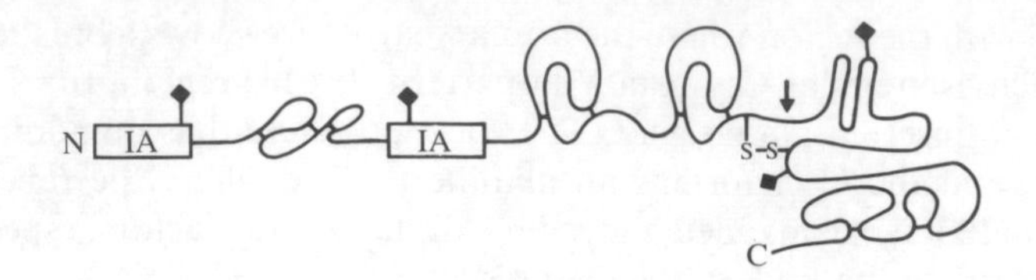

Fig. 73. The domain structure of C1r. → shows the position of cleavage by autoactivation and ◆ shows N-bound carbohydrate. (IA, interaction domains.)

```
  1 S I P I P Q K L F G E V T S P L F P K P Y P N N F E T T T V I
177 S S E L Y T E A S G Y I S S L E Y P R S Y P P D L R C N Y S I

T V P T G Y R V K L V F - Q Q F D L E P S E G - - C F Y D Y V K I
R V E R G L T L H L K F L E P F D I D D H Q Q V H C P Y D Q L Q I

S A D K K S L G R F C G Q L G S P L G N P P G K K E F M S Q G N K
Y A N G K N I G E F C G K - - - - - Q R P P D L D T S - S N A V D

M L L T F H T D F S N E E N G T I M F Y 114
- L L - F F T D E S G D S R G W K L R Y 285
```

Fig. 74. Internal homology between the IA domains of C1r. Only the positions at which amino acids are identical are outlined. The degree of amino-acid identity is 30%.

C1r has been determined completely by protein-sequencing (648, 1226, 1227) and cDNA-sequencing (646, 647) (Appendix, Fig. A33). The gene for C1r has not been characterised, but it has been located on chromosome 12 at position p13 (1228).

C1r contains 688 amino-acid residues, giving a theoretical molecular mass for the unmodified polypeptide of 78.3 kDa, and 85.7 kDa for the complete monomer including the 9.4% carbohydrate. This is attached to C1r as four chains at residues Asn108, Asn204, Asn497 and Asn564.

The polypeptide chain of C1r is organised into six domains in four classes (Fig. 73). The first of these, at the N terminus, has no known counterpart except later in the same chain and in C1s (see section 12.1.3; Fig. 74). It is called the IA domain. Thereafter comes a growth-factor domain, then a second IA domain, and then two domains of a class that we shall soon encounter frequently, the so-called short consensus repeat (SCR) domain (Fig. 75). (We have already encountered SCR domains in the b-chain of factor XIII, section 7.1.2.) Finally, the C-terminal domain of C1r is the potentially catalytic serine-proteinase domain.

A noteworthy detail in the sequence of C1r is the fact that the growth-factor domain contains a modified amino-acid residue in position 150,

```
292 C P Q P K T L D - - E F T I I Q N L Q - P Q Y Q F R D Y F I A
359 C G Q P R N L P N G D F R Y T T T M G V N T Y - - K A R I Q Y

T C K Q G Y - - - Q L I E G N Q V L H S F T A V C Q D D G T W H R
Y C H E P Y Y K M Q T R A G S R E S E Q G V Y T C T A Q G I W K N

A - - - - - M P R C 354
E Q K G E K I P R C 430
```

Fig. 75. Internal homology between the SCR domains of C1r. Only the positions at which amino acids are identical are outlined. The degree of amino-acid identity is 25%.

identified as a β-OH-asparagine residue (1229). This is the second time that we have met this modified amino-acid residue; the first time was in protein S from the coagulation system (section 8.2.3). Position 135 is dimorphic, in that both the protein sequence and the cDNA sequence have shown this position occupied variously by Ser and by Leu.

The disulphide pattern in C1r has not been demonstrated experimentally, but a qualified suggestion can be made on the basis of homology and a knowledge of the autolytic fragments (see section 12.2). The disulphide bonds proposed are: in the first IA domain, Cys54–Cys72; in the growth-factor domain, Cys129–Cys148, Cys144–Cys157 and Cys159–Cys172; in the second IA domain, Cys176–Cys203 and Cys233–Cys251; in the first SCR domain, Cys292–Cys341 and Cys321–Cys354; in the second SCR domain, Cys339–Cys412 and Cys389–Cys430; between chains Cys434–Cys560; and in the serine-proteinase domain, Cys603–Cys622 and Cys633–Cys663.

The activation of $[C1r]_2$ to $[C\bar{1}r]_2$ proceeds by the cleavage of the peptide bond between Arg446 and Ile447. The product is a non-covalent dimer of active, two-chained enzymes, with heavy chains from the N termini of the zymogens (58 kDa) and light chains, containing the active sites, from the C termini (34 kDa). The catalytic apparatus is taken to involve His39, Asp94 and Ser191 in the light chain, corresponding to His485, Asp540 and Ser637 in C1r. $[C\bar{1}r]_2$ is thus a ‘double’ serine proteinase, with two active sites.

12.1.3. *C1s*

C1s is the third protein component of C1. As was the case for C1r, there are two C1s molecules in C1. However, these do not form a dimeric structure in C1, even though isolated C1s dimerises in the presence of calcium ions.

Human C1s is a single-chained glycoprotein with 7.1% carbohydrate (1223) and a molecular mass around 85 kDa, as determined by sedimentation-equilibrium studies (1224). Its concentration in normal human serum is about 35 μg/ml ($\approx$400 nM) (1225).

Only a part of the amino-acid sequence of human C1s has been determined by protein-sequencing (1230–1232), but the sequenced region includes the entire catalytically active chain and extends into the non-catalytic chain. The complete amino-acid sequence of human C1s, plus that of a 15-residue signal peptide, have been determined from the cDNA sequence (649, 650, 1233) (Appendix, Fig. A34). The gene structure of C1s is partially known. On the basis of heteroduplex analysis, it has been suggested that the C1s gene may contain 12 exons and cover about 10.5 kbp of genomic DNA (1234). The last five exons in the C1s gene have been cloned, and, surprisingly, the final exon encodes the entire serine-proteinase domain (1234). This genomic organisation is unique for a serine-proteinase domain, but it is expected that the C1r gene will be organised in the same way. The gene for C1s is tightly linked to the gene for C1r (650, 1223) and has also been located on chromosome 12 at position p13 (1228). The genes for C1s and C1r are separated by only about 9.5 kbp, but the two genes are transcribed in opposite directions (1233).

C1s is made up of 673 amino-acid residues, corresponding to a theoretical molecular mass of 74.8 kDa, excluding carbohydrate. It has potential sites for N-glycosylation at Asn159, Asn391 and Asn667, of which at least the first two are known to be used (1232).

C1s is homologous to C1r along its entire length; about 40% of all residues are identical (649, 650, 1233) (Fig. 76), and it contains the same types of domain in the same order (Fig. 77): an IA domain, a growth-factor domain, another IA domain, two SCR domains and a serine-proteinase domain. C1s further contains β-OH-Asn (1235); its exact position is not yet known, but the asparagine residue modified in C1r is correspondingly placed in C1s (649, 650), so further homology at this point is not unlikely, especially as all β-OH-Asn residues found so far have been in growth-factor domains.

The disulphide-bond pattern in C1s has not been demonstrated by experiment, but considerations of homology suggest the following placings: in the first IA domain, Cys50–Cys68; in the growth-factor domain, Cys120–Cys132, Cys128–Cys141 and Cys143–Cys156; in the second IA domain, Cys160–Cys187 and Cys219–Cys236; in the first SCR domain, Cys279–Cys326 and Cys306–Cys339; in the second SCR domain, Cys344–Cys388 and Cys371–Cys406; between the chains Cys410–Cys534; and in the serine-proteinase domain, Cys580–Cys603 and Cys613–Cys644.

C1s is activated to give $\overline{\text{C1}}$s by cleavage of the bond between Arg422 and Ile423 (649), which results in an N-terminally derived heavy chain (58 kDa) disulphide-bound to a C-terminally derived light chain (27 kDa) that contains the catalytic apparatus. This is presumed to involve the residues His38, Asp92 and Ser195 of the light chain, corresponding to His460, Asp514 and Ser617 in C1s.

The serine-proteinase domains in both C1r and C1s are unusual in that

```
C1s       1 E P T M Y G E I L S P N Y P Q A Y P S E V E K S W D
C1r 1 S I P I P Q K L F G E V T S P L F P K P Y P N N F E T T T V

I E V P E G Y G I H L Y F T H L D I E L S E N C A Y D S V Q I I S
I T V P T G Y R V K L V F Q Q F D L E P S E G C F Y D Y V K I S A

G D T E E G R L C G Q R S S N N P H S P I V E E F Q V P Y N K L Q
D K K S L G R F C G Q L G S P L G N P P G K K E F M S Q G N K M L

V I F K S D F S N E E R F T - - - - - G F A A Y Y V A T D I N E C
L T F H T D F S N E E N G T I M F Y K G F L A Y Y Q A V D L D E C

T D F V - - - - - - - - D V P C S H F C N N F I G G Y F C S C P P E
A S R S K L G E E D P Q P Q C Q H L C H N Y V G G Y F C S C R P G

Y F L H D D M K N C G V N C S G D V F T A L I G E I A S P N Y P K
Y E L Q E D R H S C Q A E C S S E L Y T E A S G Y I S S L E Y P R

P Y P E N S R C E Y Q I R L E K G F Q V V V T L R R E D F D V E A
S Y P P D L R C N Y S I R V E R G L T L H - L K F L E P F D I D D

A D S A G N C L G D L V F V A G D R Q F G P Y C G H G F P G P L N
H Q Q V H C P Y D Q L Q I Y A N G K N I G E F C G K Q R P P D L D

I E T K S N A L D I I F Q T D L T G Q K K G W K L R Y H G D P M P
- - T S S N A V D L L F F T D E S G D S R G W K L R Y T T E I I K

C P K E D T P N S - - V W E P A K A K Y V F R D V V Q I T C L D G
C P Q P K T L D E F T I I Q N L Q P Q Y Q F R D Y F I A T C K Q G

F E V V E G R V G A T S F Y S T C Q S N C K W S N S K L K C Q P V
Y Q L I E G N Q V L H S F T A V C Q D D G T W H R A M P R C K I K

D C G I P E S I E N G K V E D P - - - E S T L F G S V I R Y T C E
D C G Q P R N L P N G D F R Y T T T M G V N T Y K A R I Q Y Y C H

E P Y Y Y M E N G G G - - - - - - G E Y H C A G N G S W V N E V L
E P Y Y K M Q T R A G S R E S E Q G V Y T C T A Q G I W K N E Q K

G P E L P K C V P V C G V P R E P F E E K Q R I I G G S D A D I K
G E K I P R C L P V C G K P V N P V E Q R Q R I I G G Q K A K M G

N F P W Q V F F D N P - W A G G A L I N E Y W V L T A A H V V E G
N F P W Q V F T N I H G R G G G A L L G D R W I L T A A H T L Y P

N R E P T M Y V G S T S V Q T S R L A - K S K M L T P E H V F I H
K E H E A Q S N A S L D V F L G H T N V E E L M K L G N H P I R R

P G W K L L E V P E G R T N F D N D I A L V R L K D P V K M G P T
V S V H P D Y R Q D E S Y N F E C D I A L L E L E N S V T L G P N

V S P I C L P G T S S D Y N L M D G D L G L I S G W G R T E K R D
L L P I C L P D N D T F Y - - - D L G L M G Y V S G F G V M E E K

R A V R L K A A R L P V A P L R K C K E V K V E K P T A D A E A Y
I A H D L R F V R L P V A N P Q A C E N W L R G K N R M D - - - -

V F T P N M I C A G G E K - G M D S C K G D S G G A F A V Q D P N
V F S Q N M F C A G H P S L K Q D A C Q G D S G G V F A V R D P N

D K T K F Y A A G L V S W G P Q C G - T Y G L Y T R V K N Y V D W
T D - R W V A T G I V S W G I G C S R G Y G F Y T K V L N Y V D W

I M K T M Q E N S T P R E D 673
I K K E M E E E D 688
```

Fig. 76. Comparison of the amino-acid sequences for C1r and C1s. Only the positions at which amino acids are identical are outlined. The degree of amino-acid identity is 37%.

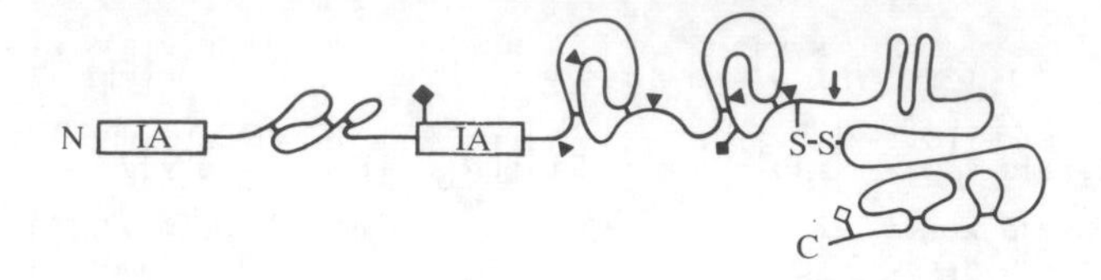

Fig. 77. The domain structure of C1s. → shows the position of cleavage by C$\overline{1}$r, ▼ the positions of the introns localised so far, ◆ the N-bound carbohydrate and ◇ the potential attachment sites for N-bound carbohydrate. (IA, interaction domains.)

they lack the two cysteine residues that normally form the so-called histidine loop in other serine proteases.

12.1.4. *Complete C1*

As described at the beginning of this chapter, C1 is a composite structure with C1q, C1r and C1s as its components. C1 has two major subunits, C1q and $[(C1r)_2(C1s)_2]$. We have examined the way in which C1q is built up, so let us now examine the structure of the $[(C1r)_2(C1s)_2]$ complex, before going on to C1 as a complete entity.

The kernel of this complex is the dimer $[C1r]_2$. It has long been known that C1r is a dimer, both under physiological conditions and even in non-physiological buffers, for example in the absence of calcium ions. As described in section 12.1.2, C1r can be divided into domains, and after activation it consists of a light and a heavy chain. The light chain (also called the B chain) is derived from the C terminus of the complete C1r polypeptide and contains little other than the serine-proteinase domain, while the heavy chain (also called the A chain), derived from the N terminus of C1r, contains all of the remaining defined domains. The A chain of C1r can be divided further into autolytic fragments, as a special property of C1r is the ability to be activated autocatalytically; this we return to at a later point.

The autolytic fragments of the C1r A chain are designated α, β and γ (1236–1238), and that is their order in the sequence (from the N terminus). The α fragment corresponds to residues 1–211, the β to residues 212–279, and the γ to residues 280–446 (648). The γ fragment is disulphide-bound to the catalytically active B chain from the zymogen's C terminus. The autocatalytic processes result in the formation of a molecule that is called C$\overline{1}$rII. This molecule is still a homodimer, and it consists of monomers containing the γ fragment of the A chain plus the entire B chain. It is thus clear that the C1r–C1r interaction takes place between these two regions. Chemical cross-linking experiments indicate that the interaction takes place between the γ fragment in one monomer and the B chain in the other

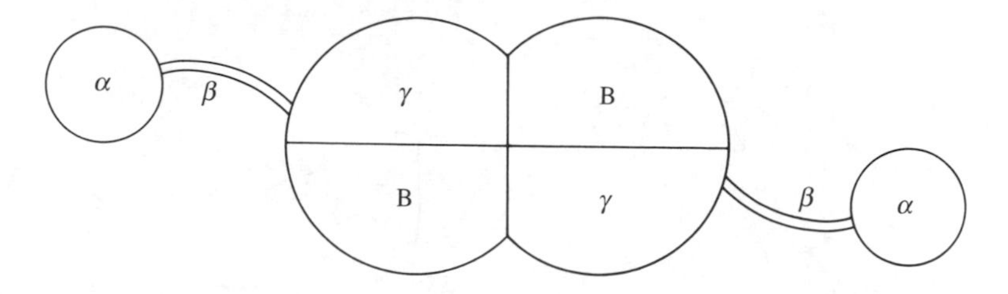

Fig. 78. A schematic model of $[\mathrm{C\bar{1}r}]_2$. α, β and γ are the autocatalytically-formed fragments of the heavy chain of C1r; B is the serine-proteinase domain. (After reference 1239.)

monomer (1239) (Fig. 78). As the amino-acid sequence shows, the γ fragment consists largely of the two SCR domains, whose function may thus be connected with the formation of dimers.

The formation of the complex between C1s and C1r is calcium-dependent (1240, 1241), just as C1s forms dimers in the presence of calcium ions (1240). Since $\mathrm{C\bar{1}rII}$ has lost the ability to bind C1s (1238) and Ca^{2+} (1242), it must be assumed that the interaction between C1r and C1s involves the two N-terminal fragments of C1r, α and β. The same procedure has been employed to find out which regions in C1s are important for complexation with C1r, with the difference that the fragments of C1s were generated not by autolysis but by plasmin-catalysed degradation (1243). The molecule $\mathrm{C\bar{1}sII}$, which corresponds largely to $\mathrm{C\bar{1}rII}$, has lost the ability to bind C1r and to undergo calcium-dependent dimerisation. On the other hand, the ability to undergo calcium-dependent dimerisation and to bind C1r is retained in a C1s fragment consisting of the whole heavy chain disulphide-bound to a short fragment from the light chain (1244). It follows from this that it must be the three domains closest to the N terminus that mediate the bindings between C1s and C1r. By analogy with the assumed ability to bind Ca^{2+} shown by other growth-factor domains, one may imagine the growth-factor domains in C1r and C1s to be the domains that mediate the Ca^{2+}-dependent complex formation between C1r and C1s.

These results, taken together, suggest a model for the complex: a linear tetramer C1s–C1r–C1r–C1s, as shown in Figure 79. The linear tetrameric structure has also been supported by the results of studies involving affinity labelling of C1s (1245). However, the existence of a linear tetramer should not be taken as implying that the domains of all these proteins are lined up like beads on a string. $[(\mathrm{C1r})_2(\mathrm{C1s})_2]$ appears most of all to resemble an asymmetric letter X (1245, 1246), a fact which is in gratifying agreement with the manner in which the two C1r chains are held together.

In both C1r and C1s we have two IA domains with as yet unexplained functions. One hypothesis is that these domains are of importance for the interaction between C1s–C1r–C1r–C1s and C1q, but there is at present no

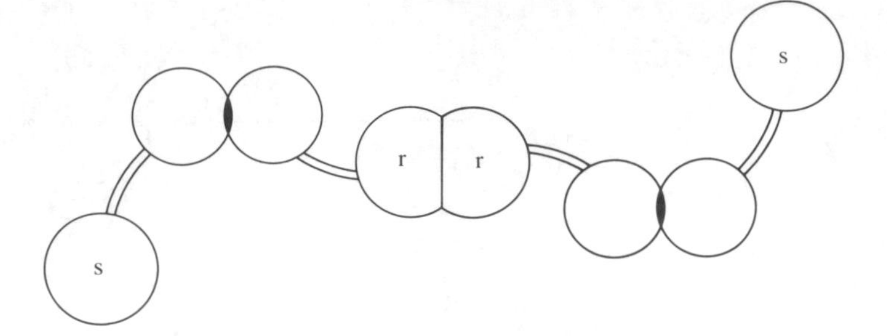

Fig. 79. A schematic model of $[(C1r)_2(C1s)_2]$. Domains labelled r and s are serine-proteinase domains in C1r or C1s respectively. (After reference 1243.)

evidence for this. Notwithstanding, the domains have been given the name 'IA' in token of their assumed function of **i**nter**a**ction. The only domains whose function or possible function we have not yet discussed are the two SCR domains in C1s, and these we return to later. Before proceeding to the complexation between $[(C1r)_2(C1s)_2]$ and C1q, it should be mentioned that C1r does not display autolysis when it is present in the tetramer with C1s (1238, 1247).

Complex formation between $[(C1r)_2(C1s)_2]$ and C1q is the final remaining step needed for the formation of functional C1. C1q forms a relatively loose complex with $[(C1r)_2(C1s)_2]$ (1223, 1241), but, although the interaction is weak, it has functional importance, as it allows the activation of C1. Furthermore, neither C1r nor C1s binds alone to C1q – at least, not to any significant degree (1240, 1241).

The relatively weak binding of $[(C1r)_2(C1s)_2]$ to C1q suggests that models for C1 should not involve structures with too intimate association between these components. As described above (section 12.1.1), the overall structure of C1q resembles a bunch of tulips; electron-microscopic studies have shown the $[(C1r)_2(C1s)_2]$ complex bound between the heads and the centre stalk (1248, 1249) – in other words, in the general region where the stalks diverge owing to the kink in their otherwise regular structure.

Several models have been suggested for the structure of C1 (1187, 1220, 1245, 1248, 1251). It is unlikely that any of them is completely correct, but they give a basis for further development. Figure 80 shows one of these models.

12.2. Activation of complement component C1

We have now seen something of the construction of C1. It is this structure that is converted to the activated form $C\overline{1}$, and it is the latter that takes part in complement activation.

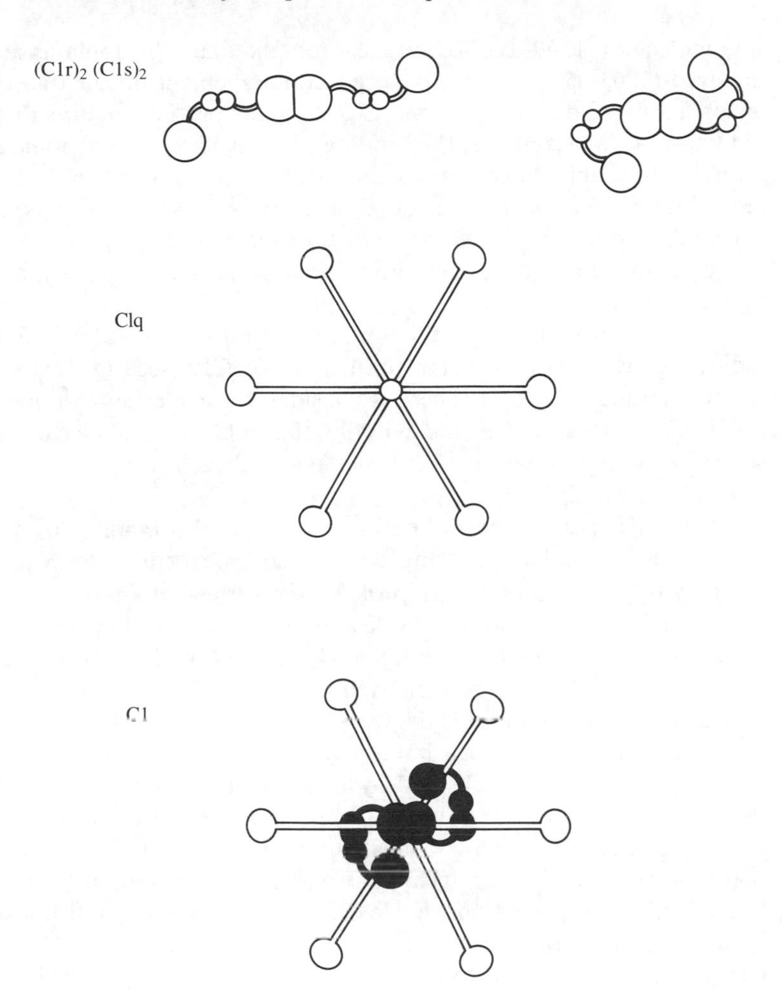

Fig. 80. A schematic model of C1. C1q is looked at from the heads of the 'tulips' such that the 'stalks' point into the paper. (Based upon references 1245 and 1251.)

C1 binds to, and is activated by, immune complexes or aggregates that contain IgG or IgM, but not complexes of of IgA, IgD or IgE (1252–1256). However, it is not only IgG or IgM complexes and aggregates that can activate C1; a whole series of other substances also possess this property (see, for example, reference 1187). In the following discussion, we shall just look at the activation of C1 as induced by immune complexes, since it is likely that the mechanism can be extrapolated to other substances that initiate the complement system's classical activation.

The part of C1 that binds to immune complexes is C1q, which thus functions as a recognition unit (1257–1259). C1q binds to C_H2 domains in

IgG molecules (1260–1262), but it is probable that C_H3 domains are also involved (1263–1265). There is no general agreement on the question of which amino-acid residues in the C_H2 domains participate directly in the binding to C1q (1266), but the binding is believed to involve ionic rather than hydrophobic interactions, and studies of murine IgG made by recombinant techniques indicate that Glu318, Lys320 and Lys322 are essential residues for this (1267). On IgM molecules, C1q recognises the C_H4 domain (1268, 1269), but there are results that suggest participation by C_H3 as well (1270).

The binding of C1q to IgG takes place with a stoichiometry of 1:6 (1253), indicating that it is the globular 'tulip' heads on C1q (see Fig. 72) that bind to the immunocomplexes. There is considerable evidence to support this (1271–1273). This implies, incidentally, that a C1q molecule can bind to several immunoglobulin molecules at once, if each 'tulip' possesses the same binding ability.

There is a considerable difference between the binding of C1q to monomeric IgG and to IgG complexes. It also appears that it is the presence of many IgG molecules, for example on the surface of an antigen, that is important for the activation of C1, and not merely the fact that the IgG molecules have bound an antigen (1274, 1275). This is termed the associative model of complement activation.

The detailed mechanism of the steps that follow is certainly not yet clear. However, it is known that the binding of C1q to activators such as surface-bound IgG molecules is followed by an autocatalytic activation of C1r, in the sense that the two C1r polypeptides convert themselves to two-chained C$\overline{1}$r molecules. Whatever its mechanism, it must require a molecular rearrangement in C1 before it can take place, since C1 is stable in plasma. It is assumed that the signal for this rearrangement is given by C1q, which by binding to a surface carrying antibodies 'locates' the site at which complement activation is to take place. The conformational signalling in C1q is presumed to be highly dependent upon the flexibility of the collagen-like domains of C1q at the point where these have a kink (see, for example, reference 1196); this is of course also the point at which the interaction of C1q with $[(C1r)_2(C1s)_2]$ is believed to take place.

Apart from the necessary rearrangement, there are two ways in which the C1r polypeptide chains in C1 can activate themselves, or each other (1276): either each C1r polypeptide chain cleaves itself, or each C1r polypeptide within a C1 molecule cleaves a bond in the other. As far as is known, both chains are activated simultaneously. At least, it is established that C$\overline{1}$r cannot activate C1r (1277). (The third conceivable possibility, that autoactivation involves collision and reaction between two separate dimers, is ruled out by the fact that dimers are bound to surface-attached C1q during the activation process and therefore cannot diffuse.) The bond cleaved in C1r is Arg446–Ile447, and normally the second hypothesis is preferred (1239). However, it should be noted that the presence of C1s in a C1 complex is essential for the autoactivation of $[\mathrm{C1r}]_2$.

After the autoactivation of $[\mathrm{C1r}]_2$ to give $[\mathrm{C\bar{1}r}]_2$, the $[\mathrm{C\bar{1}r}]_2$ activates the C1s polypeptide chains to give $\mathrm{C\bar{1}s}$ (1278, 1279) by splitting the bond Arg422–Ile 423. Figure 81 gives a model (1251) of how this might happen; note that C1q does not play any part in this model, and that others could be thought out.

Finally, it should be emphasized that the activation of C1s by $[\mathrm{C\bar{1}r}]_2$ is a very unusual reaction. This is not only because the activation of C1 is triggered off by a conformational change, but also because the substrate of the $[\mathrm{C\bar{1}r}]_2$ is never replenished. The only substrates of the $[\mathrm{C\bar{1}r}]_2$ are the two C1s polypeptides that reside in the same C1 complex. However, it is also true that $[(\mathrm{C\bar{1}r})_2(\mathrm{C\bar{1}s})_2]$ can easily dissociate from C1q and thus lead to $[(\mathrm{C\bar{1}r})_2(\mathrm{C\bar{1}s})_2]$ in bulk plasma and to free C1q. The latter can then activate more $[(\mathrm{C1r})_2(\mathrm{C1s})_2]$ in the plasma. It is by no means true that all C1q and all $[(\mathrm{C1r})_2(\mathrm{C1s})_2]$ circulate together as a complex, even though their molar concentrations are roughly the same.

In order to proceed with the classical pathway of activation of the complement system, we now introduce three more of its components: C2, C3 and C4.

12.3. **Complement component C2**

C2 is a single-chained, glycosylated serine-proteinase zymogen. Its molecular mass has been estimated by physical methods to be about 100 kDa (1280), while its carbohydrate content has been determined as 16% (1281). Its concentration in normal human serum lies around 20–25 μg/ml ($\approx$200 nM).

For human C2, only sparse sequence data have been obtained at the protein level (1282–1284), but the complete amino-acid sequence, including that of a 20-residue signal peptide, has been deduced from that of cDNA (637, 1285) (Appendix, Fig. A35). For murine C2, a partial cDNA sequence is known (1286). The gene for human C2 lies on chromosome 6 at position p21 (1287) and covers 18 kbp of genomic DNA (1192, 1288). Nothing has been reported about its genomic organisation.

Human C2 has an overall structure as shown in Figure 82, and the molecule has 732 amino-acid residues, corresponding to a molecular mass of 81 kDa, to which the weight of the carbohydrate must be added. There are eight potential attachment sites for N-acetylglucosamine-based carbohydrate on the asparagine residues at positions 9, 92, 270, 313, 447, 451, 601, and 631. Position 513 may be dimorphic, as the two published cDNA sequences code respectively for Leu (637) and Phe (1285) at this position.

When C2 is activated by $\mathrm{C\bar{1}s}$, the peptide bond between Arg223 and Lys224 is broken. The fragment from the N terminus, C2b, is the light chain

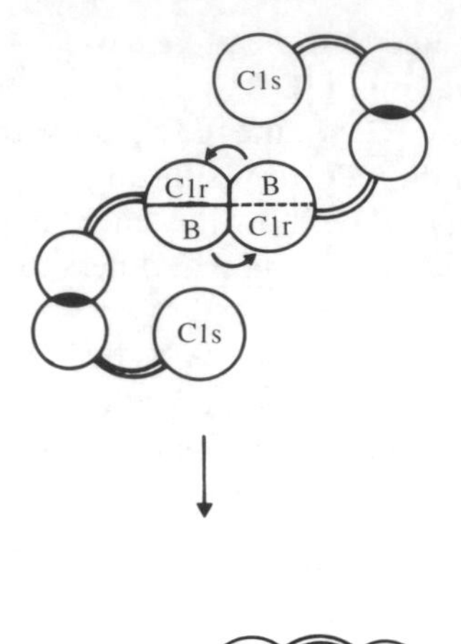

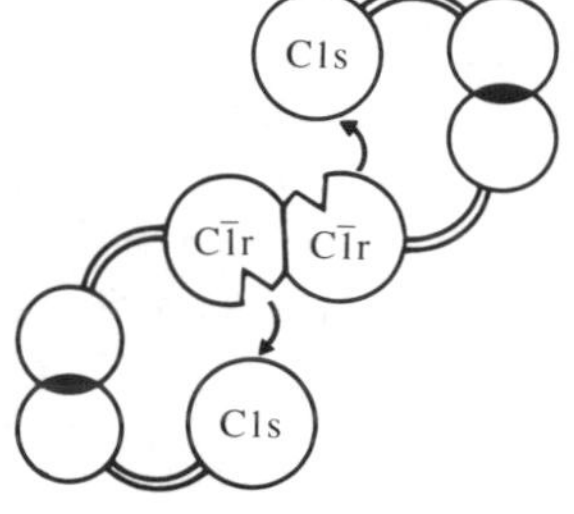

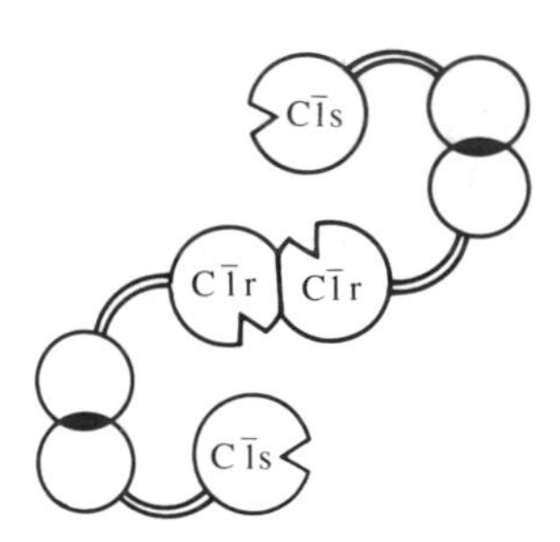

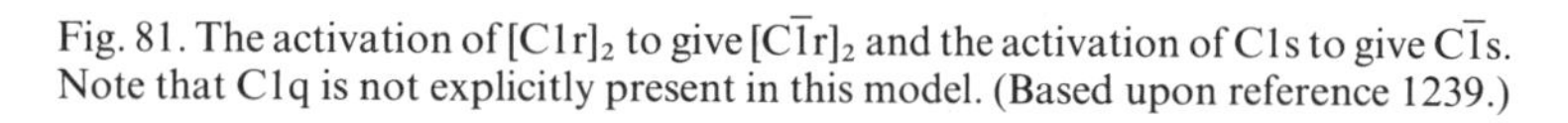

Fig. 81. The activation of $[C1r]_2$ to give $[C\bar{1}r]_2$ and the activation of C1s to give C$\bar{1}$s. Note that C1q is not explicitly present in this model. (Based upon reference 1239.)

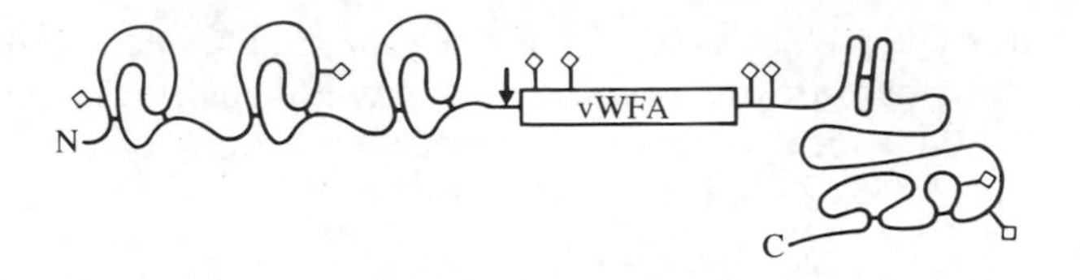

Fig. 82. The domain structure of C2. → shows the position of cleavage by C$\bar{1}$s and ◇ the potential attachment sites for N-bound carbohydrate. vWFA is the part of C2 that is homologous to the A domains defined in von Willebrand factor.

```
  4 C P Q N V N I S G G T F T L S H G - W A P G S L L T Y S C P Q G -
 69 C P A P V S F E N G I Y T P R L G S Y P V G G N V S F E C E D G F
131 C P N P G I S L - G A V R T G F R - F G H G D K V R Y R C S S N -

- L Y P S P A S - R L C K S S G Q W Q T P G A T R S L S K A V C 64
I L R G S P V - - R Q C R P N G M W D G E - - - - - - - T A V C 124
- L V L T G S S E R E C Q G N G V W S G T - - - - - - - E P I C 184
```

Fig. 83. Internally homologous SCR domains in C2. Only the positions at which amino acids are the same in all three domains are outlined. Pairwise comparison of the three domains reveals the following degrees of amino-acid identity: domains 1 and 2, 35%; domains 1 and 3, 24%; domains 2 and 3, 27%.

(30 kDa) and that from the C terminus, C2a, is the heavy chain (70 kDa). The latter contains the catalytic apparatus, which is presumed to involve His264, Asp318 and Ser436 in C2a, corresponding to His487, Asp541 and Ser659 in C2. C2a and C2b are held together by non-covalent forces only; there are no intercatenary disulphide bonds.

C2b has 223 amino-acid residues and consists mainly of three SCR domains (Fig. 83), while C2a, with 509 residues, appears at first sight to be an unusually large serine-proteinase domain – about twice as large as those we have met up to now. In fact, it is the C-terminal part of C2a that is homologous to the other serine proteases, while its N-terminal part is homologous to the A domains as defined in von Willebrand factor. C2 is homologous to complement factor B throughout its entire sequence, with a degree of amino-acid identity of 39% (see section 13.2).

The disulphide-bond pattern in C2 has not been determined, but one free cysteine residue, Cys241, has been found (1283). This free thiol group appears to play a part in the activity of C2 at a later stage: this activity is raised by oxidation, but destroyed by *p*-chloromercuribenzoate (1283, 1289). On the basis of homology, the following pattern of disulphide bonds can be proposed: in the first SCR domain, Cys4–Cys44 and Cys31–Cys64; in the second SCR domain, Cys69–Cys111 and Cys97–Cys124; in the third SCR domain, Cys131–Cys171 and Cys157–Cys184; and in the serine-proteinase domain, Cys472–Cys488, Cys618–Cys645 and Cys655–Cys685. In addition to this, there are four cysteine residues whose pairing cannot be inferred directly, but which may form S–S bridges according to the pattern Cys443–Cys561, Cys564–Cys580.

12.4. **Complement component C3**

C3 is a two-chained glycoprotein, consisting of an α chain and a β chain held together by disulphide bonds. Its carbohydrate content is 1.5%, and its molecular mass is estimated to be 185 kDa, distributed between the α chain (110 kDa) and the β chain (75 kDa) (1290, 1291). C3 is found in normal human serum at a concentration of 1–1.2 mg/ml ($\approx$6 μM).

The amino-acid sequence of human C3 and fragments thereof has been studied at the protein level (1292–1297), but only on the basis of the cDNA sequence has the complete amino-acid sequence of C3 been determined, for human (1298) and murine (1299, 1300) C3. Partial cDNA sequences have also been published for rabbit C3 (1301) and for *Xenopus* C3 (1302). The gene for human C3 has been located on chromosome 19 at position p13.2–13.3 (1303, 1304) and the genomic organisation of the α' chain has also been determined (1305). This part of the C3 gene covers 23–24 kbp, and has 24 exons. The exons vary in size from 52 bp to 213 bp, while the introns measure between 85 bp and about 4.2 kbp. The complete C3 gene is estimated to occupy some 35–40 kbp of genomic DNA.

C3 is synthesized as a single-chained polypeptide (prepro-C3) consisting of 1663 amino-acid residues (Appendix, Fig. A36), of which 22 are those of the signal peptide. The remaining 1641 residues (Fig. 84), from the N terminus, make up first the β chain (645 residues, corresponding to 71.1 kDa), followed by four arginine residues that are removed during processing, and finally the α chain (992 residues, corresponding to 112.8 kDa). There are three potential N-glycosylation sites: Asn63 in the β chain and Asn268 in the α chain are glycosylated, while it is not known whether the third, Asn946 in the α chain, is used (1297). The disulphide-bond pattern in C3 has been partially determined (1306–1308).

C3 contains an unusual post-translational modification: an internal thioester (1309–1311), formed between Cys339 and Gln342 in the α chain by deamidation of the glutamine. There is also a thioester in complement component C4 (see section 12.5) and in α_2-macroglobulin (1312). Indeed, there are yet further points of resemblance between these structures. The sequence of C3 is homologous to complement components C4 and C5, and also to α_2-macroglobulin (1313). However, C5 does not contain a thioester (section 14.1). In addition, it appears that C3 is partially phosphorylated (1314).

A common variant of C3 is C3F, which has been shown to be associated with heightened risk of certain diseases. (The normal form is termed C3S.)

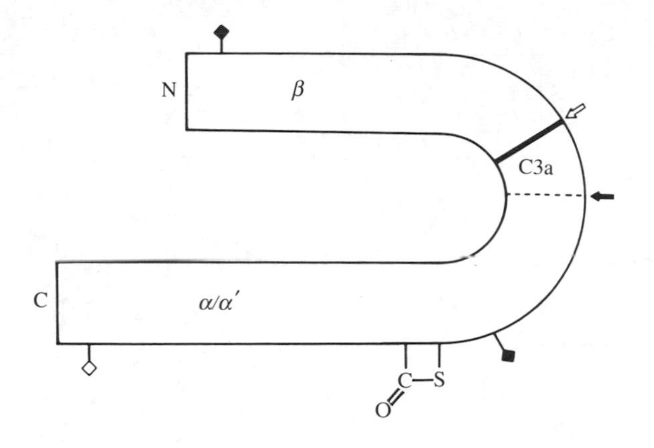

Fig. 84. The synthesis and structure of C3. → shows the position of cleavage by C3-convertases, ⇒ the positions of the bonds that are cleaved when C3 is formed from pro-C3, ◆ the N-bound carbohydrate and ◇ the potential attachment sites for N-bound carbohydrate.

The difference between the two variants lies at position 80 in proC3, corresponding to position 80 in C3 β chain, where C3S contains an Arg and C3F an Gly residue (1315).

When C3 is activated, 77 residues are removed from the N terminus of the α chain. This fragment is called C3a, and it has anaphylatoxic effects. The rest of the molecule is called C3b, and it consists of an α′ chain (915 residues) and a β chain (unaltered, with 645 residues).

12.5. Complement component C4

C4 is a three-chained glycoprotein that consists of an α, a β and a γ chain, all held together by disulphide bonds. Its carbohydrate content is about 7% (1316). The concentration of C4 in normal human serum is some 350–500 μg/ml (≈1.8–2.5 μM). The molecular mass is estimated as about 200 kDa on the basis of the molecular masses of the individual chains; 95 kDa is associated with the α chain, 75 kDa with the β chain and 30 kDa with the γ chain (1317).

At the level of protein, parts of human C4 have been sequenced (1318–1325), and the complete sequence has been deduced from that of the cDNA (1326). The complete sequence of the signal peptide has been revealed by genomic sequencing (1327), and the signal peptide for human prepro-C4 has 19 amino-acid residues. The gene for C4 lies on chromosome 6, position p21 (1287) and consists of two coupled loci, C4A and C4B. These both show a fairly high degree of polymorphism, with a total number of known alleles in excess of 40 (1328). The gene structure for C4 is only partially known. The gene that codes for C4A normally covers 22 kbp, but the gene

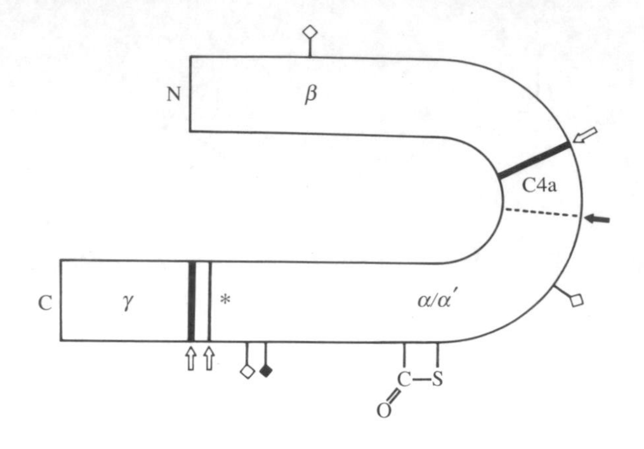

Fig. 85. The synthesis and structure of C4. → shows the position of cleavage by C$\overline{1}$s, ⇒ the positions of the bonds that are cleaved when C4 is formed from pro-C4, ◆ the N-bound carbohydrate, ◇ the potential attachment sites for N-bound carbohydrate and * the region where three tyrosine residues are sulphated.

that encodes C4B covers either 16 kbp or 22 kbp, presumably as a result of the respective presence or absence of a 6–7-kbp-long intron located 2–2.5 kbp from the 5′ end of the C4B gene (1328, 1329).

The two C4 loci are separated by approximately 10 kbp of genomic DNA (1287), but 1.5 kbp downstream (on the 3′ side) of each of the C4 loci there is a gene for cytochrome P-450 steroid 21-hydroxylase (1330, 1331). This means that the cytochrome P-450 steroid 21-hydroxylase A gene, which incidentally is a pseudogene, in fact lies between the two C4 loci. (See also section 13.2.)

Like C3, human C4 is synthesized as a single polypeptide chain, prepro-C4 (Fig. 85). It originally seemed that pro-C4 consisted of 1722 amino-acid residues (1326), but further cloning of cDNA (1327), and in particular protein-sequencing studies (1324, 1332), have shown that pro-C4 in fact has 1725 residues (Appendix, Fig. A37). At the N terminus comes the β chain (656 residues, corresponding to 71.6 kDa), followed by the sequence R-K-K-R and then the α chain (748 residues, corresponding to 82.1 kDa), after which come in turn first a 22-residue peptide, then the sequence R-R-R-R and finally the γ chain (291 residues, corresponding to 33.0 kDa) (1324, 1332). Both of the basic tetrapeptides are removed within the cell, while the last 22 amino-acid residues are probably removed from the α chain's C terminus at a later point.

The polymorphism between C4A and C4B can so far be attributed to 12 dimorphic amino-acid residues. One dimorphic position is in the β chain, one is in the γ chain and the other ten are in the α chain. The residue responsible in the β chain is number 399 (Ala/Val), and that in the γ chain is number 47 (Asp/Tyr). In the α chain, most of the dimorphic positions are found in the C-terminal part of the fragment known as C4d, which covers residues 278 to 657 in the α chain of C4. The residues concerned are numbers 394 (Asp/Gly), 441 (Pro/Leu), 442 (Cys/Ser), 445 (Leu/Ile), 446 (Asp/His), 497 (Asn/Ser), 522 (Ser/Thr), 528 (Val/Ala), 531 (Leu/Arg) and 607 (Ser/Ala). We shall later (section 12.6) see the significance of some of these dimorphic positions.

There are four potential glycosylation sites in C4. These are Asn207 in the β chain and Asn183, Asn649 and Asn712 in the α chain. Only the glycosylation of α-chain Asn649 has been demonstrated experimentally (1325). The disulphide-bond pattern in C4 has been determined partially (1308). C4 contains two further post-translational modifications. The murine C4 α chain is sulphated at one or more tyrosine residues (1333), and for human C4 there are strong indications that the α chain contains tyrosine-O-sulphate located at positions 738, 741 and 743 (1334). The C4 α chain also contains an internal thioester (1335), just as C3 and α_2-macroglobulin do. The thioester of C4 is made up of Cys331 and Gln334 in the α chain. C4 is, as mentioned in section 12.4, homologous to C3, C5 and α_2-macroglobulin (1313).

The activation of C4 is associated with the removal of the N-terminal 77 residues from the α chain. Like C3a, the fragment removed (called C4a) is anaphylatoxic. The activated molecule is termed C4b, and is made up of the rest of the α chain (α', 690 residues), together with the unmodified β and γ chains.

It should be mentioned that the situation in the mouse is somewhat different from that in humans. The complete amino-acid sequence of murine prepro-C4 has been determined by cDNA-sequencing (1336, 1337). Instead of containing two loci for C4, the mouse genome contains one locus for C4 and one for a protein called sex-limited protein (Slp) that does not possess complement activity. The amino-acid sequence of Slp has been determined on the basis of its cDNA (1338, 1339), and it has 95% identity with murine pro-C4. Slp contains a thioester (1340), but it is not activated by $C\overline{1}s$ in the way that C4 is (see section 12.6). Mutagenesis studies indicate that a deletion of amino-acid residues 8–10, on the C-terminal side of the potential $C\overline{1}s$ cleavage site in Slp, may be responsible for Slp's lack of complement activity (1341). The complete sequence of the murine C4 gene has been determined (1342). The gene covers some 16 kbp genomic DNA and possesses 41 exons between 52 bp and 241 bp in length. The introns vary in size from 75 bp to 1089 bp.

12.6. Further processes in classical complement activation

We have already described the activation of C1 to give $C\overline{1}$, for example following the binding of C1q to antigen–antibody complexes. This led to the formation of $C\overline{1}s$, the enzyme that furthers the activation of the complement system by splitting C2 and C4 (1343–1348). How does this take place?

First, C$\overline{1}$s cleaves the bond Arg77–Ala78 in the α chain of C4, releasing the anaphylatoxin C4a (Asn1–Arg77 from the α chain) to leave C4b ($\alpha'\beta\gamma$). The sulphation of the α chain, described earlier, appears to be important for the activation of C4 by C$\overline{1}$s, since non-sulphated or under-sulphated C4 is a poorer substrate for C$\overline{1}$s than correctly sulphated C4 (1349). The part of C$\overline{1}$s that interacts with C4 has been localised to the two SCR domains: a monoclonal antibody against an epitope in this region blocked the activation of C4 by C$\overline{1}$s (1350).

C4b is unstable, in that its cleavage by C$\overline{1}$s makes the internal thioester become reactive. This bond is particularly sensitive towards nucleophilic attack, and a reaction of this kind is the next step in complement activation (1351–1353). Nucleophiles that can react with the thioester in this step include amino groups, hydroxyl groups and water. If the thioester group of C4b reacts with a nucleophile attached to the surface of the activator that activated C1 to C$\overline{1}$ – either the antibody or a surface structure – then this surface becomes labelled with the C4b. If C4b's thioester reacts with water, then the C4b remains in the bulk fluid phase. In practice, it is only a minority of the C4b molecules that becomes bound to surfaces (Fig. 86).

In this connection it is especially interesting that the activated forms of the two isotypes of C4 (C4Ab and C4Bb, see section 12.5) have different preferences in respect of the nucleophile with which they react. C4Ab reacts preferentially with amino groups on peptide antigens, while C4Bb reacts preferentially with hydroxyl groups on carbohydrate antigens (1354–1356). It appears that this difference in reactivity can be ascribed to the polymorphic region around amino-acid residues 441–446 in the C4 α chain (1328), and it has been suggested that the residue at position 446 contributes specially to the specificity (1357). In this position, C4A has an aspartic acid residue, while C4B has histidine.

C4b (and C4) bind C2; the binding is dependent upon magnesium ions (1358, 1359), and there are indications that there is a C4b-binding site in the C2b part of C2 that is important for the initial binding of C2 to C4b (1359–1361). Note here that C2b is built up of three SCR domains. After it has been bound to C4b, C2 is cleaved by C$\overline{1}$s. The peptide bond cleaved is Arg223–Lys224, and this results in the formation of C2b from the N-terminal fragment and C2a from the C-terminal fragment. C2b dissociates from the complex, while C2a remains non-covalently bound to C4b. This complex is the complement system's classical 'C3-convertase', C4b2a (Fig. 87). It must be underlined that the presence of C4b is necessary in order for C$\overline{1}$s to cleave and activate C2 under physiological conditions (1362–1364). The formation of the C4bC2 complex, and thus the formation of C3-convertase, depend absolutely upon the C2 structure being intact. C4b does not form complexes either with C2a alone or with C2aC2b in a non-covalent complex.

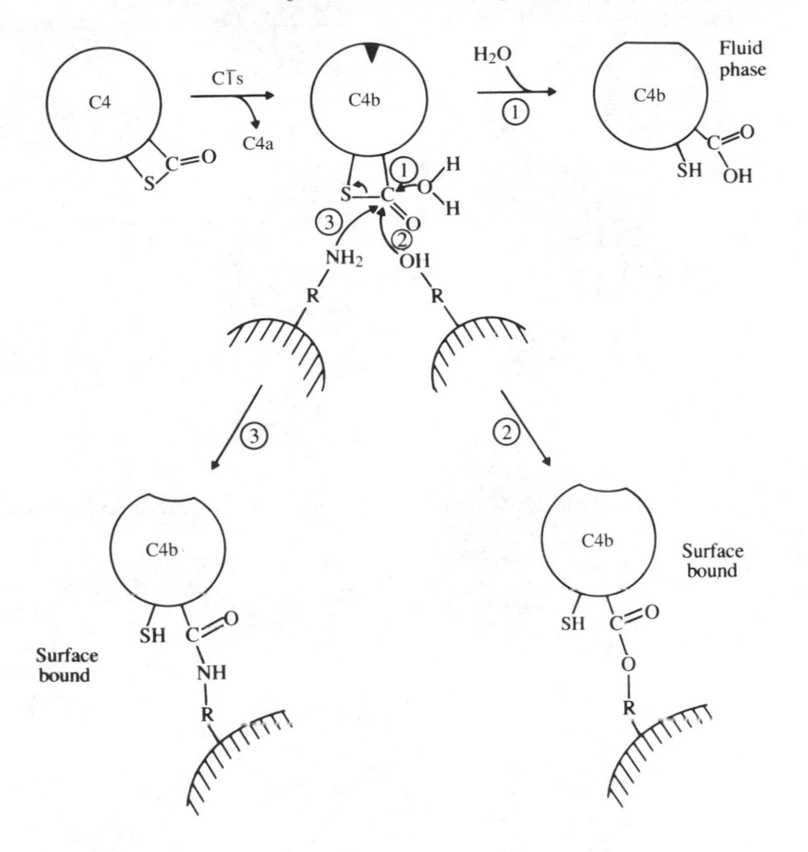

Fig. 86. Possible reaction pathways for C4b. After the cleavage of C4 by C$\bar{1}$s, the thioester reacts with a nucleophile. The reaction with H_2O (reaction 1) produces fluid-phase C4b, while the reaction with a surface-bound hydroxyl group (reaction 2) or amino group (reaction 3) produces surface-bound C4b. Reaction 1 is the predominant one; if this path is not followed, then the predominant reaction of C4Ab is reaction 3, and that of C4Bb is reaction 2.

As its name suggests, the task of C3-convertase is to cleave and thus to activate C3. The catalytically active part of the enzyme resides in C2a, which, as mentioned earlier, is a serine proteinase. C4b2a cleaves the bond Arg77–Ser78 in the α chain of C3, whereupon C3 releases the anaphylatoxin C3a (Ser1–Arg77 in the α chain) and becomes activated to C3b ($\alpha'\beta$).

Like C3b, C4b is unstable owing to the presence of a thioester. C3b can thus also react with nucleophiles (1352, 1353, 1365, 1366). The relative instability of C3b can be illustrated by comparing the half-lives of the thioester bonds in C3 and in C3b. In C3b this half-life is 60 μs and in C3 it is 230 hours (1367, 1368). Figure 88 shows the ways in which C3b can react; these are completely analogous to the reactions of C4b: some C3b is fixed to the surface of activators of the complement system, either in the vicinity of or in contact with C3-convertase, C4b2a. Within minutes, this changes the

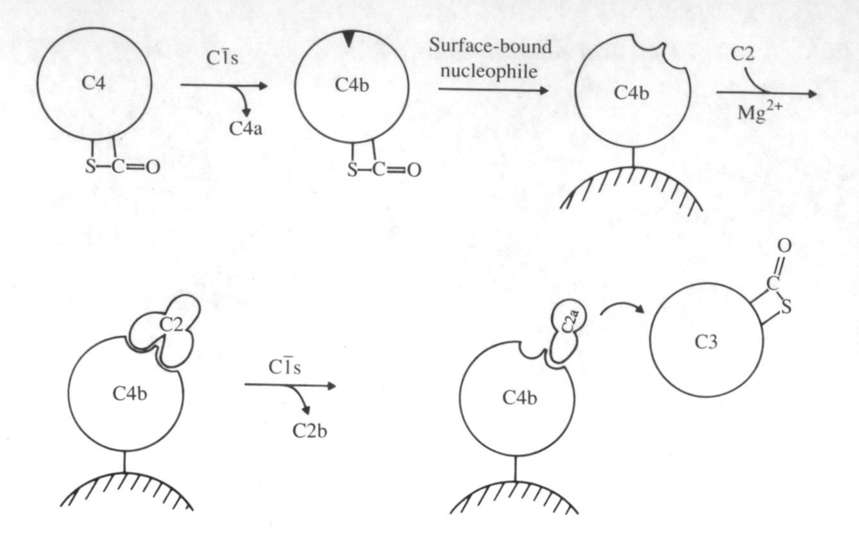

Fig. 87. The formation of the classical C3-convertase of the complement system.

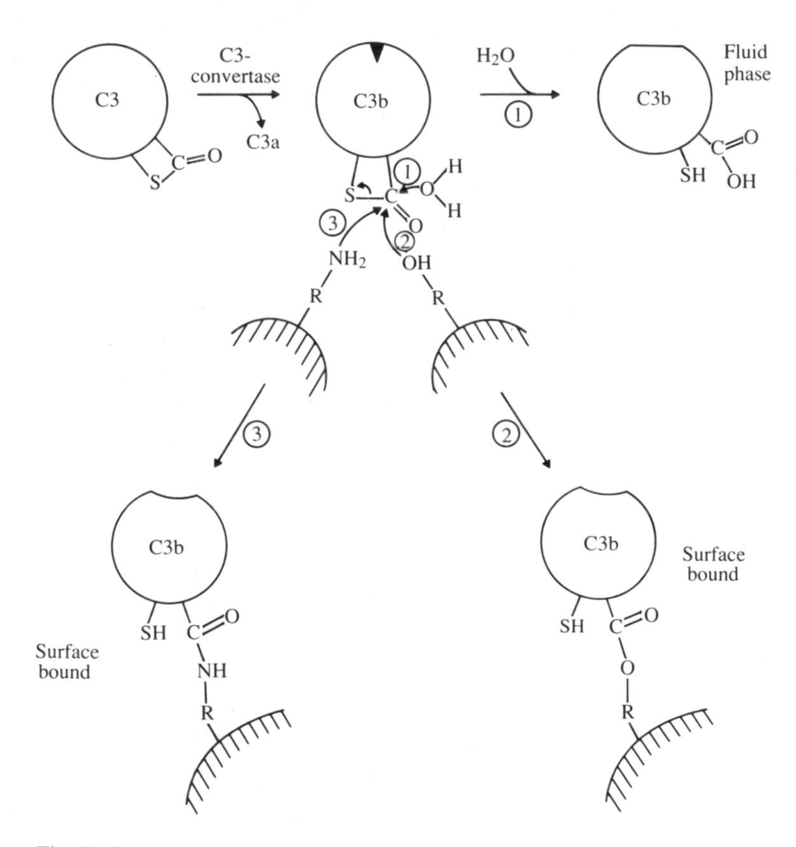

Fig. 88. Possible reaction pathways for C3b. After the cleavage of C3 by a C3-convertase, the thioester reacts with a nucleophile. The reaction with H_2O (reaction 1) produces fluid-phase C3b, while the reaction with a surface-bound nucleophile (reactions 2 and 3) produces surface-bound C3b.

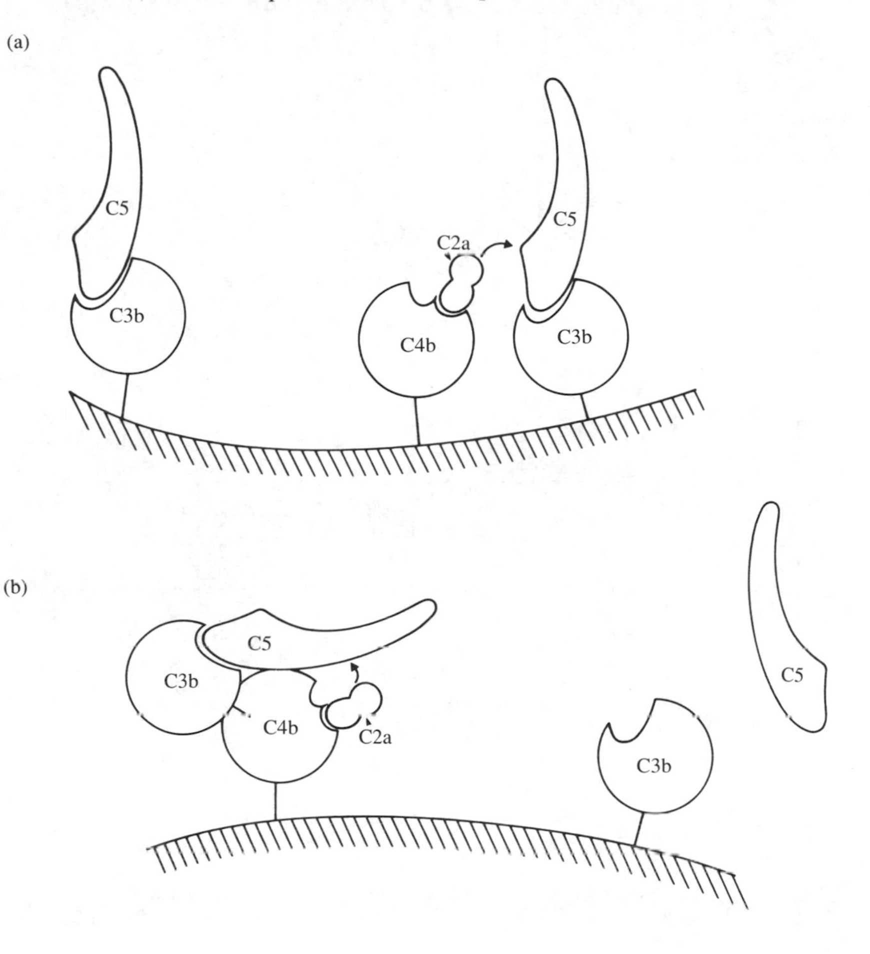

Fig. 89. The classical C5-convertase of the complement system. (a) An early model. (b) A recent model. Note that in the recent model it is the [C4b.C3b] dimer that binds C5, in contrast to the early model, in which only C3b does this.

specificity of the C3-convertase, and it turns into a C5-convertase – the complement system's classical 'C5-convertase', C4b2a3b. It is not yet clear how this change in specificity takes place, but this could be because the association between C5 and bound C3b changes the conformation of the C5 in such a way that proteolytic cleavage by C2a becomes possible (1369, 1370). This would mean that the change of C3-convertase to a C5-convertase was more a matter of activation of the substrate than of modulation of the enzyme's specificity.

It may therefore well be asked whether the random placing of C3b on the activating surface fulfils the conditions that must be met as regards the correct molecular orientation of the C3b-bound C5 in relation to C3-convertase. If C3b were placed at random on the surface, the orientation of

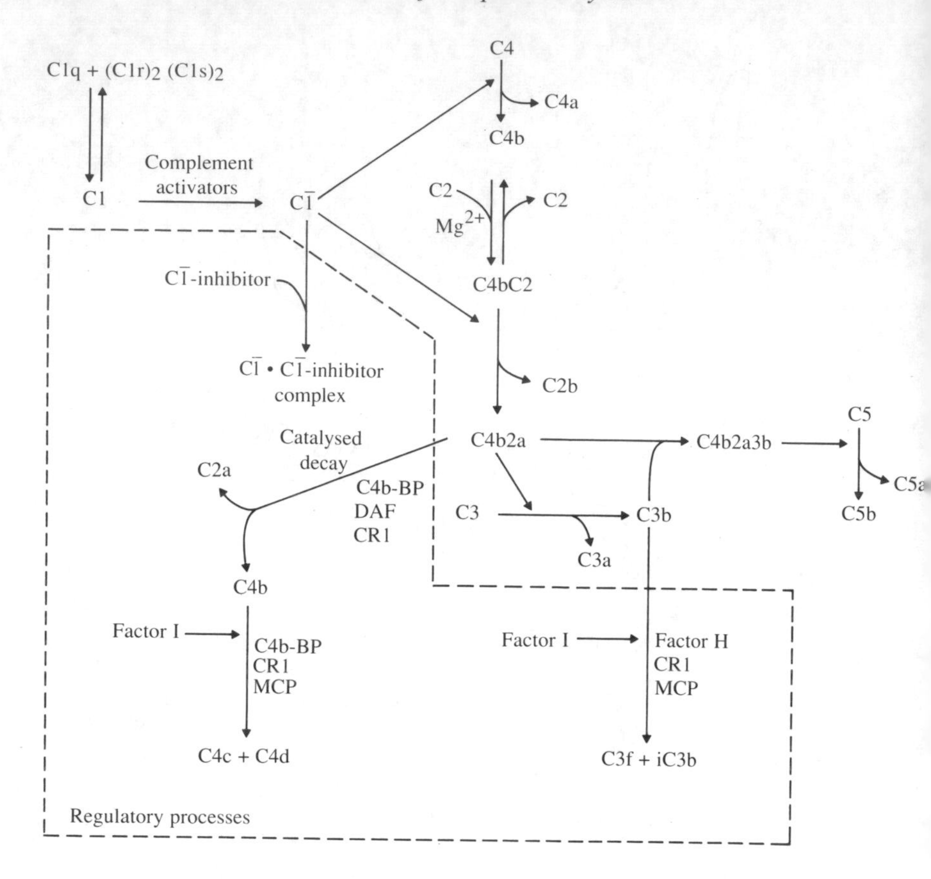

Fig. 90. The classical activation pathway of the complement system. Note that the figure also includes the regulatory processes, which have not yet been discussed (see Chapter 15).

the C3b-bound C5 in relation to the C3-convertase would be strongly dependent upon the dynamic properties of the surface and the distribution of the relevant nucleophilic acceptor groups. One possible way of overcoming this problem was discovered when it was demonstrated that C3b binds covalently to the α chain in C4b in C3-convertase (1371). In this way, C5-convertase comes to contain a covalent complex of C3b and surface-bound C4b (Fig. 89).

Figure 90 shows schematically the classical activation of the complement system.

13.

Alternative activation of the complement system

The second pathway for activation of the complement system is called alternative activation, and it is independent of antibody–antigen complexes. The proteins that are involved here are factor $\overline{D}$, factor B and C3.

13.1. Complement factor $\overline{D}$

Factor $\overline{D}$ is a single-chained serine proteinase with a molecular mass of 23 kDa (1372). The use of a bar over the D implies that (as is currently believed) this component of the complement system circulates in its active form (1373). In this respect it is unusual. Its concentration in normal human serum is about 1.5 μg/ml (≈65 nM).

The amino-acid sequence of human factor $\overline{D}$ has been determined at the protein level (1374, 1375). By the isolation and characterisation of a partial cDNA clone, it has been shown that factor $\overline{D}$ is synthesized in a *pro* form (1376), but the step in the process of synthesis and secretion at which factor D is activated to give factor $\overline{D}$ is not known. In an early study (1377), an inactive pro form of factor $\overline{D}$ in plasma was described, but this work has not yet been confirmed.

The two published amino-acid sequences of human factor $\overline{D}$ are not quite identical (Appendix, Fig. A38), consisting respectively of 218 and 222 residues, corresponding to a molecular mass just below 24 kDa. It is not yet clear whether factor $\overline{D}$ is a glycoprotein or not, and there is disagreement on this (1378–1380). In the amino-acid sequence there are no consensus sequences for N-bound carbohydrate, but of course O-bound carbohydrate can be present. If carbohydrate is present, then according to carbohydrate analysis it can only be glucose (1281), which would be very unusual.

Factor $\overline{D}$ shows sequence homology with other serine proteases, but it is of greatest interest that factor $\overline{D}$ and murine adipsin are homologous, with 61% identical amino-acid residues (1381, 1382). Adipsin is synthesized in – and secreted from – adipocytes, and purified adipsin synthesized from recombinant DNA can replace factor $\overline{D}$ in alternative complement activation *in vitro* (1381). It is therefore possible that adipsin is the murine counterpart to human factor $\overline{\mathrm{D}}$.

The disulphide bonds in factor $\overline{\mathrm{D}}$ have not been investigated by experiment, but on the basis of homology the S–S pattern in factor $\overline{\mathrm{D}}$ can be proposed to be Cys26–Cys42, Cys124–Cys184, Cys155–Cys165 and Cys174–Cys199. The catalytic apparatus of factor $\overline{\mathrm{D}}$ is presumed to involve His41, Asp88, Ser178.

Factor $\overline{\mathrm{D}}$ shows a remarkably high specificity for a single Arg–Lys bond in factor B, once the factor B has been complexed by binding to C3b (see section 13.3). This could be the explanation for the unusual fact that factor $\overline{\mathrm{D}}$ is allowed to circulate in the plasma as an active enzyme.

13.2. Complement factor B

Factor B is a single-chained glycoprotein, the zymogen of a serine proteinase. Its molecular mass is estimated to be 90–93 kDa, and the carbohydrate content to be 7–9% (1281, 1383). Its concentration in normal human serum is about 200 μg/ml (≈2 μM).

Human factor B has been partly sequenced at the levels of protein (633, 1384, 1385) and cDNA (633, 634), and combining these leads to the complete sequence (Appendix, Fig. A39). A part of the sequence for mouse factor B is also known (1386). The gene of human factor B has been characterised, but not all details are known. The gene for factor B consists of some 6 kbp of genomic DNA containing 18 exons (1387, 1388) residing on chromosome 6, position p21 (1287) and tightly linked to the C2 gene and also, less tightly, to the C4A and C4B genes. Factor B is synthesized with a 25-residue signal peptide. Figure 91 shows the overall structure of factor B.

Factor B consists of 739 amino-acid residues, corresponding to a molecular mass of 83 kDa, which, with the inclusion of the carbohydrate, rises to the value of 90–93 kDa estimated by physical methods. Factor B contains only N-bound carbohydrate, attached at Asn97, Asn117, Asn260 and Asn353. There are two common alleles of factor B, termed F and S, and the difference between these is found at position 7, which is either arginine (in the S allele) or glutamine (in the F allele) (1389).

Factor B is activated by the cleavage of the peptide bond between Arg234 and Lys235, and the two fragments are not connected by disulphide bonds. The light fragment (30 kDa), from the N terminus of factor B, is termed Ba and the heavy fragment (60 kDa), from the C terminus, is called Bb. Ba has 234 residues and, like C2b, it contains three internally homologous SCR domains (Fig. 92). Bb, with 505 residues, resembles C2a in that it appears to be a large serine proteinase, the additional size of which is due to an N-terminal region without counterpart among the 'classical' serine proteases (such as trypsin and chymotrypsin). This N-terminal

Fig. 91. The domain structure of factor B. → shows the position of cleavage by factor $\overline{D}$, ▼ the positions of the introns and ◆ the N-bound carbohydrate. vWFA is the part of factor B that is homologous to the A domains defined in von Willebrand factor.

```
 12 C S L E - G V E I K G G S F R - - L L Q E G Q A L E Y V C P S G F
 78 C P R P H D F E N G E Y W P R S P Y Y N V S D E I S F H C Y D G Y
140 C S N P - G I P I G T R K V G S - Q Y R L E D S V T Y H C S R G L

Y P Y P V Q T R T C R S T G S W S T L K T Q D Q K T V R K A E C 73
T L R G S A N R T C Q V N G R W S G - - - - - - - - - Q T A I C 133
T L R G S Q R R T C Q E G G S W S G - - - - - - - - - T E P S C 193
```

Fig. 92. Internally homologous SCR domains in factor B. Only the positions at which amino acids are the same in all three domains are outlined. Pairwise comparison of the three domains reveals the following degrees of amino-acid identity: domains 1 and 2, 21%; domains 1 and 3, 28%; domains 2 and 3, 42%.

region in Bb is homologous to the A domains as defined in von Willebrand factor. The catalytic apparatus in Bb is believed to be made up of His267, Asp317 and Ser440, corresponding to His501, Asp551 and Ser674 in factor B. Factor B is, as mentioned above (section 12.3), homologous to C2 over its entire sequence, with a degree of identity of 39% (Fig. 93).

The pattern of disulphide bridges in factor B has not been determined experimentally, but, as in C2, it possesses a free thiol group, Cys267 (1390). The following disulphide pattern has been suggested, on the basis of homology: in the first SCR domain, Cys12–Cys51 and Cys37–Cys73; in the second SCR domain, Cys78–Cys120 and Cys106–Cys133; in the third SCR domain, Cys140–Cys180 and Cys166–Cys193; and in the serine-proteinase domain, Cys486–Cys502, Cys631–Cys657 and Cys670–Cys700. There are four additional Cys residues that cannot be paired up directly, but for which the pattern Cys453–Cys571, Cys574–Cys590 seems reasonable.

The genes for C2 and factor B are, as mentioned above, located at position p21 of chromosome 6 together with the genes for C4A and C4B. The overall structure of this region is shown in Figure 94. The genes for C2 and for factor B are transcribed in the same direction, and they are only separated by the 421 bp that lie between the polyadenylation site of the C2 gene and the initiation site for the transcription of the factor B gene (1391). The C2 gene covers much more DNA (18 kbp) than does the factor B gene

```
Factor B  1 T P W S L A R P Q G S C S L E G V E I K G G S F R L L Q - - -
C2                          1 A P S C P Q N V N I S - G G T F T L S H G W A

G Q A L E Y V C P S G F Y P Y P V Q T R T C R S T G S W S T L K T Q D Q K
G S L L T Y S C P Q G L Y P S P A - S R L C K S S G Q W Q T P G A T R S L

V R K A E C R A I H C P R P H D F E N G E Y W P R S P Y Y N V S D E I S F
- - K A V C K P V R C P A P V S F E N G I Y T P R L G S Y P V G G N V S F

C Y D G Y T L R G S A N R T C Q V N G R W S G Q T A I C D N G A G Y C S N
C E D G F I L R G S P V R Q C R P N G M W D G E T A V C D N G A G H C P N

G I P I G T R K V G S Q Y R L E D S V T Y H C S R G L T L R G S Q R R T C
G I S L G A V R T G F R F G H G D K V R Y R C S S N L V L T G S S E R E C

E G G S W S G T E P S C Q D S F M Y D T P Q E V A E A F L S S L T E T I E
G N G V W S G T E P I C R Q P Y S Y D F P E D V A P A L G T S F S H M L G

V D A E D G H G P G E Q Q K R K I V L D P S G S M N I Y L V L D G S D S I
T N P T Q K T K - - E S L G R K I Q I Q R S G H L N L Y L L L D C S Q S V

A S N F T G A K K C L V N L I E K V A S Y G V K P R Y G L V T Y A T Y P K
E N D F L I F K E S A S L M V D R I F S F E I N V S V A I I T F A S E P G

W V K V S E A D S S N A D W V T K Q L N E I N Y E D H K L K S G T N T K K
L M S V L N D N S R D M T E V I S S L E N A N Y K D H E N G T G T N T Y A

L Q A V Y S M M S W P D D V - - - P P E G W N R T R H V I I L M T D G L H
L N S V Y L M M N N Q M R L L G M E T M A W Q E I R H A I I L L T D G K S

M G G D P I T V I D E I R D L L Y I G K D R K N P R E D Y L D V Y V F G V
M G G S P K T A V D H I R E I L N I N Q K - - - - R N D Y L D I Y A I G V

P L - V N Q V N I N A L A S K K D N E Q H V F K V K D M E N L E D V F Y Q
K L D V D W R E L N E L G S K K D G E R H A F I L Q D T K A L H Q V F E H

I D E S - Q S L S L C G M V W E H R K G T D Y H K Q P W Q A K I S V I R P
L D V S K L T D T I C G V G N M S A N A S D Q E R T P W H - - - V T I K P

K G H E S C M G A V V S E Y F V L T A A H C F T V D D K E H S I - K V S V
K S Q E T C R G A L I S D Q W V L T A A H C F R D G N - D H S L W R V N V

G E K - - - - R D L E I E V V L F H P N Y N I N G K K E A G I P E F Y D Y
D P K S Q W G K E L L I E K A V I S P G F D V F A K K N Q G I L E F Y G D

V A L I K L K N K L K Y G Q T I R P I C L P C T E G T T R A L R L P P T T
I A L L K L A Q K V K M S T H A R P I C L P C T M E A N L A L R R P Q G S

C Q Q Q K E E L L P A Q D I K A L F V S E E E K K L T R K E V Y I K N G D
C R D H E N E L L N K Q S V P A H F V A L N G S K L N - - - I N L K M G V

K G S C E R D A - Q Y A P G Y D K V K D I S E V V T P R F L C T G G V S P
W T S C A E V V S Q E K T M F P N L T D V R E V V T D Q F L C S G T Q E -

A D P N T C R G D S G G P L I V H K R S R F I Q V G V I S W G V V D V C K
- D E S P C K G E S G G A V F L E R R F R F F Q V G L V S W G L Y N P C L

Q K R Q K Q V - - - - - - - - P A H A R D F H I N L F Q V L P W L K E K L Q
S A D K N S R K R A P R S K V P P P R D F H I N L F R M Q P W L R Q H L G

E D L G F L 739
V - L N F L P I 732
```

Fig. 93. Comparison of the amino-acid sequences for factors B and C2. Only the positions at which amino acids are identical are outlined. The degree of identity between these is 39%.

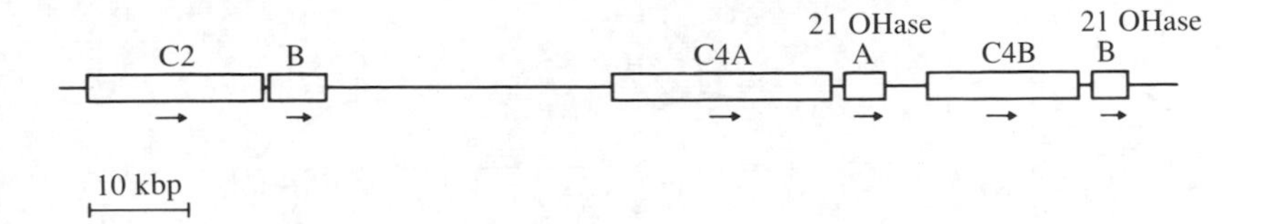

Fig. 94. The organisation of the genes for C2, factor B and C4. 21-OHase stands for the genes for cytochrome P450 steroid 21-hydroxylase. The arrows indicate the direction of transcription.

(6 kbp). The closeness of these two genes and the difference in their sizes are interesting in the context of their high degree of identity. There lies about 30 kbp of DNA between the factor B gene and the C4A gene, and about 10 kbp between the C4A and C4B genes. In the 10 kbp between the two C4 gene loci lies the gene for cytochrome P-450 steroid 21-hydroxylase A (which is in fact a pseudogene); it is placed about 1.5 kbp downstream (on the 3′ side) of the C4A gene. Similarly, the gene for cytochrome P-450 steroid 21-hydroxylase B is situated downstream from the C4B gene. These four genes are all transcribed in the same direction as the C2 and factor B genes. It further appears that a gene is transcribed from the other DNA strand, and therefore in the oppsite direction. It is not known what this gene encodes, but, unusually, its 3′ end overlaps the 3′ end of one of the cytochrome P-450 steroid 21-hydroxylase genes (1392).

13.3. Mechanism of alternative complement activation

We have just introduced factors $\overline{D}$ and B, and have earlier (section 12.4) made the acquaintance of C3. The only component still needed in the alternative activation of complement is the activator itself.

Activators of the complement system's alternative activation pathway are many and various, and they include lipopolysaccharides, viruses, bacteria, fungi, parasites, tumour cells and so forth, from which little can be concluded except that these activators have no common feature that might serve as a defining property. This point will be taken up in connection with the regulation of the complement system (Chapter 15, especially section 15.4).

The following sequence of events in the alternative activation of the complement system is generally accepted as corresponding roughly to reality.

The formation of C3b from C3 is a prerequisite for the initiation of alternative complement activation. Less clear is how this takes place, as C3b itself is a part of the alternative C3-convertase. One suggestion has been that there is a slow, perpetual enzymic formation of C3b – a 'ticking over' mechanism (1393, 1394). With certain modifications, the principle is still accepted today.

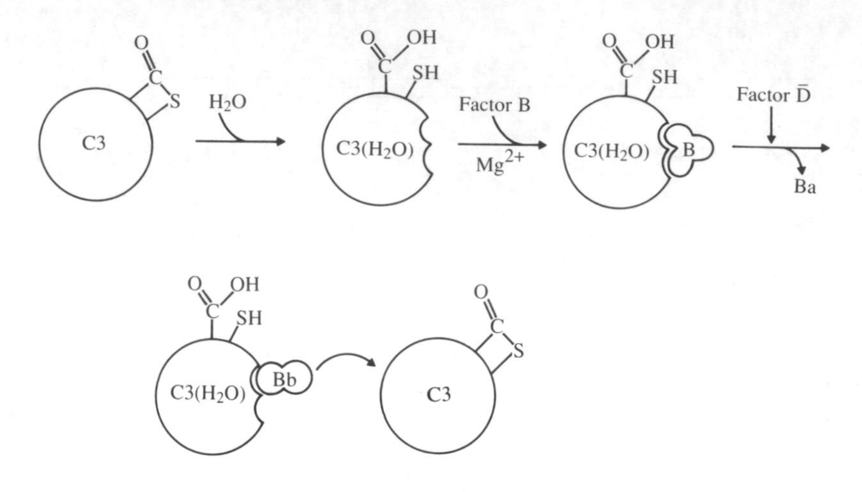

Fig. 95. The formation of fluid-phase C3-convertase in the alternative complement activation pathway.

C3 contains an unstable thioester group. The hydrolysis of this thioester normally occurs first after the cleavage of the α chain of C3, the associated release of the anaphylotoxic C3a and the formation of C3b with its unstable thioester. If the thioester in C3 is hydrolysed without prior cleavage of the α chain, then the product is called C3(H_2O); this product resembles C3b in some of its properties (1367, 1395). Studies with monoclonal antibodies support this, showing that the formation of C3(H_2O) is accompanied by the exposure of neo-epitopes that also appear when C3 is activated, but are never seen in native C3 (1396). The initial alternative C3-convertase molecules are formed around these C3(H_2O) molecules (Fig. 95) in the same way as the 'correct' alternative C3-convertases are formed around C3b.

Factor B binds to C3(H_2O); the binding is dependent upon magnesium ions. Only after this binding has occurred can factor $\overline{\text{D}}$ cleave the bond between Arg234 and Lys235 in factor B. The Ba fragment (amino-acid residues 1–234) dissociates from the complex, while the Bb fragment remains non-covalently associated with C3(H_2O), so that the C3-convertase initially formed has the structure [C3(H_2O).Bb]. Studies with 'bulk-phase' C3b have shown that factor B can bind to it independently of magnesium, and is activated by factor $\overline{\text{D}}$, so that there is not necessarily an absolute requirement for Mg^{2+} in the formation of this alternative C3-convertase (1397).

After this, [C3(H_2O).Bb] can activate C3 to give C3a and C3b, cleaving the Arg77–Ser78 bond in the α chain of C3. Again, this process occurs in the bulk phase; it is indeed only here that the enzyme is present. The fate of the C3b thus formed is either hydrolysis of the thioester by water or disruption

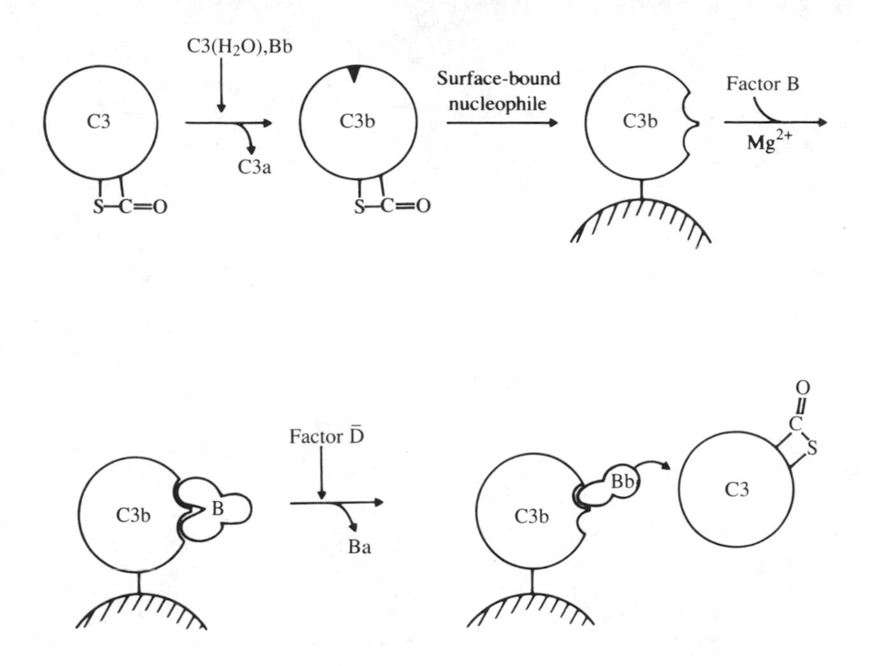

Fig. 96. The formation of surface-bound C3-convertase in the alternative complement activation pathway.

of the thioester by a surface-bound nucleophile such as a hydroxyl or amino group (Fig. 96). In the former case, the hydrolysed product cannot participate in the targeting processes leading towards localised complement activation. In the latter case, the C3b becomes bound to the surface and the complement activation process becomes localised.

The surface-bound C3b functions as a receptor for factor B, since C3b binds factor B reversibly in a manner dependent upon the presence of magnesium ions (1398–1400). After this, factor D̄ can cleave factor B into the two fragments Ba and Bb. This cleavage does not occur until factor B has bound to C3b; afterwards, the Ba leaves the complex. In this way, the alternative C3-convertase [C3b.Bb] becomes localised on the surface. Its activity reinforces the formation of C3b and thus the alternative complement activation pathway. This process is called amplification.

Details of the interaction between factor B and C3b are beginning to emerge (Fig. 97). On the basis of electron-microscopic studies, it is believed that factor B has three globular domains, of which Ba makes up one and Bb the other two (1401, 1402). It is believed that factor B binds to C3b by means of two contact points, one located in Ba and the other in Bb. Since isolated Ba binds to C3b without the involvement of magnesium ions, the Mg^{2+}-dependent binding site in factor B is taken to be in Bb (1403). This localisation is supported by the observation that the 33-kDa C-terminal

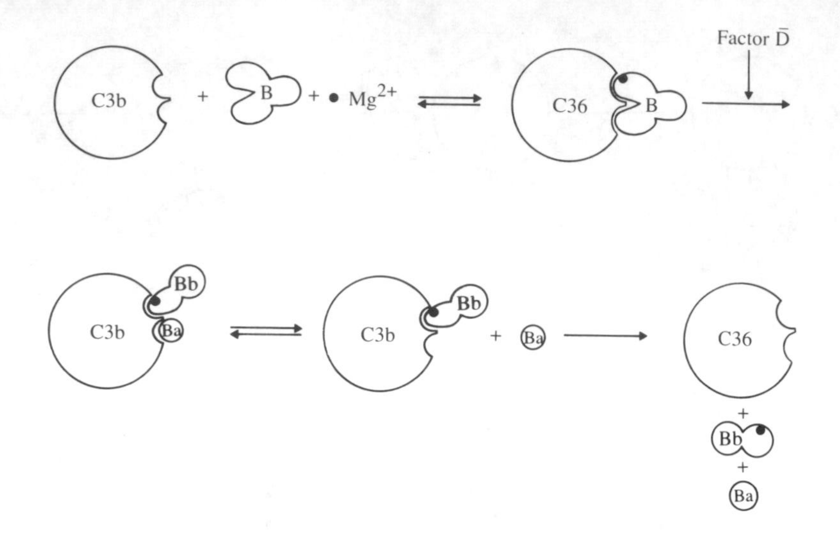

Fig. 97. The interaction between factor B and C3b. Factor B binds to C3b in a magnesium-dependent manner at two points of contact; the magnesium is important only for the conformation of the Bb fragment at its point of contact with C3b.

region of Bb shows strong and magnesium-dependent binding to C3b (1404). Further, there is only one domain in Bb that binds to C3b in the finished C3-convertase [C3b.Bb].

After factor D̄ has cleaved factor B to Ba and Bb, the affinity of the various protein components for each other is reduced. The Ba fragment leaves the complex, and this, along with the simultaneous exposure of the active serine residue in Bb, determines the appearance of active C3-convertase [C3b.Bb]. However, this is held together in a metastable manner, and if its components dissociate then they do so irreversibly. As we saw in the case of the classical C3-convertase, where activated C2 cannot form a complex with C4b, the alternative C3-convertase cannot be formed again from activated factor B and C3b.

The enhanced formation of C3b at the surface increases the deposition of C3b on the surface, and also results in a gradual change in the specificity of C3-convertase [C3b.Bb] to a C5-convertase [$(C3b)_2$.Bb] or [$(C3b)_n$.Bb].

Earlier, in connection with C5-convertase in classical complement activation (section 12.6), we discussed whether the random placing of C3b on the activating surface could fulfil the demand for correct orientation of the C3b-bound C5 with respect to the C3-convertase. The same arguments hold for the alternative activation of complement. If C3b is distributed randomly on the surface, the orientation of the C3b-bound C5 with respect to C3-convertase depends to a high degree on the properties of the surface.

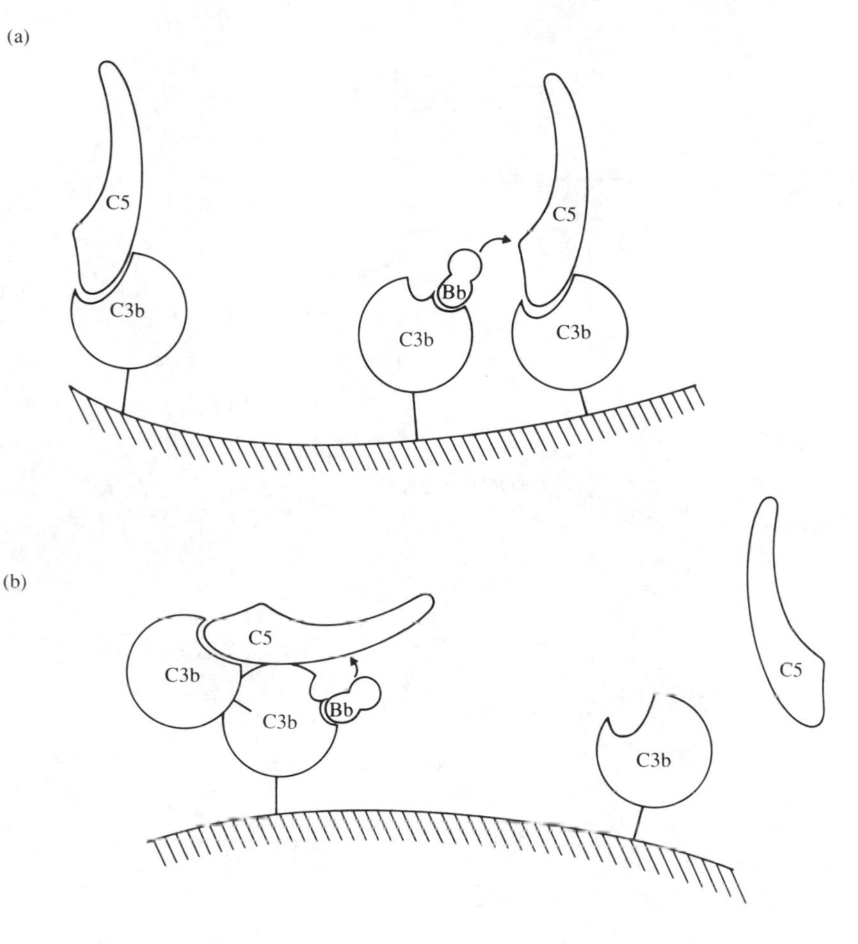

Fig. 98. The alternative C5-convertase of the complement system. (a) An early model. (b) A recent model. Note that in the recent model it is the C3b dimer that binds C5, in contrast to the early model, in which it is the C3b monomer that does this.

In the same way as it was shown that C3b can bind covalently to the α' chain of C4b in the classical C3-convertase (1371), it has been shown that C3b can bind covalently to the α' chain in C3b in the alternative C3-convertase (1405). In this way, the alternative C5-convertase comes to contain a covalent dimer of C3b (Fig. 98).

Figure 99 gives a schematic picture of alternative activation in the complement system.

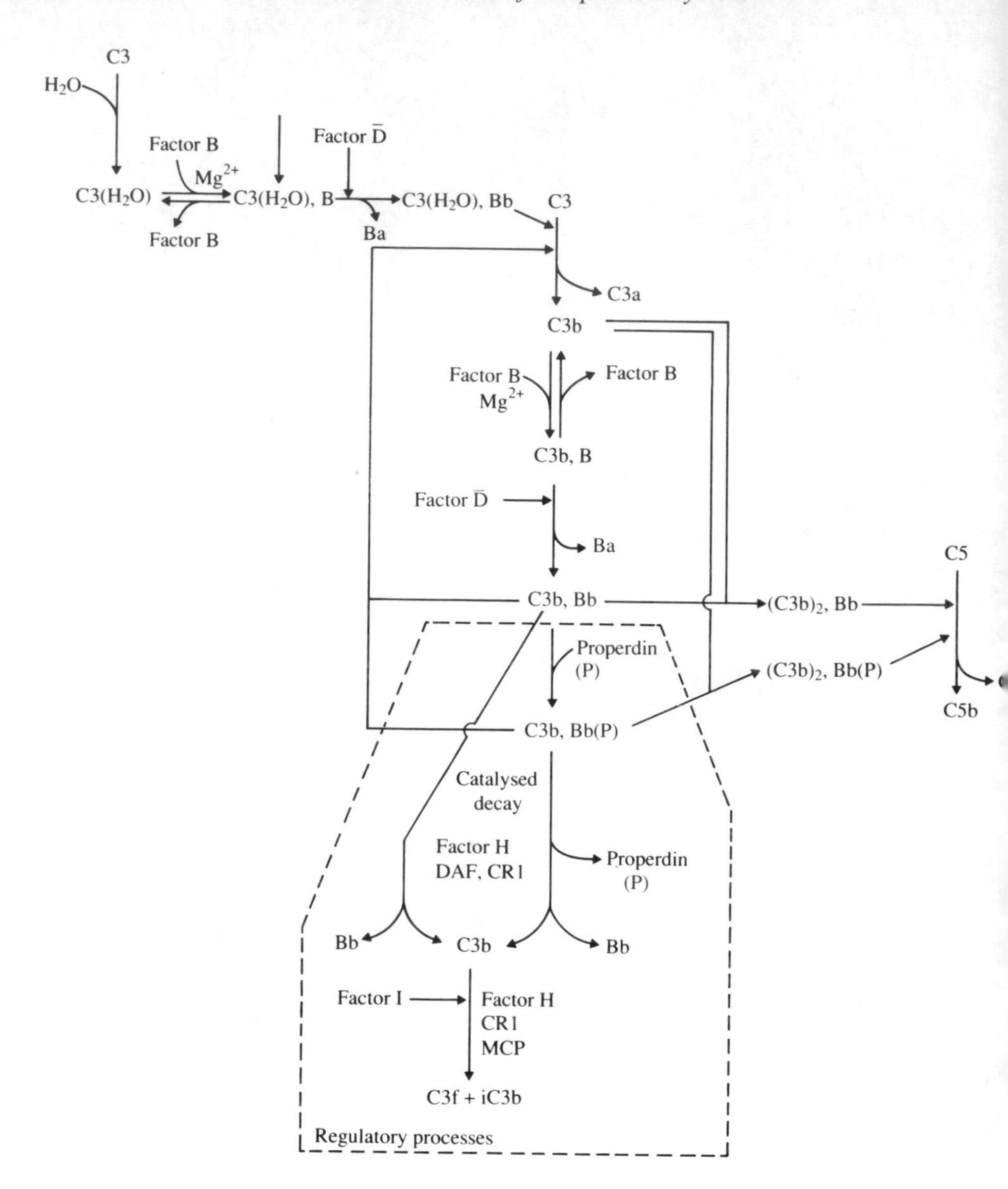

Fig. 99. The alternative activation pathway of the complement system. Note that the figure also includes the regulatory processes, which have not yet been discussed (see Chapter 15).

14.

The lytic complex (MAC)

The purpose of the complement system is, among other things, to bring about cytolysis by puncturing the outer barriers – for example, the plasma membranes – of foreign organisms. Their means of doing this consist of the membrane attack complex (MAC), a cytolytic protein complex whose formation is based upon the initiation of C5-convertase activity. The proteins that make up MAC are termed (somewhat uninspiringly) C5, C6, C7, C8 and C9.

14.1. Complement component C5

C5 is a two-chained glycoprotein made up of an α chain and a β chain connected by disulphide bridges. Its molecular mass is estimated as 191 kDa, of which 115 kDa is associated with the α chain and 75 kDa with the β chain (1406). Its carbohydrate content is about 3% (1281, 1407) and its concentration in normal human serum is approximately 70 μg/ml ($\approx$350 nM).

Only a little of the sequence of this protein has been determined at the level of protein, but the complete sequence of the anaphylatoxin C5a has been determined for human (1408), porcine (1409), bovine (1410) and rat (1411) C5. Most of the cDNA sequence of murine pro-C5 has been determined (1412), and sequencing of partial cDNA clones of human pro-C5 combined with sequencing of genomic DNA has given some 80% of the structure of this protein (1413, 1414) (Appendix, Fig. A40). The gene for human C5 is on chromosome 9 at position q32–34 (1414, 1415).

C5 is synthesized as a single-chained polypeptide with the orientation $\beta\alpha$ (Fig. 100). For human C5, the known amino-acid sequence corresponds to the 262 C-terminal residues in the β chain, plus an R-P-R-R peptide and the complete a chain of 999 residues, corresponding to 112.3 kDa. In the human C5 α chain there are four potential N-glycosylation sites: Asn64, Asn234, Asn438 and Asn953, of which only the first is known to be utilised (1408). The disulphide-bond pattern is not known.

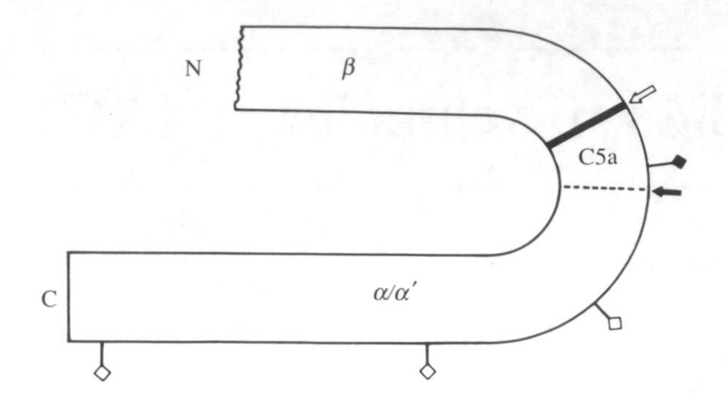

Fig. 100. The synthesis and structure of C5. → shows the position of cleavage by C5-convertases, ⇒ the positions of the bonds that are cleaved when C5 is formed from pro-C5, ◆ the N-bound carbohydrate and ◇ the potential attachment sites for N-bound carbohydrate. Note that the exact size of the β chain of C5 and the number of potential N-glycosylation sites in this chain are unknown.

As C5 has a blocked N-terminal residue, it is not known how much of the pro-C5 molecule and/or signal peptide the isolated murine cDNA clone covers. However, the known amino-acid sequence does at least cover the 634 C-terminal residues in the β chain from mouse C5, followed by a four-residue peptide (R-S-L-R) that connects this to the α chain, which has 1002 residues. In mouse pro-C5 there are five possible sites for N-glycosylation.

C5 is homologous in sequence to C3, C4 and α_2-macroglobulin, but in contrast to these it does not contain a thioester, a prediction already made on the basis of early functional studies (1351). The amino-acid sequence deduced for human C5 shows that C5 contains a serine and an alanine residue in the positions corresponding to the cysteine and glutamine residues that form the thioester in C3, C4 and α_2-macroglobulin.

The activation of C5 involves the proteolytic removal of 74 residues from the N terminus of the α chain with a C5-convertase, in the process of which the anaphylatoxin C5a, consisting of these residues, is released. The remaining part of the molecule is denoted C5b.

14.2. Complement components C6 and C7

Human C6 and C7 are both single-chained glycoproteins (1416). Estimates of their molecular mass vary from 105 kDa to 128 kDa for C6 and from 92 kDa to 121 kDa for C7, depending upon the method used (1416, 1417). The carbohydrate content is about 5%, the exact figure depending upon which molecular mass is taken; it is known that C7 contains more carbohydrate than C6: the absolute masses for the carbohydrate content are 4 kDa for C6 and 6 kDa for C7 (1417). Normal concentrations in human serum are roughly the same, about 60 μg/ml (≈600 nM).

A partial (1418) and two complete (655, 656) amino-acid sequences for human C6, including a 21-residue signal peptide, have been derived from cDNA-sequencing (Appendix, Fig. A41). The entire amino-acid sequence of human C7, including a 22-residue signal peptide, has likewise been deduced from the cDNA sequence (657) (Appendix, Fig. A42). There is homology ($\approx$33% identity) between C6 and C7 (Fig. 101), and the two are genetically tightly linked (1419), as is further documented by the finding that when one is missing, the other is too (1420). There is also some evidence of immunological similarity (1421).

Human C6 contains 913 amino-acid residues, corresponding to a calculated molecular mass of 102.4 kDa. There are two potential N-glycosylation sites in C6, at Asn303 and Asn834. Both are believed to be used (655). The molecular mass of glycosylated C6 is calculated to be 108 kDa. Position 761 is dimorphic, as amino-acid sequencing shows both Ser and Asp at this position (655). The disulphide-bond pattern in C6 is not known.

Human C7 consists of 821 amino-acid residues, corresponding to a molecular mass of 91.1 kDa. There are two potential N-glycosylation sites in C7, at Asn180 and Asn732, both of which are used (657), and the molecular mass of glycosylated C7 is calculated to be 97.1 kDa. Position 367 is dimorphic and can be either a serine or a threonine residue. The disulphide pattern of C7 is unknown.

C6 and C7 show homology with various domains that are known from other proteins (Figs. 102, 103). The most interesting aspect of this is that the 610 N-terminal residues in C6 show homology with the α and β chains of C8 and with C9, with respectively 30%, 29% and 24% identical amino acids (see Fig. 101). Likewise, the 525 N-terminal residues of C7 show homology with the α and β chains of C8 and with C9, with respectively 28%, 27% and 24% identical amino acids (see Fig. 101).

14.3. **Complement component C8**

C8 is a three-chained glycoprotein (1422) with a molecular mass of 150 kDa and assembled from an α chain (64 kDa), a β chain (64 kDa) and a γ chain (22 kDa). Unusually, the α chain and the γ chain are covalently bound by a disulphide bridge, while the β chain is bound to the other two by non-covalent forces only (1423). There are two linked genetic loci for C8 upon chromosome 1; one codes for the α and the γ chain and one for the β chain (1424, 1425). The existence of these two loci is also supported by the pattern of C8 deficiency, in which either the β chain or else both the α and the γ chains are missing (1426, 1427). In normal human serum, the concentration of C8 is about 55 μg/ml ($\approx$350 nM).

Only a part of the protein sequence of C8 has been determined directly, but the sequences of all three human C8 chains have been deduced from cDNA.

The α chain of C8 (1428) consists of 553 amino-acid residues (Appendix,

```
C6     1 C F C D H Y A W T Q W T S C S K T C N S G T Q S R H R Q I V V D K Y

Y Q E N F C E Q I C S K Q E T R E C N W Q R C P I N C L L G D F G P W S D C
C7                                       1 S S P V N C Q W D F Y A P W S E C
C8α                                1 A A T P A A V T C Q L S N W S E W T D C
C8β                            1 S V D V T L M P I D C E L S S W S S W T T C
C9         1 Q Y T T S Y D P E L T E S S G S A S H I D C R M S P W S E W S Q C

D P C I E K Q S K V R S V L R P S Q F G G Q P C T A P L V A F Q P C I P S K
N G C T K T Q T R R R S V A V Y G Q Y G G Q P C V G N A F E T Q S C E P T R
F P C Q D K K Y R H R S L L Q P N K F G G T I C S G D I W D Q A S C S S S T
D P C Q K K R Y R Y A Y L L Q P S Q F H G E P C N F S D K E V E D C V T N R
D P C L R Q M F R S R S I E V F G Q F N G K R C T D A V G D R R Q C V P T E

L C K I E E A D C K N K F R C - D S G R C I A R K L E C N G E N D C G D N S
G C P T E E G - C G E R F R C - F S G Q C I S K S L V C N G D S D C D E D S
T C V R - Q A Q C G Q D F Q C K E T G R C L K R H L V C N G D Q D C L D G S
P C G S - Q V R C - E G F V C A Q T G R C V N R R L L C N G D N D C G D Q S
P C E D A E D D C G N D F Q C - S T G R C I K M R L R C N G D N D C G D F S

D E R D - C G R T K - - - A V C T R K Y N P I P S V Q L M G N - G F H F L A
A D E D R C E D S E R - R P S C D I D - K P P P N I E L T G N - G Y N E L T
D E D D - C E D V R A I D E D C S Q - Y E P I P G S Q K A A L - G Y N I L T
D E A N - C - - - R R I Y K K C Q H E M D Q Y W G I G S L A S - G I N L F T
D E D D - C E S E P R - - P P C R D R V V E E S E L A R T A G Y G I N I L G

G E P R G E V L D N S F T G G I C K T V K S - - - - - - - - - - - - - - - -
G Q F R N R V I N T K S F G G Q C R K V F S - - - - - - - - - - - - - - - -
Q E D A Q S V Y D A S Y Y G G Q C E T V Y N G E W R E L R Y D S T C E R L Y
N S F E G P V L D H R Y Y A G G C S P H Y I L N T R - - - - - - - - - - - -
M D P L S T P F D N E F Y N G L C N R D R D - - - - - - - - - - - - - - - -

- S R T S N P Y R V P A N L E N V G F E V Q T A E D D L K T D F Y K D L T S
- G D G K D F Y R L S G N V L S Y T F Q V K - I N N D F N Y E F Y N S T W S
Y G D D E K Y F R K P Y N F L K Y H F E A L - A D T G I S S E F Y D N A N D
- - - - - - - F R K P Y N V E S Y T P Q T Q - G K Y E F I L K E Y E S Y S D
- G N T L T Y Y R R P W N V A S L I Y E T K - G E K N F R T E H Y E E Q I E

L G H N - - E N Q Q G S F S S Q G G S S F S V P I F Y - - - - - - - - - - -
Y V K H - - - - - - - - T S T E H T S S S R K R S F F - - - - - - - - - - -
L L S K V K K D K S D S F - - G V T I G I G P A G S P L - - - - - - - - - -
F E R N V T E K M A S K - - S G F S F G F K I P G I F - - - - - - - - - - -
A F K S I I Q E K T S N F N A A I S L K F T P T E T N K A E Q C C E E T A S

S S K R S E N I N H N S A F K Q A - - - - - I Q A S H K - K D S S F I R I H
R S S S S S S R S Y T S H - - - - - - - - - T N E I H K G K S Y Q L L V V E
L V G V G V S H S Q D T S F - - - - - L N E L N K Y - N E K K F I F T R I F
E L G I S S Q S D R G K H Y - - - - - I R R T K R F - S H T K S V F L H A R
S I S L H G K G S F R F S Y S K N E T Y Q L F L S Y S S K K E K M F L H V K

K V M K V L N F T T K A K D - L H L S D V F L K A L N H L P L E Y N S A L Y
N T V E V A Q F I N N N P E F L Q L A E P F W K E L S H L P S L Y D Y S A Y
T K V Q T A H F K M R K D D - I M L D E G M L Q S L M E L P D Q Y N Y G M Y
S D L E V A H Y K L K P R S - L M L H Y E F L Q R V K R L P L E Y S Y G E Y
G E I H L G R F V M R N R D - V V L T T T F V D D I K A L P T T Y E K G E Y
```

Fig. 101. Comparison of the amino-acid sequences for C6, C7, C8 α chain, C8 β chain and C9. Only the positions at which amino acids are the same in all protein chains shown are outlined. Pairwise comparison of the five chains reveals the following degrees of amino-acid identity: C6 and C7, 33%; C6 and C8 α, 30%; C6 and C8 β, 29%; C6 and C9, 29%; C7 and C8 α, 28%; C7 and C8 β, 27%; C7 and C9, 24%; C8 α and C8 β, 34%; C8 α and C9, 27%; C8 β and C9, 25%.

```
S R I F D D F G T H Y F T S G S L G G V Y D L L Y Q F S S E E L K N S G L T
R R L I D Q Y G T H Y L Q S G S L G G E Y R V L F Y V D S E K L K Q N D F N
A K F I N D Y G T H Y I T S G S M G G I Y E Y I L V I D K A K M E S L G I T
R D L F R D F G T H Y I T E A V L G G I Y E Y T L V M N K E A M E R G D Y T
F A F L E T Y G T H Y S S S G S L G G L Y E L I Y V L D K A S M K R K G V E

E E E A K H C V R I E T K K R V L F A K K T K V - - - - E H R C T T N K L S
S V E E K K C K S S G W H F V V K F S S - - - - - - - - - - H G C K E L E N A
S R D I T T C F G G S L G I Q Y E - D K I N V G G G L S G D H C K K F G G G
L N N V H A C A K N D F K I G G A I E E V Y V S L G V S V G K C R G I L N E
L K D I K R C L G Y H L D V S L A F S E I S V G A E F N K D D C V K R G E G

E K H E G S F I Q G A E K S I S L I R G G R S E Y G A A L A W - E K G S S G
L K A A S G T Q N N V L R G E P F I R G G G A G F I S G L S Y L E L D N P A
K T E R A R K A M A V E D I I S R V R G G S S G W S G G L A - Q N R S T I T
I K D R N K R D T M V E D L V V L V R G G A S E H I T T L A Y Q E L P T A D
R A V N I T S E N L I D D V V S L I R G G T R K Y A F E L K E K L L R G T V

L E E K T F S E W L E S V K E N P A V I D F E L A P I V D L V - - R N I P C
G N K R R Y S A W A E S V T N L P Q V I K Q K L T P L Y E L V - - K E V P C
- - - - - Y R S W G R S L K Y N P V V I D F E M Q P I H E C C G T Q A W A -
L - - - - M Q E W G D A V Q Y N P A I I K V K V E P L Y E L V T A T D F A Y
I D V T D F V N W A S S I N D A P V L I S Q K L S P I Y N L V P V K - M K N

A V T K R N N L R K A L Q E Y A A K F D P C Q C A P C P N N G R P T L S G T
A S V K K L Y L K W A L E E Y L D E F D P C H C R P C Q N G G L A T V E G T
S G G Q R Q N L R R A L D Q Y L M E F N A C R C G P C F N N G V P I L E G T
S S T V R Q N M K Q A L E E F Q K E V S S C H C A P C Q G N G V P V L K G S
A H L K K Q N L E R A I E D Y I N E F S V R K C H T C Q N G G T V I L M D G

E C L C V C Q S G T Y G E N C E K Q S - P D Y K S N A V D G Q W C C W S S W
H C L C H C K P Y T F G A A C E Q G V L V G N Q A G G V D G G W S C W S S W
S C R C Q C R L G S L G A A C E Q T Q - - - T E G A K A D G S W S C W S S W
R C D C I C P V G S Q G L A C E V S Y R - - - K N T P I D G K W N C W S N W
K C L C A C P F K F E G I A C E I S K Q K I S E G L P A L E F P N E K 538

S T C D A T Y K R S R T R E C N N P A P Q R G G K R C E G E K R Q E E D C T
S P C V Q G - K K T R S R E C N N P P P S G G G R S C V G E T T E S T Q C -
S V C R A G - I Q E R R R E C D N P A P Q N G G A S C P G R K V Q T Q A C 553
S S C S G R - R K T R Q R Q C N N P P P Q N G G S P C S G P A S E T L D C S537

F S I M E N N G Q P C I N D D E E M K E V D L P E - - - - - - - - I E A D S G
- - - - - - - - - - - - - - E D E E L E H L R L L E P H C F P L S L V P T E F

C P Q P V P P E N G F I R N E K Q L Y L V G E D V E I S C L T G F E T V G Y
C P S P P A L K D G F V Q D E G P M F P V G K N V V Y T C N E G Y S L I G N

Q Y F R C L P D G T W R Q G D V E C Q R T E C I K P V V Q E V L T I T P F Q
P V A R C G E D L R W L V G E M H C Q K I A C V L P V L M D G I Q S H P Q K

R L Y R I G E S I E L T C P K G F V V A G P S R Y T C Q G N - S W T P P I S
P F Y T V G E K V T V S C S G G M S L E G P S A F L C G S S L K W S P E M K

N S L T C E K D T - - L T K L K G H C Q L G Q K Q S G S E C I C M S P E E D
N A - R C V Q K E N P L T Q A V P K C Q R W E K L Q N S R C V C K M P Y E -

C S H H S E D L C V F D T D S N D Y F T S P A C K F L A E K C L N N Q Q L H
C G P - S L D V C A Q D E R S K R I L P L T V C K M H V L H C - Q G R N Y T
```

```
F L H I G S C Q D G R Q L E W G L E R T R L S S N S T K K E S C G Y D T C Y
L T G R D S C T L P A S A E - - - - - - - - - - - - - - - - K A C G A - - C P

D W E K C S A S T S K C V C L L P P Q C F K G G N Q L Y C V K M G S S T S E
L W G K C D A E S S K C V C R E A S E C E E E G - F S I C V E V - - N G K E

K T L N I C E V G T I R C A N R K M E I L H P G K C L A 913
Q T M S E C E A G A L R C R G Q S I S V T S I R P C A A E T Q 821
```

Fig. 101. (*cont.*)

Fig. A43), corresponding to a calculated molecular mass of 61.5 kDa without carbohydrate. The α chain contains carbohydrate attached to Asn13 but not to the other potential N-glycosylation site Asn407 or to any O-glycosylation site (1428). The C8 α chain is synthesized with a 30-residue extension at the N terminus. The first 20 residues are believed to be a signal peptide and the remaining 10 to be a propeptide that is also removed before the α chain becomes part of mature C8. The C8 α chain is homologous to C6, C7, the C8 β chain and C9 (see Fig. 101), with respective degrees of amino-acid identity of 30%, 28%, 34% and 27%. Locally, these values can be much higher. The homologous stretches include two domains showing homology with type I domains in thrombospondin, a domain showing homology with a well-defined domain in the LDL receptor (1429) and a growth-factor domain (Figs. 102, 103). C8 α chain cross-reacts immunologically with an antibody raised against a part of the LDL-receptor-like region in C9 (1430).

The β chain of C8 consists of either 537 (1431) or 536 (1432) amino-acid residues, giving a calculated molecular mass of 60.7 kDa (1432) (Appendix, Fig. A44). The difference is due to the existence of a stretch of either nine (1431) or eight (1432) residues in the two published sequences that bear no similarity to each other. These are residues 383–391 and 383–390 respectively. In addition, these two published sequences differ at position 63, containing either an arginine (1431) or a glycine (1432). There are three possible attachment sites for N-bound carbohydrate, at Asn47, Asn189 and Asn499/498, but which are used is not known. Like the C8 α chain, the β chain contains no O-bound carbohydrate (1432). It is not clear how the signal peptide for the β chain is organised, but the deduced amino-acid sequence on the 5′ side of the N terminus is clearly not a normal signal peptide, and it may be a propeptide, as was found on the α chain. The disulphide pattern of the C8 β chain is unknown. As mentioned earlier (above and section 14.2), the C8 β chain is homologous to C6 (29%), C7 (27%), the C8 α chain (34%) and C9 (25%) (see Fig. 101). The domain structure of the C8 β chain corresponds exactly to that of the C8 α chain (see Figs. 102, 103).

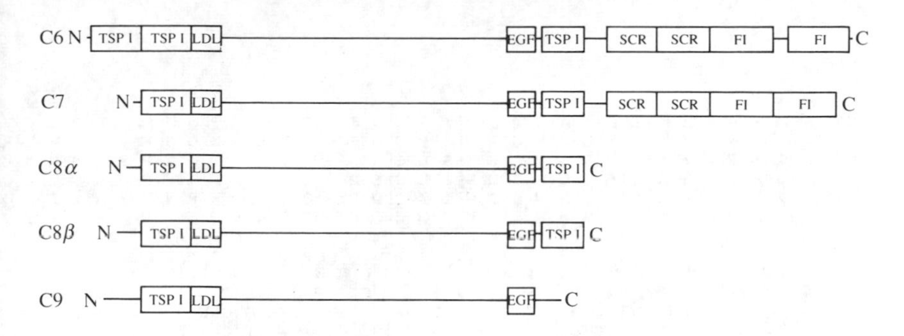

Fig. 102. Overall domain structure of C6, C7, C8 α chain, C8 β chain and C9. (TSP I, thrombospondin type I domains; LDL, LDL receptor domains; SCR, SCR domains; EGF, growth-factor domains; FI, factor I-homologous domains.) (Based upon reference 656.)

The γ chain of C8 was once assumed to be encoded by the same gene as that encoding C8 α chain, on account of their very tight genetic linkage. However, analysis of the cDNA sequence of C8 α chain showed that this is not the case, and this was confirmed by the cloning of the independent γ-chain cDNA (1433, 1434). This means that C8 is encoded by three genes in all. The C8 γ chain consists of 182 amino-acid residues (Appendix, Fig. A45), corresponding to a theoretical molecular mass of 20.3 kDa. The signal peptide is 20 residues long, and, after removal of this peptide, the N-terminal residue Gln1 is usually cyclised to give pyroglutamic acid, resulting in a blocked N terminus for the γ chain. C8 γ chain contains only three cysteines, of which Cys76 and Cys168 form an internal disulphide bond, while Cys40 therefore participates in the disulphide bridge between the α and the γ chains (1434). There is no carbohydrate attached to the C8 γ chain. The amino-acid sequence of C8 γ chain is homologous (25% identity) to the almost identical proteins α_1-microglobulin and protein HC (1434–1436), but the significance of this homology is not clear at present.

The details of the assembly of C8 from its three separate chains are not known. However, studies of rat hepatocytes in culture have shown that the C8 α–γ dimer is synthesised more rapidly than the C8 β chain (1437). Other investigations have shown that normal human serum contains free C8 α–γ dimers but no free β chain. These results suggest that it is the quantity of available C8 β chain that determines the formation of functional C8 (1438).

14.4. Complement component C9

C9 is a single-chained glycoprotein with a physically estimated molecular mass of 71 kDa and a carbohydrate content of about 8% (1439), and its

(a)

N-terminal domains

```
C6    1 - - C F C D H Y A W T Q W T S C S K T C N S G T Q S R H R Q I V V
C6   60 - - - N C L L G D F G P W S D C D P - C I E - K Q S K V R S V L R
C7    2 S P V N C Q W D F Y A P W S E C N G - C T K - T Q T R R R S V A V
C8α   5 A A V T C Q L S N W S E W T D C F P - C Q D - K K Y R H R S L L Q
C8β   7 M P I D C E L S S W S S W T T C D P - C Q K - K R Y R Y A Y L L Q
C9   18 S H I D C R M S P W S E W S Q C D P - C L R - Q M F R S R S I E V

D - K Y Y Q E N F C E Q I C S K Q E T R E C N W Q R - C P I - - 59
P - S Q F G G Q P C T A P - L V - A F Q P C I P S K L C K I E E 116
Y - G Q Y G G Q P C V G N - A F - E T Q S C E P T R G C P T E E 61
P - N K F G G T I C S G D - I W - D Q A S C S S S T T C V R - Q 63
P - S Q F H G E P C N F S - D K - E V E D C V T N R P C G S - Q 65
F - G Q F N G K R C T D A - V G - D R R Q C V P T E P C E D A E 77
```

C-terminal domains

```
C6   541 N A V D G Q W G C W S S W S T C D A T Y K R S R T R E C N N P A P
C7   475 G G V D G G W S C W S S W S P C V Q - G K K T R S R E C N N P P P
C8α  505 A K A D G S W S C W S S W S V C R A - G I Q E R R R E C D N P A P
C8β  488 T P I D G K W N C W S N W S S C S G - R R K T R Q R Q C N N P P P

Q R G G K R C E G E K R Q E E D C 590
S G G G R S C V G E T T E S T Q C 523
Q N G G A S C P G R K V Q T Q A C 553
Q N G G S P C S G P A S E T L D C 536
```

Fig. 103. Comparison of homologous domains in C6, C7, C8 α chain, C8 β chain and C9. Only the positions at which amino acids are the same in all domains are outlined.

(a) Thrombospondin type I domains, subdivided according to their appearance of N-terminal (NT) or C-terminal (CT). Pairwise comparison of the domains reveals the following degrees of amino-acid identity: C6NT-1 and C6NT-2, 19%; C6NT-1 and C7NT, 27%; C6NT-1 and C8αNT, 20%; C6NT-1 and C8βNT, 22%; C6NT-1 and C9, 20%; C6NT-2 and C7NT, 37%; C6NT-2 and C8αNT, 33%; C6NT-2 and C8βNT, 32%; C6NT-2 and C9, 32%; C7NT and C8αNT, 27%; C7NT and C8βNT, 25%; C7NT and C9, 30%; C8αNT and C8βNT, 41%; C8αNT and C9, 30%; C8βNT and C9, 34%; C6CT and C7CT, 52%; C6CT and C8αCT, 53%; C6CT and C8βCT, 47%; C7CT and C8αCT, 51%; C7CT and C8βCT, 51%; C8αCT and C8βCT, 45%; C6NT-1 and C6CT, 20%; C6NT-2 and C6CT, 28%; C7NT and C7CT, 31%; C8αNT and C8αCT, 29%; C8βNT and C8βCT, 33%. (b) SCR domains in C6 and C7. Pairwise comparison of the domains reveals the following degrees of amino-acid identity: C6SCR-1 and C6SCR-2, 30%; C6SCR-1 and C7SCR-1, 30%; C6SCR-1 and C7SCR-2, 29%; C6SCR-2 and C7SCR-1, 18%; C6SCR-2 and C7SCR-2, 29%; C7SCR-1 and C7SCR-2, 29%. (c) Factor I-homologous domains in C6 and C7. Pairwise comparison of the domains reveals the following degrees of amino-acid identity: C6FI-1 and C6FI-2, 27%; C6FI-1 and C7FI-1, 29%; C6FI-1 and C7FI-2, 23%; C6FI-2 and C7FI-1, 23%; C6FI-2 and C7FI-2, 36%; C7FI-1 and C7FI-2, 20%.

(b)

```
C6   623 C P Q P V P P E N G F I R N E K Q L Y L V G E D V E I S C L T G F
C6   683 C I K P V V Q E V L T I T P F Q R L Y R I G E S I E L T C P K G F
C7   549 C P S P P A L K D G F V Q D E G P M F P V G K N V V Y T C N E G Y
C7   609 C V L P V L M D G I Q S H P Q K P F Y T V G E K V T V S C S G G M

E T V G Y Q Y F R C L P D G T W R Q G D - - - V E C 678
V V A G P S R Y T C Q G N - S W T P P I S N S L T C 740
S L I G N P V A R C G E D L R W L V G E - - - M H C 604
S L E G P S A F L C G S S L K W S P E M K N A - R C 666
```

(c)

```
C6   745 - L T K L K G H C Q L G Q K Q S G - - S E C I C M S P E E D C S H
C6   838 K E S C G Y D T C Y D W E K C S A S T S K C V C L L P P Q - C F K
C7   672 P L T Q A V P K C Q R W E K L Q N - - S R C V C K M P Y E - C G P
C7   749 K - A C G A - - C P L W G K C D A E S S K C V C R E A S E - C E E

H S E D L - C V F D T D S N D Y F T S P A C K F L A E K C L N N Q Q L H F
G G N Q L Y C V K M G S S T S E K T L N I C E V G T I R C A N - R K M E I
- S L D V - C A Q D E R S K R I L P L T V C K M H V L H C - Q G R N Y T L
E G F S I - C V E V - - N G K E Q T M S E C E A G A L R C R G - Q S I S V

L H I G S C Q D 818
L H P G K C L A 913
T G R D S C T L 743
T S I R P C A A 818
```

concentration in normal human serum is approximately 60 μg/ml (≈850 nM).

The amino-acid sequence of human C9, including a signal peptide at least 21 residues in length, has been derived from cDNA-sequencing (1440, 1441), with minor corrections from sequencing of the C9 gene (1442). The sequencing of genomic DNA added a Val residue in position 293; human C9 therefore consists of 538 amino-acid residues (Appendix, Fig. A46), which give a calculated molecular mass of 60.8 kDa, excluding carbohydrate. There are two potential N-glycosylation sites, Asn256 and Asn394 (1441–1443), and both are used (1441). (The first published cDNA sequence (1440) placed a proline residue at position 396, destroying the consensus sequence involving Asn394. It is not known whether this is due to genetic variation (1443) or a trivial sequencing error.) C9 appears to have more carbohydrate than that which is attached at these two sites, since both of the two thrombin fragments, C9a and C9b, which can be generated *in vitro*, contain carbohydrate (1444). The N-bound carbohydrate in C9 is attached to C9b, which embraces residues 245 to 538 of C9. Carbohydrate analysis (1443) shows that C9a does not contain N-acetylglucosamine (typical of

N-glycosylation; Chapter 19), and its amino-acid sequence (residues 1–244) contains no N-glycosylation sites. The carbohydrate attached to C9a is therefore probably O-bound.

The disulphide-bond pattern of C9 has not been determined. It appears that the N terminus of C9 is blocked by the cyclisation of an N-terminal glutamine residue to give pyroglutamic acid.

The gene for human C9, covering more than 80 kbp of genomic DNA, has been partially characterised. It possesses at least 11 exons (1442) and lies at position p13 of chromosome 5 (1445, 1446). The studies of the genomic DNA for C9 also settled a disagreement between two published cDNA sequences concerning position 22 (Arg (1440) or Cys (1441)) in favour of cysteine, bringing this sequence into line with the C9 sequences of mouse and trout (1447).

As we have seen, C9 is homologous to C6, to C7 and to the C8 α and β chains, with respectively 24%, 24%, 27% and 25% identical amino-acid residues (see Fig. 101). The domain structure of C9 is shown in Figures 102 and 103.

14.5. **Formation of MAC**

The actual formation of the complement system's lytic complex (MAC) presupposes just one proteolytic cleavage; the subsequent steps are more or less spontaneous (Fig. 104).

The proteolytic cleavage needed is carried out by the membrane-bound C5-convertase, whether 'classical' or 'alternative' in origin, and it takes place only if the C5 is bound to C3b in C5-convertase (1369). The bond in C5 cleaved is the bond Arg74–Leu75 in the N-terminal region of the α chain; after the cleavage, the anaphylatoxic polypeptide C5a (Thr1–Arg74) is released from the rest of the molecule, C5b. C5b is metastable, in that it contains a metastable binding site for C6. The binding site for C6 is probably hydrophobic, exposed as a consequence of the cleavage and subsequent conformational changes. The binding site for C5b in C6 has been localised to a 34-kDa C-terminal fragment (residues 612–913). This fragment contains the two SCR domains and the two regions with homology to complement factor I (656) (see section 15.2.7). If C5b is not bound to C6, it decays rapidly at physiological temperature, 37 °C (1448). If it is bound, then the result is a stable [C5b.C6] complex with a total molecular mass around 320–330 kDa (1449, 1450).

[C5b.C6] probably remains bound loosely to C3b in C5-convertase, until the additional binding of C7 (1451). When the latter protein binds, there first forms a complex [C5b.C6.C7], also called C5b–7, with an aggregate molecular mass of 430–450 kDa. This possesses a metastable membrane-binding site that decays with a life-time of less than 10 ms (1452). This site enables the C5b–7 to insert itself into a membrane's lipid layer via

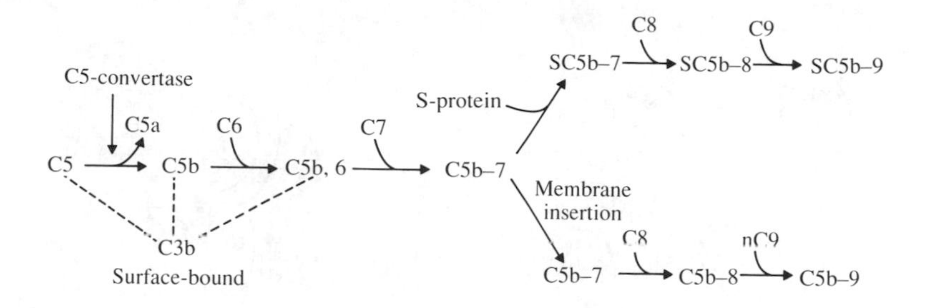

Fig. 104. Formation of MAC. Note that the figure also includes the regulatory processes, which have not yet been discussed (see Chapter 15).

hydrophobic interactions. If insertion into a membrane does not occur before decay of the binding site, then soluble aggregates of C5b–7 are formed (presumably by interaction between the same hydrophobic sites (1453)). There is good evidence that the membrane-binding site in C5b–7 is located in C7 (1453, 1454).

C5b–7 embedded in a membrane functions as a receptor for C8. C8 has in itself no affinity for membranes, so that it is C5b–7 alone that determines where the MAC is bound. The recognition site for C5b–7 in C8 has been localised to the β chain (1455), which in isolated form binds C5b–7 with the same affinity as native C8 does (1456), and it appears that it is C5b that is recognised by C8 (1457). The α chain of C8 contains various other binding domains. There is a domain that binds the C8 β chain (1458), one that binds the C8 γ chain (1459) and another that binds C9 (1460). In addition, C8 α chain contains a region, consisting of one or more domains, that after the formation of the C5b–8 complex (C5b–7 plus C8) is inserted into the membrane bilayer (1461). The C8 γ chain appears to be completely dispensable in these processes, in that C8 depleted of γ chain functions just as well as C8 does (1458). The molecular mass of the C5b–8 complex is presumed to be just below 600 kDa.

The binding of C9 to C5b–8 takes place via C8, presumably with the help of the C8 α chain, as this possesses a binding site for C9. Studies using monoclonal antibodies have shown that the interaction of C9 with C5b–8 is mediated by the C9b part of C9, which consists of amino acids 245–538 (1462) (section 14.4). C5b–8 promotes the insertion of C9 into the membrane's hydrocarbon core, and thereafter there occurs a non-enzymic polymerisation of C9 that leads to the formation of MAC. The mechanisms of the steps in which the membrane is penetrated and C9 is polymerised have not been elucidated, but they appear to involve hydrophilic–amphiphilic transitions: in other words, the hydrophilic proteins that are soluble in plasma undergo conformational changes such that they become amphiphilic (i.e., possess both hydrophilic and lipophilic character).

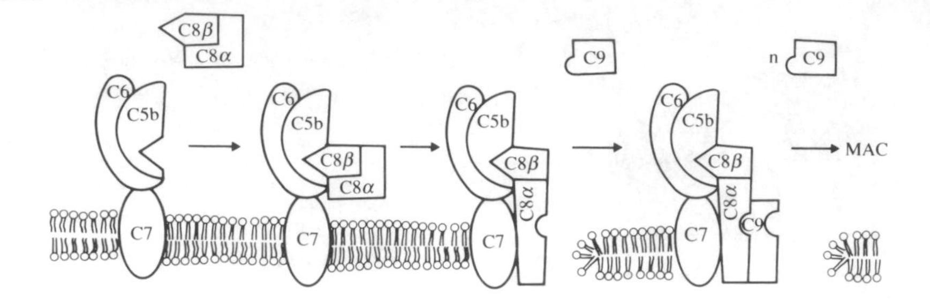

Fig. 105. A model for the formation of MAC.

The polymerisation of C9 can result either in the formation of tubular poly(C9), $(C9)_n$, attached to C5b–8 (1463), or in the formation of non-tubular oligomers of C9, likewise attached to C5b–8 (1464). Both complexes are potentially lethal for the cells upon whose membranes they are assembled.

The macromolecular structure of MAC can thus be denoted $[C5b_1.C6_1.C8_1.C9_n]$, where n can vary from 1 to 18. This means that the molecular mass of MAC can lie between 650 kDa and 1.7–1.8 MDa.

The overall structure of MAC depends upon the molar ratio between C9 and C5–8, in that the formation of tubular C9 demands a high C9:C5b–8 ratio (1464), which has interesting physiological implications. Initially, the ratio between C9 and C5b–8 will be high on account of the large excess of C9, so that tubular poly(C9) will probably dominate. Later, as the ratio becomes more even (by the arrival of more C5b–8 and the consumption of C9), it is possible that non-tubular structures will make up a larger part of the MAC.

Which form of MAC is predominant or of greater physiological importance is not fully clear. However, even the membrane-associated C5b–8 makes pores in membranes (1465), although the pores are not very large. Historically, the tubular poly(C9) molecules have always been regarded as the lethal *complement lesion*, as it clearly resembles (indeed, is) a hole. However, this visual similarity is not a rigorous argument, and recent results imply that the picture may be somewhat different. These studies have employed a thrombin-modified version of C9 (C9thr), in which the bond between His244 and Gly245 in C9 has been broken, giving [C9a.C9b]. C9 and C9thr were used *in vitro* in a system of purified MAC components (1466, 1467) to puncture erythrocyte ghosts and thereby to release radioactive marker molecules. It was found that, although C9thr does not form tubular poly(C9), it was just as active in haemolysis as C9 was (1466); it fact, it led to a more rapid efflux of the markers. In a similar experiment, an analogous result was obtained for the influx of calcium ions (1467), and such an influx is normally one of the first results of the formation of MAC

(1468). Similar conclusions were reached in studies of the MAC-induced perforation of well-defined phospholipid vesicles (1469). It can be deduced from this that the circular MAC is not necessarily the functional lesion pore in complement.

A model for the formation of MAC is shown in Figure 105.

15.

Regulation processes in the complement system

The course of events in the activation of the complement system is regulated by at least six plasma proteins, with the help of certain membrane proteins as well.

15.1. Regulation of $C\bar{1}$ with $C\bar{1}$-inhibitor

15.1.1. *$C\bar{1}$-inhibitor*

$C\bar{1}$-inhibitor is a single-chained glycoprotein with a molecular mass of approximately 104 kDa as determined by sedimentation equilibrium (1470). The carbohydrate content is about 35% (1470), but this result has varied considerably from one investigation to another. The concentration in normal human serum is about 200 μg/ml ($\approx$2 μM).

The amino-acid sequence of $C\bar{1}$-inhibitor has been determined almost fully by protein-sequencing and completed in the same study by cDNA-sequencing; the latter revealed also a 22-residue signal peptide (749) (Appendix, Fig. A47). Other groups have reported partial (1471–1473) or complete (1474, 1475) cDNA sequences. The gene for $C\bar{1}$-inhibitor, which lies on chromosome 11 at position p11.2–q13 (749, 1472), covers approximately 17 kbp of genomic DNA and contains at least seven introns (1475, 1476). The 5′ end of the $C\bar{1}$-inhibitor gene has not been found, so the first exon is not yet fully characterised. The sizes of the exons vary between 73 bp and 518 bp, while those of the introns vary between 194 bp and 5.5 kbp.

$C\bar{1}$-inhibitor consists of 478 amino-acid residues with a molecular mass of 52.9 kDa, which is only 51% of the physically estimated molecular mass. The carbohydrate found bound to $C\bar{1}$-inhibitor is both N-bound (at Asn3, Asn47, Asn59, Asn216, Asn231 and Asn330) and O-bound (at Ser42, Thr26, Thr66, Thr70 and Thr74). In addition, it is possible that threonine residues at positions 49, 61, 77, 84, 85, 89, 93, 96 and 97 also carry carbohydrate (749). The O-bound carbohydrate is largely localised in a

```
63 Glu-Pro-Thr-Thr 66
67 Gln-Pro-Thr-Thr 70
71 Glu-Pro-Thr-Thr 74
75 Gln-Pro-Thr-Ile 78
79 Gln-Pro-Thr     81
82 Gln-Pro-Thr-Thr 85
94 Gln-Pro-Thr-Thr 97
```

Fig. 106. Internal homology in the N-terminal region of C$\overline{1}$-inhibitor.

region close to the N terminus, where the amino-acid sequence Glx-Pro-Thr-Thr and variants of it are repeated seven times (Fig. 106). The disulphide-bond pattern in C$\overline{1}$-inhibitor has been determined experimentally; the pairs are Cys101–Cys406 and Cys108–Cys183 (749). The reactive site in C$\overline{1}$-inhibitor is the Arg444–Thr445 bond (1477).

Genetically caused deficiency of C$\overline{1}$-inhibitor is serious and is called hereditary angioneurotic oedema (see, for example, reference 1202). The genetic basis of C$\overline{1}$-inhibitor deficiency has been documented in several cases. Two types of reactive-site variant have been described. In one, Arg444 is replaced by His (1476, 1478, 1479), while in the other it is replaced by Cys (1476, 1479). Two further known mutations lie close to the reactive site, on the N-terminal side. One of these is Ala434→Glu (1480), and the other is Ala436→Thr (1481). The last documented cause of deficiency of C$\overline{1}$-inhibitor is the absence of exon VII, causing the deletion of residues 322–394 and introducing a frameshift error that generates a stop codon 14 codons later (1482).

As described earlier, in section 8.1.2, the C$\overline{1}$-inhibitor is a member of the serpin family, by virtue of its homology with other members.

15.1.2. *Regulation mechanisms*

As its name implies, the task of the C$\overline{1}$-inhibitor is to inhibit C$\overline{1}$. It is the only known physiological inhibitor of C$\overline{1}$r and C$\overline{1}$s (1483). The inhibition occurs by the formation of stable 1:1 complexes. When C$\overline{1}$ is inhibited by C$\overline{1}$-inhibitor, the components of C$\overline{1}$ dissociate (1484): complexes of [C$\overline{1}$s.C$\overline{1}$r.(C$\overline{1}$-inhibitor)$_2$] are released (1485), in which the stoichiometry is determined by the binding of a molecule of the inhibitor to each of the two proteinases. When bound to the immune complex, C$\overline{1}$r reacts only 25–35% as rapidly with C$\overline{1}$-inhibitor as C$\overline{1}$s does (1486).

Figure 107 shows a model for this inhibition, leading to the release of two complexes with the structure [C$\overline{1}$s.C$\overline{1}$r.(C$\overline{1}$-inhibitor)$_2$].

It has been suggested that C$\overline{1}$-inhibitor also functions as inhibitor of the activation of C1 when the C1 has bound to, for example, immune complexes. This is based upon the observation that C$\overline{1}$-inhibitor inhibits

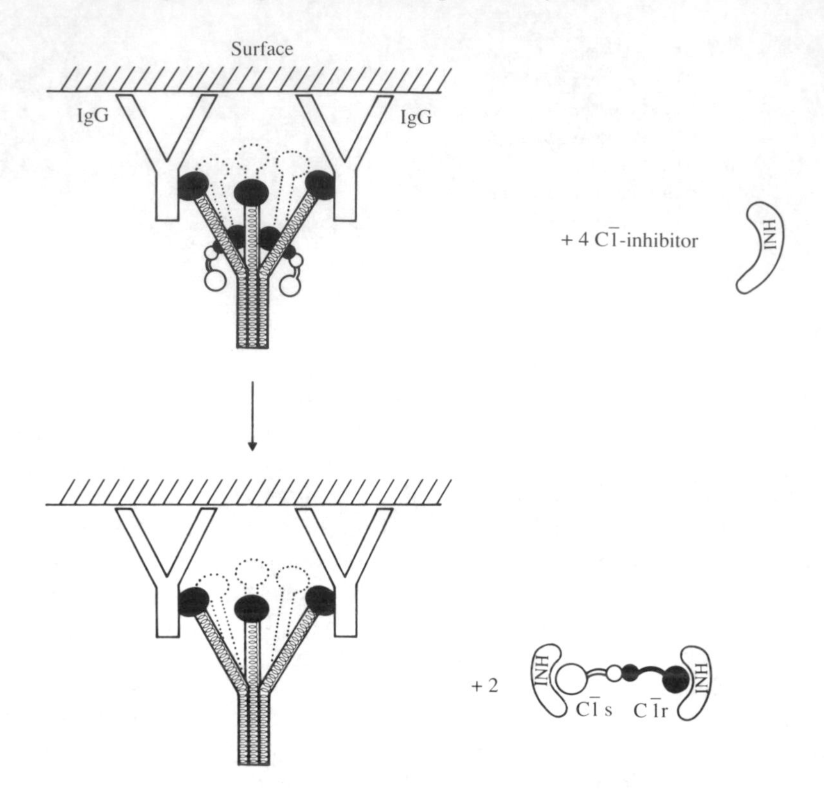

Fig. 107. A model for the control of C$\bar{1}$ by C$\bar{1}$-inhibitor. (After reference 1485.)

such activation *in vitro* when the various components are present at their physiological concentrations (1487).

C$\bar{1}$-inhibitor is the only serine-proteinase inhibitor that plays a part in the complement system, and it is thus the only agent of this kind of control.

Finally, a note should be made in this context about the possible importance of C3b and C4b for the control of the activation of C1. In systems of purified complement factors at their physiological concentrations *in vitro*, the degree of activation that can be reached with a given quantity of immune complex depends upon how many components of the system are present (1488). In an incubation mixture with C1 and C$\bar{1}$-inhibitor and including C2, C3 and C4, no activation of C1 was observed in the presence of quantities of immune complex that were sufficient to activate C1 completely if the incubation mixture contained only C1 and C$\bar{1}$-inhibitor. Reagents with the ability to inactivate C3b and C4b (nucleophiles that attack the thioester) abolished the effect of C3 and C4. In summary, it appears that, when C3b and C4b are deposited on surfaces that activate the classical pathway, they modulate the activation of C1, probably in order to prevent the complement system from being over-activated.

15.2. **Catalysed decay processes and proteolytic inactivation**

The proteins that are involved in these two control mechanisms of the complement system are the plasma proteins factor H, factor I and C4b-binding protein (C4b-BP), together with the membrane proteins decay-accelerating factor (DAF), complement receptor type 1 (CR1) and membrane co-factor protein (MCP).

15.2.1. *Complement factor H*

Factor H is a single-chained glycoprotein with a total molecular mass of about 150 kDa (1489), of which the carbohydrate content is estimated to lie between 4% and 18%. Its concentration in normal human serum is 500 μg/ml ($\approx$3 μM).

The amino-acid sequence of human factor H has been determined partially, both by direct sequencing and by deduction from cDNA (1490–1492); the complete sequence has come from cDNA (638) (Appendix, Fig. A48). The murine factor H sequence has also been determined via cDNA (639). The human and murine cDNA sequences include an 18-residue signal peptide. The gene for murine factor H has been characterised. It covers approximately 100 kbp, and contains 22 exons (1493). The sizes of the exons range from 77 bp to 210 bp and those of the introns from 86 bp to about 26 kbp.

Human factor H consists of 1213 amino-acid residues, corresponding to a molecular mass of 137.1 kDa. It is possible that the dipeptide Lys-Arg is removed from its C terminus in the bloodstream, leaving 1211 residues (638). Its chain contains nine potential N-glycosylation sites, at Asn residues in positions 199, 511, 700, 784, 804, 864, 893, 1011, 1077, of which at least Asn511, Asn784, Asn804, Asn864 and Asn893 are used (638).

The amino-acid sequences of both human and murine factors H are organised into 20 internally homologous SCR domains. The disulphide-bond pattern in factor H has not been determined experimentally. Homology with human β_2-glycoprotein I (632) suggests that, in each domain, the first and the third cysteines are connected by one S–S bridge and the second and fourth by another.

The gene for human factor H has been found on the long arm of chromosome 1 (1494), by both indirect (1494, 1495) and direct (1496) methods.

15.2.2. *C4b-binding protein*

C4b-binding protein (C4b-BP) is a many-chained glycoprotein; in non-reducing SDS-polyacrylamide gel electrophoresis it shows two bands that correspond to a molecular mass around 550 kDa (1497). Sedimentation-equilibrium analysis reveals a molecular mass for C4b-BP close to 570 kDa

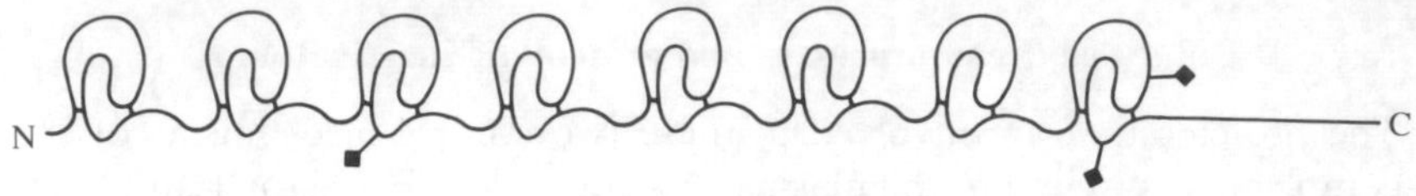

Fig. 108. The domain structure of C4b-BP. ◆ shows the N-bound carbohydrate.

(878). It was originally believed that all chains in C4b-BP were identical, with an individual molecular mass of 70 kDa (877, 1498), and that the final structure arose by the formation of disulphide bonds between seven such chains. However, it has since been shown that both molecular-weight variants of C4b-BP contain an additional chain with a molecular mass of only 45 kDa, disulphide-bound to the other C4b-BP structure (1499). Its N terminus is not related structurally to the other chain type in C4b-BP, and complete C4b-BP consists of one 45-kDa together with seven 70-kDa chains. The concentration of C4b-BP in normal human serum is about 250 μg/ml ($\approx$450 nM), and the carbohydrate content is about 12%.

The amino-acid sequence of the 70-kDa chain in human C4b-BP has been determined partially by protein-sequencing (1500) and by cDNA-sequencing (635). This has led to the complete sequence of the chain, which consists of 549 amino-acid residues (Appendix, Fig. A49). The signal peptide is not included in this cDNA sequence, which starts only at residue 32. It has since been shown that the signal peptide is probably 32 residues long (1501). The gene for this chain lies on the long arm of chromosome 1 (1494, 1496). Preliminary studies on genomic DNA have revealed one intron (1502). For the 45-kDa chain, only the N-terminal sequence has been determined (1499).

The 549 amino-acid residues in the 70-kDa polypeptide chain of C4b-BP add up to give a molecular mass of 61.5 kDa, but the attached carbohydrate raises this value to close to the observed 70 kDa. Three N-glycosylation sites are used: Asn173, Asn458 and Asn480; a fourth potential site, at Asn240, is not used, which is consistent with the fact that the second residue in the consensus sequence is proline (1500). The sequence is organised into eight internally homologous SCR domains, which start at the N terminus and continue to residue 491, after which comes a C-terminal extension 58 residues in length (Fig. 108).

The coarse three-dimensional structure of C4b-BP is interesting. Electron-microscopic pictures of C4b-BP reveal a structure reminiscent of a squashed spider, with a central 'body' and seven 'legs' (1503, 1504), or perhaps of a seven-armed nautilus (Fig. 109), as predicted by x-ray scattering (1505). Each of the seven projections is one of the 70-kDa polypeptide chains, with its N terminus at the periphery and its C terminus in the central region, held to the others by disulphide bonds (1504, 1506). It

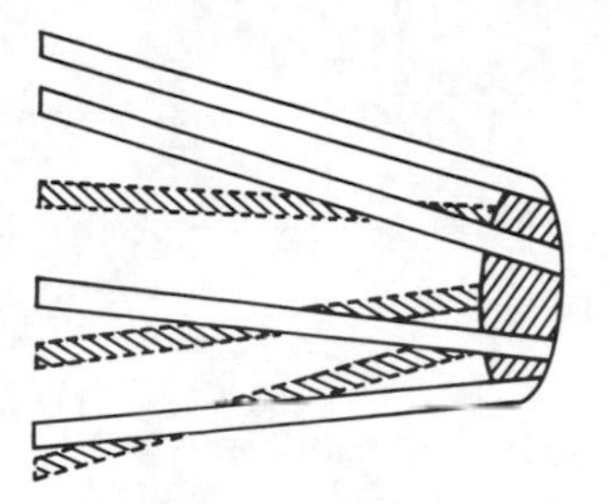

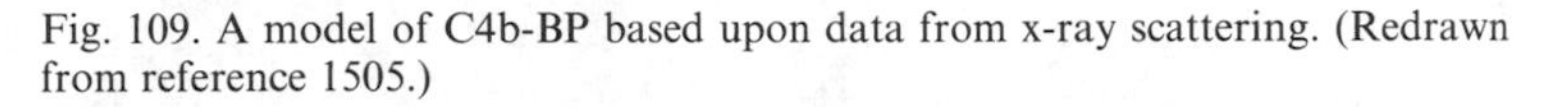

Fig. 109. A model of C4b-BP based upon data from x-ray scattering. (Redrawn from reference 1505.)

has been shown that the 45-kDa chain resides in the central body structure (1499).

It has been assumed that the two cysteine residues found in the C-terminal extension (see above) take part in the disulphide bonding that links the chains of C4b-BP (Fig. 110). This is both because the cysteines within the SCR regions are normally expected to participate in intra-domain S–S bonds, and because human C4b-BP remains intact in non-reducing, but not in reducing SDS-polyacrylamide electrophoresis, in contrast to mouse C4b-BP (see below).

Studies of the disulphide-bond pattern in the SCR domains of C4b-BP show consistency with the presumed pattern described for factor H (section 15.2.1); however, there is still some inclarity here (1507).

The murine C4b-BP chain does not contain cysteine in its C-terminal extension; accordingly, the polypeptide chains in assembled murine C4b-BP are not disulphide-bonded to one another, but are held entirely by non-covalent interactions, as shown by the instability of murine C4b-BP in non-reducing SDS electrophoresis (1508). The structure of murine C4b-BP also differs from that of human C4b-BP in that it has fewer SCR domains – six instead of eight. Exact analysis of homology indicates that the 'missing' domains are numbers 5 and 6 from the N terminus (636). The gene for murine C4b-BP has been partially characterised (1509).

The assumption that the two Cys residues in the C-terminal extension hold the 70-kDa chains together may possibly be incorrect, as indicated by studies with a monoclonal antibody against C4b-BP (1510).

It was mentioned earlier (section 8.2.3) that a certain fraction of the C4b-BP in plasma circulates in a 1:1 stoichiometric complex with the vitamin K-dependent protein S (877), which we met in connection with the regulation of coagulation. This fraction is estimated to be about 50% of the plasma C4b-BP. The binding site on C4b-BP for protein S is not the same as that for C4b, since the binding of protein S to C4b-BP does not affect the latter's function (1511). The C4b-binding region in C4b-BP has been localised to

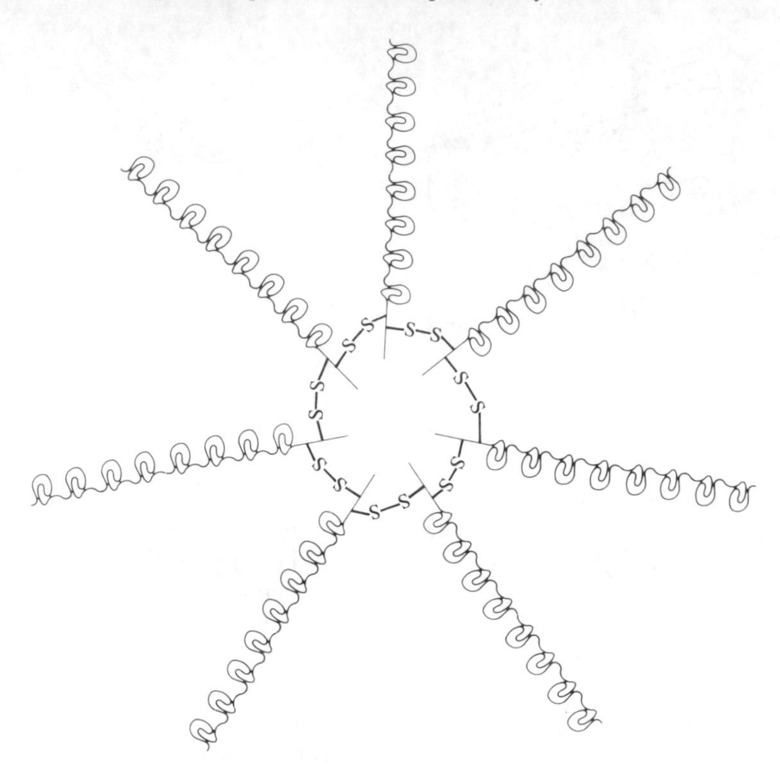

Fig. 110. The originally suggested disulphide-bond pattern in C4b-BP, with intercatenary disulphide bonds located in the C-terminal extension.

the 'arms' – the N-terminal region (1503, 1512) – while the binding site for protein S has been localised to the central 'body' (1513), possibly involving the 45-kDa subunit (1499). The 45-kDa subunit is sensitive towards cleavage by chymotrypsin; after cleavage, C4b-BP has lost its ability to bind protein S. However, treatment of C4b-BP with chymotrypsin in the presence of protein S leaves the 45-kDa subunit intact.

The function of protein S in the complex with C4b-BP is unknown, but the complex-bound protein S has no activity in coagulation (895, 915). In addition to this, it is only the high-molecular-weight form of C4b-BP that binds protein S (878). The monoclonal antibody mentioned above blocks the binding of protein S to C4b-BP and does not react at all with the low-molecular-weight form. The epitope that this antibody recognises has been found at residues 447–467 in the eighth and last SCR domain, indicating that this region is also important for the binding of protein S. Since the low-molecular-weight form is unable to bind protein S, but is held together by disulphide bonds, these intercatenary bonds should be found on the N-terminal side of residues 447–467, and (therefore) not in the C-terminal extension (1510). This question will not be resolved until the disulphide pattern has been elucidated systematically.

There is some evidence that points to a different explanation for the results obtained with this monoclonal antibody. There is no doubt that the 45-kDa chain is involved in the binding of protein S, or that this chain is found in both forms of C4b-BP (1499). Since only the high-molecular-mass form of C4b-BP binds protein S (878), there must be additional structural prerequisites for this binding. Such a prerequisite could be the epitope between position 447 and 467 that is recognised by the monoclonal antibody in question. The fact that this epitope is not found in the low-molecular-mass form of C4b-BP need not mean that the entire region of the C terminus is missing; there could well exist C4b-BP molecules that lack a part of the SCR domain 8 because of alternative splicing. No large difference in molecular mass between the 70-kDa chains in the two forms of C4b-BP could be shown by SDS-polyacrylamide gel electrophoresis (1499), although the opposite has been asserted (but not substantiated) (1510).

Another important question remains unanswered: how is the 45-kDa chain of C4b-BP disulphide-bound to the central body? At present, it is not even known how many 70-kDa chains the 45-kDa chain binds to, quite apart from the detailed disulphide pattern or its relation to the function of C4b-BP.

15.2.3. *Decay-accelerating factor*

Human decay accelerating factor (DAF, reviewed in reference 1514) is an extracellular, single-chained membrane glycoprotein with a molecular mass around 70 kDa (1515). It is bound to the plasma membrane by way of a glycophospholipid at its C terminus (1516, 1517). Originally, DAF was identified in erythrocytes (1518), but it is found on the plasma membrane of more or less all cell types in close contact with plasma (1519–1521). These include blood platelets, leucocytes and endothelial cells.

The amino-acid sequence of human DAF has been derived from the sequencing of cDNA (651, 652). The protein has 347 amino-acid residues, corresponding to a molecular mass of 46 kDa (Appendix, Fig. A50). The two inferred amino-acid sequences differ at positions 46 and 51, which have been determined respectively as Ile and Ser (651) or as Thr and Met (652). The amino-acid sequence contains only one potential N-glycosylation site, Asn61, but DAF also contains O-bound carbohydrate (1522). The enzyme-catalysed removal of the N-bound carbohydrate reduces the molecular mass by only approximately 3 kDa, while the removal of the O-bound carbohydrate reduces it by 26 kDa (1522). The signal peptide has 34 residues (651). The gene for DAF covers some 35 kbp of genomic DNA. It has not been characterised structurally, but it has been found on

```
  2 C G L P P D V P N A Q P A L E G R T S - - F P E D T V I T Y K C E
 64 C E V P T R L N S A S L K Q P Y I T Q N Y F P V G T V V E Y E C R
129 C P N P G E I R N G Q I D V P G G I L - - - - F G A T I S F S C N
191 C P A P P Q I D N G I I Q G E R D H Y G Y - - - R Q S V T Y A C N

E S F V K I P G E K D S V I C L K G - - - S Q W S D I E E F C 60
P G Y R R E P S L S P K L T C L Q N - - - L K W S T A V E F C 124
T G Y - K L F G - S T S S F C L I S G S S V Q W S D P L P E C 186
K G F - T M I G - E H S I Y C T V N N D E G E W S G P P P E C 249
```

Fig. 111. Internally homologous SCR domains in DAF. Only the positions at which amino acids are the same in all four domains are outlined. Pairwise comparison of the four domains reveals the following degrees of amino-acid identity: domains 1 and 2, 32%; domains 1 and 3, 28%; domains 1 and 4, 25%; domains 2 and 3, 22%; domains 2 and 4, 20%; domains 3 and 4, 31%.

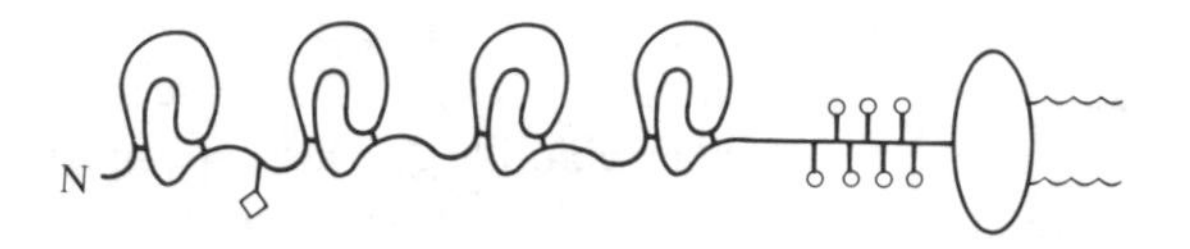

Fig. 112. The domain structure of DAF. ◇ shows the potential attachment site for N-bound carbohydrate, ○ shows the region containing potential attachment sites for O-bound carbohydrate and the ellipsoid represents the attached glycolipid.

chromosome 1 at position q31–41, and to be coupled to the genes for factor H, C4b-BP and CR1 (1523, 1524).

Human DAF contains four SCR domains close to its N terminus (Fig. 111). These are followed by a stretch rich in serine and threonine, 45 residues in length, and, in newly synthesized DAF, a hydrophobic C-terminal region (Fig. 112). The hydrophobic region is not a transmembrane peptide that anchors the protein to a membrane, but is a signal for the attachment of the glycolipid (1525–1528), and it is believed that this hydrophobic stretch is removed proteolytically during the attachment process.

It is believed that there is O-bound carbohydrate attached to the Ser/Thr-rich region. When recombinant DAF is synthesized in a cell type that needs exogenic N-acetylgalactosamine in order to carry out O-glycosylation, the product is a DAF molecule that is exported correctly to the plasma membrane but is then rapidly cleaved, so that most of the molecule dissociates from the membrane (1529). The O-bound carbohydrate thus appears to be important for the stability of DAF on the cell surface. The disulphide pattern of DAF is not known, but it is assumed that the four SCR domains show the familiar pattern (first to third and second to fourth cysteines).

Although DAF is normally bound to plasma membranes, there is also a non-membrane-bound form of DAF, for example in plasma (1530). One

```
DAF   1 D C G L P P D V P N A Q P A L E G R T S F P E D T V I T Y
K C E E S F V K I P G E K D S V I C L K G S Q W S D I E E F C N R
S C E V P T R L N S A S L K Q P Y I T Q N Y F P V G T V V E Y E C
R P G Y R R E P S L S P K L T C L Q N L K W S T A V E F C K K K S
C P N P G E I R N G Q I D V P G G I L F G A T I S F S C N T G Y K
L F G S T S S F C L I S G S S V Q W S D P L P E C R E I Y C P A P
P Q I D N G I I Q G E R D H Y G Y R Q S V T Y A C N K G F T M I G
E H S I Y C T V N N D E G E W S G P P P E C R G K S L T S K V P P
T V Q K P T T V N V P T T E V S P T S Q K T T T K T T T P N A Q A
T R S T P V S R T T K H F H E T T P N K G S G T T S G T T R L L S
G 327


DAF
328 H T C F T L T G L L G T L V T M G L L T 347

Alternative DAF C terminus
328 S R P V T Q A G M R W C D R S S L Q S R T P G F K R S F H F S
L P S S W Y Y R A H V F H V D R F A W D A S N H G L A D L A K E E
L R R K Y T Q V Y R L F L V S 406
```

Fig. 113. Synthesis of an alternative DAF molecule with an insertion of 118 bp in the codon that encodes amino-acid residue 327.

explanation for this has been suggested on the basis of the cloning of two different DAF-coding cDNA sequences: the shorter of these coded for the DAF just described, and the longer contained 118 additional base pairs inserted into the codon for amino-acid residue 327. This results in a frameshift error and consequently a much longer DAF molecule (406 residues – Fig. 113) (651) with a completely different C-terminal sequence that is no longer hydrophobic. This could possibly be the non-membrane-bound form of DAF, but this does not seem likely, as antibodies against the alternative, hydrophilic C-terminal region do not recognise the soluble DAF (1514). It is more probable that the 118 nucleotides represent a non-excised intron.

15.2.4. *Complement receptor type 1*

Human complement receptor type 1 (CR1) is an extracellular, single-chained membrane glycoprotein that was first isolated from erythrocytes (1531, 1532). Four allelic variants of CR1 have been identified; the two most common molecular masses estimated from non-reducing SDS polyacrylamide gel electrophoresis are 190 kDa (CR1-A) and 220 kDa (CR1-B) (1533–1535). Variations in the molecular mass do not appear to be due to differences in glycosylation (1536); rather, they reflect genomic polymorphism, perhaps a result of crossovers between internally

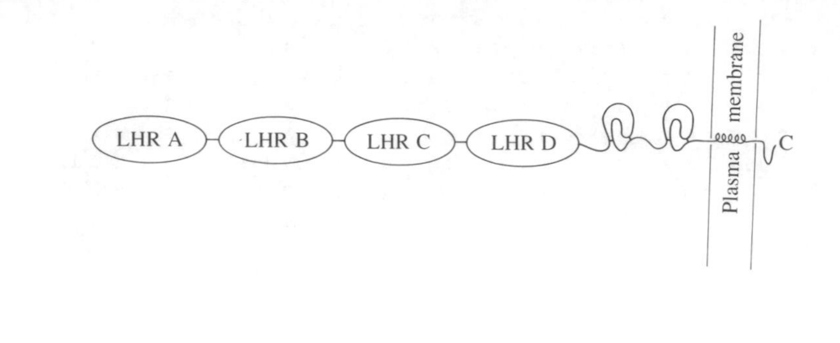

LHR =

Fig. 114. The domain structure of CR1-A. Each long homologous repeat (LHR) consists of seven SCR domains.

homologous regions (see below) (1537). The carbohydrate content is about 8% and is principally N-bound (1538); there is debate as to whether CR1 also contains O-bound carbohydrate (1536, 1538). CR1 is normally isolated, as indicated above, from erythrocytes, but it is also found, for example, on monocytes, macrophages, granulocytes and B lymphocytes. CR1 is sometimes called C3b/C4b receptor, a name that bespeaks the molecule's function.

The amino-acid sequences of peptides from human CR1 have led to the cloning of partial cDNA clones that together cover the entire sequence for CR1-A (640, 641, 1539). Pro-CR1-A has 2044 residues, of which 46 are believed to make up a signal peptide and the remaining 1998 the CR1-A (Appendix, Fig. A51). The N terminus of CR1 is blocked; if the signal peptide prediction is correct, the first amino acid is Gln, so the blocking is expected to be due to the formation of pyroglutamic acid.

The amino-acid sequence of CR1-A is organised into 30 SCR domains of about 60 residues each, a presumed transmembrane peptide of 25 residues and a 43-residue cytoplasmic domain (Fig. 114). The first 28 SCR domains from the N terminus can be divided into four groups of 7 SCR domains each, called long homologous repeats. This subdivision is based upon the greater fractions of identical amino acids (60–99%) between each of these four regions. The corresponding homology between individual SCR domains within the four regions is substantially weaker.

This organisation of CR1 into long homologous repeats, each corresponding to a molecular mass of 40–50 kDa, indicates a mechanism of unequal crossovers, which could have given rise to the different forms of CR1 (1540). The differences in molecular mass between the allelic forms correspond roughly to one long homologous repeat.

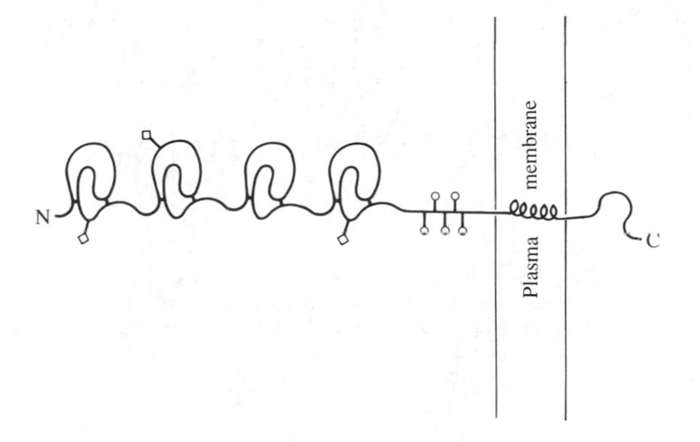

Fig. 115. The domain structure of MCP. ◇ shows potential attachment sites for N-bound carbohydrate and ○ the region containing potential attachment sites for O-bound carbohydrate.

The gene for CR1 has been localised to chromosome 1, position q32 (1495), and, depending on which allele is present, it covers 130–160 kbp of genomic DNA (1540).

15.2.5. *Membrane co-factor protein*

Human membrane co-factor protein (MCP), earlier called gp45-70, is a relatively recent discovery (1541). It is a single-chained glycoprotein, found on the surfaces of various types of leucocyte, blood platelets and endothelial cells, but not on erythrocytes (1542). Most cell types express two forms of MCP, with molecular masses of 58 kDa and 63 kDa; this is due to genetic polymorphism (1543).

The amino-acid sequence of human MCP has been derived from cDNA-sequencing (653) (Appendix, Fig. A52). It has 350 residues, corresponding to a molecular mass of 39 kDa. While MCP contains three potential N-glycosylation sites at Asn49, Asn80 and Asn239, it also contains a considerable quantity of O-bound carbohydrate (1544). The signal peptide is 34 residues long, and the gene for MCP lies on chromosome 1 at location q31–41 (653), with the result that the MCP gene is linked to the genes for DAF, C4b-BP, factor H and CR1.

The structure of MCP is reminiscent of DAF (Fig. 115), since the amino-acid sequence can be divided into four N-terminal SCR domains (Fig. 116) followed by a regions rich in serine and threonine, then a hydrophobic (probably transmembrane) domain and finally, in contrast to DAF, a short cytoplasmic domain. The pattern of disulphide bonds is not known, but this is assumed to resemble other SCR regions.

```
  1 C E E P P T F E A M E L I G K P K P Y - - - - Y E I G E R V D Y K C K
 65 C - - P Y I R - - D P L N G Q A V P A N G T - Y E F G Y Q M H F I C N
128 C T P P P K I K - - - - N G K H T F S E V E V F E Y L D A V T Y S C D
194 C R F P V V E - - - - - N G K Q I S G F G K K F Y Y K A T V M F E C D

K G Y - - - - F Y I P P L A T H T I C D R N H T - - - W L P V S D D A C 60
E G - - - - - Y Y L I G E E I - L Y C E L K G S V A I W S G K - P P I C 123
P A P G P D P F S L I G E S T - I Y C G D N S V - - - W S R A - A P E C 189
K G - - - - - F Y L D G S D T - I V C D S N S T - - - W D P P - V P K C 249
```

Fig. 116. Internally homologous SCR domains in MCP. Only the positions at which amino acids are the same in all four domains are outlined. Pairwise comparison of the four domains reveals the following degrees of amino-acid identity: domains 1 and 2, 24%; domains 1 and 3, 25%; domains 1 and 4, 30%; domains 2 and 3, 27%; domains 2 and 4, 27%; domains 3 and 4, 38%.

15.2.6. *Decay processes of C3-convertases*

Both classical and alternative C3-convertase show spontaneous and catalysed decay. Their spontaneous decay is a simple dissociation of the components of the C3-convertase. The classical convertase dissociates from C4b2a to C4b and C2a, while the alternative C3-convertase dissociates from C3b,Bb to C3b and Bb. These processes are irreversible, which means that C3-convertase activity can never be generated by the mere presence of the dissociated components. As we saw earlier (sections 12.6 and 13.3), this implies that C2b and Ba must have some property that is essential for the formation of the complex between C4b and C2 – or, respectively, between C3b and factor B.

The spontaneous decay of classical C3-convertase is accelerated by the control proteins C4b-BP (1545, 1546), DAF (1515, 1547) and CR1 (1548). That of alternative C3-convertase is catalysed by factor H (1549), by DAF (1515) and by CR1 (1531, 1532). The purpose of this is to regulate the formation of C3-convertase activity, partly in the bulk phase and partly on the surfaces at which complement activation is unwanted – those of the organism to be defended. For DAF, it has been shown that the dissociation of C3-convertase components does not occur until after the activation of C2 or factor B, respectively (1550).

The fate of C2a and Bb after their release from the C3-convertase complex is not known. However, it is known that the C3b (or C4b) released can either enter new C3-convertase complexes or else be inactivated by the serine proteinase of factor I.

15.2.7. *Factor I*

Factor I (the 'I' is a letter, not a Roman numeral) was formerly called C3b/C4b inactivator. It is a two-chained serine proteinase with a molecular mass of about 88 kDa that is distributed between a heavy (50 kDa) and a light (38 kDa) chain held together by disulphide bonds (1551, 1552). Its carbohydrate content has been found to be between 11% and 27% (1551, 1553), and its concentration in normal human serum lies around 35 μg/ml ($\approx$400 nM) (1551). Like factor $\overline{\text{D}}$, factor I is not thought to possess the zymogen form typical of the serine proteinases that we have met, but rather to circulate in the active, two-chained form. By convention, factor I has no bar over the 'I', even though it would be more logical to include the bar.

Parts of the amino-acid sequence of factor I have been determined by protein-sequencing (1554, 1554), but the complete amino-acid sequence has been deduced from that of the cDNA (654, 1556) (Appendix, Fig. A53). The cDNA sequence includes a signal peptide 18 residues in length. The structure of the gene for factor I has not been characterised, but this gene is known to be on chromosome 4 (654).

Factor I is synthesized as a single polypeptide chain of 565 amino-acid residues. The N terminus gives rise to the heavy chain and the C terminus to the light chain. There is probably a spacer tetrapeptide R-R-K-R, so that the heavy and light chains have 317 and 244 residues respectively, giving molecular masses for the polypeptides of 35.4 kDa and 27.6 kDa. Both have three potential sites for N-glycosylation. In the heavy chain these are Asn52, Asn85 and Asn159, and in the light chain these are Asn125, Asn155 and Asn197. There is protein-chemical evidence that all three sites are used in the light chain (1555).

The serine-proteinase domain of factor I is the light chain, while the N-terminally derived heavy chain has several domains, although there is disagreement over what sort and how many. One group (1556) has suggested the presence of one growth-factor domain and two domains showing homology with domains in the LDL receptor, C6, C7, C8 α and β chains and C9; another (654) has proposed the presence of three domains homologous to the LDL receptor and an SCR domain. The disulphide-bond pattern in factor I has not been determined.

It is interesting that factor I (like factor $\overline{\text{D}}$) circulates in its active form, and the explanation for this may lie in factor I's strict specificity for a very small number of peptide bonds. The catalytic apparatus in factor I is believed to involve His41, Asp90 and Ser186 in the light chain, corresponding to His362, Asp411 and Ser507 in pro-factor I.

15.2.8. *Proteolytic inactivation of C3-convertases*

The two molecules that are subject to control by proteolytic degradation and consequent inactivation are C3b and C4b, so that this mechanism functions both in the classical and the alternative activation of complement. The purpose of this, like that of the decay processes, is to regulate the formation of C3-convertase activity.

The enzymically active component in both cases is the serine proteinase factor I. However, this enzyme requires co-factors in order for it to function. The degradation of C4b requires the presence of C4b-BP (1557, 1558), of CR1 (1548) or of MCP (1559). The degradation of C3b requires the presence of factor H (1551), CR1 (1531) or MCP (1559).

It would be an over-simplification to represent the inactivation process as a simple, one-step attack on C3-convertase by cleavage of the C3b (or, respectively, C4b), because proteolytic inactivation appears always to be preceded by a decay process (section 15.2.6). Factor I only cleaves C3b or C4b if one of the co-factors mentioned above is present.

C4b is cleaved at two points by factor I (1546). Both of these are in the α' chain, Arg200–Thr201 and Arg580–Asn581 (1319, 1326). The portion of the chain from Thr201 to Arg580, from the α' chain, is called C4d, and the rest is C4c.

Factor I also cleaves C3b at two places in the α' chain (1560): at Arg555–Ser556 and at Arg572–Ser573 (1298, 1561), releasing C3f (17 residues) from the α' chain. The remainder of the molecule is termed iC3b; this undergoes further degradation by proteinases other than factor I.

Finally, it should be observed that some of the regions in C3 that interact with co-factors have been mapped out. The binding site for factor H in C3b has been located between residues 461 and 523 in the α' chain (1562), while the binding site for CR1 lies between residues 1 and 42 in the α' chain (1563). The C3b-binding site in factor H is located in a 38-kDa N-terminal fragment (1564), while the C3b-binding site in CR1 has been localised to the first two SCR domains in the second and the third long homologous repeats (641). The binding site for C4b in CR1 is within the first two SCR domains in the first long homologous repeat (641). In C4b-BP, the binding region for C4b lies between residues 332 and 395 (1565).

15.3. Properdin-stabilisation of alternative C3/C5-convertase

The reader will undoubtedly have received the impression that regulation of the complement system proceeds by mechanisms that destroy enzyme activity, such as inhibition and degradation. This is largely true. However, there is one example of positive control, associated with the protein properdin.

Human properdin was the first protein found to belong to the alternative complement activation pathway (1566). In plasma, properdin is thought to appear in the form of cyclic dimers, trimers and tetramers of identical monomeric chains, while higher polymers can also appear as purification artefacts (1567–1569). The predominant form is the trimer (1569). Properdin is a glycoprotein built up of 56-kDa chains (1570, 1571) with a carbohydrate content of about 10% (1572) and a concentration in serum of about 20 μg/ml ($\approx$350 nM for the monomer).

A large part of the amino-acid sequence of human properdin has been

```
 53 W S A W S L W G P C S V T C S - - - - - - - - E G S Q L R H R R C V G R
112 W S E W G P W G P C S V T C S - - - - - - - - K G T Q I R Q R V C D N P
169 W A S W G P W S P R S G S C L - - - G G A Q E P K E T R S R S C S A P
233 W G P W S P L S P C S V T C G - - - - - - - - L G Q T L E Q R T C D H P
291 W E A W G K W S D C S R L R M S I N C E G T P G Q Q S R S R S C G D R
354 W S Q W S T W S L C T P P C S - - - - - - P N A T R V R Q R L C T P L

G G Q C S E N V A P G T L E - W Q L Q A C E D Q P C C P E M G G 111
A P K C G G - - - H C P G E A Q Q S Q A C D T Q K T C P T H G A 168
A P S H Q P P G K P C S G P Y A E H K A C S G L P P C P V A G G 232
A P R H G G P - - F C A G D A T R N Q M C N K A V P C P V N G E 290
K F N - - - - G K P C A G K L Q D I R H C Y N I H N C I M K G S 353
L P K - Y P P T V S M V E G Q G E K N V T F W G T P R P L C E A 413
```

Fig. 117. Internally homologous thrombospondin type I domains in murine properdin.

Only the positions at which amino acids are the same in at least five of the six domains are outlined. Pairwise comparison of the six domains reveals the following degrees of amino-acid identity: domains 1 and 2, 47%; domains 1 and 3, 27%; domains 1 and 4, 31%; domains 1 and 5, 27%; domains 1 and 6, 20%; domains 2 and 3, 37%; domains 2 and 4, 47%; domains 2 and 5, 30%; domains 2 and 6, 26%; domains 3 and 4, 40%; domains 3 and 5, 30%; domains 3 and 6, 23%; domains 4 and 5, 26%; domains 4 and 6, 21%; domains 5 and 6, 13%.

determined at the protein level (1570) and a nearly complete sequence for murine properdin has been obtained from cDNA (1573) (Appendix, Fig. A54). The gene for properdin is on the X chromosome, at position p11.23–21.1 (1574, 1575).

The cDNA-derived sequence for mouse properdin corresponds to the 437 C-terminal amino-acid residues, and, as the available murine cDNA sequence starts at the point corresponding to the fifth residue of human properdin, murine properdin presumably contains 441 residues in all. Mouse properdin contains six internally homologous domains, about 60 residues in length (Fig. 117), that show homology *inter alia* with three domains in human thrombospondin and domains in human C6, C7, C8 α and β chains and C9.

By binding to the C3b part of the alternative C3/C5-convertase, properdin stabilises these enzyme complexes (1576, 1577), in that this binding reduces substantially the speed of the spontaneous and the catalysed decay reactions. Thus, properdin also affords protection against the degradation of C3b catalysed by factor I. The binding site for properdin in C3b has been localised to amino-acid residues 676–709 in the α' chain (1578).

15.4. S-protein, homologous restriction factor and the regulation of MAC formation

The formation of MAC is regulated both by the plasma protein S-protein and by the membrane protein homologous restriction factor (HRF).

S-protein, which should not be confused with protein S from blood coagulation, is a single-chained glycoprotein with a molecular mass of 83 kDa as determined by sedimentation equilibrium. Its normal concentration in plasma is between 350 μg/ml and 500 μg/ml (≈4–6 μM) (1579, 1580). S-protein is sensitive towards proteolysis; it is frequently isolated as a mixture of one- and two-chained forms, the latter containing polypeptides with respective molecular masses of 65 kDa and 12 kDa and connected by one or more disulphide bonds.

The amino-acid sequence of S-protein has been determined by cDNA-sequencing (1581) (Appendix, Fig. A55), and has shown that S-protein is identical with vitronectin, also called serum spreading factor (1582), although minor discrepancies in the sequences have not yet been resolved (1581). The cDNA sequence includes a signal peptide of 19 residues. The gene for S-protein covers about 3 kbp of genomic DNA and has eight exons. The exons vary in size from 120 bp to 334 bp and the introns from 75 bp to 459 bp (1583).

S-protein contains 459 amino-acid residues with three potential N-glycosylation sites (Asn67, Asn150 and Asn223). In addition, the presence of at least one free thiol group in S-protein has been detected (1580), and there must in fact be at least two of these, since the chain contains an even number (14) of cysteines in all (1581). The calculated molecular mass of the polypeptide is 52.4 kDa. The two-chained form of S-protein is believed to arise when the Arg379–Ala380 bond is cleaved (1581, 1580).

Of the first 44 amino-acid residues in S-protein, 42 are identical with those in a molecule called somatomedin B (1584) that originally (but erroneously) was classified as a growth factor. The two differences are perhaps due to an error in a peptide sequence, since at the point of disagreement the cDNA sequences both code for C-C-T-D (1581, 1582), while the amino-acid residues of somatomedin B were found to be N-C-T-C (1584). S-protein also contains the tripeptide sequence R-G-D (residues 45–47) that is common to a number of proteins with attachment properties, such as fibrinogen, fibronectin and von Willebrand factor. S-protein is the plasma component that becomes most heavily phosphorylated by incubation of plasma with ATP, and the phosphorylation site has been localised to Ser378 (1585). The significance of this possible phosphorylation is not certain. Furthermore, S-protein contains tyrosine-O-sulphate, but the precise location of this is not known (1586). Homology has also been claimed between S-protein and the haem-binding protein in plasma haemopexin (1587).

In the complement system, S-protein functions by binding to the exposed, metastable membrane-binding site in C5b–7, which prevents this complex from binding to a membrane (1449, 1588): the complex of S-protein and C5b–7 (denoted SC5b–7) is soluble and is found in the bulk phase. The SC5b–7 complex contains three molecules of S-protein, and is also able to bind a molecule of C8 and three molecules of C9 (1589); the C9 is bound in such a way that it does not polymerise (1590–1592), so that the formation of MAC is avoided. Studies with a synthetic peptide corresponding to residues 348–360 in S-protein have shown that it is this region that binds to the terminal complement components (1593). In C9, this synthetic peptide binds to amino-acid residues 101–111.

The function of S-protein is by no means restricted to the complement system. We have already mentioned that the sequence of S-protein is identical to that of vitronectin (1581, 1582), a protein that appears to act as a cell-adhesion and cell-spreading factor *in vitro*. It is possible that these are also the protein's natural functions *in vivo*, where they play an important part in cell differentiation, morphogenesis and metastasis. However, the exact physiological rôle of S-protein/vitronectin in this context has not been elucidated.

Another effect of S-protein/vitronectin is connected with blood coagulation: S-protein/vitronectin counteracts heparin's stimulatory effect upon the inhibition of thrombin by antithrombin III (1594–1596). The region of S-protein/vitronectin responsible for this has been pin-pointed as the heparin-binding region (1597, 1598), close to the C terminus, between residues 340 and 380 (1581, 1582), which is also (as mentioned above) the site of MAC-inhibiting activity. S-protein/vitronectin also forms a complex with the [thrombin.antithrombinIII] complex, during the process of blood coagulation (1599, 1600); this complex is found in serum, but not in plasma.

This is the place to mention briefly a recently identified human serum protein that has been named SP-40,40. It was found in the SC5b–9 complex (1601). The function of SP-40,40 is unknown, but cDNA-sequencing has shown that the precursor of SP-40,40 is very similar (77% identical amino acids) to sulphated glycoprotein-2 from the rat (1602, 1603). The latter is a principal component of rat seminal plasma, and SP-40,40 has since been found in high concentration in human seminal plasma. SP-40,40 contains a short stretch homologous to C6, C7, C8 α and β chains and C9.

An essential aspect of the complement system is *homologous restriction*. Homologous restriction is the protection of an organism's own – i.e. homologous – cells against the effects of its own complement system, which otherwise would lead to their ultimate cytolysis. (The term 'homologous' has here nothing to do with homology of sequences.) HRF, also called C8-binding protein, is also a relatively recent discovery (1604, 1605). It has a molecular mass of 65 kDa and is found anchored to most blood cells by a glycophospholipid (1606). It inhibits the perforation of the membrane by C5b–8/C5b–9 (MAC), probably by binding to the C8 α-γ subunit (1607). Thus, HRF protects the body's own (homologous) cells against MAC-induced lysis.

In principle, the other membrane co-factor proteins – CR1, DAF and MCP – fulfil a similar protective function at an earlier stage of the complement activation pathway.

It is now possible to see how the alternative activation of the complement system avoids getting out of control, in spite of the fact that it is relatively unspecific in its requirement for an activating surface. The same arguments can also be applied to the classical activation pathway. Homologous cells have a battery of inhibitors against the effects of the complement system. C3/C5-convertase is held in check by DAF, MCP and CR1, and, should there still be a tendency towards the formation of MAC, HRF is also present. At the same time, activation in the bulk phase is regulated strictly by $C\bar{1}$-inhibitor, C4b-BP, factor H and S-protein. Together, these regulatory mechanisms see to it that the activity of the complement system is directed exclusively against foreign bodies.

16.

Anaphylatoxins and anaphylatoxin inactivator

Anaphylatoxins (reviewed in reference 1608) are biologically active peptides that are released during the activation of the complement system. They come from the complement components C3, C4 and C5 and are termed C3a, C4a and C5a (see the relevant chapters in this book).

The amino-acid sequences of all the three human anaphylatoxins have been determined at the protein level (1292, 1318, 1408), and they are of course parts of the cloned sequences of C3, C4 and C5 (1298, 1326, 1414). In addition, amino-acid sequences for porcine C3a (1609) and C5a (1409), rat C3a (1610), C4a (1613) and C5a (1411), guinea-pig C3a (1611) and bovine C4a (1612) and C5a (1410) have been determined directly, while murine C3a (1299, 1300), C4a (1336, 1337) and C5a (1412) have been deduced from cDNA sequences.

Human anaphylatoxins C3a and C4a have 77 amino-acid residues and C5a has 74. All three display mutual homology (Fig. 118), with the following percentages of identical amino-acid residues: C3a and C4a, 31%; C3a and C5a, 41%; C4a and C5a, 36%. The three-dimensional structure of C3a has been determined at a resolution of 3.2 Å (0.32 nm) (1306). It shows disulphide bonds between Cys22 and Cys49, Cys23 and Cys56, and Cys36 and Cys57 (1306), and, since all these cysteines are retained in the other anaphylatoxins, the same pattern may reasonably be assumed. The C3a structure has led to the construction of a model for C5a (1614). C5a has also been studied by nuclear magnetic resonance (1615).

Chemical modification has shown that the C-terminal regions of C3a and C4a contain the functionally important amino acids, while a larger part of C5a is important (Fig. 119). There is evidence that the peptide sequences at the C termini of C3a and C4a contain sufficient structural information to allow binding to the cellular receptor that mediates the signal to the cells and thus induces the anaphylatoxic effects. It appears that the C-terminal pentapeptide from C3a makes up this anaphylatoxin's active site, and that the task of the rest of the C3a molecule is to position this active site in an

```
C3a   1 S V Q L T E K R M D K V G K Y P - K E L R K C C E D G M R
C4a   1 N V N F Q K A I N E K L G Q Y A S P T A K R C C Q D G V T
C5a       1 T L Q K K I E E I A A K Y K H S V V K K C C Y D G A C

E N P M R F S C Q R R T R F I S L G E A C K K V F L D C C N Y I T
R L P M N R S C E Q R A A R V Q Q - P D C R E P F L S C C Q F A E
V N N D E - T C E Q R A A R I S L G P R C I K A F T E C C V V A S

E L R R Q H A R A S H - - L G L A R 77
S L R K K S R D K G Q - - A G L Q R 77
Q L R - - - A N I S H K D M Q L G R 74
```

Fig. 118. Comparison of homology among the three human anaphylatoxin sequences. Only the positions at which amino acids are the same in all three sequences are outlined. Pairwise comparison of the three sequences reveals the following degrees of amino-acid identity: C3a and C4a, 31%; C3a and C5a, 41%; C4a and C5a, 36%.

optimal manner (1616–1618). The situation in C4a appears to be similar (1619).

However, for C5a, the C-terminal peptide sequence is not sufficient to specify the biological activity of the molecule, even though this peptide region is indispensible (1620, 1621). Mutagenesis studies of C5a prepared by recombinant methods indicate that Ala26 and Arg40 play an essential part in the function of C5a: Arg40 by direct interaction with the C5a receptor and Ala26 by its contribution to the protein's conformation (1622).

As mentioned above, anaphylatoxins function by binding to receptors on the surfaces of the cells that they affect. These include almost all circulating cell types (except erythrocytes) plus various types of tissue cell including mast cells, tissue macrophages and possibly endothelial cells and fibroblasts. The circulating cells that do not respond directly to anaphylatoxin still show a secondary response mediated by those cells that do respond directly.

There are four categories of cellular response to anaphylatoxins: (i) chemotactic behaviour, (ii) morphological changes, (iii) metabolic activation and (iv) release reactions.

It is assumed that all four responses follow from the binding of the anaphylatoxin to the corresponding receptor. These effects are shown most strongly by C5a and least strongly by C4a, with C3a lying midway between them. C5a is also the most thoroughly investigated anaphylatoxin. The presence of C5a receptors has been demonstrated both on polymorphonuclear granulocytes (1623) and macrophages (1624). C5a-stimulated chemotactic effects have been demonstrated for four types of leucocyte: neutrophils (1625, 1626), eosinophils (1627), basophils (1628) and monocytes (1629). The morphological changes following exposure to C5a involve the cell's overall shape and its properties of adhesion to foreign surfaces and

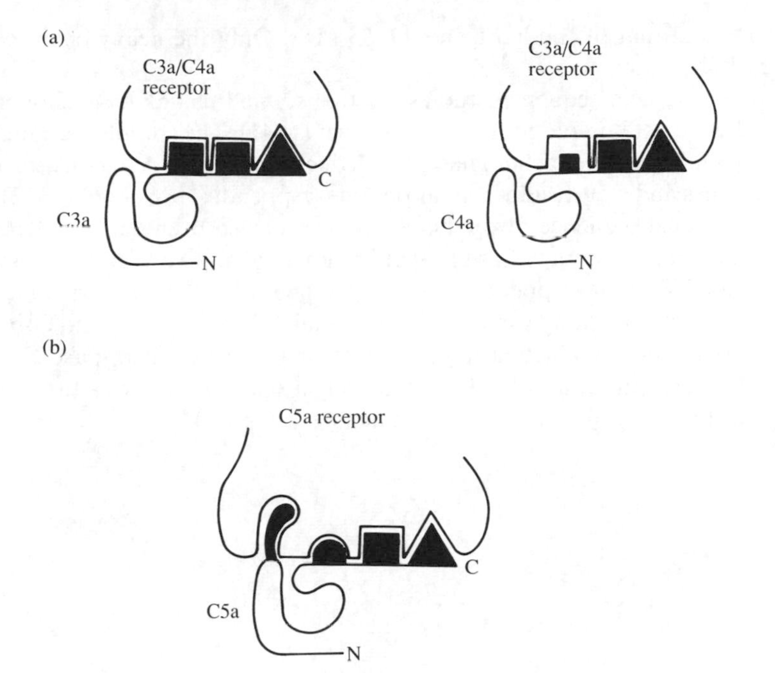

Fig. 119. The binding of anaphylatoxins to their receptors. In C3a and C4a, the C-terminal pentapeptide sequence suffices for binding to the receptor (a), while additional structures are necessary for the binding of C5a to its receptor (b).

of aggregation (1630–1632). C5a-stimulated neutrophil cells show a heightened level of oxidative metabolism (1633, 1634), and the stimulation leads to the release of enzymes (1635). If, in addition to all these primary effects, the secondary effects are taken into account, then the importance of the anaphylatoxins becomes clear. This has also been confirmed by experiments on live animals: the activation of complement *in vivo* in the presence of an inhibitor of anaphylatoxin inactivator resulted in the death of the animal (1636).

Anaphylatoxins have been called 'local hormones', and their effects are subject to strict control by anaphylatoxin inactivator, carboxypeptidase N, which removes C-terminal arginine residues (1637). If this residue is removed from anaphylatoxins, then their activity disappears.

Carboxypeptidase N is a metalloprotease found in plasma at a concentration of about 35 μg/ml ($\approx$ 100 nM). It is a tetrameric glycoprotein containing 17% carbohydrate with an estimated molecular mass around 300 kDa (1637–1640). The non-covalent tetramer consists of two identical heavy chains and two identical light ones; the latter contain the catalytic activity. The molecular mass of the heavy subunit is estimated to be 83–98 kDa and that of the light one 42–55 kDa. Only the heavy chain carries carbohydrate.

Amino-acid sequence studies on the isolated chains have shown that these are not homologous to one another (1641). The complete amino-acid sequence of the light chain has been deduced from cDNA; it consists of 438 residues and a 20-residue signal peptide (Appendix, Fig. A56) (1642). The amino-acid sequence of the heavy chain has also been deduced from the cDNA. It consists of 536 amino-acid residues, giving a calculated molecular mass of 58.9 kDa (Appendix, Fig. A57) (1643). The chain has carbohydrate that is both N-bound (up to eight potential N-binding sites) and O-bound.

In addition to inactivating anaphylatoxins, carboxypeptidase N inactivates bradykinin, which, like the anaphylatoxins, possesses a functionally essential C-terminal arginine residue (see Chapter 18).

17. Complement receptors

Nine complement receptors in all have been described. Most of them function by recognising regions of C3. These are the receptors CR1, CR2, CR3, CR4, the C3a/C4a receptor and the C3e receptor. The remaining receptors are specific for C1q, C5a and factor H. In this chapter we shall examine CR1, CR2 and CR3, on account of the scant structural information available for the others; for further information, the reader may consult the reviews in references (1182, 1190 and 1204).

17.1. Complement receptor type 1

Complement receptor type 1 (CR1) was once termed the C3b/C4b receptor. Its structure and function have already been described in section 15.2.4, in connection with regulation processes in the complement system.

As its older name suggests, CR1 binds to C3b and C4b that are already attached to foreign bodies in the bloodstream, thus binding indirectly to the foreign bodies. It also binds to the two C3b fragments iC3b and C3c, but with lower affinity. It will be recalled (section 15.2.8) that iC3b is the fragment of C3b that remains when the 17-residue fragment C3f is split off. C3c is a further cleavage product of iC3b that is released into the bulk phase, as it does not contain the thioester group. For a summary of C3 cleavage products, see Figure 120.

We have already seen that CR1 participates in the catalysed decay of C3/C5-convertase (section 15.2.6). We also noted that CR1 functions as a co-factor in the factor I-catalysed cleavage of C3b and C4b (section 15.2.8). This function is of course associated with cell surfaces to which CR1 is bound, so that it is for the most part surface-bound C3b and C4b that are affected.

Since 90% of the CR1 in the blood is found on the surfaces of erythrocytes, it may well be asked whether its main function on these cells is not homologous restriction – protection of the cells against undesired C3-convertase activity. On other cells, CR1 of course plays a further part as

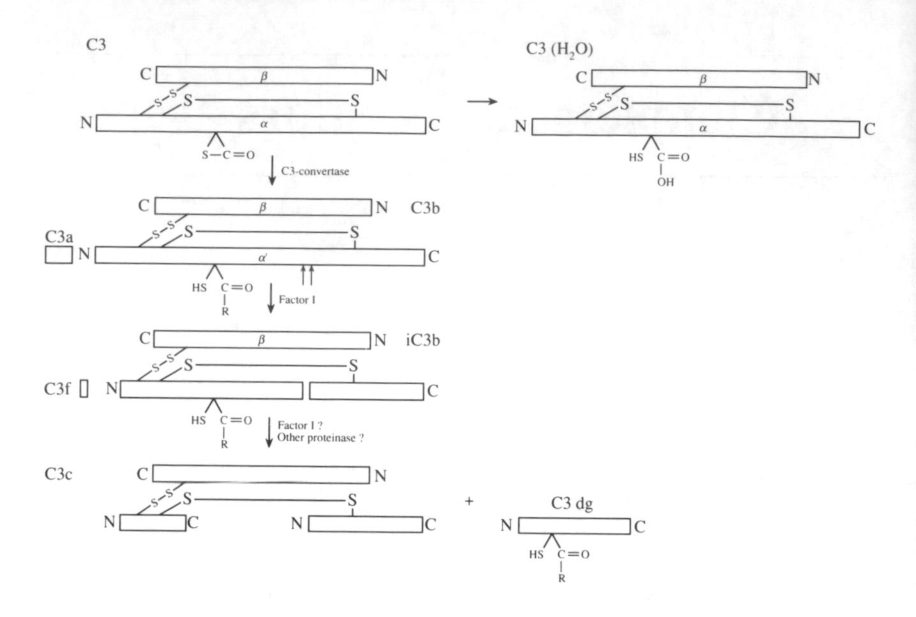

Fig. 120. Cleavage products of C3/C3b. R represents a surface-bound nucleophile. Note that the disulphide-bond pattern in C3 is not known exactly.

receptor for C3b/C4b-labelled foreign bodies that are to be removed by phagocytosis. The number of CR1 molecules on, for example, neutrophil cells rises dramatically following stimulation by C5a.

17.2. Complement receptor type 2

Complement receptor type 2 (CR2) is a single-chained membrane glycoprotein on the surface of B lymphocytes. Its molecular mass is 145 kDa (1644), of which 110 kDa is polypeptide chain and the rest carbohydrate (1645). The amino-acid sequences of tryptic peptides from CR2 have revealed strong homology between CR1 and CR2, and indeed CR1 cDNA has been used as a starting-point for the cloning of CR2 cDNA (1646).

The complete amino-acid sequence of CR2, including a 20-residue signal peptide, has been deduced from the cDNA sequence (643, 644) (Appendix, Fig. A58). The CR2 molecule can vary because of alternative exon use and also because of the presence of different alleles (1647). CR2 consists of 15 (alternatively, 16) SCR domains, a presumed transmembrane domain and a cytoplasmic domain (Fig. 121). These two forms consist of 1012 or 1067 residues with calculated molecular masses of 110 kDa and 117 kDa respectively, to which should be added carbohydrate bound at 12 (or, respectively, 14) potential N-glycosylation sites. As was the case for CR1

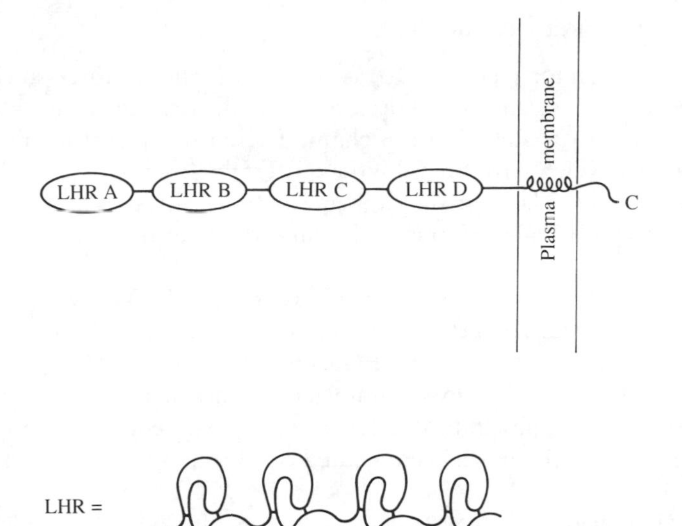

Fig. 121. Domain structure of CR2. The figure shows the form of CR2 that contains 16 SCR domains. Each long homologous repeat (LHR) contains four SCR domains.

(section 15.2.4), the SCR domains in CR2 are organised into long homologous repeats, but with the difference that in CR2 each long homologous repeat contains four SCR domains, not seven. In the shorter form of CR2, the missing SCR domain is the eleventh (644).

The gene for CR2 is linked to the gene for CR1, and they map on position q32 of chromosome 1 (1495). We have by now encountered quite a number of linked genes on chromosome 1, and these all code for proteins that consist almost entirely of SCR domains. These are the genes for CR1, CR2, DAF, MCP, factor H, C4b-BP and the b-chain of factor XIII.

CR2 functions as a receptor for the C3b fragments iC3b, C3dg and C3d (see Fig. 120), all of which are surface-bound. The affinity of CR2 for iC3b is also possessed by CR1 and CR3, and that for C3dg is also shown by CR3. In contrast, only CR2 has affinity for C3d. Its binding site for C3d has been located between residues 1209 and 1236 in C3, corresponding to residues 483 and 510 in the α' chain of C3 (1648).

It has been reported that CR2 has a weak co-factor activity in the cleavage of iC3b catalysed by factor I, but the significance of this is uncertain (1649). It should be added that CR2 shows no activity in connection with the decay of C3-convertase or in the cleavage of C3b catalysed by factor I. CR2 is also a receptor for Epstein-Barr virus.

17.3. Complement receptor type 3

Complement receptor type 3 (CR3) is one of a family of three so-called leucocyte adhesion receptors. All three are glycoproteins consisting of an α chain bound non-covalently to a β chain. These three receptors differ in their α chains, which have homologous N-terminal regions but different molecular masses. The β chains appear to be identical. The leucocyte adhesion receptors make up one of the sub-groups of the integrin family (Chapter 25).

The α chain of CR3 has a molecular mass of 165–170 kDa, while that of the β chain is 95 kDa. The α chain's amino-acid sequence has been deduced from cDNA sequences (Appendix, Fig. A59); it consists of 1136 (1650) or 1137 (1651, 1652) residues, giving a molecular mass of 125 kDa, to which should be added carbohydrate attached at up to 19 potential Asn residues. The gene for the α chain has been located on chromosome 16 at position p11–13.1 (1653, 1654).

The amino-acid sequence of the β chain has also been deduced from cDNA (1655, 1656) (Appendix, Fig. A60), but the exact size of this chain in the fully processed form cannot be determined, as the N terminus is blocked. The β chain is probably 747 residues long, corresponding to a molecular mass of 82.5 kDa. This chain has five potential N-glycosylation sites, which presumably are used, on account of the difference between the measured and the calculated molecular masses. The gene for the β chain is on chromosome 21 at position q22.1–qter (1654, 1657). The amino-acid sequences of the α and β chains of CR3 place this protein clearly in the family of receptors called integrins, all of which employ the ligand-recognition site R-G-D (see Chapter 25).

As stated above, CR3 functions as a receptor for iC3b, and the part of iC3b recognised by CR3 lies between residues 1361 and 1380 in C3, corresponding to residues 635–654 in the α' chain, a region that indeed contains the tripeptide sequence R-G-D (1658).

Part IV Special Topics

18.

The kinin system

Kinins function, among other things, as vasodilators – that is, they relax the blood vessels and thus increase their permeability, leading to a reduction in blood pressure (hypotension). It is believed that the kinins in this way participate in the regulation of microcirculation, for example in an organ. The other pharmacological properties of the kinins (causation of pain, contraction of smooth muscle and stimulation of the metabolism of arachidonic acid) indicate that the kinins are important mediators of inflammation.

The kinin system (reviewed in references 12, 1659–1661) contains two main components: kininogen molecules and kallikrein molecules. As already discussed in section 2.1.4, there are two types of kininogen in humans, HMM-kininogen and LMM-kininogen. There are also two types of kallikrein, plasma kallikrein and glandular kallikrein. Plasma kallikrein we have already met in connection with blood coagulation (section 2.1.2). The glandular kallikreins are all members of a sub-family of serine proteinases characterised by their ability to split kinins from kininogens *in vitro*. The situation *in vivo* is somewhat different, but some glandular kallikreins do release kinin from kininogen, for example in the case of renal kallikrein.

It is worth noting that there are two types of kinin, bradykinin and kallidin; these differ only in that kallidin has an additional lysine residue at its N terminus, so that it is also referred to as Lys-bradykinin (Fig. 122). Both molecules have kinin activity, but are formed from kininogens by cleavage with two different kallikreins: bradykinin is formed by cleavage with plasma kallikrein, while kallidin is formed by cleavage with glandular kallikrein.

The bradykinin sequence shown in Figure 122 has a proline residue at position 3, but it has been found that this proline can be hydroxylated, appearing as 4-hydroxy-proline (62–64). Pro3 in bradykinin is not always hydroxylated; experimental values vary, and there is disagreement as to whether HMM-kininogen and LMM-kininogen are hydroxylated to the same extent at this position.

```
Bradykinin     Arg-Pro-Pro-Gly-Phe-Ser-Pro-Phe-Arg

Kallidin   Lys-Arg-Pro-Pro-Gly-Phe-Ser-Pro-Phe-Arg
```

Fig. 122. The amino-acid sequences of bradykinin and kallidin.

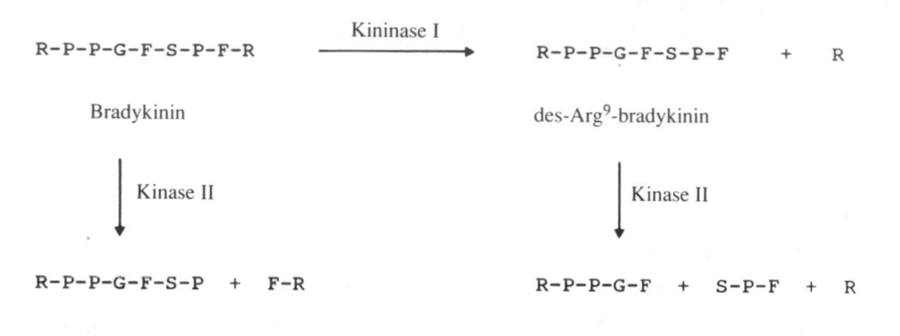

Fig. 123. The inactivation of kinins by carboxypeptidase N and angiotensin-converting enzyme. Kininase I designates carboxypeptidase N (anaphylatoxin inactivator) and kininase II designates angiotensin-converting enzyme.

Although this may some day prove to be incorrect, it appears helpful to consider the kinin system as two separate sub-systems, one consisting of HMM-kininogen and plasma kallikrein and one consisting of LMM-kininogen and glandular kallikreins. *In vitro*, plasma kallikrein clearly prefers HMM-kininogen as substrate, and even though glandular kallikrein can use both HMM-kininogen and LMM-kininogen as substrate *in vitro*, there is evidence that only LMM-kininogen is used as substrate *in vivo*. This view finds support in the fact that persons who, because of genetic defects, lack both forms of kininogen, do not excrete kinin in the urine, while persons lacking only HMM-kininogen excrete normal quantities of kinin.

Kinins can thus form in two ways. First, bradykinin can be released in connection with the coagulation of the blood, by the cleavage of HMM-kininogen by plasma kallikrein, and, secondly, kallidin can form in the tissues of various glands and organs as a consequence of the activity of kallikreins that cleave LMM-kininogen. The effects of the kinins are mediated by kinin receptors, but it is not known how the signal is passed on from the receptors.

The kinins are inactivated in plasma by carboxypeptidase N and angiotensin-converting enzyme (Fig. 123). Carboxypeptidase N we met earlier as anaphylatoxin inhibitor from the complement system. In the context of the kinin system, it is also known as kininase I, and it removes the C-terminal arginine residue from kinins, giving *des*-Arg kinins. Both kinins and *des*-Arg kinins are also inactivated by angiotensin-converting enzyme (kininase II), either by the removal of the two C-terminal amino-acid residues Phe-Arg (kinin), or by the removal of the three C-terminal amino-acid residues Ser-Pro-Phe (*des*-Arg kinin). This inactivation of kinin by angiotensin-converting enzyme has potentially interesting connotations. Angiotensin-converting enzyme is the enzyme that converts angiotensin I to angiotensin II by removing the dipeptide His-Leu from the C terminus of angiotensin I. Angiotensin I consists of 10 amino-acid residues (D-R-V-Y-I-H-P-F-H-L) that are derived from the cleavage of angiotensinogen by the enzyme renin. Angiotensin II is a potent vasoconstrictor, and the function of angiotensin-converting enzyme is thus to raise blood pressure. It is possible that the effect of the angiotensin II thus formed might be reinforced by a simultaneous inactivation of kinins, which in themselves have an opposite (vasodilatory) effect. However, this is still largely a matter for informed speculation.

19.

Glycosylation

Glycosylation is the attachment of carbohydrate, in the form of oligosaccharides, to proteins. It is a very common type of post-translational modification for proteins whose function is extracellular (reviewed in reference 1662). Glycosylation is a phenomenon so widespread among plasma proteins and membrane proteins that a short description is called for here.

For glycosylated proteins of mammalian origin, it is usual to distinguish between three manners of carbohydrate attachment, according to the amino-acid side-chain to which the carbohydrate is bound. The mode of attachment found most frequently is N-acetylglucosamine bound to asparagine residues; less frequently, N-acetylgalactosamine is found bound to serine or threonine residues; and, least frequently of all, we encounter galactose bound to 5-hydroxy-lysine residues. Most plasma glycoproteins have only Asn-bound carbohydrate, but there are some that contain more than one of the three forms.

In addition to being the most common form, asparagine-bound carbohydrate is also the most diverse. It is usual to distinguish between three types of Asn-bound oligosaccharides, called 'high mannose', 'hybrid' and 'complex' (Fig. 124). These oligosaccharides have a common nuclear structure, attached to asparagine via N-acetylglucosamine, from which comes the expression 'N-acetylglucosamine-based carbohydrate'. This kind of attachment requires that the Asn residue be part of the consensus sequence Asn-Xaa-Ser/Thr, where Xaa can be any amino-acid residue but is rarely if ever Pro or Asp. A few cases are known in which carbohydrate is attached at Asn in the sequence Asn-Xaa-Cys.

The mechanism of attachment of carbohydrate at Asn residues follows a fairly consistent pattern (Fig. 125). Attachment is carried out both in the endoplasmic reticulum and in the Golgi apparatus. The reason for this consistent pattern is doubtless that the various enzymes involved recognise particular intermediate structures.

Carbohydrates anchored at serine or threonine are much simpler

High mannose

```
Man α1,2-Man α1,6\
                  Man α1,6\
Man α1,2-Man α1,3/         \
                            Man β1,4-GlcNAc β1,4-GlcNAc β-Asn
Man α1,2-Man α1,2-Man α1,3/
```

Hybrid

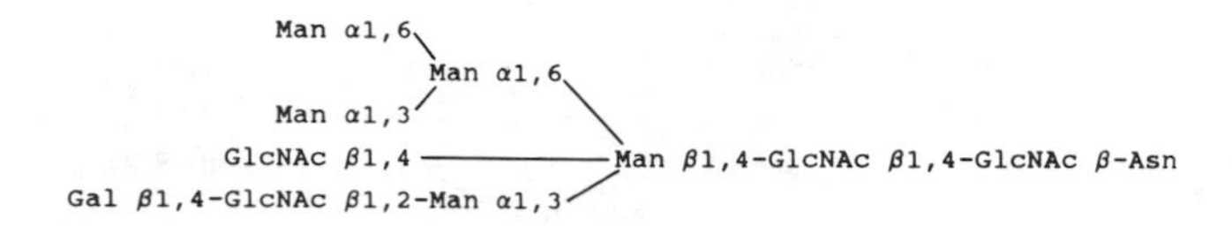

Complex

```
Sia α2,3/6-Gal β1,4-GlcNAc β1,2-Man α1,6\
                  ± GlcNAc β1,4 ---------Man β1,4-GlcNAc β1,4-GlcNAc β-Asn
Sia α2,3/6-Gal β1,4-GlcNAc β1,2-Man α1,3/              ± Fuc α1,6/
```

```
Sia α2,3/6-Gal β1,4-GlcNAc β1,2-Man α1,6\
                                         Man β1,4-GlcNAc β1,4-GlcNAc β-Asn
Sia α2,3/6-Gal β1,4-GlcNAc β1,2-Man α1,3/              ± Fuc α1,6/
Sia α2,3/6-Gal β1,4-GlcNAc β1,4/
          ± Fuc α1,3/
```

Fig. 124. Examples of the three classes of N-bound carbohydrate. (GlcNAc, N-acetylglucosamine; Man, mannose; Gal, galactose; Fuc, fucose; Sia, sialic acid.) (Based upon reference 1662.)

oligosaccharides than those we encounter bound to asparagine (Fig. 126). The latter oligosaccharides are all bound via N-acetylgalactosamine and are therefore referred to as N-acetylgalactosamine-based carbohydrate. Their synthesis is presumed to take place in the Golgi apparatus. There is no consensus sequence of residues for O-glycosylation, but O-glycosylation is often encountered in serine- and threonine-rich regions.

Carbohydrate attached to 5-hydroxy-lysine residues consists only of glucose-β-1,2-galactose, and among plasma proteins it is encountered only in the case of C1q.

Finally, the attachment of ((xylose)$_2$glucose) to serine residues in the first growth-factor domain of bovine factor VII, bovine factor IX and bovine protein Z and the attachment of (xylose-glucose) to the corresponding Ser in human factor VII, human factor IX and human protein Z (137, 138) have been described. However, for human factor VII and protein Z, the carbohydrate attached can alternatively be the trisaccharide (138).

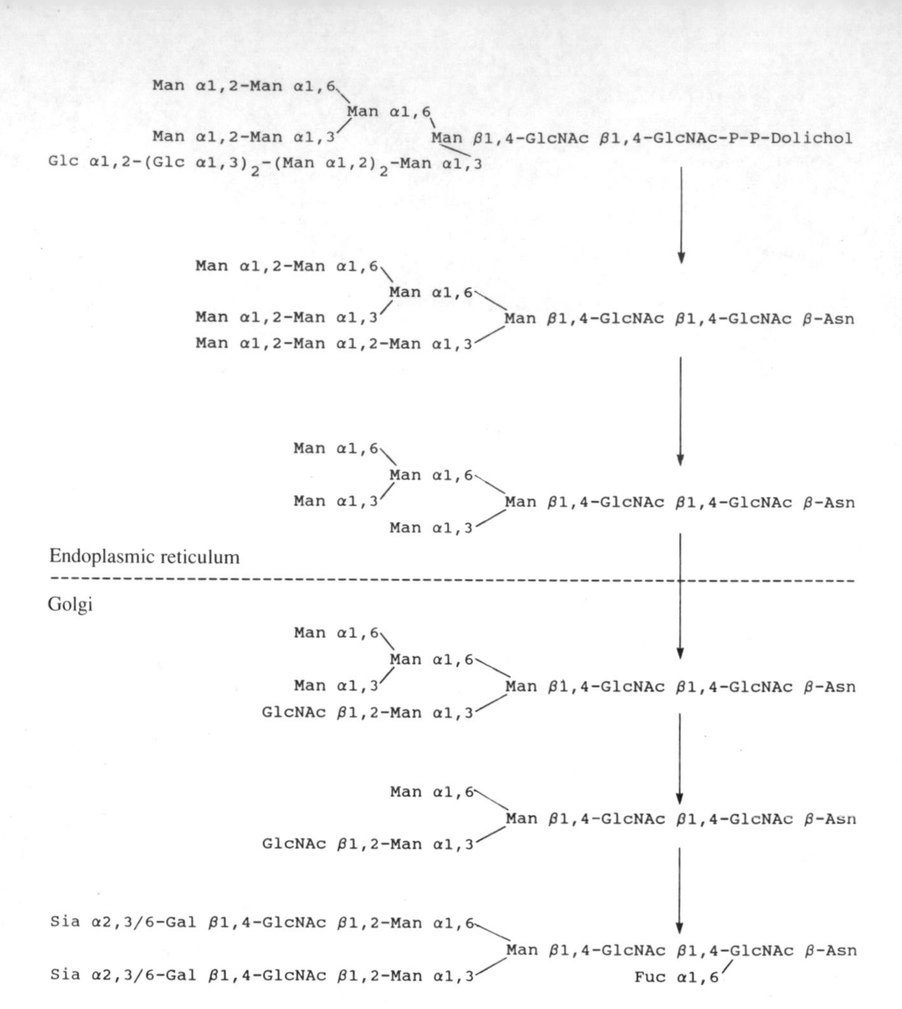

Fig. 125. A suggested synthetic pathway for N-bound carbohydrate. (GlcNAc, N-acetylglucosamine; Man, mannose; Gal, galactose; Fuc, fucose; Sia, sialic acid.) (Based upon reference 1662.)

GalNAc - Ser/Thr

Gal β1,3-GalNAc - Ser/Thr

Sia α2,3-Gal β1,3-GalNAc - Ser/Thr

Gal β1,3-GalNAc - Ser/Thr
Sia α2,6 (linked to GalNAc)

Sia α2,3-Gal β1,3-GalNAc - Ser/Thr
Sia α2,6 (linked to GalNAc)

Fig. 126. Examples of O-bound carbohydrate. (GalNAc, N-acetylgalactosamine; Gal, galactose; Sia, sialic acid.) (Based upon reference 1662.)

20.

Signal peptides

A signal peptide is normally the amino-acid sequence that starts with the methionine at which translation commences and ends immediately before the residue at the N terminus of the mature, processed protein (see, however, Chapter 21).

Virtually all proteins that traverse a membrane, and many membrane proteins, need a signal peptide, or signal sequence, in order to reach their correct location (reviewed in references 1663–1669). This is also the case for the plasma proteins that we have met, since all of them are secreted, having passed through the endoplasmic reticulum (ER) and the Golgi apparatus. The membrane traversed here is thus the ER membrane.

Quite different signals are used if other membranes are to be crossed. Each organelle has a specific target sequence, so that proteins synthesised in the cytoplasm can be transported to their correct destination.

The function of the signal peptides of the plasma proteins cannot be discussed without reference to the secretion system, so we also examine this briefly.

Let us first consider the signal peptide. Functional signal peptides vary in length from 13 amino-acid residues upwards, and most fall into the range 18–30 residues.

The N-terminal part of the signal peptide normally contains one or more charged residues, including the protonated N-terminal amino group, and these together always give a nett positive charge. There are indications that the presence of this nett positive charge is of importance for the secretion process. The length of the charged region varies considerably – from 2 to 8 residues, or even more.

The most characteristic property of the signal peptides is a stretch of uncharged, mostly hydrophobic residues, often called the hydrophobic core. This stretch follows directly after the charged stretch and is as a rule 10 to 15 residues long. The hydrophobic regions of various signal peptides have no clear homology one with another, and it is postulated that their hydrophobicity and length are determined by their function (or *vice versa*).

The amino-acid residues occurring most frequently in the hydrophobic core are leucine, alanine, valine, phenylalanine and isoleucine, but less hydrophobic residues such as threonine, serine and proline are also found there. It has been suggested that the function of the hydrophobic region is to enter the lipid membrane, while other suggestions involve more specific interactions.

The last five or so amino-acid residues in the signal peptide are more polar than those in the hydrophobic region, and they define the site of cleavage for the signal peptidase. Signal peptidase is the enzyme that removes the signal peptide from the rest of the protein. The cleavage site is defined primarily by the last three residues, of which both the ultimate and the antepenultimate residues have a small, apolar side-chain and most frequently are alanine.

Having described the signal peptide, we return to examine the translation of mRNA for a protein that is to be secreted. Not all details of this process, which will vary from case to case, are known; the following model is representative of current thinking (Fig. 127).

First, the ribosome binds to an mRNA molecule that encodes the secreted protein, and translation commences. This takes place free in the cytoplasm. When the nascent protein has grown to a polypeptide about 80 residues in length, the signal peptide begins to protrude from the ribosome. The signal peptide then binds to another particle, that is called the signal-recognition particle (SRP).

SRP is a cytoplasmic ribonucleoprotein that consists of six different polypeptide chains of molecular mass 9, 14, 19, 54, 68 and 72 kDa plus an RNA molecule of about 300 nucleotides. The sequence of the RNA molecule is known, but only three of the polypeptides (14 kDa, 19 kDa and 54 kDa) have been sequenced by inference from their cDNA. The function of the RNA in SRP is unknown; it appears at least to act as a structural skeleton, binding five of the six proteins. The sixth (54 kDa) is affixed to the 19-kDa polypeptide.

SRP recognises ribosomes that are synthesising secretory proteins, and it is the 54-kDa polypeptide that binds directly to the signal peptide. When SRP has bound to the translation complex, a blocking of translation occurs; this blocking depends upon two of the SRP chains (9 kDa and 14 kDa). It has been shown that the blocking of translation couples the protein synthesis to the extrusion process.

The next step in this process is the binding of SRP to the SRP receptor, which is also called the 'docking protein'. SRP receptor is an integral protein found on the cytoplasmic surface of the rough ER membrane. It is a two-chained integral membrane protein of rough ER. Its α chain has a molecular mass of 72 kDa and its β chain has a molecular mass of 30 kDa. The α chain has been sequenced via cDNA. Its C-terminal region shows homology with the N-terminal region of the 54-kDa polypeptide of SRP.

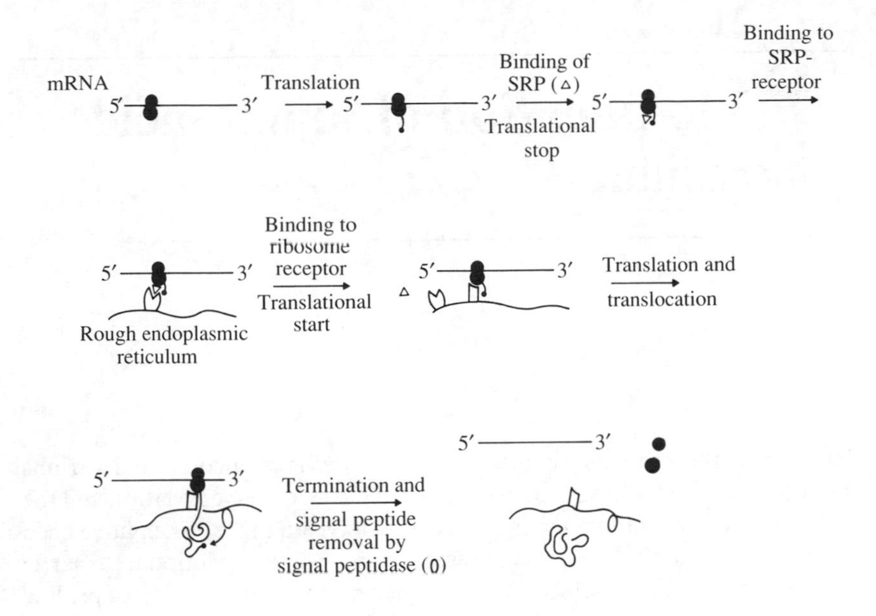

Fig. 127. A model for the secretion of proteins.

The SRP receptor binds to SRP; the binding is mediated by the 68-kDa and 72-kDa polypeptides of SRP. Thus the SRP and SRP receptor have acted as 'adaptors' between the membrane and the translation complex, anchoring the latter to the membrane through which the secreted protein is to be translated. In a GTP-dependent process, SRP dissociates from the ribosome, unblocking the translation process and allowing it to continue. At the same time, the binding of the ribosome to the rough ER membrane is taken over by other structural elements. The transport process is thus complete.

We now have a ribosome attached to the rough ER membrane through which the protein under synthesis is destined to pass. It is still a matter for speculation whether this transport is effected by isolated protein molecules or by complexes, perhaps building a pore through the membrane.

During or after translation, the enzyme signal peptidase removes the signal peptide from the nascent protein, which can then be secreted. The signal peptidase is an integral enzyme on the inner (non-cytoplasmic) side of the rough ER membrane.

21.

Modification of amino-acid residues

In most of the proteins we have discussed, we have encountered various modifications to the amino-acid residues. Apart from glycosylation and the formation of disulphide bonds, the two we have met most often have been γ-carboxylation of glutamic acid residues and β-hydroxylation of aspartic acid and asparagine residues. We have further met hydroxylation of proline and lysine residues, in connection with HMM-kininogen and C1q, and also sulphation and phosphorylation. In this book, we encounter sulphation in the cases of factor VIII, complement component C4, α_2-plasmin inhibitor and heparin co-factor II, and phosphorylation in the cases of fibrinogen, complement component C3, S-protein/vitronectin and fibronectin.

In this chapter, we examine more closely the mechanisms of γ-carboxylation and β-hydroxylation. The functions of these two modifications will be kept for Chapter 23, which deals with the binding of proteins to coagulation-active membranes.

21.1. γ-carboxylation

The significance of vitamin K for the coagulation of blood has been realised for over fifty years. It was in fact the observation of coagulation defects in chickens kept on a special diet that led Henrik Dam to the discovery of vitamin K (1670). The rôle of vitamin K, and the process it takes part in, were naturally elucidated much later.

Vitamin K functions as co-factor for the γ-carboxylation of glutamic acid (Glu) side-chains to γ-carboxyglutamic acid (Gla) in various proteins. In the coagulation system, these include factors VII, IX and X, prothrombin, protein C and protein S, but Gla residues are also found in protein Z from plasma, and in the bone proteins osteocalcin (bone Gla protein) and matrix Gla protein. (It should be mentioned that protein Z is a serine-proteinase analogue without known function.)

The mechanism of the γ-carboxylation of Glu residues has not been

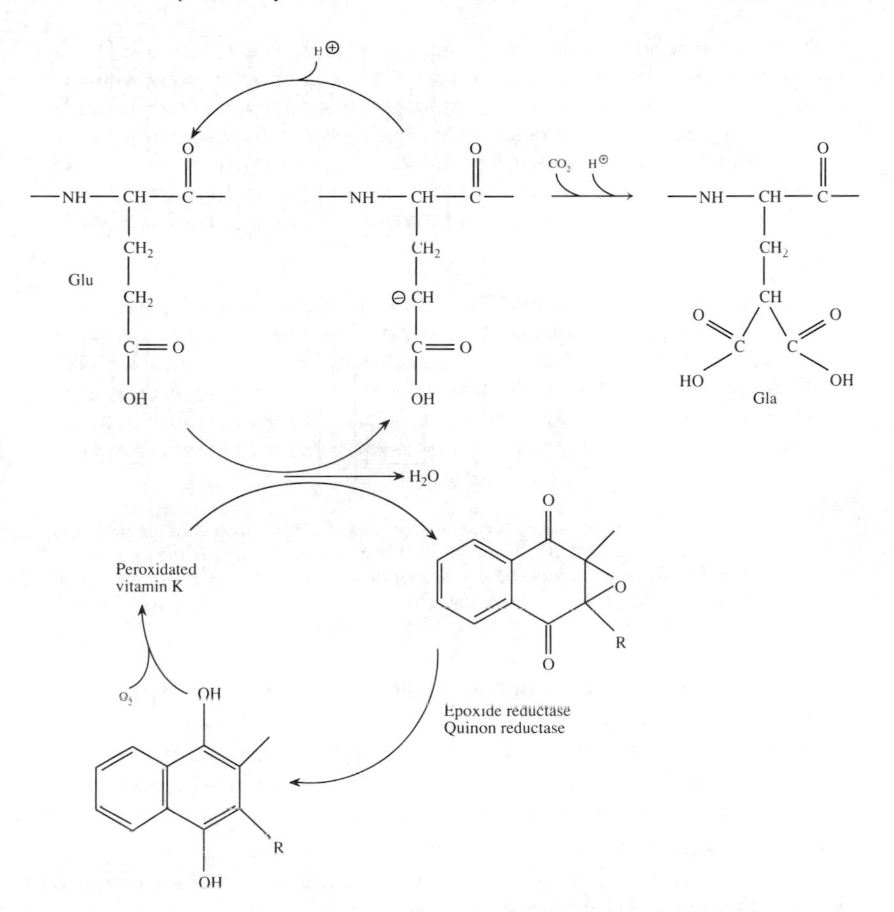

Fig. 128. A suggested mechanism for the vitamin K-dependent carboxylation of glutamic acid residues. R designates the side-chain in vitamin K. (Based upon reference 1671.)

described in detail, but it is known to require the enzyme vitamin K-dependent carboxylase, vitamin K, O_2 and CO_2. With these components, the following model can be presented for the mechanism of γ-carboxylation (reviewed in reference 1671) (Fig. 128).

The carboxylation reaction starts out from the reduced form of vitamin K. This molecule is oxygenated with molecular oxygen to give first a peroxidised intermediate which leads to another form that is both oxidised and epoxidised. It seems that, in this process, a proton is removed from the γ carbon atom in the side-chain of glutamic acid, giving a carbanion that then attacks CO_2, and the carboxylation is complete.

However, the oxidised and epoxidised forms of vitamin K must still be returned to the reduced form, which requires a vitamin K epoxide reductase

```
Bovine factor VII      ANGFLγγLLPGSLγRγCRγγLCSFγγAHγIFRNγγRTRQFWVSY
Bovine factor IX      YNSGKLγγFVRGNLγRγCKγγKCSFγγARγVFγNTγKTTγFWKQY
Bovine factor X        ANSFLγγVKQGNLγRγCLγγACSLγγARγVFγDAγQTDγFWSKY
Bovine protein C       ANSFLγγLRPGNVγRγCSγγVCτFγγARγIFQNTγDTMAFWSKY
Bovine protein S       ANTLLγγTKKGNLγRγCIγγLCNKγγARγIFγNNPγTEYFYPKY
Bovine protein Z      AGSYLLγγLFYGHLγKγCWγγICVVγγARγVFγDDγTTDγFWRTY
Bovine prothrombin    ANKGLLγγVRKGNLγRγCLγγPCSRγγAFγALγSLSATDAFWAKY

Human factor VII       ANAFLγγLRPGSLγRγCKγγQCSFγγARγIFKDAγRTKLFWISY
Human factor IX       YNSGKLγγFVQGNLγRγCMγγKCSFγγARγVFγNTγRTTγFWKQY
Human factor X         ANSFLγγMKKGHLγRγCMγγTCSYγγARγVFγDSDKTNγFWNKY
Human protein C        ANSFLγγLRHSSLγRγCIγγICDFγγALγIFQNVDDTLAFWSKH
Human protein S        ANSLLγγTKQGNLγRγCIγγLCNKγγARγVFγNDPγTDYFYPLY
Human prothrombin      ANTFLγγVRKGNLγRγCVγγTCSYγγAFγALγSSTATDVFWAKY
```

Fig. 129. The amino-acid sequences of Gla domains from bovine and human coagulation factors. Only the positions at which amino acids are the same in all Gla domains from a single species are outlined. All the bovine sequences were determined by protein-sequencing, but only the sequences for human factors VII and X and prothrombin were found by this method.

and at least two quinone reductases; the details exceed the scope of this book.

However, it should be mentioned that a vitamin K-dependent carboxylase has recently been purified from bovine liver as a single-chained, 77-kDa protein (1672). This protein possesses activity both as a vitamin K-dependent carboxylase and as a vitamin K epoxidase.

We must, however, still ask the question of what causes some proteins to be γ-carboxylated. How does the protein signal the need for such a modification? This cannot be answered with certainty, but some indications are available.

If we first examine the amino-acid sequence of the Gla-rich region in the coagulation factors and in protein Z, we find that they are highly homologous (Fig. 129), but that there is no homology with the Gla-containing bone proteins; these, however, are homologous to one another over a certain range of their sequence (Fig. 130). Therefore, if there is a common signal for γ-carboxylation, it must reside at a different point in the mature protein.

As was described in Part I (sections 3.1, 4.1.4, 5.1.2, 6.1.1, 8.2.1 and 8.2.3), the vitamin K-dependent coagulation factors are synthesised with a signal/propeptide some 40 residues long, often a little longer still. The term signal/propeptide implies that the N-terminal part of this peptide has a structure resembling an ordinary signal peptide, and this is followed by a stretch of residues that are considerably more hydrophilic and are without counterpart in normal signal peptides. Studies on mutants of factor IX have

```
Swordfish BGP  13 γ S L R γ V C τ L N V A C D G M A D T A G I V A A Y I A 40
Murine BGP     13 γ P T R γ Q C τ L N P A C D E L S D Q Y G L Y T A Y K R 40
Rat BGP        17 γ P H R γ V C τ L N P N C D E L A D H I G F Q D A Y K R 44
Bovine BGP     17 γ P K R γ V C τ L N P D C D E L A D H I G P Q E A Y R R 44
Human BGP      17 E P R R γ V C τ L N P D C D E L A D H I G F Q E A Y R R 44
Human MGP      48 γ L N R γ A C D D Y R L C E R Y A M V Y G Y N A A Y N R 75
Bovine MGP     48 γ L N R γ A C D D F K L C E R Y A M V Y G Y N A A Y D R 75
Rat MGP        48 γ I N R γ A C D D Y K L C E R Y A L I Y G Y N A A Y N R 75
```

Fig. 130. Comparison of amino-acid sequences of selected stretches from osteocalcin (bone Gla protein, BGP) and matrix Gla protein (MGP) from various species. Only the positions at which amino acids are the same in at least seven of the sequences are outlined. Note that only in swordfish BGP, bovine BGP, human BGP and bovine MGP have the Gla residues been located experimentally. The amino-acid sequences are taken from the following references: swordfish, bovine and human BGP (1910), murine and rat BGP (1911), bovine MGP (1912), human and rat MGP (1913).

shown directly that the last 18 residues of the signal/propeptide are not cleaved off by signal peptidase in the secretion process, and these must therefore make up a propeptide. The signal peptide thus stretches from the initiator Met residue in position –46 to the Cys residue in position –19. Other signal/propeptides follow the same pattern, even though the exact cleavage site for signal peptidase is not known in all cases. In protein C, the propeptide starts at the Thr residue in position –24, while in the case of factor X it is believed to start at the Leu residue in position –17.

If signal/propeptides are compared with one another, the picture shown in Figure 131 arises. Interestingly, there is clear homology between the propeptides, indicating that the propeptide stretch may play an important part in correct γ-carboxylation. There is now considerable evidence for the correctness of this conclusion (150, 151, 154, 1673–1679).

Studies of factor IX and protein C have shown that removal of the propeptides results in the abolition of γ-carboxylation, just as the mutation of Phe(–16) and Ala(–10) in factor IX also depresses this process. It is therefore interesting that one published cDNA sequence for human protein S codes for Phe at position –16 (882), while the other codes for Leu at this position (881). In bovine protein S, codon –16 encodes a Phe residue (879).

Nothing is known about the identity of the enzyme that removes the propeptide at the right point, but it is known to be arginine-specific: the mutation of Arg(–1) to Ser or Arg(–4) to Gln in factor IX prevents the removal of the propeptide and appears to reduce the extent of γ-carboxylation.

The essence of the model is thus that the propeptide part of the signal/propeptide functions as a recognition signal for vitamin K-dependent carboxylase.

```
                                 -40       -30       -20       -10        -1
Human factor VII          MVSQALRLLCLLLGLQGCLAAVFVTQEEAHGVLHRRRRA
Human factor IX     MGRVNMIMAESPGLITICLLGYLLSAECTVFLDHENANKILNRPKRY
Canine factor IX              MAEASGLVTVCLLGYLLSAECAVFLDRENATKILSRPKRY
Human factor X               MGRPLHLVLLSASLAGLLLLGESLFIRREQANNILARVTRA
Bovine factor X              MAGLLHLVLLSTALGGLLRPAGSVFLPRDQAHRVLQRARRA
Human protein C            MWQLTSLLLFVATWGISGTPAPLDSVFSSSERAHQVLRIRKRA
Bovine protein C               TSLLLFVTIWGISSTPAPPDSVFSSSQRAHQVLRIRKRA
Human protein S              MRVLGGRCGAPLACLLLVLPVSEANFLSKQQASQVLVRKRRA
Bovine protein S             MRVLGGRTGTLLACLALVLPVLEANFLSRQHASQVLIRRRRA
Human prothrombin          MAHVRGLQLPGCLALAALCSLVHSQHVFLAPQQARSLLQRVRRA
Bovine prothrombin         MARVRGPRLPGCLALAALFSLVHSQHVFLAHQQASSLLQRARRA

Rat BGP       -49 MRTLSLLTLLALTAFCLSDLAGAKPSDSESDKAFMSKQEGSKVVNRLRRY 1
Murine BGP    -49 MRTLSLLTLLALAALCLSDLTDAKPSGPESDKAFMSKQEGNKVVNRLRRY 1
Human BGP     -49 MRALTLLALLALAALCIAGQAGAKPSGPESSKAFVSKQEGSEVVKRPRRY 1
Human MGP     -19 MKSLILLAILAALAVVTLCYESHESMESYELNPFINRRNANTFISPQQRW 31
Bovine MGP    -19 MKSLLLLSILAALAVAALCYESHESLESYEINPFINRRNANSFISPQQRW 31
Rat MGP       -19 MKSLLPLAILAALAVAALCYESHESMESYEVSPFTNRRNANTFISPQQRW 31
```

Fig. 131. The amino-acid sequences of signal/propeptides for vitamin K-dependent coagulation factors and for the vitamin K-dependent bone Gla protein. Amino-acid sequences are included from matrix Gla protein, which is presumed to play the rôle of a propeptide. Positions at which amino acids are the same in all sequences are outlined. Furthermore, positions at which amino acids are identical in all the signal/propeptides from coagulation factors are outlined.

Fig. 132. Occurrence of β-OH-aspartic acid and β-OH-asparagine in domains homologous to epidermal growth factor. (a) Growth-factor domains with documented occurrence of β-OH-Asp or β-OH-Asn, the positions of which are indicated by β. In the last three stretches of bovine protein S, in human C1r and in uromodulin there occur β-OH-Asn residues, while in the other cases β-OH-Asp is found. Note that the degree of hydroxylation is not 100% in all cases. The figures in brackets give the locations of the peptide sequences shown within the amino-acid sequences of the proteins. The sequence for human uromodulin is taken from reference (1914) and the location of β-OH-Asn residues in this sequence from reference (1915). The sequence of bovine protein Z is from reference (1916). All other sequences are taken from references cited in the text. Amino-acid residues thought to be part of the consensus sequence for β-hydroxylation are outlined. (b) Amino-acid sequences of growth-factor domains of proteins (i) in which β-hydroxylated side chains are found but in which their location in the sequence is not known, (ii) in which β-hydroxylated side chains are believed to be present because the consensus sequence is found and (iii) in which β-hydroxylation is known not to occur, even though a consensus (or a very similar) sequence is present. The figures in brackets give the locations of the peptide sequences shown within the amino-acid sequences of the proteins. The amino-acid sequence of human LDL receptor is taken from reference (1429) and that for murine EGF precursor is from references (1917) and (1918). All other sequences are taken from references cited in the text. Amino-acid residues thought to be part of the consensus sequence for β-hydroxylation are outlined.

(a)

Bovine protein S	M T C K β G Q A T F T C I C K S	(91-106)
	Q I C E β T P G S Y H C S C K N	(132-147)
	A V C E β I P G D F E C E C D G	(174-189)
	Q L C V β Y P G G Y S C Y C D G	(213-228)
Bovine protein C	G K C I β G L G G F R C D C A E	(67-82)
Bovine factor IX	G M C K β D I N G Y E C W C Q A	(60-75)
Human factor IX	G S C K β D I N S Y E C W C P F	(60-75)
Bovine factor X	G H C K β G I G D Y T C T C A E	(59-74)
Human factor X	G K C K β G L G E Y T C T C L E	(59-74)
Bovine protein Z	G S C Q β S I R G Y A C T C A P	(60-75)
Bovine factor VII	G S C E β Q L R S Y I C F C P D	(59-74)
Human C1r	H L C H β Y V G G Y F C S C R P	(146-161)
Human uromodulin	S S C V β T P G S F S C V C P E	(57-72)
	A T C V β V V G S Y L C V C P A	(100-115)

(b)

(i)

Human C1s	H P C N N F I G G Y F C S C P P	(130-145)
Human protein S	M S C K D G L A S F T C T C K P	(91-106)
	Q I C P N T P G S Y H C S C K N	(132-147)
	A V C K N I L G D F E C E C P E	(174-189)
	Q L C V N Y P G G Y T C Y C D G	(213-228)
Human protein C	G T C I D G I G S F S C D C D G	(67-82)
Bovine thrombomodulin	Q R C V N T E G G F Q C H C D T	domain 3
	G Q C H N L P G T Y E C I C G P	domain 6

(ii)

Human thrombomodulin	Q R C V N T Q G G F E C H C Y P	(339-354)
	G V C H N L P G T F E C I C G P	(453-468)
Murine thrombomodulin	Q L C V N T K G G F E C F C Y D	domain 3
	S E C H N F P G S Y E C I C G P	domain 6
Human LDL-receptor	H V C N D L K I G Y E C L C P D	(306-321)
	Q L C V N L E G G Y H C Q C E E	(345-360)
Murine EGF-precursor	L G C E N T P G S Y H C T C P T	(375-390)
	S R C I N T E G G Y V C R C S E	(893-908)
	A A C T N T E G G Y N C T C A G	(934-949)

(iii)

Human factor VII	G S C K D Q L Q S Y I C F C L P	(58-73)
Human factor XII	G T C V N M P S G P H C L C P Q	(89-104)

21.2. β-hydroxylation

The other modification of amino-acid side-chains that we shall consider is the β-hydroxylation of Asp and Asn residues.

The modification of aspartic acid in this way was detected in plasma proteins at the beginning of the 1980s (139, 140, 314, 830), and more recently the same modification was found in asparagine (889). β-hydroxy-Asp and β-hydroxy-Asn residues have so far been found in bovine factor VII; in bovine and human factors IX and X; in bovine and human proteins C and S; in human C1r, C1s and uromodulin; and in bovine protein Z, thrombomodulin, LDL receptor and thrombospondin. The exact positions of the modified residues are not known in all these cases. It should be noted that the finding of β-hydroxylation in a bovine protein does not necessarily imply its presence in the human equivalent or *vice versa.*

The fundamental problem with all post-translational modifications of amino-acid side-chains is that they can only be detected at the level of protein, since modifications to the residues are not encoded directly in the cDNA. Their identification at protein level requires either direct sequencing or amino-acid analysis – preferably both.

An interesting feature of β-hydroxylation is that it always occurs at defined positions in growth-factor domains, and a comparison of amino-acid sequences shows certain structural similarities (Fig. 132). In the figure, parts of growth-factor domains from other proteins have been included so as to show other sites where β-Asp/Asn can be expected, but has not yet been demonstrated.

The hydroxylation of aspartic acid and asparagine does not require either vitamin K (involved in γ-carboxylation (1680)) or vitamin C (often involved in hydroxylation (1681)); instead, it proceeds with the help of a dioxygenase enzyme and 2-oxoglutarate (211–213).

22.

Coagulation-active surfaces *in vivo*

The major section of this book devoted to blood coagulation can hardly have left the reader in doubt as to the importance of surfaces. Coagulation-active surfaces contain negatively charged phospholipids. Experiments *in vitro* employ synthetic vesicles made of these. However, biological surfaces are more complex, so we examine these in more detail in this chapter. In particular, we shall look at two possibilities for the mobilisation *in vivo* of coagulation-active surfaces: circulating cells, especially blood platelets, and endothelial cells.

Platelets have long been known to be able to activate the mechanism of coagulation (1682–1684); the interesting fact is rather that platelets are (fortunately) not *always* coagulation-active, but first require suitable activation or stimulation. The same can be said of the endothelium, which under normal conditions does not possess coagulation activity, but in fact acts as an anticoagulant. Damage to the endothelium generates a coagulation-active surface, which probably arises by the exposure of subendothelial components.

The rôle of platelets in the activation of blood coagulation is still somewhat unclear; we summarise what is known at present.

HMM-kininogen binds both to stimulated (1685) and to unstimulated (1686) platelets, which indicates that the platelets contain some kind of a receptor for HMM-kininogen. The α granula of platelets also contain HMM-kininogen of their own, and this becomes exposed at the surface after stimulation by thrombin (1687, 1688). Optimal binding of HMM-kininogen to both stimulated and unstimulated platelets is dependent upon zinc ions, with an optimum at the physiological Zn^{2+} concentration (1685, 1686).

Factor XI binds to stimulated platelets in the presence of zinc ions, calcium ions and HMM-kininogen, while the extent of binding to unstimulated platelets is much lower (1689). It thus seems plausible that HMM-kininogen functions as a receptor for factor XI on the platelet surface, since HMM-kininogen and factor XI have affinity for one another (section 2.2.1) and circulate in a 2:1 stoichiometric complex in plasma. Interestingly, this binding of factor XI to platelets cannot be reversed competitively by the addition of plasma prekallikrein (1689), which normally also circulates in a complex with HMM-kininogen. However, the binding can be reversed by factor XIa, and this is interesting, because factor XIa was originally found to bind to platelets in a manner dependent upon HMM-kininogen and independent of zinc ions, with a binding site different from that of factor XI (1690). It thus seems that factor XIa has both a zinc-dependent and a zinc-independent mode of binding to platelets, while both modes of binding depend upon HMM-kininogen.

For platelet-bound factor XIa to have any effect, two conditions must be met: factor XIa must be formed on the blood platelets, and it must be enzymically active with respect to factor IX. Are these conditions in fact fulfilled?

Factor XIa appears to be formed in contact activation on stimulated platelets (1691), but there are contradictory results on whether platelet-bound factor XIa can activate factor IX. It has been stated that when factor XIa is bound to platelets, both its structural integrity and its ability to activate factor IX are fully preserved (1692). However, other authors (1693) have reported that the activity of factor XIa is inhibited by binding to stimulated platelets; the inhibition is thought to be reversible, carried out by a fairly large inhibitor (with a molecular mass over 650 kDa). This discrepancy can probably be explained by the difference in experimental conditions, since the experiments in reference (1693) were carried out in the absence of zinc ions or added HMM-kininogen, while the experiments described in reference (1692) included both of these and thus probably approximated better to physiological conditions. It has since been confirmed (1694) that platelets secrete an inhibitor of factor XIa, but shown that at the same time they protect factor XIa from inhibition both by this inhibitor and by α_1-proteinase inhibitor. (The latter is normally regarded as the most important inhibitor of factor XIa.)

Platelets contain factor VIII (1695), and stimulated blood platelets bind both factor IX and factor IXa in the presence of calcium ions (1696). Stimulated blood platelets support the activation of factor X by factor IXa in the same manner as synthetic phospholipid vesicles do in the presence of factor VIIIa and Ca^{2+} (355, 1697).

Factor V is found in platelets (1698), stored in the α granula, from which it can be secreted upon stimulation (1699). Furthermore, factors V and Va from plasma bind both to stimulated and to unstimulated blood platelets (1700). These possess a high-affinity binding site, for factor Va only, and also another binding site, of lower affinity, for both factors V and Va. Factor Va, localised at the platelet membrane, functions as a receptor for factor Xa (441, 442), and, together with calcium ions, these two proteins make up the platelet-bound prothrombinase complex.

The platelet surface is thus able to support all the activation steps in the intrinsic pathway. However, this is subject to the condition that the platelets are stimulated. It is not sufficient that some of the factors can bind to unstimulated platelets, if the rest of the factors cannot.

In the light of this, it is not surprising that stimulated platelets also provide a surface for the proteolytic inactivation processes catalysed by protein Ca, in which factors VIIIa and Va are degraded (1701–1703). As we saw with artificial phospholipid vesicles, protein S also acts as a co-factor for protein Ca on the platelet surface.

A question that arises at once, when platelets are considered as a coagulation-active surface, is that of whether they will be able to function as initiators of contact activation. The answer is both yes and no. Non-stimulated platelets cannot initiate contact activation of the intrinsic pathway, but the pro-cofactors necessary for this, HMM-kininogen and factors VIII and V, appear to reside on the membrane, fulfilling the requirements for activation of the intrinsic pathway. Stimulated platelets, on the other hand, can readily initiate the intrinsic pathway by contact activation.

There are therefore two further questions to be considered: (i) do platelets play a part in the initiation of blood coagulation, and (ii) how do platelets attain activity in coagulation?

The answers to both questions contain a theoretical element. Platelets can only play a part in the initiation of coagulation if they are present in stimulated form before coagulation starts. After injury, the first visible step in coagulation and haemostasis is the adhesion of platelets to the subendothelium. This may stimulate the platelets by exposing them either to collagen or to adenosine diphosphate (ADP), and if this is the case then platelets must be regarded as capable of playing a part in the initiation of coagulation, by acting as a negatively charged phospholipid for contact activation. We shall return to this point later, where arguments will be adduced to show that platelets in reality do not have this function.

There is, on the other hand, no doubt that platelets attain coagulation activity, and we shall here consider the mechanism of this. The principle is that coagulation activity is attained by the transition from the (normal) non-stimulated state to the stimulated state, and that this transition can be caused in various ways, for example by ADP, collagen or thrombin.

Stimulation has many effects, and coagulation activity is only one of these. It is claimed that the negatively charged phospholipids that we have seen are necessary for the coagulation activity of synthetic phospholipid vesicles are found almost exclusively on the inner side of the membranes of non-stimulated platelets (220) (section 4.1.1), and this is presumed to be responsible for the fact that non-stimulated platelets have no coagulation activity. The stimulation of platelets, for example by thrombin, is thought to cause a rearrangement of the membrane lipids such that some 25% of the phosphatidyl serine and phosphatidyl inositol molecules become exposed upon the outside of the membrane (1704). This exposure would provide the necessary charge (468, 1697, 1704, 1705).

Although this sounds attractive, good evidence to support it is still lacking, especially as this explanation postulates implicitly a hitherto unknown mechanism for the rapid transport of negatively charged phospholipid head-groups through the platelet membrane, a 'flip-flop' process known normally to have a high activation energy. Another possibility might be that the secretion processes that follow stimulation also cause the negatively charged lipids to traverse the membrane. It has recently been shown that so-called 'micro-particles', formed from the platelet membrane when platelets are stimulated by thrombin and collagen, contain sites that bind factor Va. Furthermore, their formation is associated with the appearance of the platelets' ability to function as surfaces for the prothrombinase enzyme complex (1706).

The stimulation of platelets by thrombin has been shown to be mediated by a receptor; this receptor has not been identified with certainty, but it may be the membrane-bound glycoprotein Ib (1707, 1708).

A final circulating cell-type with possible coagulation activity is the monocytes. These are able to initiate the coagulation of blood by way of the stimulated and regulated surface-exposure of tissue factor. When stimulated by ADP, these cells also expose a receptor with high affinity for factor X (1709). This receptor has been identified as CR3, complement receptor type 3 (1710) (section 17.3). When factor X binds to this, factor Xa is formed. The binding of factor X to CR3 is inhibited by competition with iC3b.

Another surface to which an active rôle in coagulation is frequently ascribed is the endothelial/subendothelial surface. Normally, the endothelium does not promote coagulation, but it acquires activity if damage to the tissue has occurred.

The most likely pathway for the initiation of coagulation in connection with damage to the endothelium is activation of the extrinsic pathway by tissue factor, factor VII and Ca^{2+}. Tissue factor is localised on the plasma membranes of many different kinds of tissue, including the tissue around the blood vessels. This means that even minimal damage to this tissue will cause the initiation of the extrinsic pathway towards coagulation.

The exposed, negatively charged phospholipids from endothelial/subendothelial components can also serve as components in contact activation and thus in the intrinsic pathway.

However, intact endothelium also appears to have some functions in the coagulation system. Thrombomodulin, which was a co-factor for the thrombin-catalysed activation of protein C to give protein Ca, is permanently exposed upon the endothelial surface, where its effect is to inhibit coagulation.

Intact endothelium binds HMM-kininogen (1711, 1712), factors IX/IXa (1713–1715) and factor X (1715). For factor IX/IXa, it has been shown that the binding to endothelial cells is due to a specific receptor (1716), and it appears that the binding involves regions of the Gla-domain and of the first growth-factor domain in factor IX (1717).

The intact endothelium can support many of the activation processes of blood coagulation (1718–1720). Endothelial-cell-bound factor IX can be activated with factor XIa, and the factor IXa formed can, in the presence of exogenous [von Willebrand factor.factor VIII] complex, activate factor X to give factor Xa on the endothelial cell surface. The factor Xa generated can take the coagulation process further, without exogenous factor Va, by activating prothrombin, and the thrombin formed leads in turn to the formation of fibrin.

Naturally, these observations do not show that intact endothelium is coagulation-active; what they show is that some of the components of the coagulation system are in position, ready and waiting to do their job when it is required of them.

We conclude our discussion of coagulation-active surfaces *in vivo* by examining the possibility of time-dependent, differential use of various surfaces *in vivo*.

The various kinds of surface that we have discussed all play different parts, so they naturally come into play at different points in time. The question of whether platelets participate in the first steps of the initiation of coagulation by supporting contact activation is largely hypothetical, as there is scarcely a way for them to do this: platelets must be stimulated before being used in coagulation, and this requires thrombin, collagen or ADP. Thrombin is not formed until late in the coagulation process, and, even though this formation is rapid, it can hardly overtake the first step in the activation, the step that leads to the formation of thrombin. Exposed collagen and released ADP are signs of tissue damage, and thus also of the exposure of tissue factor and the activation of the extrinsic pathway.

As already argued in section 6.2, there is every indication that the coagulation of blood is initiated primarily by tissue factor and the extrinsic pathway, leading to the formation of thrombin. Thrombin formed in this way activates platelets, factor VIII and factor V, so that the activation processes of the intrinsic pathway that are localised on the activated platelets can start. This leads on to the formation of larger quantities of thrombin, and, perhaps more important still, the site of thrombin formation moves from the endothelium out to the platelet surface.

Without repetition of the arguments for regarding the exposure of tissue factor and the subsequent initiation of coagulation as the primary events in blood coagulation (see section 6.2), it should be underlined that both the extrinsic and the intrinsic pathways make indispensable contributions to fibrin formation and to haemostasis. There is a regrettable tendency to ask which activation pathway is the more important, and this obscures both the general perspective and the fact that the coagulation of blood is an intimate interplay between both pathways.

It seems as though effective coagulation and haemostasis depend upon the separation in time and space between the two activation pathways. The exposed tissue factor will be much too localised to contribute adequately to thrombin formation in the entire haemostatic plug (see Chapter 24), and a platelet-based activation of coagulation is therefore a prerequisite for the platelets to become joined together by fibrin.

23.

The binding of proteins to coagulation-active surfaces

There are in principle two ways in which a protein can bind non-covalently to a membrane. One is the direct binding of the protein to the lipid, and the other is the binding of the protein to another protein – a receptor – that itself is bound to, or anchored in, the lipid. Both of these are encountered in blood coagulation.

We examine first the direct binding to membrane surfaces. The interactions that cause this are either hydrophobic or electrostatic.

Proteins that bind to membranes by way of hydrophobic interactions can achieve this by two means: either by being synthesized as a membrane protein or by adopting, immediately after synthesis or by later activation, a structure in which a hydrophobic domain is exposed.

There are two membrane proteins in the coagulation system, tissue factor and thrombomodulin, and both of these are typical members of their class, since towards the C terminus of their amino-acid sequence there lies a longish ($\approx$25-residue) hydrophobic sequence that can span a membrane and give a firm anchoring-point. There is no known reason why the anchor sequence should be C-terminal, but it is frequently found to be so.

The exposure of a hydrophobic domain, leading to insertion into a membrane, is a process that we have already met in the complement system. However, this process is not used in the coagulation system.

The electrostatic interactions between protein and lipid can also be divided into two categories: Gla-dependent and Gla-independent. The interaction involved is the attraction, familiar from elementary physics, between positive and negative charges. Gla-independent binding to a membrane is relatively simple in nature, and the stability of this binding depends upon physical factors such as ionic strength and pH. Factor XII/XIIa, HMM-kininogen, factor VIII/VIIIa and factor V/Va interact with negatively charged phospholipid surfaces in this way, with a positively charged protein surface binding to a negatively charged phospholipid. The lipid-binding region in HMM-kininogen is probably the histidine-rich region (101). In factor XII/XIIa, the binding site is located among the 360 amino-acid residues closest to the N terminus (104, 105), while in factors V/Va and VIII/VIIIa binding takes place via the light chains from the C-terminal region (342, 343, 411).

The Gla-dependent binding appears to be more complex, and not all data appertaining to it are easily interpreted. One thing, however, is certain: Gla-dependent lipid binding is calcium-dependent. There is no doubt that the Gla residues are needed for binding to lipid, and the association between Gla-rich proteins and negatively charged phospholipid vesicles can be observed and quantified *in vitro*. The proteins showing this kind of binding are factors VII, IX and X, prothrombin, protein C and protein S.

Two models for the interaction can be set up. In the most frequently discussed model, the negatively charged protein groups and the negatively charged lipid molecules are bridged by Ca^{2+}. In the second model, the binding of calcium to the Gla domain causes a conformational change in the whole protein such that a membrane-binding site is exposed. The first model is best developed for prothrombin. It also postulates a Ca^{2+}-dependent conformation that exposes the required amino-acid residues in the correct way (1721). This model has received support from the three-dimensional structure of the Gla domain of prothrombin in the presence of calcium ions, determined at a resolution of 2.8 Å (0.28 nm) (393). It is found that six of the ten Gla residues are positioned on the same outer surface of the Gla domain (fragment 1, see section 5.1.2), opening the possibility of a strong concerted interaction with a membrane via calcium ions.

A central question in connection with the Gla-containing proteins and their binding to membranes is that of these proteins' behaviour as substrates. When these proteins are substrates for surface-bound enzyme complexes, where do they come from – the negatively charged surface or the bulk solution phase? We have earlier (sections 4.2, 5.2 and 6.2) seen arguments to support the assertion that the surface-bound enzyme complexes (Xase, prothrombinase, and the [tissue factor.factor VIIa] complex) take their substrates from the bulk phase. It has also been noted that the issue may be an artificial one engendered by the use of an excessively high content of negatively charged lipid head-groups in the artificial vesicles used for the study of the activation processes. If these arguments are true, then they imply that the primary function of the Gla domain is in reality not to provide binding to a membrane, but rather to help generate the conformation that gives the best protein-protein interaction.

Another clue in this direction comes from the activation of protein C by thrombin/thrombomodulin. Here the binding of protein C to the enzyme complex is dependent upon calcium, yet it is independent of phospholipid and Gla residues. It thus provides yet another example of the binding of substrate directly from the bulk phase.

It is worthwhile to speculate on how the newly activated zymogens carry out the jump from the enzyme complex that activated them to themselves becoming an active part of an enzyme complex. A prerequisite for this is that the enzyme complex and the now cloven substrate lose their affinity for one another, while the substrate – itself an enzyme – develops affinity for the co-factor with which it must later co-operate in an active complex. One could imagine that the Gla domains play an important part here in retaining the enzyme in the membrane bound complex. The transition that takes place is from a mobile zymogen to an immobile enzyme.

We encounter frequently in the coagulation system the binding of one protein to a membrane with the assistance of another protein. This may seem to resemble the binding of a ligand to a receptor, but in fact the similarity is not always strong. An example of a binding that does not have this resemblance is that of factor XI and plasma prekallikrein, which bind to the membrane via HMM-kininogen, with which they circulate as a complex in plasma before the whole becomes bound to the membrane. Examples of binding resembling a ligand-receptor interaction are provided by the binding of factor IXa to factor VIIIa, that of factor X to the [factorIXa.factor VIIIa] complex, that of factor Xa to factor Va, that of prothrombin to the [factorXa.factorVa] complex, that of thrombin to thrombomodulin, that of protein C to the [thrombin.thrombomodulin] complex, that of protein Ca to membrane-bound protein S, that of factor VII/VIIa to tissue factor, that of factor X to tissue factor/factor VIIa, and possibly that of factor IX to HMM-kininogen/factor XIa.

We conclude this section by returning briefly to the high-affinity binding of calcium ions seen in some of the coagulation factors, a binding that could involve β-hydroxy aspartic acid.

The coagulation factors found to contain a Gla-independent high-affinity Ca^{2+}-binding site all contain β-OH-Asp residues. They are factor IX (206, 207), factor X (404), protein C (864) and protein S (907). Thus, it was once suspected that this modified amino-acid residue was important for the Gla-independent binding of Ca^{2+}. There was no firm evidence for this suspicion, but it was suggested by results from mutagenesis studies on human factor IX (section 3.3) in which the residue normally hydroxlated (Asp64) was replaced by other residues (210). After the discovery that the hydroxylation process is carried out by a 2-oxoglutarate-dependent dioxygenase (211–213), it became possible to make recombinant factor IX containing unmodified Asp64. This unmodified factor IX appears so far to behave just like native factor IX as regards coagulation activity (212), leaving the function of β-OH-Asp to be elucidated.

This raises again the question of the significance of the high-affinity Ca^{2+}-binding site, the current answer to which is provided by an observation on protein C. Both native protein C and protein C from which the Gla domain has been removed must bind one calcium ion at a high-affinity site in order to attain the conformation that is recognised by the [thrombin.thrombomodulin] complex (864). In support of this, conformational changes induced by strongly bound Ca^{2+} have also been detected for factor IX (206, 207) and for factor X (404); the precise meaning of this is still unclear, but it could well be imagined that for these two factors, by analogy with protein C, the strong binding of a calcium ion is needed for normal interaction with factors VIIIa and Va respectively.

24. Haemostasis

Haemostasis is a wide-ranging concept that includes many phenomena. In this chapter we shall examine the processes that lead to the adhesion of blood platelets to thrombogenic surfaces and subsequent aggregation. The processes of haemostasis actually fall outside the scope of this book, so that the topic will be covered in a relatively cursory manner in this chapter.

Up to now, we have concentrated on the processes that make blood coagulate, and this has led us up to a network of polymerised fibrin, which, once stabilised, forms a hard clot. However, the fibrin on its own is not enough to close the rupture in the endothelium that originally set the coagulation process in motion.

For this reason, while the various coagulation processes are working to produce the fibrin clot, other mechanisms are in progress, and these lead ultimately to the formation of the so-called *haemostatic plug*. The participants in these mechanisms include the subendothelium, collagen, platelets, fibrinogen, fibronectin, von Willebrand factor and thrombospondin. For once, the description of the proteins will be left to the end of the chapter, and we start by examining the mechanism of the process in which they act.

24.1. The mechanism of haemostasis

The mechanisms that are associated with the adhesion of blood platelets and their aggregation have not been documented fully, so that the following picture is still partly hypothetical.

Normally, intact endothelium is surrounded by subendothelium; the latter varies in composition from one type of blood vessel to another. For capillaries, the subendothelium consists mostly of basal membranes, while for veins and arteries it contains a heterogeneous layer of components of connective tissue, such as collagen. Surrounded by the endothelium, the blood circulates freely, its plasma containing *inter alia* fibrinogen, fibronectin and von Willebrand factor, while the platelets contain in their

α granula the glycoprotein thrombospondin plus, in connection with their plasma membranes, two other glycoprotein components called glycoprotein Ib and glycoprotein IIb-IIIa.

Haemostasis is initiated by damage to the endothelium. Von Willebrand factor from plasma binds to subendothelial components not yet defined, and thereafter it acts as a bridge between the newly exposed subendothelium and activated platelets (1722): at one end it binds to components of the subendothelium, suggested to be various types of collagen (738, 1723–1726) and microfibrils (1727, 1728), and on the other end it binds to the activated platelets. Platelets that are not activated in some way or other fail to bind von Willebrand factor to any significant extent, since, if they did, they would be associated in plasma and circulate as a complex.

By what mechanism does a blood platelet become stimulated in such a way that it can be termed activated?

Platelets can be activated by the antibiotic ristocetin, in which process a receptor for von Willebrand factor becomes exposed on the platelet surface; this receptor has been identified as glycoprotein Ib (1729). The activation of blood platelets with physiological substances such as thrombin or ADP exposes a second receptor for von Willebrand factor called glycoprotein IIb-IIIa (1730–1732). A physiological activator corresponding to ristocetin is not yet known, but there is no reason at present to believe that no such activator exists. The situation *in vivo* is thus uncertain, but there is at least a theoretical possibility of the binding of von Willebrand factor to two different receptors.

In the same way, fibronectin is thought to be able to bind both to collagen and to platelets (1726), but the rôle of fibronectin in the adhesion of platelets is unclear, and it is known that fibronectin does not play any part in the aggregation of platelets (1733).

It is, of course, not sufficient that the platelets bind to the site at which the endothelium is damaged; it is also necessary that they aggregate. The first requirement is thus for the platelets to be located correctly with respect to the point of damage to the endothelium, and the second requirement is for them to aggregate, so that they fill the lesion gap.

The aggregation of platelets is to a large extent a phenomenon that can be ascribed to fibrinogen, which binds specifically to glycoprotein IIb-IIIa (1734–1736), as fibronectin also does. Other proteins, such as von Willebrand factor, compete for binding to glycoprotein IIb-IIIa. However, the concentration of free fibrinogen in comparison with that of von Willebrand factor, fibronectin etc. ensures that most of the glycoprotein IIb-IIIa binds to fibrinogen *in vivo*. It has also been shown that von Willebrand factor does not bind to thrombin or to ADP-stimulated platelets in plasma (1737, 1738). For this reason, it seems more probable that the physiologically important receptor for von Willebrand factor is glycoprotein Ib.

The activation of platelets by thrombin releases thrombospondin, which is a multifunctional protein that can bind to a number of other proteins including fibrinogen, fibronectin and type V collagen (review in reference 1739). The rôle of thrombospondin in platelet aggregation may be to stabilise the [fibrinogen.glycoprotein IIb-IIIa] complex (1740), since fibrinogen binds specifically to thrombospondin, using a binding site that is different from the binding site for glycoprotein IIb-IIIa (1741).

The binding of proteins to glycoprotein IIb-IIIa depends upon the presence of the amino-acid sequence Arg-Gly-Asp-(Ser) in the protein. This peptide sequence has been shown to be the sequence that confers upon fibronectin the ability to bind cells (1742). Fibrinogen, fibronectin, von Willebrand factor, thrombospondin and vitronectin all contain this sequence. Studies employing synthetic peptides that contain the sequence Arg-Gly-Asp-Ser have shown that these peptides inhibit the binding of fibrinogen, fibronectin and von Willebrand factor to platelets and also inhibit the aggregation of platelets (1743, 1744). Glycoprotein IIb-IIIa can therefore be classified as a member of the class of (Arg-Gly-Asp-Ser)-specific adhesion receptors (1745), which in turn is a part of the integrin family (see Chapter 25).

The interaction between fibrinogen and glycoprotein IIb-IIIa has been studied in some detail (1746–1753), and it has been shown that not only the two Arg-Gly-Asp sequences in the Aα chain of fibrinogen (positions 95–97 and 572–574), but also the last 12 residues at the C terminus in its γ chain take part in the binding. In addition, synthetic peptides representing the various binding sites in fibrinogen inhibit each other's binding to glycoprotein IIb-IIIa, possibly because of partially overlapping sites in glycoprotein IIb-IIIa or as a consequence of binding-induced changes of conformation.

The glycoprotein IIb-IIIa complex is found in the membranes of non-stimulated platelets, but it has no ligand-binding properties before stimulation. This immediately raises two questions: (i) what it is about platelet stimulation that confers receptor properties upon glycoprotein IIb-IIIa, and (ii) what are the changes in glycoprotein IIb-IIIa that activate receptor function? A possible answer to both these questions is offered by the fact that the stimulation of platelets leads to the hydrolysis of polyphospho-inositide (PIP_2) and thus causes a heightened intracellular Ca^{2+} concentration, which in turn activates the glycoprotein IIb-IIIa receptor function via conformational changes.

Let us summarise the course of events. A lesion occurs, and von Willebrand factor acts as a bridge between subendothelial components and activated platelets, probably with the essential involvement of glycoprotein Ib. The platelets thus accumulate around the lesion, and aggregate as a result of the binding of fibrinogen/fibrin to glycoprotein IIb-IIIa, possibly involving a stabilisation by thrombospondin. The platelets are simply

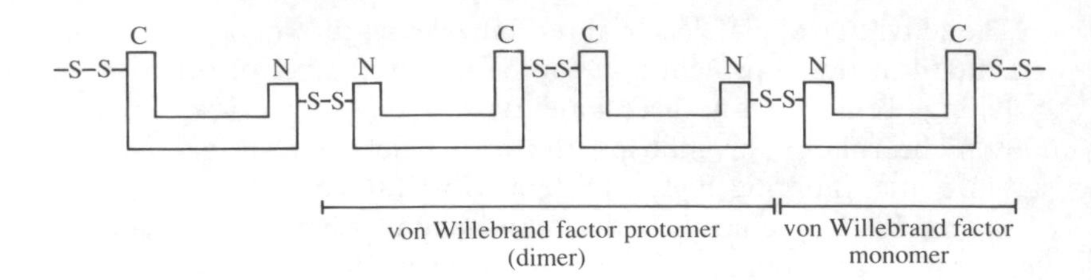

Fig. 133. The structure of the multimer of von Willebrand factor.

caught in a net consisting of fibrin and all the adhesion proteins that have been described, and this initially seals the lesion in the tissue. Thereafter, the haemostatic plug made up of fibrin and platelets contracts, owing to the contractile apparatus that the platelets possess.

The molecular processes that follow, leading to tissue repair, are considerably less clear, and no attempt will be made to describe them here in any detail. They involve various kinds of leucocytes (granulocytes, monocytes and lymphocytes) and also fibroblasts. We have earlier alluded to a possible rôle for fibronectin in tissue repair, and to the dependence of tissue repair upon cross-linking reactions catalysed by factor XIIIa that also involve both von Willebrand factor and thrombospondin. The entire fibrinolytic system is also involved, as are various growth factors.

24.2. Von Willebrand factor

Von Willebrand factor (reviews in references 221, 277, 1754–1756) is a multifunctional glycoprotein with a concentration in plasma of about 10 μg/ml (≈40 nM for the monomer). Von Willebrand factor is a multimer of disulphide-linked subunits, each of which is a dimer of identical, disulphide-linked polypeptide chains (Fig. 133). This tightly associated dimeric unit is often termed a 'protomer'. Each polypeptide chain has an estimated molecular mass of 250–270 kDa, so that the protomer of von Willebrand factor has a molecular mass of at least 500 kDa. In plasma, the observed molecular mass of von Willebrand factor varies from 550 kDa to over 12 MDa.

The primary structure of human von Willebrand factor has been determined both by protein-sequencing and by inference from cDNA (1757–1760). Its gene lies at position p12–pter of chromosome 12 (1761, 1762), covers approximately 178 kbp genomic DNA and has been characterised (1763–1765). The gene has 52 exons between 40 bp and 1379 bp in length, while the intron lengths vary from 97 bp to approximately 19.9 kbp (1765). The complete cDNA sequence of von Willebrand factor shows that this 2050-residue-long protein is synthesized with both a 22-residue signal peptide and a 741-residue propeptide, giving pro-von Willebrand factor a total length of 2791 residues (1759, 1760) (Appendix, Fig. A61).

The 2050 amino-acid residues of human von Willebrand factor monomer give a molecular mass for the polypeptide of 226 kDa, to which both N-bound and O-bound carbohydrate must be added. There are 13 potential sites for N-glycosylation, of which 11 are used, at Asn residues in positions 94, 468, 752, 811, 1460, 1527, 1594, 1637, 1783, 1822 and 2027 (1758). Asn452 and Asn1872 obey the consensus-sequence rule but are not modified, while Asn384 in the sequence Asn-Ser-Cys is glycosylated (1758). (After bovine protein C, this is the second example of Cys replacing Ser/Thr in the consensus sequence; cf. Chapter 19.) O-bound carbohydrate is attached at Ser500 and Ser723, and at threonines 485, 492, 493, 705, 714, 724, 916 and 1535 (1758). The glycosylated von Willebrand factor polypeptide thus has an estimated molecular mass of about 280 kDa. It has been reported that von Willebrand factor is sulphated, but it is not known where or in what way. Human von Willebrand factor also contains the sequence Arg1744-Gly-Asp-Ser, which may be expected to mediate binding to glycoprotein IIb-IIIa on platelets.

The propeptide in pro-von Willebrand factor has been found to be identical to a protein that earlier was called von Willebrand antigen II (1766). This consists (see above) of 741 amino-acid residues, with a molecular mass of 81 kDa. It has four potential N-glycosylation sites, at Asn77, Asn134, Asn189 and Asn644. Like von Willebrand factor, the propeptide contains the tripeptide Arg-Gly-Asp (residues 676–678), but the significance of this is not known.

The amino-acid sequence of pro-von Willebrand factor allows the definition of four types of homologous domains denoted A to D. Pro-von Willebrand factor can thus be ordered D1-D2-D′-D3-A1-A2-A3-D4-B1-B2-C1-C2, where D′ is a partial D domain and where D1 and D2 together make up the propeptide (1759). Another suggestion was based upon the amino-acid sequence of von Willebrand factor, giving a division into five internally homologous domains, as follows: E1-D1-A1-A2-A3-D2-E2-B1-B2-B3-C1-C2 (1757, 1767). The main difference between the two classifications lies in the definition of the D domains, and in the classification of the E domains as part of the D domains in reference (1759) (Fig. 134).

The propeptide part of pro-von Willebrand factor is cleaved off within the cell, giving von Willebrand factor directly, and it has been found that this peptide is necessary for the formation of multimers from the dimeric subunits, but not necessary for the formation of von Willebrand factor dimers from the monomeric polypeptide chains (1768, 1769). For participation in the assembly process, it does not matter whether the propeptide is cleaved off or still contiguous with the main chain.

The dimerisation of von Willebrand factor monomer occurs by the formation of a disulphide bond between cysteine residues in the C-terminal

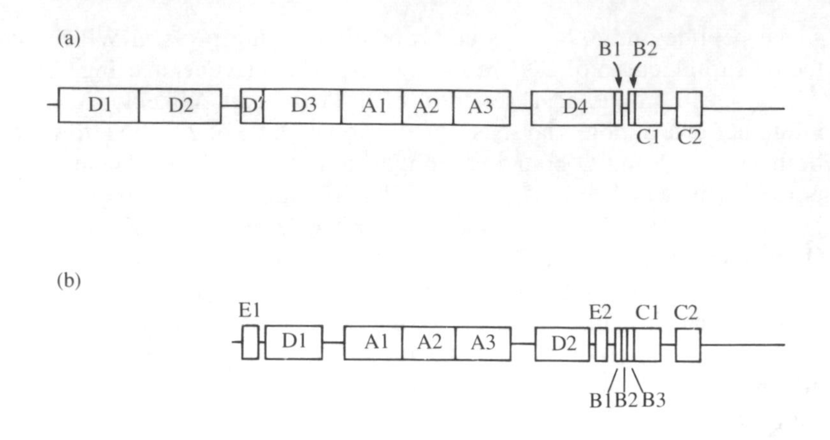

Fig. 134. The internally homologous domains of von Willebrand factor. The domains are defined on the basis of pro-von Willebrand factor (a; reference 1759) or of von Willebrand factor (b; references 1757, 1767).

region, and the union of dimers to give the complete multimer probably involves a disulphide bond in the N-terminal region. The disulphide pattern in von Willebrand factor has been partially determined (1770).

Von Willebrand factor has two main functions in the bloodstream. The first is protection of factor VIII, and the second is mediating the adhesion of platelets to thrombogenic surfaces in connection with haemostasis; accordingly, von Willebrand factor contains various functional domains. Up to now, the following functional domains have been defined.

1. The interaction between von Willebrand factor and factor VIII occurs by way of a 34-kDa region at the N terminus, Ser1–Arg272 (1771, 1772). By the combined use of monoclonal antibodies against von Willebrand factor (that prevent it from binding factor VIII) and of recombinant fragments of von Willebrand factor, it has been shown that the interaction between factor VIII and von Willebrand factor takes place by way of the first 106 residues of von Willebrand factor, probably involving the stretch Thr78–Thr96 (1773, 1774).
2. The binding between von Willebrand factor and glycoprotein Ib was first attributed to the region Val449–Lys728 (1775) and has since been narrowed down to two 15-residue stretches, Cys474–Pro488 and Leu694–Pro708 (1776). Independently of one another, each of these peptides obstructs the binding of von Willebrand factor to glycoprotein Ib, but they stimulate each other's effect if both are present (1776).
3. The binding between von Willebrand factor and glycoprotein IIb-IIIa appears to depend upon the residues Arg1744-Gly-Asp1746. This has been shown (i) by studies with oligopeptides containing Arg-Gly-Asp (1743, 1744), and (ii) by the use of monoclonal and polyclonal antibodies raised against synthetic oligopeptides based upon the amino-acid sequence from Glu1737 to Ser1750 of von Willebrand factor (1777).

4. There are two binding sites for collagen in von Willebrand factor, of which one is localised in the region Glu542–Met622 and a second in Ser948–Met998 (1778–1781).
5. Von Willebrand factor also contains a heparin-binding site located in the region between Val449 and Lys728 (1782).

Deficiency of von Willebrand factor causes von Willebrand's disease, which is a form of haemophilia (277, 1755).

24.3. Fibronectin

Fibronectin is a 450-kDa, homodimeric, multi-domain glycoprotein found both in plasma and also, in a slightly modified form, immobilised in the extracellular matrix. It is reviewed in reference (1783). The dimer's concentration in plasma is about 300 μg/ml ($\approx$700 nM). The primary structure has been determined at protein level for bovine plasma fibronectin (32, 36, 1784, 1785) and at cDNA level for human cellular fibronectin (1786).

Bovine plasma fibronectin has 2265 amino-acid residues (Appendix, Fig. A62), corresponding to a molecular mass of 249 kDa, to which must be added carbohydrate at asparagines 399, 497, 511, 846, 976, 1213, 1987 and at Thr1943/1944. (The last position cannot be located more precisely.) There are two unused potential N-glycosylation sites at Asn1205 and Asn1692. Bovine plasma fibronectin contains phospho-O-serine at position 2263 and two free cysteine residues at positions 1201 and 2015. Bovine plasma fibronectin contains two Arg-Gly-Asp sequences; the one at positions 1493–1495 has been shown to be involved in binding to outer cell membrane. The fibronectin dimer is held together in an antiparallel manner by two disulphide bonds in the C-terminal region.

By far the greatest part of fibronectin can be classified into three types of internal homology. There are 12 homologous units of type I, 2 of type II and 15 of type III, and the disulphide patterns in each type have been determined. It will be recalled that coagulation factor XII (section 2.1.1) contains a domain homologous to type I and another homologous to type II, while t-PA (section 9.3) contains one of type I. The overall classification of regions of bovine plasma fibronectin into homologous units is shown in Figure 135, which also displays the locations of the various binding domains.

24.4. Thrombospondin

Thrombospondin (reviewed in references 1738 and 1787) is a homotrimeric, 450-kDa, multi-domain glycoprotein found in the α granula of platelets. Its sequence has been determined from its cDNA (1788, 1789).

Human thrombospondin is made up of 1152 amino-acid residues

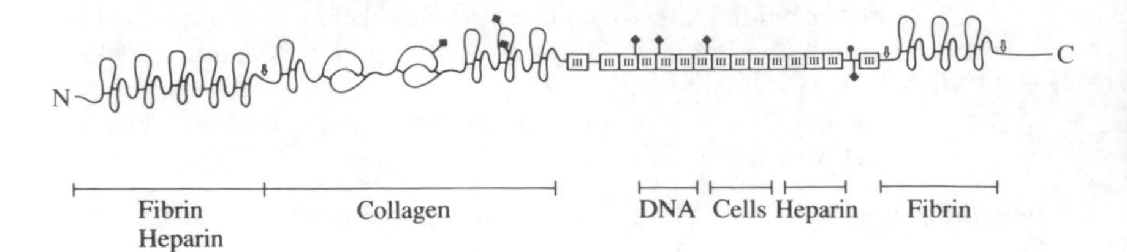

Fig. 135. The domain structure of the monomer of bovine plasma fibronectin. ⇒ shows the positions of cleavage by plasmin, ◆ N-bound carbohydrate and ● O-bound carbohydrate. (III, internally homologous domains of type III.) The different binding sites are indicated below the domain structure.

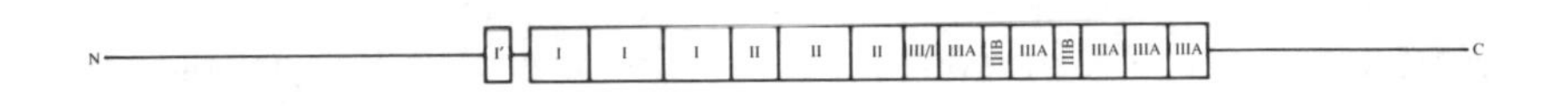

Fig. 136. The domain structure of thrombospondin. (As defined in reference 1788.)

(Appendix, Fig. A63), corresponding to a molecular mass of 127.5 kDa, to which carbohydrate, bound to one or more of the potential sites Asn230, Asn342, Asn690 and Asn1049 must be added. Thrombospondin contains the tripeptide Arg-Gly-Asp at positions 908–910.

Thrombospondin contains three types of internal homology, shown together with the placing of the various binding domains in Figure 136.

24.5. Glycoprotein Ib

Glycoprotein Ib (GPIb) is a 160-kDa heterodimeric extracellular glycoprotein of the platelet membrane made up of a 145-kDa α chain disulphide-bound to a 22-kDa β chain. Both chains are integral membrane proteins, and both are glycosylated (Fig. 137).

The amino-acid sequence of human GPIb α chain has been determined partially at the protein level (1790) and completely at the level of cDNA (1791) (Appendix, Fig. A64). The amino-acid sequence deduced from cDNA contains 626 residues, of which the first 16 make up the signal peptide and the remaining 610 the GPIb α chain. The calculated molecular mass of the polypeptide chain is 67.2 kDa, to which carbohydrate must be added. There are four potential sites for N-glycosylation, at Asn21, Asn159, Asn346 and Asn382, of which at least the first two are used (1790). The GPIb α chain also contains substantial amounts of O-bound carbohydrate (1792), but only one point of attachment has been located, at Thr292. A large part of the remaining O-bound carbohydrate is expected to be found in a Ser/Thr-rich region between residues 320 and 420.

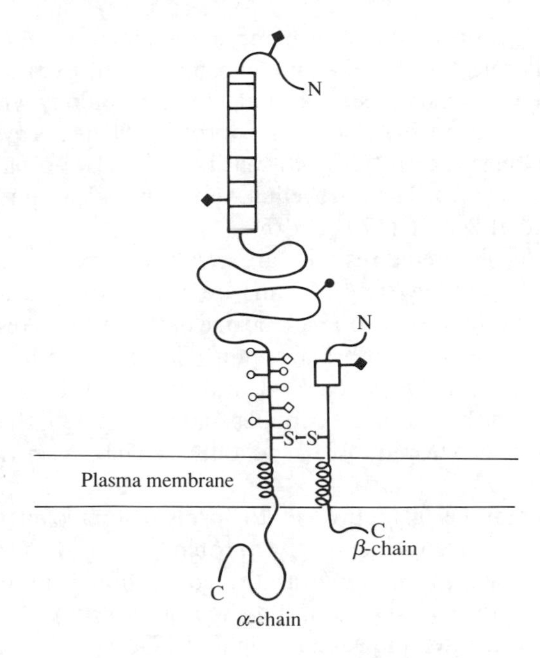

Fig. 137. A schematic model of glycoprotein Ib. ◆ shows N-bound carbohydrate, ◇ potential attachment sites for N-bound carbohydrate, ● O-bound carbohydrate and ○ the region containing potential attachment sites for O-bound carbohydrate. □ represents the leucine-rich sequences that show homology with leucine-rich α_2-glycoprotein.

The α chain of GPIb consists of an N-terminal extracellular domain of about 485 residues, a putative transmembrane segment of about 29 residues and a C-terminal cytoplasmic domain of about 100 residues. The extracellular domain contains two types of internally homologous sequence. One type, 24 residues in length, occurs six times between positions 36 and 187, with a part of a seventh at the beginning of this stretch; it shows homology with the 13 corresponding internally homologous stretches in the plasma protein leucine-rich α_2-glycoprotein (1793). The other type of internal homology, nine residues in length, with the consensus sequence T-T-X-E-P-T-P-X-P, occurs five times between residues 363 and 414, and is expected to prove of importance for the process of O-glycosylation.

The amino-acid sequence of the human GPIb β chain has been inferred from cDNA-sequencing (1794) (Appendix, Fig. A65). It contains 206 residues, of which the first 25 are the signal peptide and the remaining 181 are the β chain itself, with a calculated molecular mass of 19.3 kDa not counting the carbohydrate bound to Asn41. The sequence indicates that the extracellular region of GPIb comprises approximately the first 120 residues, which are followed by a proposed transmembrane segment of about 25 residues and a cytoplasmic domain of about 34 residues. The extracellular region contains a single 24-residue stretch that shows homology with the glycoprotein Ib α chain and leucine-rich α_2-glycoprotein. There is a cysteine residue in the intracellular region (Cys148) that is acylated with palmitic acid, to give a thioester (1795). The intracellular domain is also sometimes found phosphorylated at Ser166 (1796, 1797).

It is not known which cysteine residues are involved in the disulphide bridge, or bridges, that unite the α and β chains, but it can be expected that the residue(s) responsible in the α chain include one of the two Cys residues that occur just outside the transmembrane region. The region with affinity for von Willebrand factor lies between His1 and Arg393 in the α chain (1790, 1798). Studies with synthetic peptides have indicated that the binding of von Willebrand factor involves a peptide stretch between Ser251 and Tyr279 (1799).

In the platelet membrane, glycoprotein Ib forms a non-covalent 1:1 complex with another glycoprotein, glycoprotein IX (GPIX). The [GPIb.GPIX] complex serves *inter alia* as the connection between the platelet membrane and the cytoskeleton, but the rôle of GPIX in this is unclear. GPIX has an estimated molecular weight of 17 kDa and possesses a single polypeptide chain. The amino-acid sequence of GPIX has been inferred from its cDNA sequence (1800) (Appendix, Fig. A66) and contains 160 residues, corresponding to a molecular weight of 17.3 kDa. There is a potential N-glycosylation site at Asn 44. It is generally accepted that the hydrophobic stretch between Val135 and Ala154 makes up the transmembrane segment; this stretch contains a Cys residue acylated with palmitic acid (1795). Like the GPIb β chain, the extracellular part of GPIX contains a single 24-residue leucine-rich amino-acid sequence that is homologous to a segment of leucine-rich α_2-glycoprotein.

24.6. Glycoprotein IIb-IIIa

Glycoprotein IIb-IIIa (GPIIb-IIIa) is a non-covalent, calcium-dependent complex of molecular mass 260 kDa, made up of the two platelet-membrane glycoproteins GPIIb and GPIIIa (Fig. 138). GPIIb is a heterodimer composed of a 125-kDa α chain and a 23-kDa β chain, while GPIIIa is a single-chained 110-kDa polypeptide. These will be described in turn.

24.6.1. *GPIIb*

The complete amino-acid sequence of GPIIb has been determined from the cDNA (1801); partial cDNA sequences (1802–1804) and protein sequences (1805–1807) have also been published. The gene for GPIIb is located on

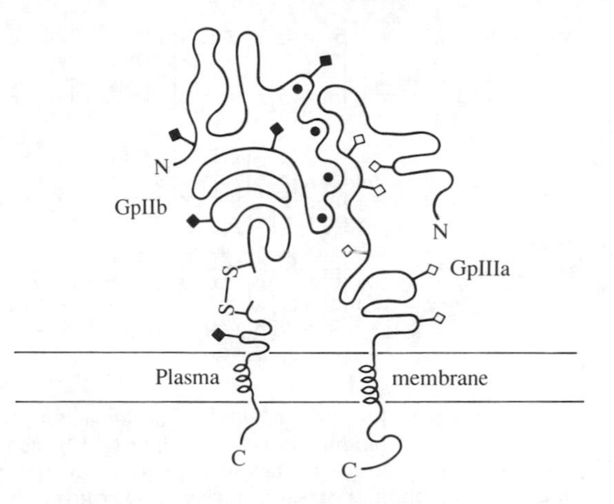

Fig. 138. Schematic model of glycoprotein IIb-IIIa. ◆ shows N-bound carbohydrate and ◇ potential attachment sites for N-bound carbohydrate. ● represents presumed calcium-binding sites.

chromosome 17 at position q21–22 (1803, 1808, 1809), and it is linked to the gene for GPIIIa (1808, 1809).

The amino-acid sequence inferred from the cDNA (Appendix, Fig. A67) has 1039 residues, of which the first 31 make up the signal peptide, leaving 1008 for the precursor of GPIIb. This precursor is processed proteolytically to give the α and β chains. The exact location of the cleavage site between the N-terminal α chain and the C-terminal β chain is not yet certain, and there could be some heterogeneity in respect of this. Direct amino-acid-sequencing has shown that the N-terminal sequence of the GPIIb β chain is identical with that of the precursor starting from the residue Leu872, implying a cleavage site between Arg871 and Leu872 (1801, 1806, 1810). However, studies with antibodies suggest a somewhat different site: antibodies against a synthetic oligopeptide with the same amino-acid sequence as residues 859–871 in the precursor react with the β chain and not with the α chain, while antibodies against an oligopeptide with the same sequence as residues 844–860 reacted with the α and not with the β chain. These results led to the suggestion (1811) of a cleavage site between the two regions named, possibly at one of the two points in the intervening sequence Lys855-Arg-Asp-Arg-Arg-Gln860 where two consecutive side-chains are positively charged. The same authors find that the N terminus of the β chain is blocked, which would be consistent with cleavage at Arg859 followed by cyclisation of the glutamine exposed at the new N terminus. If this is correct, the α chain contains 859 residues with a calculated molecular mass of 93 kDa and the β chain contains either 137 or 149 residues, with a corresponding molecular mass of 15.5 Kda or 17 kDa.

GPIIb A	233	G	Y	W	G	Y	S	V	A	-	V	G	E	F	D	G	D	L	N	T	T	E	Y	V	V	G	A	257
GPIIb B	287	S	Y	F	G	H	S	V	A	-	V	T	D	V	N	G	D	G	R	H	D	L	L	V	-	G	A	310
GPIIb C	354	G	R	F	G	S	A	I	A	P	L	G	D	L	D	R	D	G	Y	N	D	I	A	V	-	A	A	378
GPIIb D	415	S	A	F	G	F	S	L	R	G	A	V	D	I	D	D	N	G	Y	P	D	L	I	V	-	G	A	439

Calmodulin (bovine brain)														
	20	D	K	D	G	N	G	T	I	T	T	K	E	31
	56	D	A	D	G	N	G	T	I	D	F	P	E	67
	93	D	K	D	G	N	G	Y	I	S	A	A	E	104
	129	N	I	D	G	D	G	E	V	N	Y	E	E	140

Troponin C (rabbit skeletal)														
	27	D	A	D	G	G	G	D	I	S	V	K	E	38
	63	D	E	D	G	S	G	T	I	D	F	E	E	74
	103	D	R	N	A	D	G	Y	I	D	A	E	E	114
	139	D	K	N	N	D	G	R	I	D	F	D	E	150

Fig. 139. Internally homologous putative calcium-binding domains in glycoprotein IIb, compared with the calcium-binding sites of calmodulin (1812) and troponin C (1813). The positions at which amino acids are the same in all four glycoprotein IIb domains are outlined. In addition, the residues presumed to be involved in calcium binding are outlined. Pairwise comparison of the four glycoprotein IIb domains reveals the following degrees of amino-acid identity: domains A and B, 46%; domains A and C, 32%; domains A and D, 24%; domains B and C, 38%; domains B and D, 42%; domains C and D, 36%.

The α chain of GPIIb contains four repeating peptide sequences of about 25 residues in length; each includes a stretch of 12 residues that resembles the Ca^{2+}-binding β turns in Ca^{2+}-binding proteins such as calmodulin and troponin C (1812–1815) (Fig. 139). This is of course interesting because it is known that the GPIIb-IIIa complex is held together in a manner dependent upon Ca^{2+} (1816–1818): the ions bind to GPIIb, which is then able to bind to GPIIIa (1806). It should, however, be mentioned that the Ca^{2+}-binding site in classical Ca^{2+}-binding structures consists of a so-called 'EF hand', made up of a Ca^{2+}-binding β turn flanked upon either side by an α helix. Structural predictions for GPIIb made on the basis of its amino-acid sequence indicate that these α helices are absent in the α chains of GPIIb; how much their presence or absence means is not known at present.

The β chain of GPIIb contains – in contrast to the α chain – a possible transmembrane domain, 26 residues in length, between Ala963 and Trp988, so that the cytoplasmic domain is restricted to the final 20 residues at the C terminus.

N-bound carbohydrate is found at all the five potential attachment sites in GPIIb: Asn15, Asn249, Asn570 and Asn680 in the α chain and the position in the β chain, presumably Asn60, that corresponds to Asn931 in the precursor (1819). In addition, the α chain contains O-bound carbohydrate.

The α and β chains of GPIIb are held together by a disulphide bond that connects the α chain at Cys826 with the β chain at Cys9 (1810). The complete disulphide-bond pattern for GPIIb has been determined and, apart from the intercatenary bond, is: in the α chain, Cys56–Cys65,

```
                          448 C N N G N G T F E C G V C R C G P G W - L G
486 C S P R E G Q P V C S - Q R G E C L C G Q C V C H S S D - F G
527 C V R Y K G E - M C S - G H G Q C S C G D C L C D S D W - T G
567 C M S S N G L - L C S - G R G K C E C G S C V C I Q P G S Y G

- - - - S Q C E - C 473
K I T G K Y C E - C 522
- - - - Y Y C N - C 560
- - - - D T C E K C 601
```

Fig. 140. Internally homologous domains in glycoprotein IIIa. Only the positions at which amino acids are the same in all four domains are outlined. The first domain is incomplete. Pairwise comparison of the three complete domains reveals the following degrees of amino-acid identity: domains 2 and 3, 45%; domains 2 and 4, 43%; domains 3 and 4, 42%.

Cys107–Cys130, Cys146–Cys167, Cys473–Cys484, Cys490–Cys545, Cys602–Cys608 and Cys674–Cys687; and in the β chain, Cys14–Cys19 (1810, 1819).

24.6.2. *GPIIIa*

The complete amino-acid sequence of GPIIIa has been derived from the sequences of cDNA from endothelial cells and human erythroleukaemia (HEL) cells (1820–1822), and scattered partial sequences have also been derived at the protein level from platelet GPIIIa (1807, 1820, 1821). The gene for GPIIIa has been localised on chromosome 17, position q21–22 (1808, 1809, 1821), and, as mentioned above, it is linked to the gene for GPIIb.

The amino-acid sequence inferred from the cDNA (Appendix, Fig. A68) contains 788 residues, of which the first 26 are the signal peptide. The remaining 762 residues correspond to a molecular mass of 84.5 kDa, of which by far the greater part is extracellular, since the presumed transmembrane segment is close to the C terminus. There is disagreement as to the exact demarcation of this transmembrane segment, but it seems sufficient to assign the hydrophobic stretch Ile693–Trp715 to this.

On this basis, the extracellular region is 692 residues long and contains six potential N-glycosylation sites, at Asn99, Asn320, Asn371, Asn452, Asn559 and Asn654, leaving the 47 C-terminal residues to make up the intracellular domain. There is also a consensus sequence N-X-S/T involving Asn756, but the placing of this residue within the intracellular domain excludes the possibility of its glycosylation. On account of homology with other sequences, Tyr747 is regarded as a candidate for phosphorylation by tyrosine kinases, but there is no evidence for such phosphorylation at present. GPIIIa contains four mutually homologous 40-residue-long regions rich in cysteine (Fig. 140) with a repeated C-X-C-G-X-C-X-C motif.

Another form of GPIIIa is known; the only difference between this and the form just described lies in the intracellular domain at the C terminus. The amino-acid sequence of the second form diverges after Thr741, and thereafter has only 13 instead of the usual 21 residues (1823). It is known that the second form arises as a result of alternative splicing, but the significance of the existence of two forms of GPIIIa can only be guessed at.

25.

Integrins

Integrins (reviewed in references 1824–1826) is a collective name for an entire class of adhesion receptors with a common overall structure. The integrins are also termed cytoadhesins. The name 'integrin' was originally suggested for a complex of glycoproteins that make up a part of the transmembrane connection between the extracellular matrix and the intracellular cytoskeleton in chickens (1827). Since then, there have been rapid developments.

The members of the integrin family have two conspicuous common features. First, these receptors identify in part their ligands by recognising the Arg-Gly-Asp tripeptide. Secondly, they are all non-covalently bound heterodimers, all glycoproteins and all membrane-bound. In each case, the individual chains are designated α and β. Each integrin has a characteristic α subunit, but many share the same β subunit. The integrins can thus be grouped according to their β subunit.

Group I consists of the fibronectin receptor (found on fibroblasts *inter alia*), six different VLA antigens ('very late' antigens, found in many places, and originally described for stimulated T lymphocytes) and the CSAT antigen (described for chicken fibroblasts).

Group II consists of the leucocyte adhesion receptors LFA-1, Mac-1 (also called complement receptor type 3 – CR3) and p150,95; all of these, as their names suggest, are found on leucocytes.

Group III consists of glycoprotein IIb-IIIa (on platelets) and the vitronectin receptor, found on endothelial cells and elsewhere.

All the α subunits for which sequences are known are homologous to one another; the same applies for all β subunits of known sequence. In contrast, there is no detectable homology between α and β subunits. Many integrin α subunits consist of a light and a heavy chain joined by a disulphide bond. In these cases, the two chains are derived from a common precursor by partial proteolysis after translation.

25.1. Group I

Of the first integrin group, two amino-acid sequences have been determined. These are the fibronectin receptor (VLA-5) and a collagen receptor (VLA-2).

25.1.1. *Fibronectin receptor (VLA-5)*

The fibronectin receptor (reviewed in reference 1749) is a heterodimeric, non-covalently linked complex; the subunits are termed α and β.

The α subunit has a molecular mass of 160 kDa, as measured by SDS-polyacrylamide gel electrophoresis, and probably consists of a heavy and a light chain bound by S–S bonds and derived from the N and C termini, respectively, of a precursor. The complete sequence of the α-subunit precursor has been inferred from cDNA (1828); it includes a 41-residue signal peptide (Appendix, Fig. A69). The gene is located on chromosome 2 (1808). The α-subunit precursor is made up of 1008 amino acids, corresponding to a calculated molecular mass of 110 kDa, to which carbohydrate at the 14 potential N-attachment sites may perhaps be added. A hydrophobic, presumably transmembrane segment is found between Gly952 and Tyr980, implying that most of the α subunit is extracellular, with some 30 intracellular residues at the C terminus. The cleavage site separating the heavy and light chains has not been determined by experiment, but its homology with glycoprotein IIb and with the vitronectin receptor α subunit suggests that it is Arg853–Glu854. The membrane-spanning segment is thus located in the light chain. The amino-acid sequence contains five potential Ca^{2+}-binding domains, located between residues 239 and 432, in agreement with the observation that calcium ions are necessary both for receptor function and for structural integrity (stable association between chains).

The β subunit is single-chained and has a molecular mass of 140 kDa, as measured by SDS-polyacrylamide gel electrophoresis. Its complete amino-acid sequence has been deduced from that of its cDNA (1828); it includes a signal peptide of 20 residues (Appendix, Fig. A70). The β subunit contains 778 residues, giving a calculated molecular mass of 82 kDa, and there is carbohydrate bound to up to 12 potentially glycosylated asparagine residues. A presumed transmembrane segment is found between Ile709 and Trp731, giving 708 extracellular and 47 intracellular residues. The gene for the β subunit lies on chromosome 10 (1829). The amino-acid sequences for β subunits in this group of integrins have also been inferred from the cDNA sequences from chicken (1827) and *Xenopus* (1830). Both of these show 85% identical amino-acid residues with the human β subunit.

25.1.2. *VLA-2*

VLA-2, a magnesium-dependent collagen receptor, resembles the other integrins in being a heterodimeric, non-covalent complex of an α and a β subunit.

The α subunit, also called platelet glycoprotein Ia, has an electrophoretically estimated molecular mass of 160 kDa. Its complete amino-acid sequence has been inferred from the cDNA (1831) and includes a 23-residue-long signal peptide (Appendix, Fig. A71). The α subunit contains 1152 residues, corresponding to a molecular mass of 126 kDa, to which carbohydrate at up to ten potentially glycosylated asparagine residues is attached. A hydrophobic, presumed transmembrane segment is found between Gly1104 and Trp1125, so that, again, most of the molecule is extracellular. Three potential binding sites for bivalent cations are also found in the sequence.

The β subunit in VLA-2 is, as mentioned earlier, identical with the β subunit in the fibronectin receptor.

25.2. **Group II**

The known members of group II of the integrins are all leucocyte adhesion receptors and mediate cell-cell interactions. The three members are called LFA-1, Mac-1 and p150,95, and all three are heterodimers of an α subunit and a common β subunit. As Mac-1 is identical with CR3 (complement receptor type 3), the reader is referred to section 17.3 for a description of the Mac-1 α subunit and of the common β subunit.

The α subunit of LFA-1 has a physically estimated molecular mass of 180 kDa, and its complete amino-acid sequence, including that of a 25-residue signal peptide, has been deduced from the cDNA sequence (1832) (Appendix, Fig. A72). The gene for the α subunit is found on chromosome 16, position p11–13.1 (1653). The α chain has 1145 residues and a calculated molecular mass of 126 kDa, to which carbohydrate bound to up to 12 potentially glycosylated Asn residues must be added. The presumed transmembrane segment is found between Met1064 and Phe1092. As in the other integrins, the α subunit of LFA-1 contains several potential binding sites for bivalent cations.

The α subunit of p150,95 has a physically estimated molecular mass of 150 kDa, and its complete amino-acid sequence, including that of a 19-residue signal peptide, has been deduced from the cDNA sequence (1833) (Appendix, Fig. A73). The gene for the α subunit is found on chromosome 16, position p11–13.1 (1653). The α chain has 1144 residues and a calculated molecular mass of 126 kDa, to which carbohydrate bound to up to ten glycosylatable Asn residues must be added. The presumed transmembrane segment is found between Leu1089 and Tyr1109, and the extracellular part of the α subunit of p150,95 contains several potential binding sites for bivalent cations.

A common feature of all α subunits in this integrin group distinguishes them from the other integrin α subunits. In the extracellular part of the subunit, there is an additional stretch of about 200 residues that shows homology with the A domain as defined for von Willebrand factor. It should also be mentioned that the genes for all these three α subunits are located at position p11–13.1 on chromosome 16.

25.3. Group III

The third group of integrins consists of glycoprotein IIb-IIIa and the vitronectin receptor. Their common β subunit is glycoprotein IIIa. Glycoprotein IIb-IIIa has been described elsewhere (section 24.6), so it will suffice here to describe the α subunit of the vitronectin receptor.

The α subunit of the vitronectin receptor is two-chained, with a heavy (125 kDa) chain disulphide-bound to a light (25 kDa) chain (1745, 1834). The complete amino-acid sequence of the α-subunit precursor has been deduced from that of the cDNA sequence (1835–1837); it includes a signal peptide of 30 residues (Appendix, Fig. A74). The gene for the α subunit of the vitronectin receptor lies on chromosome 2 (1808). The inferred amino-acid sequence shows that the α-subunit precursor is synthesized as a 1018-residue polypeptide, of which the 860 N-terminal residues become the heavy chain and the remaining 158 residues become the light chain after cleavage of the bond Arg860–Asp861 (1835, 1836).

The heavy chain of the α subunit contains four potential binding sites for bivalent cations, and these are believed to mediate the effect of calcium ions, which are essential for the functioning of vitronectin receptor in adhesion. However, unlike glycoprotein IIb-IIIa, vitronectin receptor does not dissociate into its subunits in the absence of calcium (1838, 1839). The light chain in the α subunit contains a hydrophobic segment (Ala98–Tyr126, corresponding to Ala958–Tyr986 in the precursor) presumed to span the membrane; this implies that the cytoplasmic domain is only 32 residues long. The two chains in the α subunit are linked by a disulphide bridge, and it may be assumed that the complete disulphide pattern can be taken over from glycoprotein IIb. There are ten potential N-glycosylation sites in the heavy chain and three in the light chain.

26.

Serpins

The family of proteins known as the serpins – the name alluding to the function **ser**ine-**p**roteinase **in**hibitor – has been mentioned at several places in the main text of this book, not least because all of the serine-proteinase inhibitors that we have encountered, with only two exceptions (extrinsic pathway inhibitor and α_2-macroglobulin), belong to this family.

As for all so-called 'families' of proteins, the serpins are defined purely on the basis of the homology of their amino-acid sequences. Not all members are inhibitors of serine proteinases, although most are. The mechanism of such inhibition has been described in the context of the regulation of coagulation (see section 8.1.1), so here we concentrate upon the family as a whole.

Tables 11 and 12 give a summary of currently known serpins.

The following discussion of the serpins will be restricted to those of human origin. Even so, there are 13 of these, of which 10 have been characterised as serine-proteinase inhibitors. The other three are known to have completely different functions. At the end of this chapter we shall take a brief look at the six human serpins that have not been described in this book so far.

The 'target' enzymes of the serine-proteinase inhibitors are often implied in their names, and most of them we have already met. Those we have not are: heparin co-factor II, which inhibits thrombin; α_1-proteinase inhibitor, which inhibits leucocyte elastase; and α_1-antichymotrypsin, which inhibits chymotrypsin-like proteases such as neutrophil cathepsin G and mast-cell chymase.

The serpins that actually are serine-proteinase inhibitors reveal several common characteristics in addition to their amino-acid homology.

1. Their reactive site, i.e., the bond that is cleaved after complex formation with a proteinase, is located 33–40 residues from the C terminus of the inhibitor. (The only exception to this is the α_2-plasmin inhibitor, where this distance is 88 residues.)

Table 11. *Serpins with known serine-proteinase inhibitor activity*

Inhibitor	Figure in Appendix	Reactive site	References
Antithrombin III	17	Arg393→Ser394	(747, 748)
C$\bar{1}$-inhibitor	47	Arg444→Thr445	(749)
Protein Ca inhibitor	22	Arg354→Ser355	(919)
α_2-plasmin inhibitor (α_2-PI)	26	Arg364→Met365	(1004)
Plasminogen activator inhibitor 1 (PAI-1)	27	Arg346→Met347	(1059–1061)
Plasminogen activator inhibitor 2 (PAI-2)	28	Arg380→Thr381	(1079)
Protease nexin	29	Arg345→Ser346	(1087, 1088) (also from rat (1089))
Heparin co-factor II	75	Leu444→Ser445	(1841, 1842)
α_1-protease inhibitor	76	Met358→Ser359	(745, 746) (also from baboon (1843), mouse (1844) and rat (1845))
α_1-antichymotrypsin	77	Leu358→Ser359	(1846) (also from mouse (1844))

2. The sequence homology is much less clear in the immediate vicinity of the reactive site than in the rest of the region containing it.
3. From the N terminus, there is always a stretch of up to 120 amino-acid residues showing no homology with the rest of the serpin family.

It is the second and third of these points that are believed to contain the clue to the specificity of these inhibitors. The primary source of specificity is the reactive site, because the inhibitor is directed towards a specific proteinase. However, there must be other factors that contribute to this specificity, and the observations in point 3 suggest that these could well reside in diversity of the N-terminal regions.

Proteins with mutually homologous amino-acid sequences are generally taken to have a common evolutionary origin, and the degree of their similarity is used as a measure of how tightly two proteins are related to one another in an evolutionary sense. This leads to the construction of phylogenetic trees for protein families.

For protein families for which there are enough data to allow phylogenetic investigation, there are two other properties well worth examining: the gene structure and the chromosomal location. If the homologous proteins have a common gene structure, as evidenced by intron–exon junctions occurring at homologous points in the amino-acid sequence, then a close relationship is indicated. The gene structures of most

Table 12. *Serpins without known serine proteinase inhibitor activity*

Inhibitor	Figure in Appendix	References
Angiotensinogen	78	(1847) (also from rat (1848))
Thyroxine-binding globulin	79	(1849)
Corticosteroid-binding globulin	80	(1850)
λ-Spi-1 and λSpi-2 (from rat)	–	(1851)
ORF-1 (from rabbit)	–	(1852)
Ovalbumin (from chicken)	81	(1853)
Chicken gene X and chicken gene Y	–	(1854, 1855)
Barley protein Z	–	(1856)
38-kDa protein from cowpox virus	–	(1857)
Serpin from tobacco hornworm *Manduca sexta*	–	(1858)

of the human serpins have now been investigated, but no single picture has emerged with respect to their intron–exon junctions. The serpins investigated up to now have been human antithrombin III (764), human α_1-proteinase inhibitor (746), human α_1-antichymotrypsin (1859), human $C\overline{1}$-inhibitor (1475, 1476), human PAI-1 (1065–1067), human PAI-2 (1083), human α_2-plasmin inhibitor (1044), human heparin co-factor II (1860), rat angiotensin (1861) and chicken ovalbumin (1862). The most interesting cases turn out to be human α_1-antichymotrypsin and α_1-proteinase inhibitor: for both of these, the genes not only contain intron–exon junctions in almost exactly the same positions but also lie on chromosome 14. Their level of identical residues is 39%. The same intron–exon pattern is found in human heparin co-factor II (1860) and rat angiotensin (1861), so that these four proteins make up a sub-family within the serpin family. It is also noteworthy that the genomic organisation of PAI-2 and that of ovalbumin are almost identical (1083).

The chromosomal locations have also been determined for $C\overline{1}$-inhibitor (chromosome 11), antithrombin III (chromosome 1), PAI-1 (chromosome 7), PAI-2 (chromosome 18), α_2-plasmin inhibitor (chromosome 18), heparin co-factor II (chromosome 22) and thyroxine-binding globulin (the X chromosome).

As mentioned above, it is a surprising fact that the serpin family contains some proteins that (as far as is known) are not serine-proteinase inhibitors. It is true that angiotensinogen, which participates in the regulation of blood pressure by being the precursor for angiotensins I and II, shows the weakest homology with the rest of the family. However, thyroxine-binding globulin contains 37% identical residues compared both with α_1-proteinase inhibitor and with α_1-antichymotrypsin, while corticosteroid-binding globulin has respectively 43% and 44% identity compared with these two. The non-proteinase inhibitor members of the serpin family can therefore possess degrees of identity with the genuine serine-proteinase inhibitors that are equal to, or even greater than, the degrees of identity among the proteinase inhibitors themselves.

26.1. α_1-proteinase inhibitor

Human α_1-proteinase inhibitor (α_1-PI; reviewed in, for example, reference 1863) is a single-chained, 52-kDa glycoprotein, whose concentration in normal human plasma lies between 1.1 mg/ml and 2.6 mg/ml ($\approx$20–50 μM). Its main function is the inhibition of neutrophil elastase, but, as we have seen (section 8.1.2), α_1-PI also inhibits coagulation factor XIa.

The amino-acid sequence of α_1-PI has been determined by protein-sequencing (745) and cDNA-sequencing (746). The latter showed that α_1-PI is synthesized with a 24-residue signal peptide. The gene lies at position q31–32.3 of chromosome 14 (1864) and covers about 12 kbp of genomic DNA. The first report, employing cDNA synthesized from liver mRNA, stated that the α_1-PI gene had five exons (746). It was later found that macrophages synthesize α_1-PI from a longer mRNA, which differs from liver mRNA at its 5′ end; this led to the definition of two more exons in the α_1-PI gene (1865). The gene for α_1-PI thus consists of seven exons, but the first two are not used for the synthesis of α_1-PI in liver cells. The sizes of the exons lie between 50 bp and 650 bp, and of the introns between 823 bp and 5310 bp. α_1-PI has 394 amino-acid residues (Appendix, Fig. A76), corresponding to a molecular mass of 44.3 kDa, in agreement with the results of electrophoretic determination when account has been taken of the carbohydrate bound at Asn46 and Asn247. The reactive site is the bond between Met358 and Ser359 (1866).

The three-dimensional structure of the α_1-PI molecule, cleaved at the reactive site, has been determined at a resolution of 3 Å (0.3 nm) (1867). α_1-PI is a globular protein with 30% α-helix structure and 40% β-sheet structure; there are nine helices (A–I) and three β sheets (A–C) with respectively six, six and three strands. The most remarkable feature of this structure is that Met358 and Ser359 are seen at different ends of the molecule. Since before cleavage they cannot be separated by more than the length of one peptide bond, the cleavage must be followed by major structural rearrangement. The structure of the intact inhibitor is not known, but it has been suggested that its conformation could be identical to the conformation of the cloven inhibitor, except in that strand 4A is not associated with sheet A but instead forms an unusually taut strand in which the Met358–Ser359 bond is exposed (1867).

There are four frequently occurring alleles of the α_1-PI gene, whose products are termed respectively M1(Ala213) (1868), M1(Val213) (745), M3 (corresponding to M1(Val213, Asp376)) (1869) and M2 (corresponding to M1(His101, Val213, Asp376)) (1869). In addition, there are several rarer alleles with undisturbed physiological function, of which two – α_1-$PI_{Christchurch}$ and α_1-PI X – have been partially characterised by protein-sequencing and have the respective substitutions Glu363→Lys (1870) and Glu204→Lys (1871).

The lack of α_1-PI has serious consequences (see, for example, reference 1869). It leads to the development of lung emphysema at the age of 30–40, as a consequence of deficient control of neutrophil elastase. The genetic basis for these deficiencies has to some extent been elucidated (Table 13).

The classical α_1-PI-deficiency allele is called Z, while the most frequent deficiency allele is called S. Among variants of α_1-PI there is a surprising example of a mutation in the reactive site, called α_1-$PI_{Pittsburg}$. In this variant, Met358 at the reactive site is replaced by an arginine residue, which turns α_1-$PI_{Pittsburg}$ into a powerful thrombin inhibitor (1875) that additionally inhibits β-factor XIIa, factor XIa and plasma kallikrein (1881, 1882).

26.2. α_1-antichymotrypsin

Human α_1-antichymotrypsin is a single-chained, 68-kDa glycoprotein, whose concentration in normal human plasma is 250 μg/ml ($\approx$4 μM). Its main function is the inhibition of chymotrypsin-like proteases such as neutrophil cathepsin G and mast-cell chymase. The amino-acid sequence of α_1-antichymotrypsin has been determined by cDNA-sequencing (1846); the α_1-antichymotrypsin molecule is synthesized with a 25-residue signal peptide. Its gene is linked to the gene for α_1-proteinase inhibitor and lies at position q31–32.3 of chromosome 14 (1864); it covers about 12 kbp of genomic DNA and has five exons (1859). The sizes of the exons lie between 57 bp and 649 bp. α_1-antichymotrypsin has 408 amino-acid residues (Appendix, Fig. A77), corresponding to a molecular mass of nearly 46 kDa. Potential glycosylation sites are found at Asn8, Asn68, Asn161, Asn246 and Asn394. The reactive site is the bond between Leu358 and Ser359 (1883).

26.3. Heparin co-factor II

Heparin co-factor II is a single-chained glycoprotein with a molecular mass of 65 kDa and a concentration in normal human plasma of approximately 65 μg/ml ($\approx$1 μM). The physiological importance of heparin co-factor II is unknown, but it is thought likely to be the inhibition of thrombin (1884). The inhibitory effect of heparin co-factor II is raised about 1000-fold by certain glucosaminoglycans: these are dermatan sulphate, heparin and

Table 13. *Characterised defects in α_1-proteinase inhibitor*

Name	Normal allele	Nucleotide substitution	Amino-acid substitution	Reference
Z	M1(Ala213)	G→A	Glu342→Lys	(1873)
S	M1(Val213)	A→T	Glu264→Val	(746)
Procida	M1(Val213)	T→C	Leu41→Pro	(1874)
Malton	M2	deletion of codon 52	Pro52 deleted	(1875)
Pittsburg	unknown	unknown	Met358→Arg	(1876)
Heerlen	M1(Ala213)	C→T	Pro369→Leu	(1877)
$Null_{Granite\ Falls}$	M1(Ala213)	deletion of C in codon 160	Tyr160→stop	(1878)
$Null_{Bellingham}$	M1(Val213)	A→T	Lys217→stop	(1879)
$Null_{Hong\ Kong}$	M2	deletion of TC in codon 318	Leu318–Lys394→ Leu318–stop334[a]	(1880)
$Null_{Mattawa}$	M1(Val213)	insertion of T in codon 353	Leu353–Lys394→ Phe353–stop376[b]	(1881)

[a] Because of a reading-frame shift, the normal C-terminal sequence is replaced by a 16-residue peptide.
[b] Because of a reading-frame shift, the normal C-terminal sequence is replaced by a 23-residue peptide.

heparan sulphate (1885). The amino-acid sequence of heparin co-factor II has been determined by cDNA-sequencing (1841, 1842), and the molecule is synthesized with a signal peptide of 19 residues. The two cDNA sequences imply different amino acids at position 218, either lysine (1841) or arginine (1842). The gene, located on chromosome 22 (1842), covers 14.5 kbp of chromosomal DNA and has five exons (1860). The size of the first exon is estimated to be between 25 bp and 120 bp, while the others vary from 145 bp to 904 bp. The introns vary between approximately 0.7 kbp and 5.2 kbp.

Heparin co-factor II contains 480 residues (Appendix, Fig. A75), corresponding to a molecular mass of 55 kDa. Potential carbohydrate attachment sites are found at Asn30, Asn169 and Asn368, and there are two tyrosine-O-sulphate residues at positions 60 and 73 (1886). The reactive site is at the bond between Leu444 and Ser445 (1887). The presence of a Leu residue at the reactive site of heparin co-factor II is remarkable, both because thrombin is otherwise strictly arginine-specific, and also because thrombin is the only enzyme that heparin co-factor II inhibits. A possible explanation for its strict specificity for thrombin can perhaps be found in the discovery of a thrombin-binding site in heparin co-factor II, consisting of the residues 54–75 (1888). A synthetic peptide corresponding to this peptide stretch inhibits the cleavage of fibrinogen by thrombin. Residues 54–75 bear a strong negative charge and include two nearly identical heptapeptide stretches (E-D-D-D-Y-X-D, where X = L or I) in which both Tyr residues are sulphated. Their cDNA sequences indicate different amino-acid residues at position 218, either a Lys residue (1841) or an Arg residue (1842).

There exists a variant form of heparin co-factor II that has a reduced affinity for dermatan sulphate; however, this has no clinical consequences. The genetic alteration involved is a mutation (Arg→His) at position 189 (1889). Mutagenesis studies have shown that Lys185 also is a part of the glucosaminoglycan-binding site in heparin co-factor II (1890).

26.4. Thyroxine-binding globulin

Human thyroxine-binding globulin (TBG) is the most important thyroid-hormone-binding protein in plasma, and it circulates as a single-chained, 54-kDa glycoprotein. TBG is not a proteinase inhibitor.

The amino-acid sequence of human TBG has been derived from the cDNA sequence (1849), and the molecule is synthesized with a signal peptide containing 20 residues. The gene, found on the X chromosome at position q21–22 (1891), appears to possess four exons, but details have not been published (1892). TBG contains 395 amino-acid residues (Appendix, Fig. A79), corresponding to a molecular mass of 44.2 kDa, and four potential glycosylation sites at Asn16, Asn79, Asn145 and Asn391, although Asn391 has a proline residue in the second position of the consensus tripeptide and is therefore probably not used. A variant TBG in which Ile96 is replaced by Asn has reduced hormone-binding affinity (1893). Another variant with reduced hormone affinity has been found among Australian aborigines and characterised genetically (1892): two substitutions were found, Ala191→Thr and Leu283→Phe (1892). The latter substitution has also been found alone in another person and did not change the affinity of TBG for hormone, so the mutation at position 191 was concluded to be responsible for the altered phenotype.

26.5. Corticosteroid-binding globulin

Corticosteroid-binding globulin (CBG) is the most important transport protein for corticosteroids in human plasma. It is not a proteinase inhibitor. It circulates as a single-chained glycoprotein of molecular mass 58 kDa.

The amino-acid sequence of CBG has been inferred from its cDNA (1850); it is synthesized with a 22-residue signal peptide. CGB contains 383 amino-acid residues (Appendix, Fig. A80), corresponding to a molecular mass of 42.6 kDa, to which carbohydrate must be added. Potential N-glycosylation sites are found at positions 9, 74, 154, 238, 308 and 347.

26.6. **Angiotensinogen**

Angiotensinogen is a single-chained, 61-kDa glycoprotein found in plasma. It functions as precursor for angiotensins I and II; the latter is the main vasoconstricting peptide. The amino-acid sequence of human angiotensinogen has been deduced from the sequence of the cDNA (1847, 1894), and the two reported cDNA sequences are identical except for position 359, which may be Leu (1847) or Met (1894). Angiotensinogen is synthesized with a 33-residue signal peptide. The protein itself has 452 amino-acid residues (Appendix, Fig. A78), corresponding to a molecular mass of 49.8 kDa, to which carbohydrate at up to four potentially glycosylated asparagine side-chains in positions 14, 137, 271 and 295 may be added. Protein-chemical studies have shown that at least Asn14 is glycosylated (1895). Angiotensinogen is cleaved by the enzyme renin at the peptide bond Leu10–Val11, giving angiotensin I (D-R-V-Y-I-H-P-F-H-L). The rest of the molecule has no known function. Angiotensin I is converted to angiotensin II by removal of the C-terminal H-L dipeptide with angiotensin-converting enzyme (see, for example, reference 1896).

27.

Some evolutionary considerations

All proteins evolve by gradual structural change consequent upon mutations in the DNA that encodes them. In general, a mutation that improves the function of a protein will become established, while detrimental mutations will disappear. The evolution of the protein molecules thus makes its contribution to 'macro-evolution', the evolution of species.

Molecular evolution can be approached analytically from two directions. One of these is the comparison of the genetic material of various species, and the other is the detailed analysis of genetic material from a single species.

The comparison of different species contributes to the understanding of how the various life forms have evolved; for example, the establishment of the phylogenetic family tree of the great apes shows in what relationship we are placed with respect to other primates.

The analysis of the genetic material of an individual species can give valuable insight into the evolution of different biochemical systems within a particular branch of the tree, and may help to answer questions such as how structures in the coagulation system, the fibrinolytic system or the complement system developed.

These two approaches should, of course, not be understood as representing two independent processes within evolution. The underlying mechanisms are always the same. The two approaches are simply two different ways of looking at observable facts and are in principle ultimately convergent.

In this section it is mainly the latter approach that we shall take, since this approach is the more immediately related to the way in which we have examined many protein components of the systems of interest, with main emphasis upon the components in the single species *Homo sapiens*. For further literature, the reader may consult references (1897–1903). This approach precludes, for example, the construction of family trees; for this purpose, we should have needed to examine a single protein, or a few proteins, that have been characterised in many organisms (for example, haemoglobin).

As we have seen, the amount of data available on primary structures of proteins has grown dramatically in the last few years. At the time of writing, complete amino-acid sequences have been determined for all the known components of the coagulation and fibrinolytic systems, and for practically all the components of the complement system. Knowledge of the amino-acid sequences of proteins allows us to make comparisons and to look for similarities. The more closely two proteins resemble each other, the closer they are expected to be related in evolution.

We have seen that proteins can resemble one another in two ways: (i) by having recognisable homology over their entire length (for example, plasma kallikrein and factor XI; C1r and C1s; C2 and factor B), or (ii) by possessing common domains (a good example is t-PA and C1r; however, all multi-domain proteins resemble others in this way). To account for these patterns of resemblance, one can imagine two different mechanisms. In discussing these, it should be remembered that selection pressure operates at the level of protein (more exactly: the level of function), since a 'better' mutant gives a 'greater' possibility of survival and reproduction – by definition, in fact.

One of the two mechanisms involves gene duplication and, thereafter, divergent evolution. If a gene is duplicated, we first have two identical copies of the gene. As long as one of these continues to generate their gene product, the other will be able to mutate more freely, so that it ultimately may come to code for a new product with a new function. Which of the two copies comes to deviate from its original function is primarily a random choice, and of course both continue to evolve; but the crux of the matter is that an organism with duplicate genes has more freedom to 'experiment' with new mutations that one without.

Immediately after gene duplication, at least one, perhaps both, of the duplicate genes will evolve more rapidly than before the duplication, simply because of the reduced selection pressure. As each gene approaches optimisation of its new function, its evolution will slow down. It may be supposed that the gene with the new function will in general evolve more rapidly than the other, unless the new and the old functions are tightly coupled, or unless the precursor (before duplication) carried out both functions in a less specialised – less optimised – manner.

Examples of this kind of evolution can be found in both coagulation and complement systems. The clearest are perhaps those of: (i) factor XI and plasma prekallikrein; (ii) coagulation factors VII, IX and X along with protein C and protein Z; (iii) complement components C2 and factor B; (iv) complement components C1r and C1s; (v) the α, β and γ chains of fibrinogen; (vi) the A, B and C chains of C1q; (vii) factors V and VIII; (viii) C3, C4, C5 and α_2-macroglobulin. In all these instances, the proteins concerned display homology along more or less their entire amino-acid sequences.

The second mechanism of protein evolution is called *exon-shuffling*. It is observed that very many proteins are multi-domain proteins – that is to say, they appear to have been assembled from domains collected from many sources, but they need not show homology over their entire sequence with any other protein. Thus the domains appear frequently to be 'shuffled', like cards. If this mechanism is at work, it can in general be expected that the domains from which such proteins are built up will be found encoded in separate exons, whence the term 'exon-shuffling' arises. If exon-shuffling really takes place, it should lead to more rapid evolution than simple mutation, since all the domains' functions have been optimised, or at least made operational, before their re-assembly in a new protein.

It is believed by some that there is in reality only a restricted number of domains, and that these are merely shuffled, like cards, in the process of evolution. Possible examples of products of exon-shuffling that we have seen in the proteins reviewed in this book include factor XII, HMM-kininogen, the Gla-containing coagulation factors, plasminogen, urokinase, t-PA, C1r, C1s, C2, factor B and complement components C6, C7, C8 and C9.

As noted, the exon-shuffling hypothesis implies that the functional domains lie upon separate exons. Does this accord with observation? In fact, it does so quite well. The growth-factor domains in factors VII, IX, X and XII, protein C, t-PA and urokinase are encoded in separate exons. The Gla domains are on separated exons in factors VII, IX and X, prothrombin and protein C. The fibronectin type I domains in factor XII and t-PA also have separate exons. The fibronectin type II domain in factor XII likewise has its own exon. Serine-proteinase domains are distributed over several exons, and some of the intervening introns lie in homologous positions: this applies for factors VII, IX and X and protein C (Fig. 141). With a single exception, the same is true of the serine-proteinase domains of factors XI and XII, urokinase and t-PA (Fig. 142). The only other serine-proteinase domain in which the intron–exon junctions have been determined completely is that of complement factor B. Closer examination of these patterns allows a more detailed classification of the serine proteinases.

An important prerequisite for exon-shuffling, and thus for the validity of the exon-shuffling model, is that the junctions at both ends of all potentially 'shufflable' exons must be in phase with each other. If they are not, the insertion, duplication or removal of an exon will cause a shift of reading frame and thus lead to the expression of a useless protein. An exon to be used in shuffling must therefore contain a number of base pairs that is a

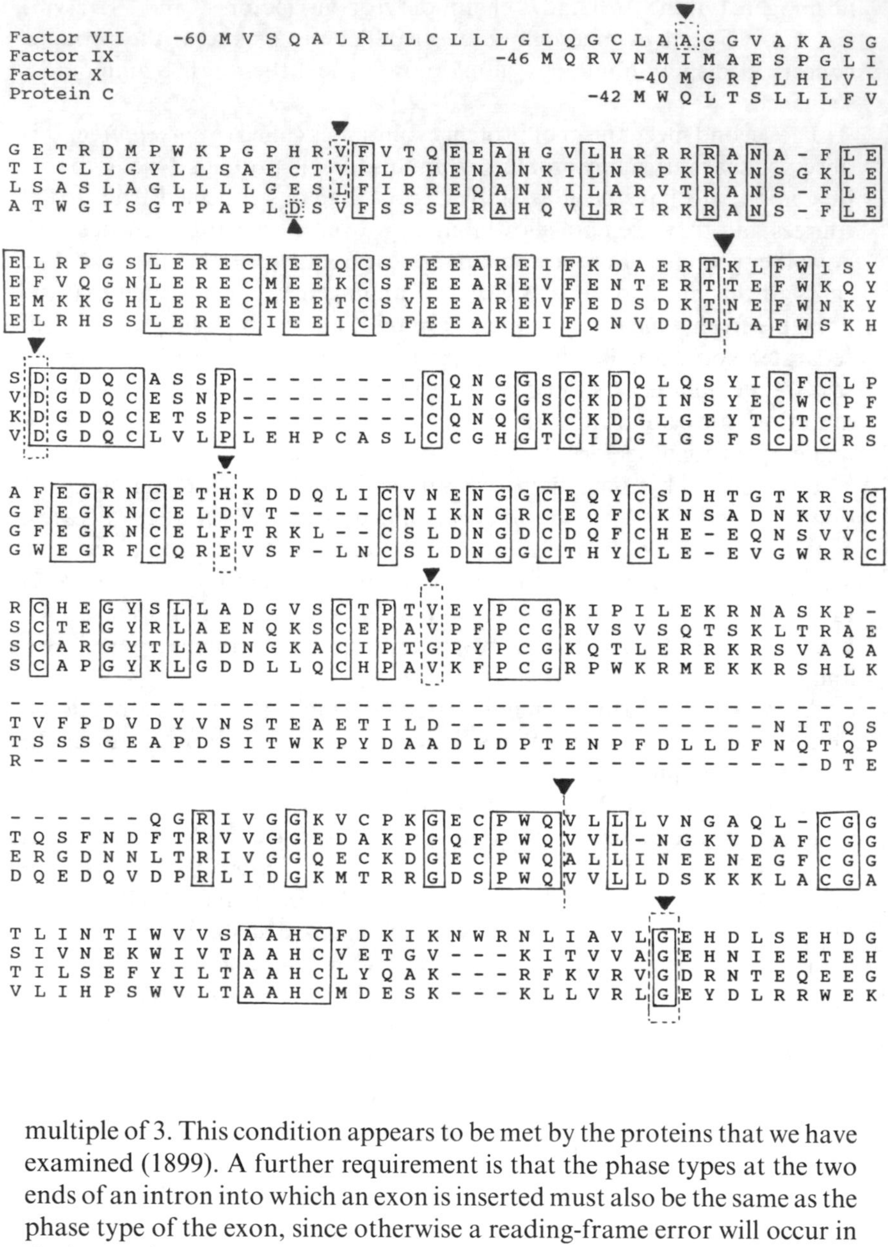

```
Factor VII   -60 M V S Q A L R L L C L L L G L Q G C L A A G G V A K A S G
Factor IX                                    -46 M Q R V N M I M A E S P G L I
Factor X                                                 -40 M G R P L H L V L
Protein C                                          -42 M W Q L T S L L L F V

G E T R D M P W K P G P H R V F V T Q E E A H G V L H R R R R A N A - F L E
T I C L L G Y L L S A E C T V F L D H E N A N K I L N R P K R Y N S G K L E
L S A S L A G L L L L G E S L F I R R E Q A N N I L A R V T R A N S - F L E
A T W G I S G T P A P L D S V F S S S E R A H Q V L R I R K R A N S - F L E

E L R P G S L E R E C K E E Q C S F E E A R E I F K D A E R T K L F W I S Y
E F V Q G N L E R E C M E E K C S F E E A R E V F E N T E R T T E F W K Q Y
E M K K G H L E R E C M E E T C S Y E E A R E V F E D S D K T N E F W N K Y
E L R H S S L E R E C I E E I C D F E E A K E I F Q N V D D T L A F W S K H

S D G D Q C A S S P - - - - - - - - - C Q N G G S C K D Q L Q S Y I C F C L P
V D G D Q C E S N P - - - - - - - - - C L N G G S C K D D I N S Y E C W C P F
K D G D Q C E T S P - - - - - - - - - C Q N Q G K C K D G L G E Y T C T C L E
V D G D Q C L V L P L E H P C A S L C C G H G T C I D G I G S F S C D C R S

A F E G R N C E T H K D D Q L I C V N E N G G C E Q Y C S D H T G T K R S C
G F E G K N C E L D V T - - - - C N I K N G R C E Q F C K N S A D N K V V C
G F E G K N C E L F T R K L - - C S L D N G D C D Q F C H E - E Q N S V V C
G W E G R F C Q R E V S F - L N C S L D N G G C T H Y C L E - E V G W R R C

R C H E G Y S L L A D G V S C T P T V E Y P C G K I P I L E K R N A S K P -
S C T E G Y R L A E N Q K S C E P A V P F P C G R V S V S Q T S K L T R A E
S C A R G Y T L A D N G K A C I P T G P Y P C G K Q T L E R R K R S V A Q A
S C A P G Y K L G D D L L Q C H P A V K F P C G R P W K R M E K K R S H L K

- - - - - - - - - - - - - - - - - - - - - - - - - - - - - - - - - - - - - -
T V F P D V D Y V N S T E A E T I L D - - - - - - - - - - - - - - N I T Q S
T S S S G E A P D S I T W K P Y D A A D L D P T E N P F D L L D F N Q T Q P
R - - - - - - - - - - - - - - - - - - - - - - - - - - - - - - - - - - D T E

- - - - - - Q G R I V G G K V C P K G E C P W Q V L L L V N G A Q L - C G G
T Q S F N D F T R V V G G E D A K P G Q F P W Q V V L - N G K V D A F C G G
E R G D N N L T R I V G G Q E C K D G E C P W Q A L L I N E E N E G F C G G
D Q E D Q V D P R L I D G K M T R R G D S P W Q V V L L D S K K K L A C G A

T L I N T I W V V S A A H C F D K I K N W R N L I A V L G E H D L S E H D G
S I V N E K W I V T A A H C V E T G V - - - K I T V V A G E H N I E E T E H
T I L S E F Y I L T A A H C L Y Q A K - - - R F K V R V G D R N T E Q E E G
V L I H P S W V L T A A H C M D E S K - - - K L L V R L G E Y D L R R W E K
```

multiple of 3. This condition appears to be met by the proteins that we have examined (1899). A further requirement is that the phase types at the two ends of an intron into which an exon is inserted must also be the same as the phase type of the exon, since otherwise a reading-frame error will occur in the inserted exon (Fig. 143).

```
D E Q S R R V A Q V I I P - S T Y V P G - - T T N H D I A L L R L H Q P V V
T E Q - K R N V I R I I P H H N Y N A A I N K Y N H D I A L L E L D E P L V
G E A V H E V E V V I K - - H N R F T K - E T Y D F D I A V L R L K T P I T
W E L D L D I K E V F V - H P N Y S K S - - T T D N D I A L L H L A Q P A T

L T D H V V P L C L P E R T F S E R T L A F - V R F S L V S G W G Q L L D R
L N S Y V T P I C I A D K E Y T N I F L K F - G S - G Y V S G W G R V F H K
F R M N V A P A C L P E R D W A E S T L M T - Q K T G I V S G F G R T H E K
L S Q T I V P I C L P D S G L A E R E L N Q A G Q E T L V T G W G Y H S S R

G A T - - - - - A L E L M V L N V P R L M T Q D C L Q Q S R K V G D S P N I
G R S - - - - - A L V L Q Y L R V P L V D R A T C L R S T K F T - - - - - I
G R Q - - - - - S T R L K M L E V P Y V D R N S C K L S S S F I - - - - - I
E K E A K R N R T F V L N F I K I P V V P H N E C S E V M S N M - - - - - V

T E Y M F C A G Y S D G S L D S C K G D S G G P H A T H Y R G T W Y L T G I
Y N N M F C A G F H E G G R D S C Q G D S G G P H V T E V E G T S F L T G I
T Q N M F C A G Y D T K Q E D A C Q G D S G G P H V T R F K D T Y F V T G I
S E N M L C A G I L G D R Q D A C E G D S G D P M V A S F H G T W F L V G L

V S W G Q G C A T V G H F G V Y T R V S Q Y I E W L Q K L M R S E P R P G V
I S W G E E C A M K G K Y G I Y T K V S R Y V N W I K E K T K L T 415
V S W G E G C A R K G K Y G I Y T K V T A F L K W I D R S M K T R G L P K A
V S W G E G C G L L H N Y G V Y T K V S R Y L D W I H G H I R D K E A P Q -

L L R A P F P 406

K S H A P E V I T S S P L K 448
K S W A P 419
```

Fig. 141. The locations of the introns in factors VII, IX and X and in protein C. Only the positions at which amino acids are the same in all four domains are outlined. The positions of the introns are indicated by ▼ and the corresponding amino-acid residues are outlined with dashed lines.

```
Factor XII  354 V V G G L V A L R G A H P Y I A A L - - - - - - -
Factor XI   370 I V G G T A S V R G E W P W Q V T L - - - H T T S
u-PA        159 I I G G E F T T I E N Q P W F A A I Y - R R H R G
t-PA        279 I K G G L F A D I A S H P W Q A A I F A K H R R S

Y W G H S F C A G S L I A P C W V L T A A H C L Q D R P A P E D L
P T Q R H L C G G S I I G N Q W I L T A A H C F Y G V E S P K I L
G S V T Y V C G G S L I S P C W V I S A T H C F I D Y P K K E D Y
P G E R F L C G G I L I S S C W I L S A A H C F Q E R F P P H H L

T V V L G Q E R R N H S C E P C Q T L A V R S Y R L H E A F S P V
R V Y S G I L N Q S E I K E D T S F F G V Q E I I I H D Q Y K M A
I V Y L G R S R L N S N T Q G E M K F E V E N L I L H K D Y S A D
T V I L G R T Y R V V P G E E E Q K F E V E K Y I V H K E F D D D

S Y Q H - - D L A L L R L Q E D A D G S C A L L S P Y V Q P V C L
E S G Y - - D I A L L K L E T T V N Y T D S Q R - - - - - P I C L
T L A H H N D I A L L K I R S K - E G R C A Q P S R T I Q T I C L
T Y D - - N D I A L L Q L K S D - S S R C A Q E S S V V R T V C L

P S G A A R P S E T T L C Q V A G W G H Q F E G A E E Y A S F L Q
P S K G D R N V I Y T D C W V T G W G Y R K L R D K I Q N T - L Q
P S M Y N D P Q F G T S C E I T G F G K E N S T D Y L Y P E Q L K
P P A D L Q L P D W T E C E L S G Y G K H E A L S P F Y S E R L K

E A Q V P F L S L E R C S A P D V H G S S I L P G M L C A G - F L
K A K I P L V T N E E C Q K R - Y R G H K I T H K M I C A G - Y R
M T V V K L I S H R E C Q Q P H Y Y G S E V T T K M L C A A D P Q
E A H V R L Y P S S R C T S Q H L L N R T V T D N M L C A G D T R

E G G T - - - - - D A C Q G D S G G P L V C E D Q A A E R R L T L
E G G K - - - - - D A C K G D S G G P L S C K H N E V - - - W H L
W - - - - - - - K T D S C Q G D S G G P L V C S L Q G - - - R M T L
S G G P Q A N L H D A C Q G D S G G P L V C L N D G - - - R M T L

Q G I I S W G S G C G D R N K P G V Y T D V A Y Y L A W I R E H T
V G I T S W G E G C A Q R E R P G V Y T N V V E Y V D W I L E K T
T G I V S W G R G C A L K D K P G V Y T R V S H F L P W I R S H T
V G I I S W G L G C G Q K D V P G V Y T K V T N Y L D W I R D N M

V S 596
Q A V 607
K E E N G L A L 411
R P 530
```

Fig. 142. The locations of the introns in the serine-proteinase domains from factors XII and XI, urokinase (u-PA) and t-PA. Only the positions at which amino acids are the same in all four domains are outlined. The position of the introns are indicated by ▼ and the corresponding amino-acid residues are outlined with dashed lines.

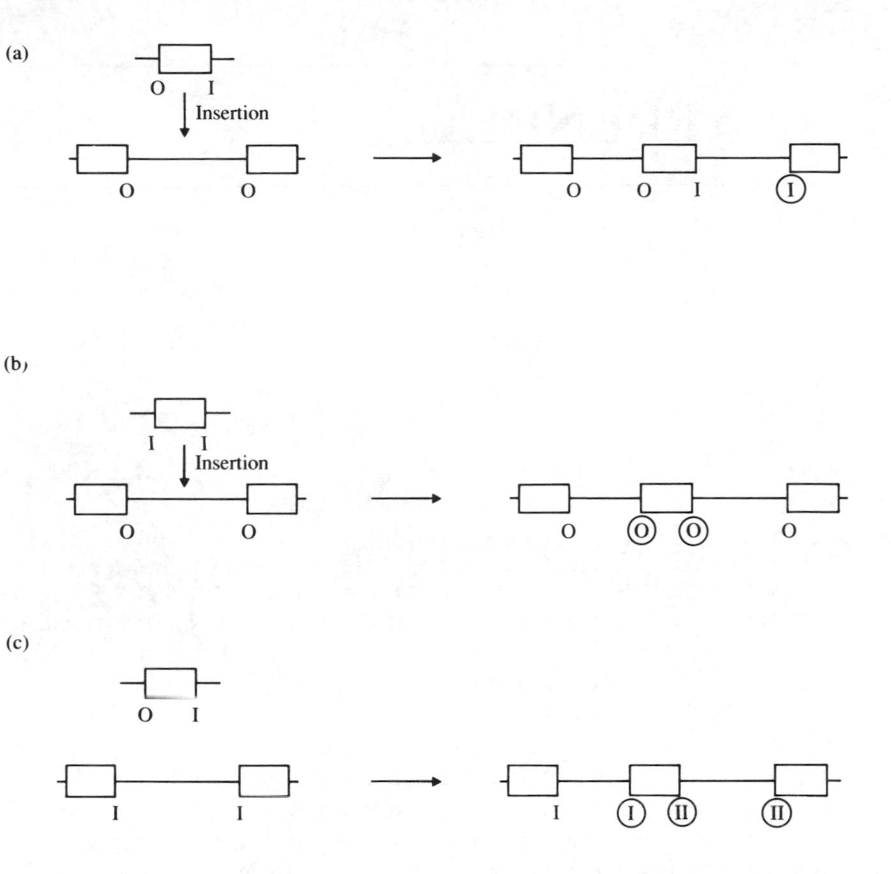

Fig. 143. The significance of the phase class of exon–intron junctions for the exon-shuffling model. Phase class 0 means that the intron is located between two codons, while phase classes I and II mean that the intron splits the codon after its first or second nucleotide, respectively. (a) shows that the phase class of an inserted exon must be the same at both ends. If that is not the case, exons located downstream will be translated with a reading-frame shift. (b) shows that the phase classes of an inserted exon must be the same as the phase classes of the intron–exon junctions between which the exon is inserted. If that is not the case, the inserted exon will be translated in another reading frame. (c) shows a combination.

APPENDIX

Factor XII

```
M R A L L L L G F L L V S L E S T L S I P P W E A P K E H K Y K A
E E H T V V L T V T G E P C H F P F Q Y H R Q L Y H K C T H K G R
P G P Q P W C A T T P N F D Q D Q R W G Y C L E P K K V K D H C S
K H S P C Q K G G T C V N M P S G P H C L C P Q H L T G N H C Q K
E K C F E P Q L L R F F H K N E I W Y R T E Q A A V A R C Q C K G
P D A H C Q R L A S Q A C R T N P C L H G G R C L E V E G H R L C
H C P V G Y T G P F C D V D T K A S C Y D G R G L S Y R G L A R T
T L S G A P C Q P W A S E A T Y R N V T A E Q A R N W G L G G H A
F C R N P D N D I R P W C F V L N R D R L S W E Y C D L A Q C Q T
P T Q A A P P T P V S P R L H V P L M P A Q P A P P K P Q P T T R
T P S Q S Q T P G A L P A K R E Q P P S L T R N G P L S C G Q R L
R K S L S S M T R V V G G L V A L R G A H P Y I A A L Y W G H S F
C A G S L I A P C W V L T A A H C L Q D R P A P E D L T V V L G Q
E R R N H S C E P C Q T L A V R S Y R L H E A F S P V S Y Q H D L
A L L R L Q E D A D G S C A L L S P Y V Q P V C L P S G A A R P S
E T T L C Q V A G W G H Q F E G A E E Y A S F L Q E A Q V P F L S
L E R C S A P D V H G S S I L P G M L C A G F L E G G T D A C Q G
D S G G P L V C E D Q A A E R R L T L Q G I I S W G S G C G D R N
K P G V Y T D V A Y Y L A W I R E H T V S
```

Fig. A1. Amino-acid sequence of human factor XII. The signal peptide is outlined. The residues presumed to be involved in the catalytic triad are encircled. → shows the position at which cleavage by plasma kallikrein leads to α-factor XIIa, ⇒ the positions at which extended cleavage by the same enzyme leads to β-factor XIIa, ◆ the N-bound carbohydrate and ● the O-bound carbohydrate.

Plasma prekallikrein

M I L F K Q A T Y F I S L F A T V S C G C L T Q L Y E N A F F R G

G D V A S M Y T P N A Q Y C Q M R C T F H P R C L L F S F L P A S

S I N D M E K R F G C F L K D S V T G T L P K V H R T G A V S G H

S L K Q C C H Q I S A C H R D I Y K G V D M R G V N F N V S K V S

S V E E C Q K R C T N N I R C Q F F S Y A T Q T F H K A E Y R N N

C L L K Y S P G G T P T A I K V L S N V E S G F S L K P C A L S E

I G C H M N I F Q H L A F S D V D V A R V L T P D A F V C R T I C

T Y H P N C L F F T F Y T N V W K I E S Q R N V C L L K T S E S G

T P S S S T P Q E N T I S G Y S L L T C K R T L P E P C H S K I Y

P G V D F G G E E L N V T F V K G V N V C Q E T C T K M I R C Q F

F T Y S L L P E D C K E E K C K C F L R L S M D G S P T R I A Y G

T Q G S S G Y S L R L C N T G D N S V C T T K T S T R I V G G T N

S S W G E W P W Q V S L Q V K L T A Q R H L C G G S L I G H Q W V

L T A A H C F D G L P L Q D V W R I Y S G I L N L S D I T K D T P

F S Q I K E I I I H Q N Y K V S E G N H D I A L I K L Q A P L N Y

T E F Q K P I C L P S K G D S T T I Y T N C W V T G W G F S K E K

G E I Q N I L Q K V N I P L V T N E E C Q K R Y Q D Y K I T Q R M

V C A G Y K E G G K D A C K G D S G G P L V C K H N G M W R L V G

I T S W G E G C A R R E Q P G V Y T K V A E Y M D W I L E K T Q S

S D G K A Q M Q S P A

Fig. A2. Amino-acid sequence of human plasma prekallikrein. The signal peptide is outlined. The residues presumed to be involved in the catalytic triad are encircled. → shows the position of cleavage by factor XIIa and ◆ the N-bound carbohydrate. Position 124 (underlined) is dimorphic and can be N or S.

Factor XI

```
M I F L Y Q V V H F I L F T S V S G E C V T Q L L K D T C F E G G
D I T T V F T P S A K Y C Q V V C T Y H P R C L L F T F T A E S P
S E D P T R W F T C V L K D S V T E T L P R V N R T A A I S G Y S
F K Q C S H Q I S A C N K D I Y V D L D M K G I N Y N S S V A K S
A Q E C Q E R C T D D V H C H F F T Y A T R Q F P S L E H R N I C
L L K H T Q T G T P T R I T K L D K V V S G F S L K S C A L S N L
A C I R D I F P N T V F A D S N I D S V M A P D A F V C G R I C T
H H P G C L F F T F F S Q E W P K E S Q R N L C L L K T S E S G L
P S T R I K K S K A L S G F S L Q S C R H S I P V F C H S S F Y H
D T D F L G E E L D I V A A K S H E A C Q K L C T N A V R C Q F F
T Y T P A Q A S C N E G K G K C Y L K L S S N G S P T K I L H G R
G G I S G Y T L R L C K M D N E C T T K I K P R I V G G T A S V R
G E W P W Q V T L H T T S P T Q R H L C G G S I I G N Q W I L T A
A H C F Y G V E S P K I L R V Y S G I L N Q S E I K E D T S F F G
V Q E I I I H D Q Y K M A E S G Y D I A L L K L E T T V N Y T D S
Q R P I C L P S K G D R N V I Y T D C W V T G W G Y R K L R D K I
Q N T L Q K A K I P L V T N E E C Q K R Y R G H K I T H K M I C A
G Y R E G G K D A C K G D S G G P L S C K H N E V W H L V G I T S
W G E G C A Q R E R P G V Y T N V V E Y V D W I L E K T Q A V
```

Fig. A3. Amino-acid sequence of human factor XI (monomer). The signal peptide is outlined. The residues presumed to be involved in the catalytic triad are encircled. → shows the position of cleavage by factor XIIa and ◇ the potential attachment sites for N-bound carbohydrate.

HMM-kininogen

```
M K L I T I L F L C S R L L L S L T Q E S Q S E E I D C N D K D L
F K A V D A A L K K Y N S Q N Q S N N Q F V L Y R I T E A T K T V
G S D T F Y S F K Y E I K E G D C P V Q S G K T W Q D C E Y K D A
A K A A T G E C T A T V G K R S S T K F S V A T Q T C Q I T P A E
G P V V T A Q Y D C L G C V H P I S T Q S P D L E P I L R H G I Q
Y F N N N T Q H S S L F M L N E V K R A Q R Q V V A G L N F R I T
Y S I V Q T N C S K E N F L F L T P D C K S L W N G D T G E C T D
N A Y I D I Q L R I A S F S Q N C D I Y P G K D F V Q P P T K I C
V G C P R D I P T N S P E L E E T L T H T I T K L N A E N N A T F
Y F K I D N V K K A R V Q V V A G K K Y F I D F V A R E T T C S K
E S N E E L T E S C E T K K L G Q S L D C N A E V Y V V P W E K K
I Y P T V N C Q P L G M I S L M K R P P G F S P F R S S R I G E I
K E E T T V S P P H T S M A P A Q D E E R D S G K E Q G H T R R H
D W G H E K Q R K H N L G H G H K H E R D Q G H G H Q R G H G L G
H G H E Q Q H G L G H G H K F K L D D D L E H Q G G H V L D H G H
K H K H G H G H G K H K N K G K K N G K H N G W K T E H L A S S S
E D S T T P S A Q T Q E K T E G P T P I P S L A K P G V T V T F S
D F Q D S D L I A T M M P P I S P A P I Q S D D D W I P D I Q T D
P N G L S F N P I S D F P D T T S P K C P G R P W K S V S E I N P
T T Q M K E S Y Y F D L T D G L S
```

Fig. A4. Amino-acid sequence of human HMM-kininogen. The signal peptide is outlined. → shows the positions of cleavage by plasma kallikrein; their use results in the formation of bradykinin. ◆ shows N-bound carbohydrate and ● O-bound carbohydrate.

Factor IX

```
M Q R V N M I M A E S P G L I T I C L L G Y L L S A E C T V F L D
H E N A N K I L N R P K R Y N S G K L γ γ F V Q G N L γ R γ C M γ
γ K C S F γ γ A R γ V F γ N T γ R T T γ F W K Q Y V D G D Q C E S
N P C L N G G S C K β D I N S Y E C W C P F G F E G K N C E L D V
T C N I K N G R C E Q F C K N S A D N K V V C S C T E G Y R L A E
N Q K S C E P A V P F P C G R V S V S Q T S K L T R↓A E T V F P D
V D Y V N S T E A E T I L D N I T Q S T Q S F N D F T R↓V V G G E
D A K P G Q F P W Q V V L N G K V D A F C G G S I V N E K W I V T
A A(H)C V E T G V K I T V V A G E H N I E E T E H T E Q K R N V I
R I I P H H N Y N A A I N K Y N H(D)I A L L E L D E P L V L N S Y
V T P I C I A D K E Y T N I F L K F G S G Y V S G W G R V F H K G
R S A L V L Q Y L R V P L V D R A T C L R S T K F T I Y N N M F C
A G F H E G G R D S C Q G D(S)G G P H V T E V E G T S F L T G I I
S W G E E C A M K G K Y G I Y T K V S R Y V N W I K E K T K L T
```

Fig. A5. Amino-acid sequence of human factor IX. The signal/propeptide is outlined. The residues presumed to be involved in the catalytic triad are encircled. → shows the positions of cleavage by factor XIa, ◇ the potential attachment sites for N-bound carbohydrate and ● the O-bound carbohydrate. Glutamic acid residues presumed to be modified to γ-carboxy-glutamic acid residues are shown as γ, while the partially hydroxylated aspartic acid residue is shown as β. Position 148 (underlined) is dimorphic and can be T or A.

Factor VIII

M Q I E L S T C F F L C L L R F C F S A T R R Y Y L G A V E L S W

D Y M Q S D L G E L P V D A R F P P R V P K S F P F N T S V V Y K

K T L F V E F T D H L F N I A K P R P P W M G L L G P T I Q A E V

Y D T V V I T L K N M A S H P V S L H A V G V S Y W K A S E G A E

Y D D Q T S Q R E K E D D K V F P G G S H T Y V W Q V L K E N G P

M A S D P L C L T Y S Y L S H V D L V K D L N S G L I G A L L V C

R E G S L A K E K T Q T L H K F I L L F A V F D E G K S W H S E T

K N S L M Q D R D A A S A R A W P K M H T V N G Y V N R S L P G L

I G C H R K S V Y W H V I G M G T T P E V H S I F L E G H T F L V

R N H R Q A S L E I S P I T F L T A Q T L L M D L G Q F L L F C H

I S S H Q H D G M E A Y V K V D S C P E E P Q L R M K N N E E A E

D Y D D D L T D S E M D V V R F D D D N S P S F I Q I R S V A K K

H P K T W V H Y I A A E E E D W D Y A P L V L A P D D R S Y K S Q

Y L N N G P Q R I G R K Y K K V R F M A Y T D E T F K T R E A I Q

H E S G I L G P L L Y G E V G D T L L I I F K N Q A S R P Y N I Y

P H G I T D V R P L Y S R R L P K G V K H L K D F P I L P G E I F

K Y K W T V T V E D G P T K S D P R C L T R Y Y S S F V N M E R D

L A S G L I G P L L I C Y K E S V D Q R G N Q I M S D K R N V I L

F S V F D E N R S W Y L T E N I Q R F L P N P A G V Q L E D P E F

Q A S N I M H S I N G Y V F D S L Q L S V C L H E V A Y W Y I L S

I G A Q T D F L S V F F S G Y T F K H K M V Y E D T L T L F P F S

G E T V F M S M E N P G L W I L G C H N S D F R N R G M T A L L K

V S S C D K N T G D Y Y E D S Y E D I S A Y L L S K N N A I E P R

S F S Q N S R H P S T R Q K Q F N A T T I P E N D I E K T D P W F

A H R T P M P K I Q N V S S S D L L M L L R Q S P T P H G L S L S

D L Q E A K Y E T F S D D P S P G A I D S N N S L S E M T H F R P

Q L H H S G D M V F T P E S G L Q L R L N E K L G T T A A T E L K

K L D F K V S S T S N N L I S T I P S D N L A A G T D N T S S L G

P P S M P V H Y D S Q L D T T L F G K K S S P L T E S G G P L S L

S E E N N D S K L L E S G L M N S Q E S S W G K N V S S T E S G R

L F K G K R A H G P A L L T K D N A L F K V S I S L L K T N K T S

N N S A T N R K T H I D G P S L L I E N S P S V W Q N I L E S D T

E F K K V T P L I H D R M L M D K N A T A L R L N H M S N K T T S

S K N M E M V Q Q K K E G P I P P D A Q N P D M S F F K M L F L P

Fig. A6. Amino-acid sequence of human factor VIII. The signal peptide is outlined. → shows the positions of cleavage by thrombin, ⇒ the position of the bond cleaved when two-chained factor VIII is formed and ◇ the potential attachment sites for N-bound carbohydrate. Sulphated tyrosine residues are encircled. The positions underlined can be dimorphic: 56 (D/V) and 1241 (D/E).

E S A R W I Q R T H G K N S L N S G Q G P S P K Q L V S L G P E K

S V E G Q N F L S E K N K V V V G K G E F T K D V G L K E M V F P

S S R N L F L T N L D N L H E N N T H N Q E K K I Q E E I E K K E

T L I Q E N V V L P Q I H T V T G T K N F M K N L F L L S T R Q N

V E G S Y D G A Y A P V L Q D F R S L N D S T N R T K K H T A H F

S K K G E E E N L E G L G N Q T K Q I V E K Y A C T T R I S P N T

S Q Q N F V T Q R S K R A L K Q F R L P L E E T E L E K R I I V D

D T S T Q W S K N M K H L T P S T L T Q I D Y N E K E K G A I T Q

S P L S D C L T R S H S I P Q A N R S P L P I A K V S S F P S I R

P I Y L T R V L F Q D N S S H L P A A S Y R K K D S G V Q E S S H

F L Q G A K K N N L S L A I L T L E M T G D Q R E V G S L G T S A

T N S V T Y K K V E N T V L P K P D L P K T S G K V E L L P K V H

I Y Q K D L F P T E T S N G S P G H L D L V E G S L L Q G T E G A

I K W N E A N R P G K V P F L R V A T E S S A K T P S K L L D P L

A W D N H Y G T Q I P K E E W K S Q E K S P E K T A F K K K D T I

L S L N A C E S N H A I A A I N E G Q N K P E I E V T W A K Q G R

T E R L C S Q N P P V L K R H Q R E I T R T T L Q S D Q E E I D Y

D D T I S V E M K K E D F D I Y D E D E N Q S P R S F Q K K T R H

Y F I A A V E R L W D Y G M S S S P H V L R N R A Q S G S V P Q F

K K V V F Q E F T D G S F T Q P L Y R G E L N E H L G L L G P Y I

R A E V E D N I M V T F R N Q A S R P Y S F Y S S L I S Y E E D Q

R Q G A E P R K N F V K P N E T K T Y F W K V Q H H M A P T K D E

F D C K A W A Y F S D V D L E K D V H S G L I G P L L V C H T N T

L N P A H G R Q V T V Q E F A L F F T I F D E T K S W Y F T E N M

E R N C R A P C N I Q M E D P T F K E N Y R F H A I N G Y I M D T

L P G L V M A Q D Q R I R W Y L L S M G S N E N I H S I H F S G H

V F T V R K K E E Y K M A L Y N L Y P G V F E T V E M L P S K A G

I W R V E C L I G E H L H A G M S T L F L V Y S N K C Q T P L G M

A S G H I R D F Q I T A S G Q Y G Q W A P K L A R L H Y S G S I N

A W S T K E P F S W I K V D L L A P M I I H G I K T Q G A R Q K F

S S L Y I S Q F I I M Y S L D G K K W Q T Y R G N S T G T L M V F

F G N V D S S G I K H N I F N P P I I A R Y I R L H P T H Y S I R

S T L R M E L M G C D L N S C S M P L G M E S K A I S D A Q I T A

S S Y F T N M F A T W S P S K A R L H L Q G R S N A W R P Q V N N

P K E W L Q V D F Q K T M K V T G V T T Q G V K S L L T S M Y V K

E F L I S S S Q D G H Q W T L F F Q N G K V K V F Q G N Q D S F T

P V V N S L D P P L L T R Y L R I H P Q S W V H Q I A L R M E V L

G C E A Q D L Y

Factor X

```
M G R P L H L V L L S A S L A G L L L L G E S L F I R R E Q A N N
I L A R V T R A N S F L γ γ M K K G H L γ R γ C M γ γ T C S Y γ γ
A R γ V F γ D S D K T N γ F W N K Y K D G D Q C E T S P C Q N Q G
K C K β G L G E Y T C T C L E G F E G K N C E L F T R K L C S L D
N G D C D Q F C H E E Q N S V V C S C A R G Y T L A D N G K A C I
P T G P Y P C G K Q T L E R R K R S V A Q A T S S S G E A P D S I
T W K P Y D A A D L D P T E N P F D L L D F N Q T Q P E R G D N N
L T R I V G G Q E C K D G E C P W Q A L L I N E E N E G F C G G T
I L S E F Y I L T A A H C L Y Q A K R F K V R V G D R N T E Q E E
G G E A V H E V E V V I K H N R F T K E T Y D F D I A V L R L K T
P I T F R M N V A P A C L P E R D W A E S T L M T Q K T G I V S G
F G R T H E K G R Q S T R L K M L E V P Y V D R N S C K L S S S F
I I T Q N M F C A G Y D T K Q E D A C Q G D S G G P H V T R F K D
T Y F V T G I V S W G E G C A R K G K Y G I Y T K V T A F L K W I
D R S M K T R G L P K A K S H A P E V I T S S P L K
```

Fig. A7. Amino-acid sequence of human factor X. The signal/propeptide is outlined. The residues presumed to be involved in the catalytic triad are encircled. ↓ shows the position of cleavage by factor IXa, ⇩ the positions of the bonds that are cleaved when two-chained factor X is formed and ◇ the potential attachment sites for N-bound carbohydrate. Glutamic acid residues modified to γ-carboxyglutamic acid residues are shown as γ, while the hydroxylated aspartic acid residue is shown as β. An R-G-D sequence is outlined with dashed lines.

Factor V

```
M F P G C P R L W V L V V L G T S W V G W G S Q G T E A A Q L R Q
F Y V A A Q G I S W S Y R P E P T N S S L N L S V T S F K K I V Y
R E Y E P Y F K K E K P Q S T I S G L L G P T L Y A E V G D I I K
V H F K N K A D K P L S I H P Q G I R Y S K L S E G A S Y L D H T
F P A E K M D D A V A P G R E Y T Y E W S I S E D S G P T H D D P
P C L T H I Y Y S H E N L I E D F N S G L I G P L L I C K K G T L
T E G G T Q K T F D K Q I V L L F A V F D E S K S W S Q S S S L M
Y T V N G Y V N G T M P D I T V C A H D H I S W H L L G M S S G P
E L F S I H F N G Q V L E Q N H H K V S A I T L V S A T S T T A N
M T V G P E G K W I I S S L T P K H L Q A G M Q A Y I D I K N C P
K K T R N L K K I T R E Q R R H M K R W E Y F I A A E E V I W D Y
A P V I P A N M D K K Y R S Q H L D N F S N Q I G K H Y K K V M Y
T Q Y E D E S F T K H T V N P N M K E D G I L G P I I R A Q V R D
T L K I V F K N M A S R P Y S I Y P H G V T F S P Y E D E V N S S
F T S G R N N T M I R A V Q P G E T Y T Y K W N I L E F D E P T E
N D A Q C L T R P Y Y S D V D I M R D I A S G L I G L L L I C K S
R S L D R R G I Q R A A D I E Q Q A V F A V F D E N K S W Y L E D
N I N K F C E N P D E V K R D D P K F Y E S N I M S T I N G Y V P
E S I T T L G F C F D D T V Q W H F C S V G T Q N E I L T I H F T
G H S F I Y G K R H E D T L T L F P M R G E S V T V T M D N V G T
W M L T S M N S S P R S K K L R L K F R D V K C I P D D D E D S Y
E I F E P P E S T V M A T R K M H D R L E P E D E E S D A D Y D Y
Q N R L A A A L G I R S F R N S S L N Q E E E E F N L T A L A L E
N G T E F V S S N T D I I V G S N Y S S P S N I S K F T V N N L A
E P Q K A P S H Q Q A T T A G S P L R H L I G K N S V L N S S T A
E H S S P Y S E D P I E D P L Q P D V T G I R L L S L G A G E F R
S Q E H A K R K G P K V E R D Q A A K H R F S W M K L L A H K V G
R H L S Q D T G S P S G M R P W E D L P S Q D T G S P S R M R P W
E D P P S D L L L L K Q S N S S K I L V G R W H L A S E K G S Y E
I I Q D T D E D T A V N N W L I S P Q N A S R A W G E S T P L A N
K P G K Q S G H P K F P R V R H K S L Q V R Q D G G K S R L K K S
Q F L I K T R K K K K E K H T H H A P L S P R T F H P L R S E A Y
N T F S E R R L K H S L V L H K S N E T S L P T D L N Q T L P S M
D F G W I A S L P D H N Q N S S N D T G Q A S C P P G L Y Q T V P
```

Fig. A8. Amino-acid sequence of human factor V. The signal peptide is outlined. → shows the positions of cleavage by thrombin and ◇ the potential attachment sites for N-bound carbohydrate. Position 1257 (underlined) may be dimorphic (L/I).

P E E H Y Q T F P I Q D P D Q M H S T S D P S H R S S S P E L S E
M L E Y D R S H K S F P T D I S Q M S P S S E H E V W Q T V I S P
D L S Q V T L S P E L S Q T N L S P D L S H T T L S P E L I Q R N
L S P A L G Q M P I S P D L S H T T L S P D L S H T T L S L D L S
Q T N L S P E L S Q T N L S P A L G Q M P L S P D L S H T T L S L
D F S Q T N L S P E L S H M T L S P E L S Q T N L S P A L G Q M P
I S P D L S H T T L S L D F S Q T N L S P E L S Q T N L S P A L G
Q M P L S P D P S H T T L S L D L S Q T N L S P E L S Q T N L S P
D L S E M P L F A D L S Q I P L T P D L D Q M T L S P D L G E T D
L S P N F G Q M S L S P D L S Q V T L S P D I S D T T L L P D L S
Q I S P P P D L D Q I F Y P S E S S Q S L L L Q E F N E S F P Y P
D L G Q M P S P S S P T L N D T F L S K E F N P L V I V G L S K D
G T D Y I E I I P K E E V Q S S E D D Y A E I D Y V P Y D D P Y K
T D V R T N I N S S R D P D N I A A W Y L R S N N G N R R N Y Y I
A A E E I S W D Y S E F V Q R E T D I E D S D D I P E D T T Y K K
V V F R K Y L D S T F T K R D P R G E Y E E H L G I L G P I I R A
E V D D V I Q V R F K N L A S R P Y S L H A H G L S Y E K S S E G
K T Y E D D S P E W F K E D N A V Q P N S S Y T Y V W H A T E R S
G P E S P G S A C R A W A Y Y S A V N P E K D I H S G L I G P L L
I C Q K G I L H K D S N M P V D M R E F V L L F M T F D E K K S W
Y Y E K K S R S S W R L T S S E M K K S H E F H A I N G M I Y S L
P G L K M Y E Q E W V R L H L L N I G G S Q D I H V V H F H G Q T
L L E N G N K Q H Q L G V W P L L P G S F K T L E M K A S K P G W
W L L N T E V G E N Q R A G M Q T P F L I M D R D C R M P M G L S
T G I I S D S Q I K A S E F L G Y W E P R L A R L N N G G S Y N A
W S V E K L A A E F A S K P W I Q V D M Q K E V I I T G I Q T Q G
A K H Y L K G C Y T T E F Y V A Y S S N Q I N W Q I F K G N S T R
N V M Y F N G N S D A S T I K E N Q F D P P I V A R Y I R I S P T
R A Y N R P T L R L E L Q G C E V N G C S T P L G M E N G K I E N
K Q I T A S S F K K S W W G D Y W E P F R A R L N A Q G R V N A W
Q A K A N N N K Q W L E I D L L K I K K I T A I I T Q G C K S L S
S E M Y V K S Y T T H Y S E Q G V E W K P Y R L K S S M V D K I F
E G N T N T K G H V K N F F N P P I I S R F I R V I P K T W N Q S
I T L R L E L F G C D I Y

Prothrombin

```
M A H V R G L Q L P G C L A L A A L C S L V H S Q H V F L A P Q Q
A R S L L Q R V R R A N T F L γ γ V R K G N L γ R γ C V γ γ T C S
Y γ γ A F γ A L γ S S T A T D V F W A K Y T A C E T A R T P R D K
L A A C L E G N C A E G L G T N Y R G H V N I T R S G I E C Q L W
R S R Y P H K P E I N S T T H P G A D L Q E N F C R N P D S S N T
G P W C Y T T D P T V R R Q E C S I P V C G Q D Q V T V A M T P R
S E G S S V N L S P P L E Q C V P D R G Q Q Y Q G R L A V T T H G
L P C L A W A S A Q A K A L S K H Q D F N S A V Q L V E N F C R N
P D G D E E G V W C Y V A G K P G D F G Y C D L N Y C E E A V E E
E T G D G L D E D S D R A I G G R T A T S E Y Q T F F N P R T F G
S G E A D C G L R P L F E K K S L E D K T E R E L L E S Y I D G R
I V E G S D A E I G M S P W Q V M L F R K S P Q E L L C G A S L I
S D R W V L T A A H C L L Y P P W D K N F T E N D L L V R I G K H
S R T R Y E R N I E K I S M L E K I Y I H P R Y N W R E N L D R D
I A L M K L K K P V A F S D Y I H P V C L P D R E T A A S L L Q A
G Y K G R V T G W G N L K E T W T A N V G K G Q P S V L Q V V N L
P I V E R P V C K D S T R I R I T D N M F C A G Y K P D E G K R G
D A C E G D S G G P F V M K S P F N N R W Y Q M G I V S W G E G C
D R D G K Y G F Y T H V F R L K K W I Q K V I D Q F G E
```

Fig. A9. Amino-acid sequence of human prothrombin. The signal peptide is outlined. The residues presumed to be involved in the catalytic triad are encircled. → shows the positions of cleavage by factor Xa and ◆ the N-bound carbohydrate. Glutamic acid residues modified to γ-carboxy-glutamic acid residues are shown as γ. An R-G-D sequence is outlined with dashed lines.

Factor VII

M V S Q A L R L L C L L L G L Q G C L A A G G V A K A S G G E T R
D M P W K P G P H R V F V T Q E E A H G V L H R R R R A N A F L γ
γ L R P G S L γ R γ C K γ γ Q C S F γ γ A R γ I F K D A γ R T K L
F W I S Y S D G D Q C A S S P C Q N G G S C K D Q L Q S Y I C F C
L P A F E G R N C E T H K D D Q L I C V N E N G G C E Q Y C S D H
T G T K R S C R C H E G Y S L L A D G V S C T P T V E Y P C G K I
P I L E K R N A S K P Q G R I V G G K V C P K G E C P W Q V L L L
V N G A Q L C G G T L I N T I W V V S A A H C F D K I K N W R N L
I A V L G E H D L S E H D G D E Q S R R V A Q V I I P S T Y V P G
T T N H D I A L L R L H Q P V V L T D H V V P L C L P E R T F S E
R T L A F V R F S L V S G W G Q L L D R G A T A L E L M V L N V P
R L M T Q D C L Q Q S R K V G D S P N I T E Y M F C A G Y S D G S
L D S C K G D S G G P H A T H Y R G T W Y L T G I V S W G Q G C A
T V G H F G V Y T R V S Q Y I E W L Q K L M R S E P R P G V L L R
A P F P

Fig. A10. Amino-acid sequence of human factor VII. The signal peptide is outlined. The residues presumed to be involved in the catalytic triad are encircled. → shows the position of cleavage by factor Xa, ◆ the N-bound carbohydrate and ● the O-bound carbohydrate (see text). Glutamic acid residues modified to γ-carboxy-glutamic acid residues are shown as γ.

Tissue factor

M E T P A W P R V P R P E T A V A R T L L L G W V F A Q V A G A S
G T T N T V A A Y N L T W K S T N F K T I L E W E P K P V N Q V Y
T V Q I S T K S G D W K S K C F Y T T D T E C D L T D E I V K D V
K Q T Y L A R V F S Y P A G N V E S T G S A G E P L Y E N S P E F
T P Y L E T N L G Q P T I Q S F E Q V G T K V N V T V E D E R T L
V R R N N T F L S L R D V F G K D L I Y T L Y Y W K S S S S G K K
T A K T N T N E F L I D V D K G E N Y C F S V Q A V I P S R T V N
R K S T D S P V E C M G Q E K G E F R E I F Y I I G A V V F V V I
I L V I I L A I S L H K C R K A G V G Q S W K E N S P L N V S

Fig. A11. Amino-acid sequence of human tissue factor. The signal peptide is outlined but may occur in two different lengths, as indicated by the dashed outline. ◆ shows N-bound carbohydrate and ◇ the potential attachment sites for N-bound carbohydrate. The esterified cysteine residue is encircled. The presumed transmembrane stretch is underlined.

Fibrinogen Aα-chain

M F S M R I V C L V L S V V G T A W T A D S G E G D F L A E G G G
V R G P R V V E R H Q S A C K D S D W P F C S D E D W N Y K C P S
G C R M K G L I D E V N Q D F T N R I N K L K N S L F E Y Q K N N
K D S H S L T T N I M E I L R G D F S S A N N R D N T Y N R V S E
D L R S R I E V L K R K V I E K V Q H I Q L L Q K N V R A Q L V D
M K R L E V D I D I K I R S C R G S C S R A L A R E V D L K D Y E
D Q Q K Q L E Q V I A K D L L P S R D R Q H L P L I K M K P V P D
L V P G N F K S Q L Q K V P P E W K A L T D M P Q M R M E L E R P
G G N E I T R G G S T S Y G T G S E T E S P R N P S S A G S W N S
G S S G P G S T G N R N P G S S G T G G T A T W K P G S S G P G S
A G S W N S G S S G T G S T G N Q N P G S P R P G S T G T W N P G
S S E R G S A G H W T S E S S V S G S T G Q W H S E S G S F R P D
S P G S G N A R P N N P D W G T F E E V S G N V S P G T R R E Y H
T E K L V T S K G D K E L R T G K E K V T S G S T T T T R R S C S
K T V T K T V I G P D G H K E V T K E V V T S E D G S D C P E A M
D L G T L S G I G T L D G F R H R H P D E A A F F D T A S T G K T
F P G F F S P M L G E F V S E T E S R G S E S G I F T N T K E S S
S H H P G I A E F P S R G K S S S Y S K Q F T S S T S Y N R G D S
T F E S K S Y K M A D E A G S E A D H E G T H S T K R G H A K S R
P V R G I H T S P L G K P S L S P

Fig. A12. Amino-acid sequence of the Aα chain of human fibrinogen. The signal peptide is outlined. → shows the position of cleavage by thrombin and ⇒ the apparent position of cleavage during synthesis. The two R-G-D sequences are outlined with dashed lines. Phosphorylated serine residues are encircled. The positions underlined are dimorphic: 47 (S/T), 296 (T/A) and 312 (A/T).

Fibrinogen Bβ-chain

[M K R M V S W S F H K L K T M K H L L L L L L C V F L V K S] Q G V

N D N E E G F F S A R ↓ G H R P L D K K R E E A P S L R P A P P P I

S G G G Y R A R P A K A A A T Q K K V E R K A P D A G G C L H A D

P D L G V L C P T G C Q L Q E A L L Q Q E R P I R N S V D E L N N

N V E A V S Q T S S S S F Q Y M Y L L K D L W Q K R Q K Q V K D N

E N V V N E Y S S E L E K H Q L Y I D E T V N S N I A T N L R V L

R S I L E N L R S K I Q K L E S D V S A Q M E Y C R T P C T V S C

N I P V V S G K E C E E I I R K G G E T S E M Y L I Q P D S S V K

P Y R V Y C D M N T E N G G W T V I Q N R Q D G S V D F G R K W D

P Y K Q G F G N V A T N T D G K N Y C G L P G E Y W L G N D K I S

Q L T R M G P T E L L I E M E D W K G D K V K A H Y G G F T V Q N

E A N K Y Q I S V N K Y R G T A G N A L M D G A S Q L M G E N◆ R T

M T I H N G M F F S T Y D R D N D G W L T S D P R K Q C S K E D G

G G W W Y N R C H A A N P N G R Y Y W G G Q Y T W D M A K H G T D

D G V V W M N W K G S W Y S M R K M S M K I R P F F P Q Q

Fig. A13. Amino-acid sequence of the Bβ chain of human fibrinogen. The signal peptide is outlined. → shows the position of cleavage by thrombin and ◆ the N-bound carbohydrate. The positions underlined are dimorphic: 162 (A/P), 296 (N/D) and 448 (R/K).

Fibrinogen γ-chain

[M S W S L H P R N L I L Y F Y A L L F L S S T C V A] Y V A T R D N

C C I L D E R F G S Y C P T T C G I A D F L S T Y Q T K V D K D L

Q S L E D I L H Q V E N◆ K T S E V K Q L I K A I Q L T Y N P D E S

S K P N M I D A A T L K S R K M L E E I M K Y E A S I L T H D S S

I R Y L Q E I Y N S N N Q K I V N L K E K V A Q L E A Q C Q E P C

K D T V Q I H D I T G K D C Q D I A N K G A K Q S G L Y F I K P L

K A N Q Q F L V Y C E I D G S G N G W T V F Q K R L D G S V D F K

K N W I Q Y K E G F G H L S P T G T T E F W L G N E K I H L I S T

Q S A I P Y A L R V E L E D W N G R T S T A D Y A M F K V G P E A

D K Y R L T Y A Y F A G G D A G D A F D G F D F G D D P S D K F F

T S H N G M Q F S T W D N D N D K F E G N C A E Q D G S G W W M N

K C H A G H L N G V Y Y Q G G T Y S K A S T P N G Y D N G I I W A

T W K T R W Y S M K K T T M K I I P F N R L T I G E G Q Q H H L G

G A K Q A G D V

Fig. A14. Amino-acid sequence of the γ chain of human fibrinogen. The signal peptide is outlined. ◆ shows N-bound carbohydrate. Position 88 (underlined) is dimorphic and can be K or I.

Factor XIII a-chain

```
[M] S E T S R T A F G G R R A V P P N N S N A A E D D L P T V E L Q
G V V P R↓G V N L Q E F L N V T S V H L F K E R W D T N K V D H H
T D K Y E N N K L I V R R G Q S F Y V Q I D L S R P Y D P R R D L
F R V E Y V I G R Y P Q E N K G T Y I P V P I V S E L Q S G K W G
A K I V M R E D R S V R L S I Q S S P K C I V G K F R M Y V A V W
T P Y G V L R T S R N P E T D T Y I L F N P W C E D D A V Y L D N
E K E R E E Y V L N D I G V I F Y G E V N D I K T R S W S Y G Q F
E D G I L D T C L Y V M D R A Q M D L S G R G N P I K V S R V G S
A M V N A K D D E G V L V G S W D N I Y A Y G V P P S A W T G S V
D I L L E Y R S S E N P V R Y G Q (C) W V F A G V F N T F L R C L G
I P A R I V T N Y F S A H D N D A N L Q M D I F L E E D G N V N S
K L T K D S V W N Y H C W N E A W M T R P D L P V G F G G W Q A V
D S T P Q E N S D G M Y R C G P A S V Q A I K H G H V C F Q F D A
P F V F A Q V N S D L I Y I T A K K D G T H V V E N V D A T H I G
K L I V T K Q I G G D G M M D I T D T Y K F Q E G Q E E E R L A L
E T A L M Y G A K K P L N T E G V M K S R S N V D M D F E V E N A
V L G K D F K L S I T F R N N S H N R Y T I T A Y L S A N I T F Y
T G V P K A E F K K E T F D V T L E P L S F K K E A V L I Q A G E
Y M G Q L L E Q A S L H F F V T A R I N E T R D V L A K Q K S T V
L T I P E I I I K V R G T Q V V G S D M T V T V Q F T N P L K E T
L R N V W V H L D G P G V T R P M K K M F R E I R P N S T V Q W E
E V C R P W V S G H R K L I A S M S S D S L R H V Y G E L D V Q I
Q R R P S M
```

Fig. A15. Amino-acid sequence of the a-chain of human plasma factor XIII. The initiator methionine residue preceding Ser1 is outlined. → shows the position of cleavage by thrombin. The catalytically active Cys residue is encircled. The positions underlined may be dimorphic: 77 (R/G), 78 (R/K), 88 (L/F), 650 (V/I) and 651 (Q/E).

Factor XIII b-chain

```
R L K N L T F I I I L I I S G E L Y A E E K P C G F P H V E N G R
I A Q Y Y Y T F K S F Y F P M S I D K K L S F F C L A G Y T T E S
G R Q E E Q T T C T T E G W S P E P R C F K K C T K P D L S N G Y
I S D V K L L Y K I Q E N M H Y G C A S G Y K T T G G K D E E V V
                                                        ◇
Q C L S D G W S S Q P T C R K E H E T C L A P E L Y N G N Y S T T
Q K T F K V K D K V Q Y E C A T G Y Y T A G G K K T E E V E C L T
Y G W S L T P K C T K L K C S S L R L I E N G Y F H P V K Q T Y E
E G D V V Q F F C H E N Y Y L S G S D L I Q C Y N F G W Y P E S P
V C E G R R N R C P P P P L P I N S K I Q T H S T T Y R H G E I V
H I E C E L N F E I H G S A E I R C E D G K W T E P P K C I E G Q
E K V A C E E P P F I E N G A A N L H S K I Y Y N G D K V T Y A C
K S G Y L L H G S N E I T C N R G K W T L P P E C V E N N E N C K
H P P V V M N G A V A D G I L A S Y A T G S S V E Y R C N E Y Y L
L R G S K I S R C E Q G K W S S P P V C L E P C T V N V D Y M N R
N N I E M K W K Y E G K V L H G D L I D F V C K Q G Y D L S P L T
P L S E L S V Q C N R G E V K Y P L C T R K E S K G M C T S P P L
                              ◆
I K H G V I I S S T V D T Y E N G S S V E Y R C F D H H F L E G S
R E A Y C L D G M W T T P P L C L E P C T L S F T E M E K N N L L
L K W D F D N K P H I L H G E V T E F I C R G D T Y P A E L Y I T
G S I L R M Q C D R G Q L K Y P R C I P R Q S T L S Y Q E P L R T
```

Fig. A16. Amino-acid sequence of the b-chain of human plasma factor XIII. The partially known signal peptide is outlined. ◆ shows N-bound carbohydrate and ◇ potential attachment sites for N bound carbohydrate. An R-G-D sequence is outlined with dashed lines.

Antithrombin III

```
M Y S N V I G T V T S G K R K V Y L L S L L L I G F W D C V T C H
G S P V D I C T A K P R D I P M N P M C I Y R S P E K K A T E D E
G S E Q K I P E A T N R R V W E L S K A N S R F A T T F Y Q H L A
D S K N D N D N I F L S P L S I S T A F A M T K L G A C N D T L Q
Q L M E V F K F D T I S E K T S D Q I H F F F A K L N C R L Y R K
A N K S S K L V S A N R L F G D K S L T F N E T Y Q D I S E L V Y
G A K L Q P L D F K E N A E Q S R A A I N K W V S N K T E G R I T
D V I P S E A I N E L T V L V L V N T I Y F K G L W K S K F S P E
N T R K E L F Y K A D G E S C S A S M M Y Q E G K F R Y R R V A E
G T Q V L E L P F K G D D I T M V L I L P K P E K S L A K V E K E
L T P E V L Q E W L D E L E E M M L V V H M P R F R I E D G F S L
K E Q L Q D M G L V D L F S P E K S K L P G I V A E G R D D L Y V
S D A F H K A F L E V N E E G S E A A A S T A V V I A G R S L N P
N R V T F K A N R P F L V F I R E V P L N T I I F M G R V A N P C
V K
```

Fig. A17. Amino-acid sequence of human antithrombin III. The signal peptide is outlined. → shows the reactive site and ◆ the N-bound carbohydrate.

EPI/LACI

```
M I Y T M K K V H A L W A S V C L L L N L A P A P L N A D S E E D
E E H T I I T D T E L P P L K L M H S F C A F K A D D G P C K A I
M K R F F F N I F T R Q C E E F I Y G G C E G N Q N R F E S L E E
C K K M C T R D N A N R I I K T T L Q Q E K P D F C F L E E D P G
I C R G Y I T R Y F Y N N Q T K Q C E R F K Y G G C L G N M N N F
E T L E E C K N I C E D G P N G F Q V D N Y G T Q L N A V N N S L
T P Q S T K V P S L F E F H G P S W C L T P A D R G L C R A N E N
R F Y Y N S V I G K C R P F K Y S G C G G N E N N F T S K Q E C L
R A C K K G F I Q R I S K G G L I K T K R K R K K Q R V K I A Y E
E I F V K N M
```

Fig. A18. Amino-acid sequence of human extrinsic pathway inhibitor (EPI/LACI). The signal peptide is outlined. ◇ shows potential attachment sites for N-bound carbohydrate. The presumed reactive sites are underlined.

Protein C

```
M W Q L T S L L L F V A T W γ I S G T P A P L D S V F S S S E R A
H Q V L R I R K R A N S F L γ γ L R H S S L γ R γ C I γ γ I C D F
γ γ A K γ I F Q N V D D T L A F W S K H V D G D Q C L V L P L E H
P C A S L C C G H G T C I β G I G S F S C D C R S G W E G R F C Q
R E V S F L N C S L D N G G C T H Y C L E E V G W R R C S C A P G
Y K L G D D L L Q C H P A V K F P C G R P W K R M E K K R S H L K
R D T E D Q E D Q V D P R L I D G K M T R R G D S P W Q V V L L D
S K K K L A C G A V L I H P S W V L T A A H C M D E S K K L L V R
L G E Y D L R R W E K W E L D L D I K E V F V H P N Y S K S T T D
N D I A L L H L A Q P A T L S Q T I V P I C L P D S G L A E R E L
N Q A G Q E T L V T G W G Y H S S R E K E A K R N R T F V L N F I
K I P V V P H N E C S E V M S N M V S E N M L C A G I L G D R Q D
A C E G D S G D P M V A S F H G T W F L V G L V S W G E G C G L L
H N Y G V Y T K V S R Y L D W I H G H I R D K E A P Q K S W A P
```

Fig. A19. Amino-acid sequence of human protein C. The signal peptide is outlined. The residues presumed to be involved in the catalytic triad are encircled. → shows the position of cleavage by thrombin, ⇒ the positions of the bonds that are cleaved when two-chained protein C is formed and ◇ the potential attachment sites for N-bound carbohydrate. Glutamic acid residues presumed to be modified to γ-carboxy-glutamic acid residues are shown as γ, while the aspartic acid residue believed to be hydroxylated is shown as β. An R-G-D sequence is outlined with dashed lines.

Thrombomodulin

```
[M L G V L V L G A L A L A G L G F P] A P A E P Q P E G S Q C V E H
                          ◇
 D C F A L Y P G P A T F L N A S Q I C D G L R G H L M T V R S S V
 A A D V I S L L L N G D G G V G R R R L W I G L Q L P P G C G D P
                               ◇ ◇
 K R L G P L R G F Q W V T G D N N T S Y S R W A R L D L N G A P L
 C G P L C V A V S A A E A T V P S E P I W E E Q Q C E V K A D G F
 L C E F H F P A T C R P L A V E P G A A A A A V S I T Y G T P F A
 A R G A D F Q A L P V G S S A A V A P L G L Q L M C T A P P G A V
 Q G H W A R E A P G A W D C S V E N G G C E H A C N A I P G A P R
 C Q C P A G A A L Q A D G R S C T A S A T Q S C N D L C E H F C V
 P N P D Q P G S Y S C M C E T G Y R L A A D Q H R C E D V D D C I
 L E P S P C P Q R C V N T Q G G F E C H C Y P N Y D L V D G E C V
                                     ◇
 E P V D P C F R A N C E Y Q C Q P L N Q T S Y L C V C A E G F A P
                         ◇
 I P H E P H R C Q M F C N Q T A C P A D C D P N T Q A S C E C P E
 G Y I L D D G F I C T D I D E C E N G G F C S G V C H N L P G T F
 E C I C G P D S A L V R H I G T D C D S G K V D G G D S G S G E P
 P P S P T P G S T L T P P A V G L V H S G L L I G I S I A S L C L
                                         -------------------------
 V V A L L A L L C H L R K K Q G A A R A K M E Y K C A A P S K E V
 ---------------------
 V L Q H V R T E R T P Q R L
```

Fig. A20. Amino-acid sequence of human thrombomodulin. The signal peptide is outlined. ◇ shows potential attachment sites for N-bound carbohydrate. The presumed transmembrane stretch is underlined.

Protein S

```
M R V L G G R C G A P L A γ L L L V L P V S E A N F L S K Q Q A S
Q V L V R K R R A N S L L γ γ T K Q G N L γ R γ C I γ γ L C N K γ
γ A R γ V F γ N D P γ T D Y F Y P K Y L V C L R S F Q T G L F T A
A R Q S T N A Y P D L R S C V N A I P D Q C S P L P C N E D G Y M
S C K β G K A S F T C T C K P G W Q G E K C E F D I N E C K D P S
N I N G G C S Q I C D β T P G S Y H C S C K N G F V M L S N K K D
C K D V D E C S L K P S I C G T A V C K β I P G D F E C E C P E G
Y R Y N L K S K S C E D I D E C S E N M C A Q L C V β Y P G G Y T
C Y C D G K K G F K L A Q D Q K S C E V V S V C L P L N L D T K Y
E L L Y L A E Q F A G V V L Y L K F R L P E I S R F S A E F D F R
T Y D S E G V I L Y A E S I D H S A W L L I A L R G G K I E V Q L
K N E H T S K I T T G G D V I N N G L W N M V S V E E L E H S I S
I K I A K E A V M D I N K P G P L F K P E N G L L E T K V Y F A G
F P R K V E S E L I K P I N P R L D G C I R S W N L M K Q G A S G
I K E I I Q E K Q N K H C L V T V E K G S Y Y P G S G I A Q F H I
D Y N N V S S A E G W H V N V T L N I R P S T G T G V M L A L V S
G N N T V P F A V S L V D S T S E K S Q D I L L S V E N T V I Y R
I Q A L S L C S D Q Q S H L E F R V N R N N L E L S T P L K I E T
T S H E D L Q R Q L A V L D K A M K A K V A T Y L G G L P D V P F
S A T P V N A F Y N G C M E V N I N G V Q L D L D E A I S K H N D
I R A H S C P S V W K K T K N S
```

Fig. A21 Amino-acid sequence of human protein S. The signal peptide is outlined. ⇒ shows the position presumed to be sensitive to cleavage by thrombin and ◇ the potential attachment sites for N-bound carbohydrate. Glutamic acid residues presumed to be modified to γ-carboxy glutamic acid residues are shown as γ, while the aspartic acid and asparagine residues believed to be hydroxylated are shown as β. Position 460 (underlined) is dimorphic and can be S or P.

Protein Ca inhibitor

```
M Q L F L L L C L V L L S P Q G A S L H R H H P R E M K K R V E D
L H V G A T V A P S S R R D F T F D L Y R A L A S A A P S Q N I F
F S P V S I S M S L A M L S L G A G S S T K M Q I L E G L G L N L
Q K S S E K E L H R G F Q Q L L Q E L N Q P R D G F Q L S L G N A
L F T D L V V D L Q D T F V S A M K T L Y L A D T F P T N F R D S
A G A M K Q I N D Y V A K Q T K G K I V D L L K N L D S N A V V I
M V N Y I F F K A K W E T S F N H K G T Q E Q D F Y V T S E T V V
                                  ◇                           ◇
R V P M M S R E D Q Y H Y L L D R N L S C R V V G V P Y Q G N A T
A L F I L P S E G K M Q Q V E N G L S E K T L R K W L K M F K K R
Q L E L Y L P K F S I E G S Y Q L E K V L P S L G I S N V F T S H
              ◇
A D L S R I S N H S N I Q V S E M V H K A V V E V D E S G T R A A
                  ↓
A A T G T I F T F R S A R L N S Q R L V F N R P F L M F I V D N N
I L F L G K V N R P
```

Fig. A22. Amino-acid sequence of human protein Ca inhibitor. The signal peptide is outlined. → shows the reactive site and ◇ the potential attachment sites for N-bound carbohydrate.

Plasminogen

```
M E H K E V V L L L L L F L K S G Q G E P L D D Y V N T Q G A S L
F S V T K K Q L G A G S I E E C A A K C E E D E E F T C R A F Q Y
H S K E Q Q C V I M A E N R K S S I I I R M R D V V L F E K K V Y
L S E C K T G N G K N Y R G T M S K T K N G I T C Q K W S S T S P
H R P R F S P A T H P S E G L E E N Y C R N P D N D P Q G P W C Y
T T D P E K R Y D Y C D I L E C E E E C M H C S G E N Y D G K I S
K T M S G L E C Q A W D S Q S P H A H G Y I P S K F P N K N L K K
N Y C R N P D R E L R P W C F T T D P N K R W E L C D I P R C T T
P P P S S G P T Y Q C L K G T G E N Y R G N V A V T V S G H T C Q
H W S A Q T P H T H N R T P E N F P C K N L D E N Y C R N P D G K
R A P W C H T T N S Q V R W E Y C K I P S C D S S P V S T E Q L A
P T A P P E L T P V V Q D C Y H G D G Q S Y R G T S S T T T T G K
K C Q S W S S M T P H R H Q K T P E N Y P N A G L T M N Y C R N P
D A D K G P W C F T T D P S V R W E Y C N L K K C S G T E A S V V
A P P P V V L L P D V E T P S E E D C M F C N G K G Y R G K R A T
T V T G T P C Q D W A A Q E P H R H S I F T P E T N P R A G L E K
N Y C R N P D G D V G G P W C Y T T N P R K L Y D Y C D V P Q C A
A P S F D C G K P Q V E P K K C P G R V V G G C V A H P H S W P W
Q V S L R T R F G M H F C G G T L I S P E W V L T A A H C L E K S
P R P S S Y K V I L G A H Q E V N L E P H V Q E I E V S R L F L E
P T R K D I A L L K L S S P A V I T D K V I P A C L P S P N Y V V
A D R T E C F I T G W G E T Q G T F G A G L L K E A Q L P V I E N
K V C N R Y E F L N G R V Q S T E L C A G H L A G G T D S C Q G D
S G G P L V C F E K D K Y I L Q G V T S W G L G C A R P N K P G V
Y V R V S R F V T W I E G V M R N N
```

Fig. A23. Amino-acid sequence of human plasminogen. The signal peptide is outlined. The residues presumed to be involved in the catalytic triad are encircled. → shows the position of cleavage by plasminogen activators, ⇒ the position of cleavage by plasmin, ◆ the N-bound carbohydrate and ● the O-bound carbohydrate. Plasminogen I carries both N-bound and O-bound carbohydrate, while plasminogen II carries only O-bound carbohydrate.

Pro-urokinase

```
M R A L L A R L L L C V L V V S D S K G S N E L H Q V P S N C D C
L N G G T C V S N K Y F S N I H W C N C P K K F G G Q H C E I D K
S K T C Y E G N G H F Y R G K A S T D T M G R P C L P W N S A T V
L Q Q T Y H A H R S D A L Q L G L G K H N Y C R N P D N R R R P W
C Y V Q V G L K P L V Q E C M V H D C A D G K K P S S P P E E L K
F Q C G Q K T L R P R F K↓I I G G E F T T I E N Q P W F A A I Y R
R H R G G S V T Y V C G G S L I S P C W V I S A T(H)C F I D Y P K
K E D Y I V Y L G R S R L N S N T Q G E M K F E V E N L I L H K D
Y S A D T L A H H N(D)I A L L K I R S K E G R C A Q P S R T I Q T
I C L P S M Y N D P Q F G T S C E I T G F G K E ◆N S T D Y L Y P E
Q L K M T V V K L I S H R E C Q Q P H Y Y G S E V T T K M L C A A
D P Q W K T D S C Q G D(S)G G P L V C S L Q G R M T L T G I V S W
G R G C A L K D K P G V Y T R V S H F L P W I R S H T K E E N G L
A L
```

Fig. A24. Amino-acid sequence of human pro-urokinase (u-PA). The signal peptide is outlined. The residues presumed to be involved in the catalytic triad are encircled. → shows the position of cleavage by plasmin and ◆ the N-bound carbohydrate.

t-PA

M D A M K R G L C C V L L L C G A V F V S P S Q E I H A R F R R G

A R S Y Q V I C R D E K T Q M I Y Q Q H Q S W L R P V L R S N R V

E Y C W C N S G R A Q C H S V P V K S C S E P R C F N G G T C Q Q

A L Y F S D F V C Q C P E G F A G K C C E I D T R A T C Y E D Q G

I S Y R G T W S T A E S G A E C T N W N S S A L A Q K P Y S G R R

P D A I R L G L G N H N Y C R N P D R D S K P W C Y V F K A G K Y

S S E F C S T P A C S E G N S D C Y F G N G S A Y R G T H S L T E

S G A S C L P W N S M I L I G K V Y T A Q N P S A Q A L G L G K H

N Y C R N P D G D A K P W C H V L K N R R L T W E Y C D V P S C S

T C G L R Q Y S Q P Q F R I K G G L F A D I A S H P W Q A A I F A

K H R R S P G E R F L C G G I L I S S C W I L S A A H C F Q E R F

P P H H L T V I L G R T Y R V V P G E E E Q K F E V E K Y I V H K

E F D D D T Y D N D I A L L Q L K S D S S R C A Q E S S V V R T V

C L P P A D L Q L P D W T E C E L S G Y G K H E A L S P F Y S E R

L K E A H V R L Y P S S R C T S Q H L L N R T V T D N M L C A G D

T R S G G P Q A N L H D A C Q G D S G G P L V C L N D G R M T L V

G I I S W G L G C G Q K D V P G V Y T K V T N Y L D W I R D N M R

P

Fig. A25. Amino-acid sequence of human tissue plasminogen activator (t-PA). The signal plus propeptide (see section 9.3) is outlined. The residues presumed to be involved in the catalytic triad are encircled. → shows the position of cleavage by plasmin, ◆ the N-bound carbohydrate and ◇ the potential attachment site for N-bound carbohydrate. Position 4 is dimorphic (S/G; underlined).

α_2-plasmin inhibitor

```
M A L L W G L L V L S W S C L Q G P C S V F S P V S A M E P L G W
Q L T S G P N Q E Q V S P L T L L K L G N Q E P G G Q T A L K S P
P G V C S R D P T P E Q T H R L A R A M M A F T A D L F S L V A Q
T S T C P N L I L S P L S V A L A L S H L A L G A Q N H T L Q R L
Q Q V L H A G S G P C L P H L L S R L C Q D L G P G A F R L A A R
M Y L Q K G F P I K E D F L E Q S E Q L F G A K P V S L T G K Q E
D D L A N I N Q W V K E A T E G K I Q E F L S G L P E D T V L L L
L N A I H F Q G F W R N K F D P S L T Q R D S F H L D E Q F T V P
V E M M Q A R T Y P L R W F L L E Q P E I Q V A H F P F K N N M S
F V V L V P T H F E W N V S Q V L A N L S W D T L H P P L V W E R
P T K V R L P K L Y L K H Q M D L V A T L S Q L G L Q E L F Q A P
D L R G I S E Q S L V V S G V Q H Q S T L E L S E V G V E A A A A
T S I A M S R M S L S S F S V N R P F L F F I F E D T T G L P L F
V G S V R N P N P S A P R E L K E Q Q D S P G N K D F L Q S L K G
F P R G D K L F G P D L K L V P P M E E D Y P Q F G S P K
```

Fig. A26. Amino-acid sequence of human α_2-plasmin inhibitor. The signal peptide is outlined. → shows the reactive site and ◆ the N-bound carbohydrate. The sulphated tyrosine residue is encircled. An R-G-D sequence is outlined with dashed lines.

PAI-1

```
M Q M S P A L T C L V L G L A L V F G E G S A V H H P P S Y V A H
L A S D F G V R V F Q Q V A Q A S K D R N V V F S P Y G V A S V L
A M L Q L T T G G E T Q Q Q I Q A A M G F K I D D K G M A P A L R
H L Y K E L M G P W N K D E I S T T D A I F V Q R D L K L V Q G F
M P H F F R L F R S T V K Q V D F S E V E R A R F I I N D W V K T
H T K G M I S N L L G K G A V D Q L T R L V L V N A L Y F N G Q W
K T P F P D S S T H R R L F H K S D G S T V S V P M M A Q T N K F
N Y T E F T T P D G H Y Y D I L E L P Y H G D T L S M F I A A P Y
E K E V P L S A L T N I L S A Q L I S H W K G N M T R L P R L L V
L P K F S L E T E V D L R K P L E N L G M T D M F R Q F Q A D F T
S L S D Q E P L H V A Q A L Q K V K I E V N E S G T V A S S S T A
V I V S A R M A P E E I I M D R P F L F V V R H N P T G T V L F M
G Q V M E P
```

Fig. A27. Amino-acid sequence of human plasminogen activator inhibitor-1 (PAI-1). The signal peptide is outlined, but may occur in two different lengths, as indicated by the dashed outline. → shows the reactive site and ◇ the potential attachment sites for N-bound carbohydrate.

PAI-2

```
M E D L C V A N T L F A L N L F K H L A K A S P T Q N L F L S P W
S I S S T M A M V Y M G S R G S T E D Q M A K V L Q F N E V G A N
A V T P M T P E N F T S C G F M Q Q I Q K G S Y P D A I L Q A Q A
A D K I H S S F R S L S S A I N A S T G D Y L L E S V N K L F G E
K S A S F R E E Y I R L C Q K Y Y S S E P Q A V D F L E C A E E A
R K K I N S W V K T Q T K G K I P N L L P E G S V D G D T R M V L
V N A V Y F K G K W K T P F E K K L N G L Y P F R V N S A Q R T P
V Q M M Y L R E K L N I G Y I E D L K A Q I L E L P Y A G D V S M
F L L L P D E I A D V S T G L E L L E S E I T Y D K L N K W T S K
D K M A E D E V E V Y I P Q F K L E E H Y E L R S I L R S M G M E
D A F N K G R A N F S G M S E R N D L F L S E V F H Q A M V D V N
E E G T E A A A G T G G V M T G R T G H G G P Q F V A D H P F L F
L I M H K I T K C I L F F G R F C S P
```

Fig. A28. Amino-acid sequence of human plasminogen activator inhibitor-2 (PAI-2). PAI-2 is synthesized without signal peptide. → shows the reactive site and ◇ the potential attachment sites for N-bound carbohydrate. The positions underlined may be dimorphic: 120 (D/N), 404 (K/D) and 413 (C/S).

Protease nexin

```
M N W H L P L F L L A S V T L P S I C S H F N P L S L E E L G S N
T G I Q V F N Q I V K S R P H D N I V I S P H G I A S V L G M L Q
L G A D G R T K K Q L A M V M R Y G V N G V G K I L K K I N K A I
V S K K N K D I V T V A N A V F V K N A S E I E V P F V T R N K D
V F Q C E V R N V N F E D P A S A C D S I N A W V K N E T R D M I
D N L L S P D L I D C V L T R L V L V N A V Y F K G L W K S R F Q
P E N T K K R T F V A A D G K S Y Q V P M L A Q L S V F R C G S T
S A P N D L W Y N F I E L P Y H G E S I S M L I A L P T E S S T P
L S A I I P H I S T K T I D S W M S I M V P K R V Q V I L P K F T
A V A Q T D K L E P L K V L G I T D M F D S S K A N F A K I T R S
E N L H V S H I L Q K A K I E V S E D G T K A S A A T T A I L I A
R S S P P W F I V D R P F L F F I R H N P T G A V L F M G Q I N K
P
```

Fig. A29. Amino-acid sequence of human protease nexin. The signal peptide is outlined. → shows the reactive site and ◇ the potential attachment sites for N-bound carbohydrate. There are two forms of protease nexin, owing to an insertion of three nucleotides in the codon for Arg310 (underlined), which changes this codon to two codons for Thr-Gly.

C1q A-chain

```
E D L C R A P D G K K G E A G R Ⓟ G R R G R Ⓟ G L K G E Q G E Ⓟ G
A Ⓟ G I R T G I Q G L K G D Q G E Ⓟ G P S G N Ⓟ G K V G Y Ⓟ G P S
G P L G A R G I K G I K G T Ⓟ G S P G N I K D Q P R P A F S A I R
R N P P M G G N V V I F D T V I T N Q E E P Y Q N H S G R F V C T
V P G Y Y Y F T F Q V L S Q W E I N L S I V S W S R G Q V R R S L
G F C D T T N K G L F Q V V S G G M V L Q L Q Q G D Q V W V E K D
P K K G H I Y Q G S E A D S V F S G F I L P G F S A
```

Fig. A30. Amino-acid sequence of the A chain of human C1q. The signal peptide is not known, so the sequence begins with Glu1. ◆ shows N-bound carbohydrate and ● carbohydrate bound to 5-hydroxy-lysine residues. 5-hydroxy-lysine residues are underlined and 4-hydroxy-proline residues are encircled.

C1q B-chain

```
M M M K I P W G S I P V L M L L L L L G L I D I S Q A Q L S C T G
P Ⓟ A I Ⓟ G I Ⓟ G I P G T P G P D G Q Ⓟ G T Ⓟ G I K G E K G L Ⓟ G
L A G D H G E F G E K G D P G I Ⓟ G D Ⓟ G K V G P K G P M G P K G
G Ⓟ G A Ⓟ G A Ⓟ G P K G E S G D Y K A T Q K I A F S A T R T I N V
P L R R D Q T I R P D H V I T N M N N N Y E P R S G K F T C K V P
G L Y Y F T Y H A S S R G N L C V N L M R G R E R A Q K V V T F C
D Y A Y N T F Q V T T G G M V L K L E Q G E N V F L Q A T D K N S
L L G M E G A N S I F S G F L L F P D M E A
```

Fig. A31. Amino-acid sequence of the B chain of human C1q. The signal peptide is outlined. ● shows carbohydrate bound to 5-hydroxy-lysine residues. 5-hydroxy-lysine residues are underlined and 4-hydroxy-proline residues are encircled.

C1q C-chain

```
N T G C Y G I Ⓟ G M Ⓟ G L Ⓟ G A Ⓟ G K D G Y D G L Ⓟ G P Ⓟ G E P G
I Ⓟ A I K G I R G P Ⓟ G Q K G E P G L Ⓟ G H K G K D G P N G P Ⓟ G
M Ⓟ G V Ⓟ G P M G I Ⓟ G E P G E E G R Y K Q K F Q S
```

Fig. A32. Partial amino-acid sequence of the C chain of human C1q. The signal peptide is not known, so the sequence begins with Asn1. ● shows carbohydrate bound to 5-hydroxy-lysine residues. 5-hydroxy-lysine residues are underlined and 4-hydroxy-proline residues are encircled.

C1r

```
M W L L Y L L L V P A L F C R A G G S I P I P Q K L F G E V T S P L
F P K P Y P N N F E T T T V I T V P T G Y R V K L V F Q Q F D L E
P S E G C F Y D Y V K I S A D K K S L G R F C G Q L G S P L G N P
P G K K E F M S Q G N K M L L T F H T D F S N E E N G T I M F Y K
G F L A Y Y Q A V D L D E C A S R S K L G E E D P Q P Q C Q H L C
H β Y V G G Y F C S C R P G Y E L Q E D R H S C Q A E C S S E L Y
T E A S G Y I S S L E Y P R S Y P P D L R C N Y S I R V E R G L T
L H L K F L E P F D I D D H Q Q V H C P Y D Q L Q I Y A N G K N I
G E F C G K Q R P P D L D T S S N A V D L L F F T D E S G D S R G
W K L R Y T T E I I K C P Q P K T L D E F T I I Q N L Q P Q Y Q F
R D Y F I A T C K Q G Y Q L I E G N Q V L H S F T A V C Q D D G T
W H R A M P R C K I K D C G Q P R N L P N G D F R Y T T T M G V N
T Y K A R I Q Y Y C H E P Y Y K M Q T R A G S R E S E Q G V Y T C
T A Q G I W K N E Q K G E K I P R C L P V C G K P V N P V E Q R Q
R I I G G Q K A K M G N F P W Q V F T N I H G R G G G A L L G D R
W I L T A A H T L Y P K E H E A Q S N A S L D V F L G H T N V E E
L M K L G N H P I R R V S V H P D Y R Q D E S Y N F E G D I A L L
E L E N S V T L G P N L L P I C L P D N D T F Y D L G L M G Y V S
G F G V M E E K I A H D L R F V R L P V A N P Q A C E N W L R G K
N R M D V F S Q N M F C A G H P S L K Q D A C Q G D S G G V F A V
R D P N T D R W V A T G I V S W G I G C S R G Y G F Y T K V L N Y
V D W I K K E M E E E D
```

Fig. A33. Amino-acid sequence of human C1r. The signal peptide is outlined. The residues presumed to be involved in the catalytic triad are encircled. → shows the position of cleavage by autoactivation and ◆ the N-bound carbohydrate. Position 135 (underlined) is dimorphic and can be L or S. The aspartic acid residue that is hydroxylated is shown as β.

C1s

```
M W C I V L F S L L A W V Y A E P T M Y G E I L S P N Y P Q A Y P
S E V E K S W D I E V P E G Y G I H L Y F T H L D I E L S E N C A
Y D S V Q I I S G D T E E G R L C G Q R S S N N P H S P I V E E F
Q V P Y N K L Q V I F K S D F S N E E R F T G F A A Y Y V A T D I
N E C T D F V D V P C S H F C N N F I G G Y F C S C P P E Y F L H
D D M K N C G V N C S G D V F T A L I G E I A S P N Y P K P Y P E
N S R C E Y Q I R L E K G F Q V V V T L R R E D F D V E A A D S A
G N C L G D L V F V A G D R Q F G P Y C G H G F P G P L N I E T K
S N A L D I I F Q T D L T G Q K K G W K L R Y H G D P M P C P K E
D T P N S V W E P A K A K Y V F R D V V Q I T C L D G F E V V E G
R V G A T S F Y S T C Q S N G K W S N S K L K C Q P V D C G I P E
S I E N G K V E D P E S T L F G S V I R Y T C E E P Y Y Y M E N G
G G G E Y H C A G N G S W V N E V L G P E L P K C V P V C G V P R
E P F E E K Q R I I G G S D A D I K N F P W Q V F F D N P W A G G
A L I N E Y W V L T A A H V V E G N R E P T M Y V G S T S V Q T S
R L A K S K M L T P E H V F I H P G W K L L E V P E G R T N F D N
D I A L V R L K D P V K M G P T V S P I C L P G T S S D Y N L M D
G D L G L I S G W G R T E K R D R A V R L K A A R L P V A P L R K
C K E V K V E K P T A D A E A Y V F T P N M I C A G G E K G M D S
C K G D S G G A F A V Q D P N D K T K F Y A A G L V S W G P Q C G
T Y G L Y T R V K N Y V D W I M K T M Q E N S T P R E D
```

Fig. A34. Amino-acid sequence of human C1s. The signal peptide is outlined. The residues presumed to be involved in the catalytic triad are encircled. → shows the position of cleavage by C$\overline{1}$r, ◆ the N-bound carbohydrate and ◇ the potential attachment sites for N-bound carbohydrate.

C2

```
M G P L M V L F C L L F L Y P G L A D S A P S C P Q N V N I S G G
T F T L S H G W A P G S L L T Y S C P Q G L Y P S P A S R L C K S
S G Q W Q T P G A T R S L S K A V C K P V R C P A P V S F E N G I
Y T P R L G S Y P V G G N V S F E C E D G F I L R G S P V R Q C R
P N G M W D G E T A V C D N G A G H C P N P G I S L G A V R T G F
R F G H G D K V R Y R C S S N L V L T G S S E R E C Q G N G V W S
G T E P I C R Q P Y S Y D F P E D V A P A L G T S F S H M L G A T
N P T Q K T K E S L G R K I Q I Q R S G H L N L Y L L L D C S Q S
V S E N D F L I F K E S A S L M V D R I F S F E I N V S V A I I T
F A S E P G V L M S V L N D N S R D M T E V I S S L E N A N Y K D
H E N G T G T N T Y A A L N S V Y L M M N N Q M R L L G M E T M A
W Q E I R H A I I L L T D G K S N M G G S P K T A V D H I R E I L
N I N Q K R N D Y L D I Y A I G V G K L D V D W R E L N E L G S K
K D G E R H A F I L Q D T K A L H Q V F E H M L D V S K L T D T I
C G V G N M S A N A S D Q E R T P W H V T I K P K S Q E T C R G A
L I S D Q W V L T A A H C F R D G N D H S L W R V N V G D P K S Q
W G K E L L I E K A V I S P G F D V F A K K N Q G I L E F Y G D D
I A L L K L A Q K V K M S T H A R P I C L P C T M E A N L A L R R
P Q G S T C R D H E N E L L N K Q S V P A H F V A L N G S K L N I
N L K M G V E W T S C A E V V S Q E K T M F P N L T D V R E V V T
D Q F L C S G T Q E D E S P C K G E S G G A V F L E R R F R F F Q
V G L V S W G L Y N P C L G S A D K N S R K R A P R S K V P P P R
D F H I N L F R M Q P W L R Q H L G D V L N F L P I
```

Fig. A35. Amino-acid sequence of human C2. The signal peptide is outlined. The residues presumed to be involved in the catalytic triad are encircled. → shows the position of cleavage by $C\bar{1}s$ and ◇ the potential attachment sites for N-bound carbohydrate. Position 513 (underlined) is dimorphic and can be L or F.

C3

```
M G P T S G P S L L L L L L T H L P L A L G S P M Y S I I T P N I
L R L E S E E T M V L E A H D A Q G D V P V T V T V H D F P G K K
L V L S S E K T V L T P A T N H M G N V T F T I P A N R E F K S E
K G R N K F V T V Q A T F G T Q V V E K V V L V S L Q S G Y L F I
Q T D K T I Y T P G S T V L Y R I F T V N H K L L P V G R T V M V
N I E N P E G I P V K Q D S L S S Q N Q L G V L P L S W D I P E L
V N M G Q W K I R A Y Y E N S P Q Q V F S T E F E V K E Y V L P S
F E V I V E P T E K F Y Y I Y N E K G L E V T I T A R F L Y G K K
V E G T A F V I F G I Q D G E Q R I S L P E S L K R I P I E D G S
G E V V L S R K V L L D G V Q N L R A E D L V G K S L Y V S A T V
I L H S G S D M V Q A E R S G I P I V T S P Y Q I H F T K T P K Y
F K P G M P F D L M V F V T N P D G S P A Y R V P V A V Q G E D T
V Q S L T Q G D G V A K L S I N T H P S Q K P L S I T V R T K K Q
E L S E A E Q A T R T M Q A L P Y S T V G N S N N Y L H L S V L R
T E L R P G E T L N V N F L L R M D R A H E A K I R Y Y T Y L I M
N K G R L L K A G R Q V R E P G Q D L V V L P L S I T T D F I P S
F R L V A Y Y T L I G A S G Q R E V V A D S V W V D V K D S C V G
S L V V K S G Q S E D R Q P V P G Q Q M T L K I E G D H G A R V V
L V A V D K G V F V L N K K N K L T Q S K I W D V V E K A D I G C
T P G S G K D Y A G V F S D A G L T F T S S S S Q Q T A Q R A E L
Q C P Q P A A R R R R S V Q L T E K R M D K V G K Y P K E L R K C
C E D G M R E N P M R F S C Q R R T R F I S L G E A C K K V F L D
C C N Y I T E L R R Q H A R A S H L G L A R S N L D E D I I A E E
N I V S R S E F P E S W L W N V E D L K E P P K N G I S T K L M N
I F L K D S I T T W E I L A V S M S D K K G I C V A D P F E V T V
M Q D F F I D L R L P Y S V V R N E Q V E I R A V L Y N Y R Q N Q
E L K V R V E L L H N P A F C S L A T T K R R H Q Q T V T I P P K
S S L S V P Y V I V P L K T G L Q E V E V K A A V Y H H F I S D G
V R K S L K V V P E G I R M N K T V A V R T L D P E R L G R E G V
Q K E D I P P A D L S D Q V P D T E S E T R I L L Q G T P V A Q M
T E D A V D A E R L K H L I V T P S G C G E Q N M I G M T P T V I
A V H Y L D E T E Q W E K F G L E K R Q G A L E L I K K G Y T Q Q
L A F R Q P S S A F A A F V K R A P S T W L T A Y V V K V F S L A
```

Fig. A36. Amino-acid sequence of human C3. The signal peptide is outlined. → shows the position of cleavage by C3-convertases, ⇒ the positions of the bonds that are cleaved when C3 is formed from pro-C3, ◆ the N-bound carbohydrate and ◇ the potential attachment sites for N-bound carbohydrate. The Cys and Gln residues forming the internal thioester are encircled. An R-G-D sequence is outlined with dashed lines. Position 80 (underlined) is dimorphic and can be R (C3S) or G (C3F).

V N L I A I D S Q V L C G A V K W L I L E K Q K P D G V F Q E D A

P V I H Q E M I G G L R N N N E K D M A L T A F V L I S L Q E A K

D I C E E Q V N S L P G S I T K A G D F L E A N Y M N L Q R S Y T

V A I A G Y A L A Q M G R L K G P L L N K F L T T A K N K N R W E

D P G K Q L Y N V E A T S Y A L L A L L Q L K D F D F V P P V V R

W L N E Q R Y Y G G G Y G S T Q A T F M V F Q A L A Q Y Q K D A P

D H Q E L N L D V S L Q L P S R S S K I T H R I H W E S A S L L R

S E E T K E N E G F T V T A E G K G Q G T L S V V T M Y H A K A K

D Q L T C N K F D L K V T I K P A P E T E K R P Q D A K N T M I L

E I C T R Y R G D Q D A T M S I L D I S M M T G F A P D T D D L K

Q L A N G V D R Y I S K Y E L D K A F S D R N T L I I Y L D K V S

H S E D D C L A F K V H Q Y F N V E L I Q P G A V K V Y A Y Y N L

E E S C T R F Y H P E K E D G K L N K L C R D E L C R C A E E N C

F I Q K S D D K V T L E E R L D K A C E P G V D Y V Y K T R L V K

V Q L S N D F D E Y I M A I E Q T I K S G S D E V Q V G Q Q R T F

◇

I S P I K C R E A L K L E E K K H Y L M W G L S S D F W G E K P N

L S Y I I G K D T W V E H W P E E D E C Q D E E N Q K Q C Q D L G

A F T E S M V V F G C P N

C4

```
M R L L W G L I W A S S F F T L S L Q K P R L L L F S P S V V H L
G V P L S V G V Q L Q D V P R G Q V V K G S V F L R N P S R N N V
P C S P K V D F T L S S E R D F A L L S L Q V P L K D A K S C G L
H Q L L R G P E V Q L V A H S P W L K D S L S R T T N I Q G I N L
L F S S R R G H L F L Q T D Q P I Y N P G Q R V R Y R V F A L D Q
K M R P S T D T I T V M V E N S H G L R V R K K E V Y M P S S I F
Q D D F V I P D I S E P G T W K I S A R F S D G L E S N S T T Q F
E V K K Y V L P N F E V K I T P G K P Y I L T V P G H L D E M Q L
D I Q A R Y I Y G K P V Q G V A Y V R F G L L D E D G K K T F F R
G L E S Q T K L V N G Q S H I S L S K A E F Q D A L E K L N M G I
T D L Q G L R L Y V A A A I I E S P G G E M E E A E L T S W Y F V
S S P F S L D L S K T K R H L V P G A P F L L Q A L V R E M S G S
P A S G I P V K V S A T V S S P G S V P E A Q D I Q Q N T D G S G
Q V S I P I I I P Q T I S E L Q L S V S A G S P H P A I A R L T V
A A P P S G G P G F L S I E R P D S R P P R V G D T L N L N L A R
V G S G A T F S H Y Y Y M I L S R G Q I V F M N R E P K R T L T S
V S V F V D H H L A P S F Y F V A F Y Y H G D H P V A N S L R V D
V Q A G A C E G K L E L S V D G A K Q Y R N G E S V K L H L E T D
S L A L V A L G A L D T A L Y A A G S K S H K P L N M G K V F E A
M N S Y D L G C G P G G G D S A L Q V F Q A A G L A F S D G D Q W
T L S R K R L S C P K E K T T R K K R N V N F Q K A I N E K L G Q
Y A S P T A K R C C Q D G V T R L P M N R S C E Q R A A R V Q Q P
D C R E P F L S C C Q F A E S L R K K S R D K G Q A G L Q R A L E
I L Q E E D L I D E D D I P V R S F F P E N W L W R V E T V D R F
Q I L T L W L P D S L T T W E I H G L S L S K T K G L C V A T P V
Q L R V F R E F H L H L R L P M S V R R F E Q L E L R P V L Y N Y
L D K N L T V S V H V S P V E G L C L A G G G G L A Q Q V L V P A
G S A R P V A F S V V P T A A A A V S L K V V A R G S F E F P V G
D A V S K V L Q I E K E G A I H R E E L V Y E L N P L D H R G R T
L E I P G N S D P N M I P D G D F N S Y V R V T A S D P L D T L G
S E G A L S P G G V A S L L R L P R G C G E Q T M I Y L A P T L A
A S R Y L D K T E Q W S T L P P E T K D H A V D L I Q K G Y M R I
Q Q F R K A D G S Y A A W L S R D S S T W L T A F V L K V L S L A
```

Fig. A37. Amino-acid sequence of human C4. The signal peptide is outlined. → shows the position of cleavage by C$\bar{1}$s, ⇒ the positions of the bonds that are cleaved when C4 is formed from pro-C4, ◆ the N-bound carbohydrate and ◇ the potential attachment sites for N-bound carbohydrate. The Cys and Gln residues forming the internal thioester are encircled, while the three sulphated tyrosine residues are outlined with dashed lines. The positions underlined are dimorphic: 399 (A/V), 1054 (D/G), 1101 (P/L) 1102 (C/S), 1105 (L/I), 1106 (D/H), 1157 (N/S), 1182 (S/T), 1188 (V/A), 1191 (L/R), 1267 (S/A) and 1481 (D/Y).

Q E Q V G G S P E K L Q E T S N W L L S Q Q Q A D G S F Q D P C P

V L D R S M Q G G L V G N D E T V A L T A F V T I A L H H G L A V

F Q D E G A E P L K Q R V E A S I S K A N S F L G E K A S A G L L

G A H A A A I T A Y A L S L T K A P V D L L G V A H N N L M A M A

Q E T G D N L Y W G S V T G S Q S N A V S P T P A P R N P S D P M

P Q A P A L W I E T T A Y A L L H L L L H E G K A E M A D Q A S A

W L T R Q G S F Q G G F R S T Q D T V I A L D A L S A Y W I A S H

T T E E R G L N V T L S S T G R N G F K S H A L Q L N N R Q I R G

L E E E L Q F S L G S K I N V K V G G N S K G T L K V L R T Y N V

L D M K N T T C Q D L Q I E V T V K G H V E Y T M E A N E D Y E D

Y E Y D E L P A K D D P D A P L Q P V T P L Q L F E G R R N R R R

R E A P K V V E E Q E S R V H Y T V C I W R N G K V G L S G M A I

A D V T L L S G F H A L R A D L E K L T S L S D R Y V S H F E T E

G P H V L L Y F D S V P T S R E C V G F E A V Q E V P V G L V Q P

A S A T L Y D Y Y N P E R R C S V F Y G A P S K S R L L A T L C S

A E V C Q C A E G K C P R Q R R A L E R G L Q D E D G Y R M K F A

C Y Y P R V E Y G F Q V K V L R E D S R A A F R L F E T K I T Q V

L H F T K D V K A A A N Q M R N F L V R A S C R L R L E P G K E Y

L I M G L D G A T Y D L E G H P Q Y L L D S N S W I E E M P S E R

L C R S T R Q R A A C A Q L N D F L Q E Y G T Q G C Q V

Factor $\overline{D}$

I L G G R E A E A H A R P Y M A S V Q L N G A H L C G G V L V A E

Q W V L S A A H C L E D A A D G K V Q V L L G A T H L P Q P E P X

X X I T I E V L R A V P H P D S Q P D T I D H D L L L L Q L S E K

A T L G P A V R P L P W Q R V D R D V A P G T L C D V A G W G I V

N H A G R R P D S L Q H V L L P V L D R A T C R L Y D V L R L M C

A E S N R R D S C K G D S G G P L V C G G V L E G V V T S G S R V

C G N R K K P G I Y T R V A T Y A A W I D H V L

Fig. A38. Amino-acid sequence of human factor $\overline{D}$. The signal peptide is not known, so the sequence begins with Ile1. The residues presumed to be involved in the catalytic triad are encircled. The sequence shown is from reference (1375) and X designates undetermined residues.

Factor B

```
M G S N L S P Q L C L M P F I L G L L S G G V T T T P W S L A R P
Q G S C S L E G V E I K G G S F R L L Q E G Q A L E Y V C P S G F
Y P Y P V Q T R T C R S T G S W S T L K T Q D Q K T V R K A E C R
A I H C P R P H D F E N G E Y W P R S P Y Y N V S D E I S F H C Y
D G Y T L R G S A N R T C Q V N G R W S G Q T A I C D N G A G Y C
S N P G I P I G T R K V G S Q Y R L E D S V T Y H C S R G L T L R
G S Q R R T C Q E G G S W S G T E P S C Q D S F M Y D T P Q E V A
E A F L S S L T E T I E G V D A E D G H G P G E Q Q K R K I V L D
P S G S M N I Y L V L D G S D S I G A S N F T G A K K C L V N L I
E K V A S Y G V K P R Y G L V T Y A T Y P K I W V K V S E A D S S
N A D W V T K Q L N E I N Y E D H K L K S G T N T K K A L Q A V Y
S M M S W P D D V P P E G W N R T R H V I I L M T D G L H N M G G
D P I T V I D E I R D L L Y I G K D R K N P R E D Y L D V Y V F G
V G P L V N Q V N I N A L A S K K D N E Q H V F K V K D M E N L E
D V F Y Q M I D E S Q S L S L C G M V W E H R K G T D Y H K Q P W
Q A K I S V I R P S K G H E S C M G A V V S E Y F V L T A A H C F
T V D D K E H S I K V S V G G E K R D L E I E V V L F H P N Y N I
N G K K E A G I P E F Y D Y D V A L I K L K N K L K Y G Q T I R P
I C L P C T E G T T R A L R L P P T T T C Q Q Q K E E L L P A Q D
I K A L F V S E E E K K L T R K E V Y I K N G D K K G S C E R D A
Q Y A P G Y D K V K D I S E V V T P R F L C T G G V S P Y A D P N
T C R G D S G G P L I V H K R S R F I Q V G V I S W G V V D V C K
N Q K R Q K Q V P A H A R D F H I N L F Q V L P W L K E K L Q D E
D L G F L
```

Fig. A39. Amino-acid sequence of human factor B. The signal peptide is outlined. The residues presumed to be involved in the catalytic triad are encircled. → shows the position of cleavage by factor $\overline{\mathrm{D}}$ and ◆ the N-bound carbohydrate. An R-G-D sequence is outlined with dashed lines. Position 7 (underlined) is dimorphic and can be R or Q.

Fig. A40. Partial amino-acid sequence of human C5. The sequence shown starts not at the N terminus but in the β chain, and proceeds to the C terminus. → shows the position of cleavage by C5-convertases, ⇒ the positions of the bonds that are cleaved when C5 is formed from pro-C5, ◆ the N-bound carbohydrate and ◇ the potential attachment sites for N-bound carbohydrate.

C5

R V D D G V A S F V L N L P S G V T V L E F N V K T D A P D L P E
E N Q A R E G Y R A I A Y S S L S Q S Y L Y I D W T D N H K A L L
V G E H L N I I V T P K S P Y I D K I T H Y N Y L I L S K G K I I
H F G T R E K F S D A S Y Q S I N T P V T Q N M V P S S R L L V Y
Y I V T G E Q T A E L V S D S V W L N I E E K C G N Q L Q V H L S
P D A D A Y S P G Q T V S L N M A T G M D S W V A L A A V D S A V
Y G V Q R G A K K P L E R V F Q F L E K S D L G C G A G G G L N N
A N V F H L A G L T F L T N A N A D D S Q E N D E P C K E I L R P
R R T L Q K K I E E I A A K Y K H S V V K K C C Y D G A C V N N D
E T C E Q R A A R I S L G P R C I K A F T E C C V V A S Q L R A N
I S H K D M Q L G R L H M K T L L P V S K P E I R S Y F P E S W L
W E V H L V P R R K Q L Q F A L P D S L T T W E I Q G I G I S N T
G I C V A D T V K A K V F K D V F L E M N I P Y S V V R G E Q I Q
L K G T V Y N Y R T S G M Q F C V K M S A V E G I C T S E S P V I
D H Q G T K S S K C V R Q K V E G S S S H L V T F T V L P L E I G
L H N I N F S L E T W F G K E I L V K T L R V V P E G V K R E S Y
S G V T L D P R G I Y G T I S R R K E F P Y R I P L D L V P K T E
I K R I L S V K G L L V G E I L S A V L S Q E G I N I L T H L P K
G S A E A E L M S V V P V F Y V F H Y L E T G N H W N T F H S D P
L I E K Q K L K K K L K E G M L S I M S Y R N A D Y S Y S V W K G
G S A S T W L T A F A L R V L G Q V N K Y V E Q N Q N S I C N S L
L W L V E N Y Q L D N G S F K E N S Q Y Q P I K L Q G T L P V E A
R E N S L Y L T A F T V I G T R K A F D I C P L V K I D T A L I K
A D N F L L E N T L P A Q S T F T L A I S A Y A L S L G D K T H P
Q F R S I V S A L K R E A L V K C N P P I Y R F W K D N L Q H K D
S S V P N T G T A R M V E T T A Y A L L T S L N L K D I N Y V N P
V I K W L S E E Q R Y G G G F Y S T Q D T I N A I E G L T E Y S L
L V K Q L R L S M D I D V S Y K H K G A L H N Y K M T D K N F L G
R P V E V L L N D D L I V S T G F G S G L A T V H V T T V V H K T
S T S E E V C S F Y L K I D T Q D I E A S H Y R G Y G N S D Y K R
I V A C A S Y K P S R E E S S S G S S H A V M D I S L P T G I S A
N E E D L K A L V E G V D Q L F T D Y Q I K D G H V I L Q L N S I
P S S D F L C V R F R I F E L F E V G F L S P A T F T V Y E Y H R
P D K Q C T M F Y S T S N I K I Q K V C E G A A C K C V E A D C G
Q M Q E E L D L T I S A E T R K Q T A C K P E I A Y A Y K V S I T
S I T V E N V F V K Y K A T L L D I Y K T G E A V A E K D S E I T
F I K K V T C T N A E L V K G R Q Y L I M G K E A L Q I K Y N F S
F R Y I Y P L D S L T W I E Y W P R D T T C S S C Q A F L A N L D
E F A E D I F L N G C

C6

```
M A R R S V L Y F I L L N A L I N K G Q A C F C D H Y A W T Q W T
S C S K T C N S G T Q S R H R Q I V V D K Y Y Q E N F C E Q I C S
K Q E T R E C N W Q R C P I N C L L G D F G P W S D C D P C I E K
Q S K V R S V L R P S Q F G G Q P C T A P L V A F Q P C I P S K L
C K I E E A D C K N K F R C D S G R C I A R K L E C N G E N D C G
D N S D E R D C G R T K A V C T R K Y N P I P S V Q L M G N G F H
F L A G E P R G E V L D N S F T G G I C K T V K S S R T S N P Y R
V P A N L E N V G F E V Q T A E D D L K T D F Y K D L T S L G H N
E N Q Q G S F S S Q G G S S F S V P I F Y S S K R S E N I N H N S
                                                    ◇
A F K Q A I Q A S H K K D S S F I R I H K V M K V L N F T T K A K
D L H L S D V F L K A L N H L P L E Y N S A L Y S R I F D D F G T
H Y F T S G S L G G V Y D L L Y Q F S S E E L K N S G L T E E E A
K H C V R I E T K K R V L F A K K T K V E H R C T T N K L S E K H
E G S F I Q G A E K S I S L I R G G R S E Y G A A L A W E K G S S
G L E E K T F S E W L E S V K E N P A V I D F E L A P I V D L V R
N I P C A V T K R N N L R K A L Q E Y A A K F D P C Q C A P C P N
N G R P T L S G T E C L C V C Q S G T Y G E N C E K Q S P D Y K S
N A V D G Q W G C W S S W S T C D A T Y K R S R T R E C N N P A P
Q R G G K R C E G E K R Q E E D C T F S I M E N N G Q P C I N D D
E E M K E V D L P E I E A D S G C P Q P V P P E N G F I R N E K Q
L Y L V G E D V E I S C L T G F E T V G Y Q Y F R C L P D G T W R
Q G D V E C Q R T E C I K P V V Q E V L T I T P F Q R L Y R I G E
S I E L T C P K G F V V A G P S R Y T C Q G N S W T P P I S N S L
T C E K D T L T K L K G H C Q L G Q K Q S G S E C I C M S P E E D
C S H H S E D L C V F D T D S N D Y F T S P A C K F L A E K C L N
                                                          ◇
N Q Q L H F L H I G S C Q D G R Q L E W G L E R T R L S S N S T K
K E S C G Y D T C Y D W E K C S A S T S K C V C L L P P Q C F K G
G N Q L Y C V K M G S S T S E K T L N I C E V G T I R C A N R K M
E I L H P G K C L A
```

Fig. A41. Amino-acid sequence of human C6. The signal peptide is outlined. ◇ shows the potential attachment sites for N-bound carbohydrate. Position 761 (underlined) is dimorphic and can be S or D.

C7

```
M K V I S L F I L V G F I G E F Q S F S S A S S P V N C Q W D F Y
A P W S E C N G C T K T Q T R R R S V A V Y G Q Y G G Q P C V G N
A F E T Q S C E P T R G C P T E E G C G E R F R C F S G Q C I S K
S L V C N G D S D C D E D S A D E D R C E D S E R R P S C D I D K
P P P N I E L T G N G Y N E L T G Q F R N R V I N T K S F G G Q C
R K V F S G D G K D F Y R L S G N V L S Y T F Q V K I N N D F N Y
      ◇
E F Y N S T W S Y V K H T S T E H T S S S R K R S F F R S S S S S
S R S Y T S H T N E I H K G K S Y Q L L V V E N T V E V A Q F I N
N N P E F L Q L A E P F W K E L S H L P S L Y D Y S A Y R R L I D
Q Y G T H Y L Q S G S L G G E Y R V L F Y V D S E K L K Q N D F N
S V E E K K C K S S G W H F V V K F S S H G C K E L E N A L K A A
S G T Q N N V L R G E P F I R G G G A G F I S G L S Y L E L D N P
A G N K R R Y S A W A E S V T N L P Q V I K Q K L T P L Y E L V K
E V P C A S V K K L Y L K W A L E E Y L D E F D P C H C R P C Q N
G G L A T V E G T H C L C H C K P Y T F G A A C E Q C V L V G N Q
A G G V D G G W S C W S S W S P C V Q G K K T R S R E C N N P P P
S G G G R S C V G E T T E S T Q C E D E E L E H L R L L E P H C F
P L S L V P T E F C P S P P A L K D G F V Q D E G P M F P V G K N
V V Y T C N E G Y S L I G N P V A R C G E D L R W L V G E M H C Q
K I A C V L P V L M D G I Q S H P Q K P F Y T V G E K V T V S C S
G G M S L E G P S A F L C G S S L K W S P E M K N A R C V Q K E N
P L T Q A V P K C Q R W E K L Q N S R C V C K M P Y E C G P S L D
                                                      ◇
V C A Q D E R S K R I L P L T V C K M H V L H C Q G R N Y T L T G
R D S C T L P A S A E K A C G A C P L W G K C D A E S S K C V C R
E A S E C E E E G F S I C V E V N G K E Q T M S E C E A G A L R C
R G Q S I S V T S T R P C A A E T Q
```

Fig. A42. Amino-acid sequence of human C7. The signal peptide is outlined. ◇ shows the potential attachment sites for N-bound carbohydrate. Position 367 (underlined) is dimorphic and can be S or T.

C8 α-chain

```
[M F A V V F F I L S L M T C Q P G V T A Q E K V N Q R V R R] A A T
                     ◆
P A A V T C Q L S N W S E W T D C F P C Q D K K Y R H R S L L Q P
N K F G G T I C S G D I W D Q A S C S S S T T C V R Q A Q C G Q D
F Q C K E T G R C L K R H L V C N G D Q D C L D G S D E D D C E D
V R A I D E D C S Q Y E P I P G S Q K A A L G Y N I L T Q E D A Q
S V Y D A S Y Y G G Q C E T V Y N G E W R E L R Y D S T C E R L Y
Y G D D E K Y F R K P Y N F L K Y H F E A L A D T G I S S E F Y D
N A N D L L S K V K K D K S D S F G V T I G I G P A G S P L L V G
V G V S H S Q D T S F L N E L N K Y N E K K F I F T R I F T K V Q
T A H F K M R K D D I M L D E G M L Q S L M E L P D Q Y N Y G M Y
A K F I N D Y G T H Y I T S G S M G G I Y E Y I L V I D K A K M E
S L G I T S R D I T T C F G G S L G I Q Y E D K I N V G G G L S G
D H C K K F G G G K T E R A R K A M A V E D I I S R V R G G S S G
W S G G L A Q N R S T I T Y R S W G R S L K Y N P V V I D F E M Q
P I H E C C G T Q A W A S G G Q R Q N L R R A L D Q Y L M E F N A
C R C G P C F N N G V P I L E G T S C R C Q C R L G S L G A A C E
Q T Q T E G A K A D G S W S C W S S W S V C R A G I Q E R R R E C
D N P A P Q N G G A S C P G R K V Q T Q A C
```

Fig. A43. Amino-acid sequence of the α chain of human C8. The signal plus propeptide (see section 14.3) is outlined. ◆ shows N-bound carbohydrate.

C8 β-chain

```
S Q C D R T W A W R A P V E L F L L C A A L G C L S L P G S R G E
R P H S F G S N A V N K S F A K S R Q M R S V D V T L M P I D C E
L S S W S S W T T C D P C Q K K R Y R Y A Y L L Q P S Q F H G E P
C N F S D K E V E D C V T N R P C G S Q V R C E G F V C A Q T G R
C V N R R L L C N G D N D C G D Q S D E A N C R R I Y K K C Q H E
M D Q Y W G I G S L A S G I N L F T N S F E G P V L D H R Y Y A G
G C S P H Y I L N T R F R K P Y N V E S Y T P Q T Q G K Y E F I L
K E Y E S Y S D F E R N V T E K M A S K S G F S F G F K I P G I F
E L G I S S Q S D R G K H Y I R R T K R F S H T K S V F L H A R S
D L E V A H Y K L K P R S L M L H Y E F L Q R V K R L P L E Y S Y
G E Y R D L F R D F G T H Y I T E A V L G G I Y E Y T L V M N K E
A M E R G D Y T L N N V H A C A K N D F K I G G A I E E V Y V S L
G V S V G K C R G I L N E I K D R N K R D T M V E D L V V L V R G
G A S E H I T(T L A Y Q E L P T)A D L M Q E W G D A V Q Y N P A I
I K V K V E P L Y E L V T A T D F A Y S S T V R Q N M K Q A L E E
F Q K E V S S C H C A P C Q G N G V P V L K G S R C D C I C P V G
S Q G L A C E V S Y R K N T P I D G K W N C W S N W S S C S G R R
K T R Q R Q C N N P P P Q N G G S P C S G P A S E T L D C S
```

Fig. A44. Amino-acid sequence of the β chain of human C8. The presumed signal plus propeptide (see section 14.3) is outlined. ◇ shows the potential attachment sites for N-bound carbohydrate. The sequence in brackets is from reference (1431) and is in reference (1432) substituted with P-G-I-P-G-A-A-D. This difference is due to the deletion of one nucleotide in each of the three codons 382, 383 and 391. An R-G-D sequence is outlined with dashed lines. Position 63 (underlined) may be dimorphic (R/G).

C8 γ-chain

```
M L P P G T A T L L T L L L A A G S L G Q K P Q R P R R P A S P I
S T I Q P K A N F D A Q Q F A G T W L L V A V G S A C R F L Q E Q
G H R A E A T T L H V A P Q G T A M A V S T F R K L D G I C W Q V
R Q L Y G D T G V L G R F L L Q A R G A R G A V H V V V A E T D Y
Q S F A V L Y L E R A G Q L S V K L Y A R S L P V S D S V L S G F
E Q R V Q E A H L T E D Q I F Y F P K Y G F C E A A D Q F H V L D
E V R R
```

Fig. A45. Amino-acid sequence of the γ chain of human C8. The signal peptide is outlined.

C9

S M S A C R S F A V A I C I L E I S I L T A Q Y T T S Y D P E L T
E S S G S A S H I D C R M S P W S E W S Q C D P C L R Q M F R S R
S I E V F G Q F N G K R C T D A V G D R R Q C V P T E P C E D A E
D D C G N D F Q C S T G R C I K M R L R C N G D N D C G D F S D E
D D C E S E P R P P C R D R V V E E S E L A R T A G Y G I N I L G
M D P L S T P F D N E F Y N G L C N R D R D G N T L T Y Y R R P W
N V A S L I Y E T K G E K N F R T E H Y E E Q I E A F K S I I Q E
K T S N F N A A I S L K F T P T E T N K A E Q C C E E T A S S I S
L H G K G S F R F S Y S K N E T Y Q L F L S Y S S K K E K M F L H
V K G E I H L G R F V M R N R D V V L T T T F V D D I K A L P T T
Y E K G E Y F A F L E T Y G T H Y S S S G S L G G L Y E L I Y V L
D K A S M K R K G V E L K D I K R C L G Y H L D V S L A F S E I S
V G A E F N K D D C V K R G E G R A V N I T S E N L I D D V V S L
I R G G T R K Y A F E L K E K L L R G T V I D V T D F V N W A S S
I N D A P V L I S Q K L S P I Y N L V P V K M K N A H L K K Q N L
E R A I E D Y I N E F S V R K C H T C Q N G G T V I L M D G K C L
C A C P F K F E G I A C E I S K Q K I S E G L P A L E F P N E K

Fig. A46. Amino-acid sequence of human C9. The partially or fully known signal peptide is outlined. ◆ shows N-bound carbohydrate.

C1̄-inhibitor

M A S R L T L L T L L L L L L A G D R A S S N P N A T S S S S Q D
P E S L Q D R G E G K V A T T V I S K M L F V E P I L E V S S L P
T T N S T T N S A T K I T A N T T D E P T T Q P T T E P T T Q P T
I Q P T Q P T T Q L P T D S P T Q P T T G S F C P G P V T L C S D
L E S H S T E A V L G D A L V D F S L K L Y H A F S A M K K V E T
N M A F S P F S I A S L L T Q V L L G A G Q N T K T N L E S I L S
Y P K D F T C V H Q A L K G F T T K G V T S V S Q I F H S P D L A
I R D T F V N A S R T L Y S S S P R V L S N N S D A N L E L I N T
W V A K N T N N K I S R L L D S L P S D T R L V L L N A I Y L S A
K W K T T F D P K K T R M E P F H F K N S V I K V P M M N S K K Y
P V A H F I D Q T L K A K V G Q L Q L S H N L S L V I L V P Q N L
K H R L E D M E Q A L S P S V F K A I M E K L E M S K F Q P T L L
T L P R I K V T T S Q D M L S I M E K L E F F D F S Y D L N L C G
L T E D P D L Q V S A M Q H Q T V L E L T E T G V E A A A A S A I
S V A R T L L V F E V Q Q P F L F V L W D Q Q H K F P V F M G R V
Y D P R A

Fig. A47. Amino-acid sequence of human C1̄-inhibitor. The signal peptide is outlined. → shows the reactive site. ◆ shows N-bound carbohydrate, ● O-bound carbohydrate and ○ potential attachment sites for O-bound carbohydrate.

Factor H

M R L L A K I I C L M L W A I C V A E D C N E L P P R R N T E I L

T G S W S D Q T Y P E G T Q A I Y K C R P G Y R S L G N V I M V C

R K G E W V A L N P L R K C Q K R P C G H P G D T P F G T F T L T

G G N V F E Y G V K A V Y T C N E G Y Q L L G E I N Y R E C D T D

G W T N D I P I C E V V K C L P V T A P E N G K I V S S A M E P D

R E Y H F G Q A V R F V C N S G Y K I E G D E E M H C S D D G F W

S K E K P K C V E I S C K S P D V I N G S P I S Q K I I Y K E N E

R F Q Y K C N M G Y E Y S E R G D A V C T E S G W R P L P S C E E

K S C D N P Y I P N G D Y S P L R I K H R T G D E I T Y Q C R N G

F Y P A T R G N T A K C T S T G W I P A P R C T L K P C D Y P D I

K H G G L Y H E N M R R P Y F P V A V G K Y Y S Y Y C D E H F E T

P S G S Y W D H I H C T Q D G W S P A V P C L R K C Y F P Y L E N

G Y N Q N H G R K F V Q G K S I D V A C H P G Y A L P K A Q T T V

T C M E N G W S P T P R C I R V K T C S K S S I D I E N G F I S E

S Q Y T Y A L K E K A K Y Q C K L G Y V T A D G E T S G S I R C G

K D G W S A Q P T C I K S C D I P V F M N A R T K N D F T W F K L

N D T L D Y E C H D G Y E S N T G S T T G S I V C G Y N G W S D L

P I C Y E R E C E L P K I D V H L V P D R K K D Q Y K V G E V L K

F S C K P G F T I V G P N S V Q C Y H F G L S P D L P I C K E Q V

Q S C G P P P E L L N G N V K E K T K E E Y G H S E V V E Y Y C N

P R F L M K G P N K I Q C V D G E W T T L P V C I V E E S T C G D

I P E L E H G W A Q L S S P P Y Y Y G D S V E F N C S E S F T M I

G H R S I T C I H G V W T Q L P Q C V A I D K L K K C K S S N L I

I L E E H L K N K K E F D H N S N I R Y R C R G K E C W I H T V C

I N G R W D P E V N C S M A Q I Q L C P P P P Q I P N S H N M T T

T L N Y R D G E K V S V L C Q E N Y L I Q E G E E I T C K D G R W

Q S I P L C V E K I P C S Q P P Q I E H G T I N S S R S S Q E S Y

A H G T K L S Y T T E G G F R I S E E N E T T C Y M G K W S S P P

Q C E G L P C K S P P E I S H G V V A H M S D S Y Q Y G E E V T Y

K C F E G F G I D G P A I A K C L G E K W S H P P S C I K T D C L

S L P S F E N A I P M G E K K D V Y K A G E Q V T Y T C A T Y Y K

M D G A S N V T C I N S R W T G R P T C R D T S C V N P P T V Q N

A Y I V S R Q M S K Y P S G E R V R Y Q C R S P Y E M F G D E E V

M C L N G N W T E P P Q C K D S T G K C G P P P P I D N G D I T S

F P L S V Y A P A S S V E Y Q C Q N L Y Q L E G N K R I T C R N G

Q W S E P P K C L H P C V I S R E I M E N Y N I A L R W T A K Q K

L Y S R T G E S V E F V C K R G Y R L S S R S H T L R T T C W D G

K L E Y P T C A K R

Fig. A48. Amino-acid sequence of human factor H. The signal peptide is outlined. ◆ shows the N-bound carbohydrate and ◇ the potential attachment sites for N-bound carbohydrate. An R-G-D sequence is outlined with dashed lines.

C4b-BP

N C G P P P T L S F A A P M D I T L T E T R F K T G T T L K Y T C

L P G Y V R S H S T Q T L T C N S D G E W V Y N T F C I Y K R C R

H P G E L R N G Q V E I K T D L S F G S Q I E F S C S E G F F L I

G S T T S R C E V Q D R G V G W S H P L P Q C E I V K C K P P P D

I R N G R H S G E E N F Y A Y G F S V T Y S C D P R F S L L G H A

S I S C T V E N E T I G V W R P S P P T C E K I T C R K P D V S H

G E M V S G F G P I Y N Y K D T I V F K C Q K G F V L R G S S V I

H C D A D S K W N P S P P A C E P N S C I N L P D I P H A S W E T

Y P R P T K E D V Y V V G T V L R Y R C H P G Y K P T T D E P T T

V I C Q K N L R W T P Y Q G C E A L C C P E P K L N N G E I T Q H

R K S R P A N H C V Y F Y G D E I S F S C H E T S R F S A I C Q G

D G T W S P R T P S C G D I C N F P P K I A H G H Y K Q S S S Y S

F F K E E I I Y E C D K G Y I L V G Q A K L S C S Y S H W S A P A

P Q C K A L C R K P E L V N G R L S V D K D Q Y V E P E N V T I Q

C D S G Y G V V G P Q S I T C S G N R T W Y P E V P K C E W E T P

E G C E Q V L T G K R L M Q C L P N P E D V K M A L E V Y K L S L

E I E Q L E L Q R D S A R Q S T L D K E L

Fig. A49. Amino-acid sequence of human C4b-binding protein (C4b-BP). The signal peptide is not known, so the sequence begins with Asn1. ◆ shows the N-bound carbohydrate.

DAF

M T V A R P S V P A A L P L L G E L P R L L L L V L L C L P A V W

G D C G L P P D V P N A Q P A L E G R T S F P E D T V I T Y K C E

E S F V K I P G E K D S V I C L K G S Q W S D I E E F C N R S C E

V P T R L N S A S L K Q P Y I T Q N Y F P V G T V V E Y E C R P G

Y R R E P S L S P K L T C L Q N L K W S T A V E F C K K K S C P N

P G E I R N G Q I D V P G G I L F G A T I S F S C N T G Y K L F G

S T S S F C L I S G S S V Q W S D P L P E C R E I Y C P A P P Q I

D N G I I Q G E R D H Y G Y R Q S V T Y A C N K G F T M I G E H S

I Y C T V N N D E G E W S G P P P E C R G K S L T S K V P P T V Q

K P T T V N V P T T E V S P T S Q K T T T K T T T P N A Q A T R S

T P V S R T T K H F H E T T P N K G S G T T S G T T R L L S G H T

C F T L T G L L G T L V T M G L L T

Fig. A50. Amino-acid sequence of human decay-accelerating factor (DAF). The signal peptide is outlined. ◇ shows the potential attachment sites for N-bound carbohydrate. The positions underlined may be dimorphic: 46 (I/T) and 51 (S/M).

CR1-A

M C L G R M G A S S P R S P E P V G P P A P G L P F C C G G S L L

A V V V L L A L P V A W G Q C N A P E W L P F A R P T N L T D E F

E F P I G T Y L N Y E C R P G Y S G R P F S I I C L K N S V W T G

A K D R C R R K S C R N P P D P V N G M V H V I K G I Q F G S Q I

K Y S C T K G Y R L I G S S S A T C I I S G D T V I W D N E T P I

C D R I P C G L P P T T T N G D F I S T N R E N F H Y G S V V T Y

R C N P G S G G R K V F E L V G E P S I Y C T S N D D Q V G I W S

G P A P Q C I I P N K C T P P N V E N G I L V S D N R S L F S L N

E V V E F R C Q P G F V M K G P R R V K C Q A L N K W E P E L P S

C S R V C Q P P P D V L H A E R T Q R D K D N F S P G Q E V F Y S

C E P G Y D L R G A A S M R C T P Q G D W S P A A P T C E V K S C

D D F M G Q L L N G R V L F P V N L Q L G A K V D F V C D E G F Q

L K G S S A S Y C V L A G M E S L W N S S V P V C E Q I F C P S P

P V I P N G R H T G K P L E V F P F G K A V N Y T C D P H P D R G

T S F D L I G E S T I R C T S D P Q G N G V W S S P A P R C G I L

G H C Q A P D H F L F A K L K T Q T N A S D F P I G T S L K Y E C

R P E Y Y G R P F S I T C L D N L V W S S P K D V C K R K S C K T

P P D P V N G M V H V I T D I Q V G S R I N Y S C T T G H R L I G

H S S A E C I L S G N A A H W S T K P P I C Q R I P C G L P P T T

A N G D F I S T N R E N F H Y G S V V T Y R C N P G S G G R K V F

E L V G E P S I Y C T S N D D Q V G I W S G P A P Q C I I P N K C

T P P N V E N G I L V S D N R S L F S L N E V V E F R C Q P G F V

M K G P R R V K C Q A L N K W E P E L P S C S R V C Q P P P D V L

H A E R T Q R D K D N F S P G Q E V F Y S C E P G Y D L R G A A S

M R C T P Q G D W S P A A P T C E V K S C D D F M G Q L L N G R V

L F P V N L Q L G A K V D F V C D E G F Q L K G S S A S Y C V L A

G M E S L W N S S V P V C E Q I F C P S P P V I P N G R H T G K P

L E V F P F G K A V N Y T C D P H P D R G T S F D L I G E S T I R

C T S D P Q G N G V W S S P A P R C G I L S H C Q A P D H F L F A

K L K T Q T N A S D F P I G T S L K Y E C R P E Y Y G R P F S I T

C L D N L V W S S P K D V C K R K S C K T P P D P V N G M V H V I

T D I Q V G S R I N Y S C T T G H R L I G H S S A E C I L S G N T

A H W S T K P P I C Q R I P C G L P P T T A N G D F I S T N R E N

Fig. A51. Amino-acid sequence of human complement receptor type 1-A (CR1-A). The presumed signal peptide is outlined. ◇ shows the potential attachment sites for N-bound carbohydrate. The presumed transmembrane stretch is underlined.

F H Y G S V V T Y R C N L G S R G R K V F E L V G E P S I Y C T S

N D D Q V G I W S G P A P Q C I I P N K C T P P N V E N G I L V S

D N R S L F S L N E V V E F R C Q P G F V M K G P R R V K C Q A L

N K W E P E L P S C S R V C Q P P P E I L H G E H T P S H Q D N F

S P G Q E V F Y S C E P G Y D L R G A A S L H C T P Q G D W S P E

A P R C A V K S C D D F L G Q L P H G R V L F P L N L Q L G A K V

S F V C D E G F R L K G S S V S H C V L V G M R S L W N N S V P V

C E H I F C P N P P A I L N G R H T G T P S G D I P Y G K E I S Y

T C D P H P D R G M T F N L I G E S T I R C T S D P H G N G V W S

S P A P R C E L S V R A G H C K T P E Q F P F A S P T I P I N D F

E F P V G T S L N Y E C R P G Y F G K M F S I S C L E N L V W S S

V E D N C R R K S C G P P P E P F N G M V H I N T D T Q F G S T V

N Y S C N E G F R L I G S P S T T C L V S G N N V T W D K K A P I

C E I I S C E P P P T T S N G D F Y S N N R T S F H N G T V V V T Y

Q C H T G P D G E Q L F E L V G E R S I Y C T S K D D Q V G V W S

S P P P R C I S T N K C T A E N V E N A I R V P G N R S F F S L T

E I I R F R C Q P G F V M V G S H T V Q C Q T N G R W G P K L P H

C S R V C Q P P P E I L H G E H T L S H Q D N F S P G Q E V F Y S

C E P S Y D L R G A A S L H C T P Q G D W S P E A P R C T V K S C

D D F L G Q L P H G R V L L P L N L Q L G A K V S F V C D E G F R

L K G R S A S H C V L A G M K A L W N S S V P V C E Q I F C P N P

P A I L N G R H T G T P F G D I P Y G K E I S Y A C D T H P D R G

M T F N L I G E S S I R C T S D P Q G N G V W S S P A P R C E L S

V P A A C P H P P K I Q N G H Y I G G H V S L Y L P G M T I S Y T

C D P G Y L L V G K G F I F C T D Q G I W S Q L D H Y C K E V N C

S F P L F M N G I S K E L E M K K V Y H Y G D Y V T L K C E D G Y

T L E G S P W S Q C Q A D D R W D P P L A K C T S R A H D A L I V

G T L S G T I F F I L L I I F L S W I I L K H R K G N N A H E N P

K E V A I H L H S Q G G S S V H P R T L Q T N E E N S R V L P

MCP

M E P P G R R E C P F P S W R F P G L L L A A M V L L L Y S F S D

A C E E P P T F E A M E L I G K P K P Y Y E I G E R V D Y K C K K

G Y F Y I P P L A T H T I C D R N H T W L P V S D D A C Y R E T C

P Y I R D P L N G Q A V P A N G T Y E F G Y Q M H F I C N E G Y Y

L I G E E I L Y C E L K G S V A I W S G K P P I C E K V L C T P P

P K I K N G K H T F S E V E V F E Y L D A V T Y S C D P A P G P D

P F S L I G E S T I Y C G D N S V W S R A A P E C K V V K C R F P

V V E N G K Q I S G F G K K F Y Y K A T V M F E C D K G F Y L D G

S D T I V C D S N S T W D P P V P K C L K V S T S S T T K S P A S

S A S G P R P T Y K P P V S N Y P G Y P K P E E G I L D S L D V W

V I A V I V I A I V V G V A V I C V V P Y R Y L Q R R K K K G K A

D G G A E Y A T Y Q T K S T T P A E Q R G

Fig. A52. Amino-acid sequence of human membrane co-factor protein (MCP). The signal peptide is outlined. ◇ shows the potential attachment sites for N-bound carbohydrate. The presumed transmembrane stretch is underlined.

Factor I

M K L L H V F L L F L C F H L R F C K V T Y T S Q E D L V E K K C

L A K K Y T H L S C D K V F C Q P W Q R C I E G T C V C K L P Y Q

C P K N G T A V C A T N R R S F P T Y C Q Q K S L E C L H P G T K

F L N N G T C T A E G K F S V S L K H G N T D S E G I V E V K L V

D Q D K T M F I C K S S W S M R E A N V A C L D L G F Q Q G A D T

Q R R F K L S D L S I N S T E C L H V H C R G L E T S L A E C T F

T K R R T M G Y Q D F A D V V C Y T Q K A D S P M D D F F Q C V N

G K Y I S Q M K A C D G I N D C G D Q S D E L C C K A C Q G K G F

H C K S G V C I P S Q Y Q C N G E V D C I T G E D E V G C A G F A

S V A Q E E T E I L T A D M D A E R R R I K S L L P K L S C G V K

N R M H I R R K R I V G G K R A Q L G D L P W Q V A I K D A S G I

T C G G I Y I G G C W I L T A A H C L R A S K T H R Y Q I W T T V

V D W I H P D L K R I V I E Y V D R I I F H E N Y N A G T Y Q N D

I A L I E M K K D G N K K D C E L P R S T P A C V P W S P Y L F Q

P N D T C I V S G W G R E K D N E R V F S L Q W G E V K L I S N C

S K F Y G N R F Y E K E M E C A G T Y D G S I D A C K G D S G G P

L V C M D A N N V T Y V W G V V S W G E N C G K P E F P G V Y T K

V A N Y F D W I S Y H V G R P F I S Q Y N V

Fig. A53. Amino-acid sequence of human factor I. The signal peptide is outlined. The residues presumed to be involved in the catalytic triad are encircled. ⇒ shows the positions of the bonds that are cleaved when factor I is formed from pro-factor I, ◆ the N-bound carbohydrate and ◇ the potential attachment sites for N-bound carbohydrate.

Murine properdin (partial)

C F T Q Y E E S S G R C K G L L G R D I R V E D C C L N A A Y A F
Q E H D G G L C Q A C R S P Q W S A W S L W G P C S V T C S E G S
Q L R H R R C V G R G G Q C S E N V A P G T L E W Q L Q A C E D Q
P C C P E M G G W S E W G P W G P C S V T C S K G T Q I R Q R V C
D N P A P K C G G H C P G E A Q Q S Q A C D T Q K T C P T H G A W
A S W G P W S P R S G S C L G G A Q E P K E T R S R S C S A P A P
S H Q P P G K P C S G P Y A E H K A C S G L P P C P V A G G W G P
W S P L S P C S V T C G L G Q T L E Q R T C D H P A P R H G G P F
C A G D A T R N Q M C N K A V P C P V N G E W E A W G K W S D C S
R L R M S I N C E G T P G Q Q S R S R S C G D R K F N G K P C A G
K L Q D I R H C Y N I H N C I M K G S W S Q W S T W S L C T P P C
S P N A T R V R Q R L C T P L L P K Y P P T V S M V E G Q G E K N
V T F W G T P R P L C E A L Q G Q K L V V E E K R S C L H V P V C
K D P E E K K P

Fig. A54. Partial amino-acid sequence of murine properdin. The sequence probably starts with Cys5 and covers the rest of the molecule. ◇ shows the potential attachment sites for N-bound carbohydrate.

S-protein

M A P L R P L L I L A L L A W V A L A D Q E S C K G R C T E G F N
V D K K C Q C D E L C S Y Y Q S C C T D Y T A E C K P Q V T R G D
V F T M P E D E Y T V Y D D G E E K N N A T V H E Q V G G P S L T
S D L Q A Q S K G N P E Q T P V L K P E E E A P A P E V G A S K P
E G I D S R P E T L H P G R P Q P P A E E E L C S G K P F D A F T
D L K N G S L F A F R G Q Y C Y E L D E K A V R P G Y P K L I R D
V W G I E G P I D A A F T R I N C Q G K T Y L F K G N Q Y W R F E
D G V L D P D Y P R N I S D G F D G I P D N V D A A L A L P A H S
Y S G R E R V Y F F K G K Q Y W E Y Q F Q H Q P S Q E E C E G S S
L S A V F E H F A M M Q R D S W E D I F E L L F W G R T S A G T R
Q P Q F I S R D W H G V P G Q V D A A M A G R I Y I S G M A P R P
S L A K K Q R F R H R N R K G Y R S Q R G H S R G R N Q N S R R P
S R A M W L S L F S S E E S N L G A N N Y D D Y R M D W L V P A T
C E P I Q S V F F F S G D K Y Y R V N L R T R R V D T V D P P Y P
R S I A Q Y W L G C P A P G H L

Fig. A55. Amino-acid sequence of human S-protein. The signal peptide is outlined. ◇ shows the potential attachment sites for N-bound carbohydrate. The R-G-D sequence is outlined with dashed lines.

Carboxypeptidase N (light chain)

M S D L L S V F L H L L L L F K L V A P V T F R H H R Y D D L V R

T L Y K V Q N E C P G I T R V Y S I G R S V E G R H L Y V L E F S

D H P G I H E P L E P E V K Y V G N M H G N E A L G R E L M L Q L

S E F L C E E F R N R N Q R I V Q L I Q D T R I H I L P S M N P D

G Y E V A A A Q G P N K P G Y L V G R N N A N G V D L N R N F P D

L N T Y I Y Y N E K Y G G P N H H L P L P D N W K S Q V E P E T R

A V I R W M H S F N F V L S A N L H G G A V V A N Y P Y D K S F E

H R V R G V R R T A S T P T P D D K L F Q K L A K V Y S Y A H G W

M F Q G W N C G D Y F P D G I T N G A S W Y S L S K G M Q D F N Y

L H T N C F E I T L E L S C D K F P P E E E L Q R E W L G N R E A

L I Q F L E Q V H Q G I K G M V L D E N Y N N L A N A V I S V S G

I N H D V T S G D H G D Y F R L L L P G I Y T V S A T A P G Y D P

E T V T V T V G P A E P T L V N F H L K R S I P Q V S P V R R A P

S R R H G V R A K V Q P Q A R K K E M E M R Q L Q R G P A

Fig. A56. Amino-acid sequence of the light chain of human carboxypeptidase N. The signal peptide is outlined.

Carboxypeptidase N (heavy chain)

Q P M G C D C F V Q E V F C S D E E L A T V P L D I P P Y T K N I

I F V E T S F T T L E T R A F G S N P N L T K V V F L D T Q L C Q

F R P D A F G G L P R L E D L E V T G S S F L N L S T N I F S N L

T S L G K L T L N F N M L E A L P E G L F Q H L A A L E S L H L Q

G N Q L Q A L P R R L F Q P L T H L K T L N L A Q N L L A Q L P E

E L F H P L T S L Q T L K L S N N A L S G L P Q G V F G K L G S L

Q E L F L D S N N I S E L P P Q V F S Q L F C L E R L W L Q R N A

I T H L P L S I F A S L G N L T F L S L Q W N M L R V L P A G L F

A H T P C L V G L S L T H N Q L E T V T E G T F A H L S N L R S L

M L S Y N A I T H L P A G I F R D L E E L V K L Y L G S N N L T A

L H P A L F Q N L S K L E L L S L S K N Q L T T L P E A S S T P T

T T C S T W P C T V T P G S A T A P G L P L Q L A A A V H R S A P

E H P D L L R C P A Y L K G Q V V H A L N E K Q L V S V T R D H L

G F Q V T W P D E S K A G G S W D L A V Q E R A A R S Q C T Y S N

P E G T V V L A C D Q A Q C R W L N V Q L S P R Q G S L G L Q Y N

A S Q E W D L R R A A V L C G S P C L S R L G Q Q G P S S S H T G

A G E G G L W G

Fig. A57. Amino-acid sequence of the heavy chain of human carboxypeptidase N. ◇ shows the potential attachment sites for N-bound carbohydrate.

CR2

M G A A G L L G V F L A L V A P G V L G I S C G S P P P I L N G R
I S Y Y S T P I A V G T V I R Y S C S G T F R L I G E K S L L C I
T K D K V D G T W D K P A P K C E Y F N K Y S S C P E P I V P G G
Y K I R G S T P Y R H G D S V T F A C K T N F S M N G N K S V W C
Q A N N M W G P T R L P T C V S V F P L E C P A L P M I H N G H H
T S E N V G S I A P G L S V T Y S C E S G Y L L V G E K I I N C L
S S G K W S A V P P T C E E A R C K S L G R F P N G K V K E P P I
L R V G V T A N F F C D E G Y R L Q G P P S S R C V I A G Q G V A
W T K M P V C E E I F C P S P P P I L N G R H I G N S L A N V S Y
G S I V T Y T C D P D P E E G V N F I L I G E S T L R C T V D S Q
K T G T W S G P A P R C E L S T S A V Q C P H P Q I L R G R M V S
G Q K D R Y T Y N D T V I F A C M F G F T L K G S K Q I R C N A Q
G T W E P S A P V C E K E C Q A P P N I L N G Q K E D R H M V R F
D P G T S I K Y S C N P G Y V L V G E E S I Q C T S E G V W T P P
V P Q C K V A A C E A T G R Q L L T K P Q H Q F V R P D V N S S C
G E G Y K L S G S V Y Q E C Q G T I P W F M E I R L C K E I T C P
P P P V I Y N G A H T G S S L E D F P Y G T T V T Y T C N P G P E
R G V E F S L I G E S T I R C T S N D Q E R G T W S G P A P L C K
L S L L A V Q C S H V H I A N G Y K I S G K E A P Y F Y N D T V T
F K C Y S G F T L K G S S Q I R C K R D N T W D P E I P V C E K G
C Q P P P G L H H G R H T G G N T V F F V S G M T V D Y T C D P G
Y L L V G N K S I H C M P S G N W S P S A P R C E E T C Q H V R Q
S L Q E L P A G S R V E L V N T S C Q D G Y Q L T G H A Y Q M C Q
D A E N G I W F K K I P L C K V I H C H P P P V I V N G K H T G M
M A E N F L Y G N E V S Y E C D Q G F Y L L G E K N C S A E V I L
K A W I L E R A F P Q C L R S L C P N P E V K H G Y K L N K T H S
A Y S H N D I V Y V D C N P G F I M N G S R V I R C H T D N T W V
P G V P T C I K K A F I G C P P P P K T P N G N H T G G N I A R F
S P G M S I L Y S C D Q G Y L V V G E P L L L C T H E G T W S Q P
A P H C K E V N C S S P A D M D G I Q K G L E P R K M Y Q Y G A V
V T L E C E D G Y M L E G S P Q S Q C Q S D H Q W N P P L A V C R
S R S L A P V L C G I A A G L I L L T F L I V I T L Y V I S K H R
E R N Y Y T D T S Q K E A F H L E A R E V Y S V D P Y N P A S

Fig. A58. Amino-acid sequence of human complement receptor type 2 (CR2). The presumed signal peptide is outlined. The sequence shown is from reference (643). ◇ shows the potential attachment sites for N-bound carbohydrate. The presumed transmembrane stretch is underlined.

CR3 α-chain

M A L R V L L L T A L T L C H G F N L D T E N A M T F Q E N A R G
F G Q S V V Q L Q G S R V V V G A P Q E I V A A N Q R G S L Y Q C
D Y S T G S C E P I R L Q V P V E A V N M S L G L S L A A T T S P
P Q L L A C G P T V H Q T C S E N T Y V K G L C F L F G S N L R Q
Q P Q K F P E A L R G C P Q E D S D I A F L I D G S G S I I P H D
F R R M K E F V S T V M E Q L K K S K T L F S L M Q Y S E E F R I
H F T F K E F Q N N P N P R S L V K P I T Q L L G R T H T A T G I
R K V V R E L F N I T N G A R K N A F K I L V V I T D G E K F G D
P L G Y E D V I P E A D R E G V I R Y V I G V G D A F R S E K S R
Q E L N T I A S K P P R D H V F Q V N N F E A L K T I Q N Q L R E
K I F A I E G T Q T G S S S S F E H E M S Q E G F S A A I T S N G
P L L S T V G S Y D W A G G V F L Y T S K E K S T F I N M T R V D
S D M N D A Y L G Y A A A I I L R N R V Q S L V L G A P R Y Q H I
G L V A M F R Q N T G M W E S N A N V K G T Q I G A Y F G A S L C
S V D V D S N G S T D L V L I G A P H Y Y E Q T R G G Q V S V C P
L P R G Q R A R W Q C D A V L Y G E Q G Q P W G R F G A A L T V L
G D V N G D K L T D V A I G A P G E E D N R G A V Y L F H G T S G
S G I S P S H S Q R I A G S K L S P R L Q Y F G Q S L S G G Q D L
T M D G L V D L T V G A Q G H V L L L R S Q P V L R V K A I M E F
N P R E V A R N V F E C N D Q V V K G K E A G E V R V C L H V Q K
S T R D R L R E G Q I Q S V V T Y D L A L D S G R P H S R A V F N
E T K N S T R R Q T Q V L G L T Q T C E T L K L Q L P N C I E D P
V S P I V L R L N F S L V G T P L S A F G N L R P V L A E D A Q R
L F T A L F P F E K N C G N D N I C Q D D L S I T F S F M S L D C
L V V G G P R E F N V T V T V R N D G E D S Y R T Q V T F F F P L
D L S Y R K V S T L Q N Q R S Q R S W R L A C E S A S S T E V S G
A L K S T S C S I N H P I F P E N S E V T F N I T F D V D S K A S
L G N K L L L K A N V T S E N N M P R T N K T E F Q L E L P V K Y
A V Y M V V T S H G V S T K Y L N F T A S E N T S R V M Q H Q Y Q
V S N L G Q R S P P I S L V F L V P V R L N Q T V I W V R P Q V T
F S E N L S S T C H T K E R L P S H S D F L A E L R K A P V V N C
S I A V C Q R I Q C D I P F F G I Q E E F N A T L K G N L S F D W
Y I K T S H N H L L I V S T A E I L F N D S V F T L L P G Q G A F
V R S Q T E T K V E P E E V P N P L P L I V G S S V G G L L L L A
L I T A A L T K L G F F K R Q Y K D M M S E G G P P G A E P Q

Fig. A59. Amino-acid sequence of the α subunit of human complement receptor type 3 (CR3). The signal peptide is outlined. The sequence shown is the one containing 1137 residues in the mature protein (1651, 1652). The encircled residue is the one that is absent in the sequence determined in reference (1650). ◇ shows the potential attachment sites for N-bound carbohydrate. The presumed transmembrane stretch is underlined.

CR3 β-chain

```
M L G L R P P L L A L V G L L S L G C V L S Q E C T K F K V S S C
R E C I E S G P G C T W C Q K L N F T G P G D P D S I R C D T R P
Q L L M R G C A A D D I M D P T S L A E T Q E D H N G G Q K Q L S
P Q K V T L Y L R P G Q A A A F N V T F R R A K G Y P I D L Y Y L
M D L S Y S M L D D L R N V K K L G G D L L R A L N E I T E S G R
I G F G S F V D K T V L P F V N T H P D K L R N P C P N K E K E C
Q P P F A F R H V L K L T N N S N Q F Q T E V G K Q L I S G N L D
A P E G G L D A M M Q V A A C P E E I G W R N V T R L L V F A T D
D G F H F A G D G K L G A I L T P N D G R C H L E D N L Y K R S N
E F D Y P S V G Q L A H K L A E N N I Q P I F A V T S R M V K T Y
E K L T E I I P K S A V G E L S E D S S N V V H L I K N A Y N K L
S S R V F L D H N A L P D T L K V T Y D S F C S N G V T H R N Q P
R G D C D G V Q I N V P I T F Q V K V T A T E C I Q E Q S F V I R
A L G F T D I V T V Q V L P Q C E C R C R D Q S R D R S L C H G K
G F L E C G I C R C D T G Y I G K N C E C Q T Q G R S S Q E L E G
S C R K D N N S I I C S G L G D C V C G Q C L C H T S D V P G K L
I Y G Q Y C E C D T I N C E R Y N G Q V C G G P G R G L C F C G K
C R C H P G F E G S A C Q C E R T T E G C L N P R R V E C S G R G
R C R C N V C E C H S G Y Q L P L C Q E C P G C P S P C G K Y I S
C A E C L K F E K G P F G K N C S A A C P G L Q L S N N P V K G R
T C K E R D S E G C W V A Y T L E Q Q G D M D R Y L I Y V D E S R
E C V A G P N I A A I V G G T V A G I V L I G I L L L V I W K A L
I H L S D L R E Y R R F E K E K L K S Q W N N D N P L F K S A T T
T V M N P K F A E S
```

Fig. A60. Amino-acid sequence of the β subunit of human complement receptor type 3 (CR3). The presumed signal peptide is outlined. ◇ shows the potential attachment sites for N-bound carbohydrate. The presumed transmembrane stretch is underlined. An R-G-D sequence is outlined with dashed lines.

pro-von Willebrand factor

```
M I P A R F A G V L L A L A L I L P G T L C A E G T R G R S S T A
R C S L F G S D F V N T F D G S M Y S F A G Y C S Y L L A G G C Q
K R S F S I I G D F Q N G K R V S L S V Y L G E F F D I H L F V N
G T V T Q G D Q R V S M P Y A S K G L Y L E T E A G Y Y K L S G E
A Y G F V A R I D G S G N F Q V L L S D R Y F N K T C G L C G N F
N I F A E D D F M T Q E G T L T S D P Y D F A N S W A L S S G E Q
W C E R A S P P S S S C N I S S G E M Q K G L W E Q C Q L L K S T
S V F A R C H P L V D P E P F V A L C E K T L C E C A G G L E C A
C P A L L E Y A R T C A Q E G M V L Y G W T D H S A C S P V C P A
G M E Y R Q C V S P C A R T C Q S L H I N E M C Q E R C V D G C S
C P E G Q L L D E G L C V E S T E C P C V H S G K R Y P P G T S L
S R D C N T C I C R N S Q W I C S N E E C P G E C L V T G Q S H F
K S F D N R Y F T F S G I C Q Y L L A R D C Q D H S F S I V I E T
V Q C A D D R D A V C T R S V T V R L P G L H N S L V K L K H G A
G V A M D G Q D V Q L P L L K G D L R I Q R T V T A S V R L S Y G
E D L Q M D W D G R G R L L V K L S P V Y A G K T C G L C G N Y N
G N Q G D D F L T P S G L A E P R V E D F G N A W K L H G D C Q D
L Q K Q H S D P C A L N P R M T R F S E E A C A V L T S P T F E A
C H R A V S P L P Y L R N C R Y D V C S C S D G R E C L C G A L A
S Y A A A C A G R G V R V A W R E P G R C E L N C P K G Q V Y L Q
C G T P C N L T C R S L S Y P D E E C N E A C L E G C F C P P G L
Y M D E R G D C V P K A Q C P C Y Y D G E I F Q P E D I F S D H H
T M C Y C E D G F M H C T M S G V P G S L L P D A V L S S P L S H
R S K R S L S C R P P M V K L V C P A D N L R A E G L E C A K T C
Q N Y D L E C M S M G C V S G C L C P P G M V R H E N R C V A L E
R C P C F H Q G K E Y A P G E T V K I G C N T C V C R D R K W N C
T D H V C D A T C S T I G M A H Y L T F D G L K Y L F P G E C Q Y
V L V Q D Y C G S N P G T F R I L V G N K G C S H P S V K C K K R
V T I L V E G G E I E L F D G E V N V K R P M K D E T H F E V V E
S G R Y I I L L L G K A L S V V W D R H L S I S V V L K Q T Y Q E
K V C G L C G N F D G I Q N N D L T S S N L Q V E E D P V D F G N
S W K V S S Q C A D T R K V P L D S S P A T C H N N I M K Q T M V
D S S C R I L T S D V F Q D C N K L V D P E P Y L D V C I Y D T C
```

Fig. A61. Amino-acid sequence of human pro-von Willebrand factor. The signal peptide is outlined. ⇒ shows the positions of the bonds that are cleaved when von Willebrand factor is formed from pro-von Willebrand factor, ◆ the N-bound carbohydrate, ◇ the potential attachment sites for N-bound carbohydrate and ● O-bound carbohydrate. The two R-G-D sequences are outlined with dashed lines. Position 767 (underlined) is dimorphic and can be A or T.

S C E S I G D C A C F C D T I A A Y A H V C A Q H G K V V T W R T
A T L C P Q S C E E R N L R E N G Y E C E W R Y N S C A P A C Q V
T C Q H P E P L A C P V Q C V E G C H A H C P P G K I L D E L L Q
T C V D P E D C P V C E V A G R R F A S G K K V T L N P S D P E H
C Q I C H C D V V N L T C E A C Q E P G G L V V P P T D A P V S P
T T L Y V E D I S E P P L H D F Y C S R L L D L V F L L D G S S R
L S E A E F E V L K A F V V D M M E R L R I S Q K W V R V A V V E
Y H D G S H A Y I G L K D R K R P S E L R R I A S Q V K Y A G S Q
V A S T S E V L K Y T L F Q I F S K I D R P E A S R I A L L L M A
S Q E P Q R M S R N F V R Y V Q G L K K K K V I V I P V G I G P H
A N L K Q I R L I E K Q A P E N K A F V L S S V D E L E Q Q R D E
I V S Y L C D L A P E A P P P T L P P D M A Q V T V G P G L L G V
S T L G P K R N S M V L D V A F V L E G S D K I G E A D F N R S K
E F M E E V I Q R M D V G Q D S I H V T V L Q Y S Y M V T V E Y P
F S E A Q S K G D I L Q R V R E I R Y Q G G N R T N T G L A L R Y
L S D H S F L V S Q G D R E Q A P N L V Y M V T G N P A S D E I K
R L P G D I Q V V P I G V G P N A N V Q E L E R I G W P N A P I L
I Q D F E T L P R E A P D L V L Q R C C S G E G L Q I P T L S P A
P D C S Q P L D V I L L L D G S S S F P A S Y F D E M K S F A K A
F I S K A N I G P R L T Q V S V L Q Y G S I T T I D V P W N V V P
E K A H L L S L V D V M Q R E G G P S Q I G D A L G F A V R Y L T
S E M H G A R P G A S K A V V I L V T D V S V D S V D A A A D A A
R S N R V T V F P I G I G D R Y D A A Q L R I L A G P A G D S N V
V K L Q R I E D L P T M V T L G N S F L H K L C S G F V R I C M D
E D G N E K R P G D V W T L P D Q C H T V T C Q P D G Q T L L K S
H R V N C D R G L R P S C P N S Q S P V K V E E T C G C R W T C P
C V C T G S S T R H I V T F D G Q N F K L T G S C S Y V L F Q N K
E Q D L E V I L H N G A C S P G A R Q G C M K S I E V K H S A L S
V E L H S D M E V T V N G R L V S V P Y V G G N M E V N V Y G A I
M H E V R F N H L G H I F T F T P Q N N E F Q L Q L S P K T F A S
K T Y G L C G I C D E N G A N D F M L R D G T V T T D W K T L V Q
E W T V Q R P G Q T C Q P I L E E Q C L V P D S S H C Q V L L L P
L F A E C H K V L A P A T F Y A I C Q Q D S C H Q E Q V C E V I A
S Y A H L C R T N G V C V D W R T P D F C A M S C P P S L V Y N H
C E H G C P R H C D G N V S S C G D H P S E G C F C P P D K V M L
E G S C V P E E A C T Q C I G E D G V Q H Q F L E A W V P D H Q P
C Q I C T C L S G R K V N C T T Q P C P T A K A P T C G L C E V A
R L R Q N A D Q C C P E Y E C V C D P V S C D L P P V P H C E R G
L Q P T L T N P G E C R P N F T C A C R K E E C K R V S P P S C P

```
P H R L P T L R K T Q C C D E Y E C A C N C V N S T V S C P L G Y
L A S T A T N D C G C T T T T C L P D K V C V H R S T I Y P V G Q
F W E E G C D V C T C T D M E D A V M G L R V A Q C S Q K P C E D
S C R S G F T Y V L H E G E C C G R C L P S A C E V V T G S P R G
D S Q S S W K S V G S Q W A S P E N P C L I N E C V R V K E E V F
I Q Q R N V S C P Q L E V P V C P S G F Q L S C K T S A C C P S C
R C E R M E A C M L N G T V I G P G K T V M I D V C T T C R C M V
Q V G V I S G F K L E C R K T T C N P C P L G Y K E E N N T G E C
C G R C L P T A C T I Q L R G G Q I M T L K R D E T L Q D G C D T
H F C K V N E R G E Y F W E K R V T G C P P F D E H K C L A E G G
K I M K I P G T C C D T C E E P E C N D I T A R L Q Y V K V G S C
K S E V E V D I H Y C Q G K C A S K A M Y S I D I N D V Q D Q S C
C C S P T R T E P M Q V A L H C T N G S V V Y H E V L N A M E C K
C S P R K C S K
```

Bovine plasma fibronectin

Q E Q Q I V Q P Q S P L T V S Q S K P G C Y D N G K H Y Q I N Q Q
W E R T Y L G S A L V C T C Y G G S R G F N C E S K P E P E E T C
F D K Y T G N T Y R V G D T Y E R P K D S M I W D C T C I G A G R
G R I S C T I A N R C H E G G Q S Y K I G D T W R R P H E T G G Y
M L E C V C L G N G K G E W T C K P I A E K C F D Q A A G T S Y V
V G E T W E K P Y Q G W M M V D C T C L G E G S G R I T C T S R N
R C N D Q D T R T S Y R I G D T W S K K D N R G N L L Q C I C T G
N G R G E W K C E R H T S L Q T T S A G S G S F T D V R T A I Y Q
P Q P H P Q P P P Y G H C V T D S G V V Y S V G M Q W L K T Q G N
K Q M L C T C L G N G V S C Q E T A V T Q T Y G G N S N G E P C V
L P F T Y N G K T F Y S C T T E G R Q D G H L W C S T T S N Y E Q
D Q K Y S F C T D H T V L V Q T R G G N S N G A L C H F P F L Y N
N H N Y T D C T S E G R R D N M K W C G T T Q N Y D A D Q K F G F
C P M A A H E E I C T T N E G V M Y R I G D Q W D K Q H D M G H M
M R C T C V G N G R G E W T C V A Y S Q L R D Q C I V D G I T Y N
V N D T F H K R H E E G H M L N C T C F G Q G R G R W K C D P V D
Q C Q D S E T R T F Y Q I G D S W E K Y L Q G V R Y Q C Y C Y G R
G I G E W A C Q P L Q T Y P D T S G P V Q V I I T E T P S Q P N S
H P I Q W S A P E S S H I S K Y I L R W K P K N S P D R W K E A T
I P G H L N S Y T I K G L R P G V V Y E G Q L I S V Q H Y G Q R E
V T R F D F T T T S T S P A V T S N T V T G E T T P L S P V V A T
S E S V T E I T A S S F V V S W V S A S D T V S G F R V E Y E L S
E E G D E P Q Y L D L P S T A T S V N I P D L L P G R K Y T V N V
Y E I S E E G E Q N L I L S T S Q T T A P D A P P D P T V D Q V D
D T S I V V R W S R P R A P I T G Y R I V Y S P S V E G S S T E L
N L P E T A N S V T L S D L Q P G V Q Y N I T I Y A V E E N Q E S
T P V F I Q Q E T T G V P R S D K V P P P R D L Q F V E V T D V K
I T I M W T P P E S P V T G Y R V D V I P V N L P G E H G Q R L P
V S R N T F A E V T G L S P G V T Y H F K V F A V N Q G R E S K P
L T A Q Q A T K L D A P T N L Q F I N E T D T T V I V T W T P P R
A R I V G Y R L T V G L T R G G Q P K Q Y N V G P A A S Q Y P L R
N L Q P G S E Y A V S L V A V K G N Q Q S P R V T G V F T T L Q P
L G S I P H Y N T E V T E T T I V I T W T P A P R I G F K L G V R

Fig. A62. Amino-acid sequence of bovine plasma fibronectin. The signal peptide is not known, so the sequence begins with Gln1 that is cyclised. ◆ shows the N-bound carbohydrate and ● the O-bound carbohydrate. It is not known which of the two adjacent Thr residues carries the O-bound carbohydrate. The phosphorylated Ser residues are encircled. Two R-G-D sequences are outlined with dashed lines.

P S Q G G E A P R E V T S E S G S I V V S G L T P G V E Y V Y T I
S V L R D G Q E R D A P I V K K V V T P L S P P T N L H L E A N P
D T G V L T V S W E R S T T P D I T G Y R I T T T P T N G Q Q G Y
S L E E V V H A D Q S S C T F E N L S P G L E Y N V S V Y T V K D
D K E S V P I S D T I I P A V P P P T D L R F T N V G P D T M R V
T W A P P S S I E L T N L L V R Y S P V K N E E D V A E L S I S P
S D N A V V L T N L L P G T E Y L V S V S S V Y E Q H E S I P L R
G R Q K T A L D S P S G I D F S D I T A N S F T V H W I A P R A T
I T G Y R I R H H P E N M G G R P R E D R V P P S R N S I T L T N
L N P G T E Y V V S I V A L N S K E E S L P L V G Q Q S T V S D V
P R D L E V I A A T P T S L L I S W D A P A V T V R Y Y R I T Y G
E T G G S S P V Q E F T V P G S K S T A T I S G L K P G V D Y T I
T V Y A V T G R G D S P A S S K P V S I N Y R T E I D K P S Q M Q
V T D V Q D N S I S V R W L P S S S P V T G Y R V T T A P K N G P
G P S K T K T V G P D Q T E M T I E G L Q P T V E Y V V S V Y A Q
N Q N G E S Q P L V Q T A V T T I P A P T N L K F T Q V T P T S L
T A Q W T A P N V Q L T G Y R V R V T P K E K T G P M K E I N L A
P D S S S V V V S G L M V A T K Y E V S V Y A L K D T L T S R P A
Q G V V T T L E N V S P P R R A R V T D A T E T T I T I S W R T K
T E T I T G F Q V D A I P A N G Q T P I Q R T I R P D V R S Y T I
T G L Q P G T D Y K I H L Y T L N D N A R S S P V V I D A S T A I
D A P S N L R F L A T T P N S L L V S W Q P P R A R I T G Y I I K
Y E K P G S P P R E V V P R P R P G V T E A T I T G L E P G T E Y
T I Q V I A L K N N Q K S E P L I G R K K T D E L P Q L V T L P H
P N L H G P E I L D V P S T V Q K T P F I T N P G Y D T G N G I Q
L P G T S G Q Q P S L G Q Q M I F E E H G F R R T T P P T T A T P
V R H R P R P Y P P N V N E E I Q I G H V P R G D V D H H L Y P H
V V G L N P N A S T G Q E A L S Q T T I S W T P F Q E S S E Y I I
S C H P V G I D E E P L Q F R V P G T S A S A T L T G L T R G A T
Y N I I V E A V K D Q Q R Q K V R E E V V T V G N S V D Q G L S Q
P T D D S C F D P Y T V S H Y A I G E E W E R L S D S G F K L S C
Q C L G F C S C H F R C D S S K W C H D N G V N Y K I G E K W D R
Q G E N G Q M M S C T C L G N G K G E F K C D P H E A T C Y D D G
K T Y H V E E Q W Q K E Y L G A I C S C T C F G G Q R G W R C D N
C R R P G A E P G N E G S T A H S Y N Q Y S Q R Y H Q R T N T N V
N C P I E C F M P L D V Q A D R E D S R E

Thrombospondin

```
M G L A W G L G V L F L M H V C G T N R I P E S G G D N S V F D I
F E L T G A A R K G S G R R L V K G P D P S S P A F R I E D A N L
I P P V P D D K F Q D L V D A V R T E K G F L L L A S L R Q M K K
T R G T L L A L E R K D H S G Q V F S V V S N G K A G T L D L S L
T V Q G K Q H V V S V E E A L L A T G Q W K S I T L F V Q E D R A
Q L Y I D C E K M E N A E L D V P I Q S V F T R D L A S I A R L R
I A K G G V N D N F Q G V L Q N V R F V F G T T P E D I L R N K G
                                ◇
C S S S T S V L L T L D N N V V N G S S P A I R T N Y I G H K T K
D L Q A I C G I S C D E L S S M V L E L R G L R T I V T T L Q D S
I R K V T E E N K E L A N E L R R P P L C Y H N G V Q Y R N N E E
                                                    ◇
W T V D S C T E C H C Q N S V T I C K K V S C P I M P C S N A T V
P D G E C C P R C W P S D S A D D G W S P W S E W T S C S T S C G
N G I Q Q R G R S C D S L N N R C E G S S V Q T R T C H I Q E C D
K R F K Q D G G W S H W S P W S S C S V T C G D G V I T R I R L C
N S P S P Q M N G K P C E G E A R E T K A C K K D A C P I N G G W
G P W S P W D I C S V T C G G G V Q K R S R L C N N P T P Q F G G
K D C V G D V T E N Q I C N K Q D C P I D G C L S N P C F A G V K
C T S Y P D G S W K C G A C P P G Y S G N G I Q C T D V D E C K E
V P D A C F N H N G E H R C E N T D P G Y N C L P C P P R F T G S
Q P F G Q G V E H A T A N K Q V C K P R N P C T D G T H D C N K N
A K C N Y L G H Y S D P M Y R C E C K P G Y A G N G I I C G E D T
                            ◇
D L D G W P N E N L V C V A N A T Y H C K K D N C P N L P N S G Q
E D Y D K D G I G D A C D D D D D N D K I P D D R D N C P F H Y N
P A Q Y D Y D R D D V G D R C D N C P Y N H N P D Q A D T D N N G
E G D A C A A D I D G D G I L N E R D N C Q Y V Y N V D Q R D T D
M D G V G D Q C D N C P L E H N P D Q L D S D S D R I G D T C D N
N Q D I D E D G H Q N N L D N C P Y V P N A N Q A D H D K D G K G
D A C D H D D D N D G I P D D K D N C R L V P N P D Q K D S D G D
G R G D A C K D D F D H D S V P D I D D I C P E N V D I S E T D F
R R F Q M I P L D P K G T S Q N P D N W V V R H Q G K E L V Q T V
N C D P G L A V G Y D E F N A V D F S G T F F I N T E R D D D Y A
G F V F G Y Q S S S R F Y V V M W K Q V T Q S Y V D T N P T R A Q
                    ◇
G Y S G L S V K V V N S T T G P G E H L R N A L W H T G N T P G Q
V R T L W H D P R H I G W K D F T A Y R W R L S H R P K T G F I R
V V M Y E G K K I M A D S G P I Y D K T Y A G G R L G L F V F S Q
E M V F F S D L K Y E C R D P
```

Fig. A63. Amino-acid sequence of human thrombospondin. The signal peptide is outlined. ◇ shows the potential attachment sites for N-bound carbohydrate. An R-G-D sequence is outlined with dashed lines.

Glycoprotein Ib α-chain

```
M P L L L L L L L P S P L H P H P I C E V S K V A S H L E V N C
D K R N L T A L P P D L P K D T T I L H L S E N L L Y T F S L A T
L M P Y T R L T Q L N L D R C E L T K L Q V D G T L P V L G T L D
L S H N Q L Q S L P L L G Q T L P A L T V L D V S F N R L T S L P
L G A L R G L G E L Q E L Y L K G N E L K T L P P G L L T P T P K
L E K L S L A N N N L T E L P A G L L N G L E N L D T L L L Q E N
S L Y T I P K G F F G S H L L P F A F L H G N P W L C N C E I L Y
F R R W L Q D N A E N V Y V W K Q G V D V K A M T S N V A S V Q C
D N S D K F P V Y K Y P G K G C P T L G D E G D T D L Y D Y Y P E
E D T E G D K V R A T R T V V K F P T K A H T T P W G L F Y S W S
T A S L D S Q M P S S L H P T Q E S T K E Q T T F P P R W T P N F
T L H M E S I T F S K T P K S T T E P T P S P T T S E P V P E P A
P N M T T L E P T P S P T T P E P T S E P A P S P T T P E P T P I
P T I A T S P T I L V S A T S L I T P K S T F L T T T K P V S L L
E S T K K T I P E L D Q P P K L R G V L Q G H L E S S R N D P F L
H P D F C C L L P L G F Y V L G L F W L L F A S V V L I L L L S W
V G H V K P Q A L D S G Q G A A L T T A T Q T T H L E L Q R G R Q
V T V P R A W L L F L R G S L P T F R S S L F L W V R P N G R V G
P L V A G R R P S A L S Q G R C Q D L L S T V S I R Y S G H S L
```

Fig. A64. Amino-acid sequence of the α chain of human glycoprotein Ib. The signal peptide is outlined. ◆ shows the N-bound carbohydrate, ◇ the potential attachment sites for N-bound carbohydrate and ● O-bound carbohydrate. The presumed transmembrane stretch is underlined.

Glycoprotein Ib β-chain

```
M G S G P R G A L S L L L L L L A P P S R P A A G C P A P C S C A
G T L V D C G R R G L T W A S L P T A F P V D T T E L V L T G N N
L T A L P P G L L D A L P A L R T A H L G A N P W R C D C R L V P
L R A W L A G R P E R A P Y R D L R C V A P P A L R G R L L P Y L
A E D E L R A A C A P G P L C W G A L A A Q L A L L G L G L L H A
L L L V L L L C R L R R L R A R A R A R A A A R L S L T D P L V A
E R A G T D E S
```

Fig. A65. Amino-acid sequence of the β chain of human glycoprotein Ib. The signal peptide is outlined. ◆ shows the N-bound carbohydrate. The phosphorylated serine residues are doubly underlined, and the esterified Cys residue is encircled. The presumed transmembrane stretch is underlined.

Glycoprotein IX

```
T K D C P S P C T C R A L E T M G L W V D C R G H G L T A L P A L
                    ◇
P A R T R H L L L A N N S L Q S V P P G A F D H L P Q L Q T L D V
T Q N P W H C D C S L T Y L R L W L E D R T P E A L L Q V R C A S
P S L A A H G P L R L T G Y Q L G S C G W Q L Q A S W V R P G V L
W D V A L V A V A A L G L A L L A G L L (C) A T T E A L D
    ‾‾‾‾‾‾‾‾‾‾‾‾‾‾‾‾‾‾‾‾‾‾‾‾‾‾‾‾‾‾‾‾‾‾‾‾‾‾‾‾
```

Fig. A66. Amino-acid sequence of human glycoprotein IX. The signal peptide is not known, so the sequence begins with Thr1. ◇ shows the potential attachment site for N-bound carbohydrate, and the esterified Cys residue is encircled. The presumed transmembrane stretch is underlined.

Glycoprotein IIb

```
[M A R A L C P L Q A L W L L E W V L L L L G P C A A P P A W A] L N
                        ◆
L D P V Q L T F Y A G P N G S Q F G F S L D F H K D S H G R V A I
V V G A P R T L G P S Q E E T G G V F L C P W R A E G G Q C P S L
L F D L R D E T R N V G S Q T L Q T F K A R Q G L G A S V V S W S
D V I V A C A P W Q H W N V L E K T E E A E K T P V G S C F L A Q
P E S G R R A E Y S P C R G N T L S R I Y V E N D F S W D K R Y C
E A G F S S V V T Q A G E L V L G A P G G Y Y F L G L L A Q A P V
A D I F S S Y R P G I L L W H V S S Q S L S F D S S N P E Y F D G
                              ◆
Y W G Y S V A V G E F D G D L N T T E Y V V G A P T W S W T L G A
V E I L D S Y Y Q R L H R L R G E Q M A S Y F G H S V A V T D V N
G D G R H D L L V G A P L Y M E S R A D R K L A E V G R V Y L F L
Q P R G P H A L G A P S L L L T G T Q L Y G R F G S A I A P L G D
L D R D G Y N D I A V A A P Y G G P S G R G Q V L V F L G Q S E G
L R S R P S Q V L D S P F P T G S A F G F S L R G A V D I D D N G
Y P D L I V G A Y G A N Q V A V Y R A Q P V V K A S V Q L L V Q D
S L N P A V K S C V L P Q T K T P V S C F N I Q M C V G A T G H N
I P Q K L S L N A E L Q L D R Q K P R Q G R R V L L L G S Q Q A G
T T L N L D L G G K H S P I C H T T M A F L R D E A D F R D K L S
            ◆
P I V L S L N V S L P P T E A G M A P A V V L H G D T H V Q E Q T
R I V L D S G E D D V C V P Q L Q L T A S V T G S P L L V G A D N
V L E L Q M D A A N E G E G A Y E A E L A V H L P Q G A H Y M R A
                                  ◆
L S N V E G F E R L I C N Q K K E N E T R V V L C E L G N P M K K
N A Q I G I A M L V S V G N L E E A G E S V S F Q L Q I R S K N S
Q N P N S K I V L L D V P V R A E A Q V E L R G N S F P A S L V V
A A E E G E R E Q N S L D S W G P K V E H T Y E L H N N G P G T V
N G L H L S I H L P G Q S Q P S D L L Y I L D I Q P Q G G L Q C F
                                                          (⇩)
P Q P P V N P L K V D W G L P I P S P S P I H P A H H K R D R R Q
                     ⇩
I F L P E P E Q P S R L Q D P V L V S C D S A P C T V V Q C D L Q
E M A R G Q R A M V T V L A F L W L P S L Y Q R P L D Q F V L Q S
        ◆
H A W F N V S S L P Y A V P P L S L P R G E A Q V W T Q L L R A L
E E R A I P I W W V L V G V L G G L L L L T I L V L A M W K V G F
      ‾‾‾‾‾‾‾‾‾‾‾‾‾‾‾‾‾‾‾‾‾‾‾‾‾‾‾‾‾‾‾‾‾‾‾‾‾‾‾‾‾‾‾‾‾‾‾‾‾‾‾‾‾
F K R N R P P L E E D D E E G E
```

Glycoprotein IIIa

M R A R P R P R P L W V T V L A L G A L A G V G V G G P N I C T T
R G V S S C Q Q C L A V S P M C A W C S D E A L P L G S P R C D L
K E N L L K D N C A P E S I E F P V S E A R V L E D R P L S D K G
S G D S S Q V T Q V S P Q R I A L R L R P D D S K ◇N F S I Q V R Q
V E D Y P V D I Y Y L M D L S Y S M K D D L W S I Q N L G T K L A
T Q M R K L T S N L R I G F G A F V D K P V S P Y M Y I S P P E A
L E N P C Y D M K T T C L P M F G Y K H V L T L T D Q V T R F N E
E V K K Q S V S R N R D A P E G G F D A I M Q A T V C D E K I G W
R N D A S H L L V F T T D A K T H I A L D G R L A G I V Q P N D G
Q C H V G S D N H Y S A S T T M D Y P S L G L M T E K L S Q Q N I
N L I F A V T E N V V N L Y Q ◇N Y S E L I P G T T V G V L S M D S
S N V L D L I V D A Y G K I R S K V E L E V R D L P E E L S L S F
◇N A T C L N N E V I P G L K S C M G L K I G D T V S F S I E A K V
R G C P Q E K E K S F T I K P V G F K D S L I V Q V T F D C D C A
C Q A Q A E P N S H R C N N G ◇N G T F E C G V C R C G P G W L G S
Q C E C S E E D Y R P S Q Q D E C S P R E G Q P V C S Q R G E C L
C G Q C V C H S S D F G K I T G K Y C E C D D F S C V R Y K G E M
C S G H G Q C S C G D C L C D S D W T G Y Y C ◇N C T T R T D T C M
S S N G L L C S G R G K C E C G S C V C I Q P G S Y G D T C E K C
P T C P D A C T F K K E C V E C K K F D R E P Y M T E N T C N R Y
C R D E I E S V K E L K D T G K D A V ◇N C T Y K N E D D C V V R F
Q Y Y E D S S G K S I L Y V V E E P E C P K G P D I L V V L L S V
M G A I L L I G L A A L L I W K L L I T I H D R K E F A K F E E E
R A R A K W D T A N N P L Y K E A T S T F T N I T Y R G T

Fig. A68. Amino-acid sequence of human glycoprotein IIIa. The signal peptide is outlined. ◇ shows the potential attachment sites for N-bound carbohydrate. The presumed transmembrane stretch is underlined.

Fig. A67. Amino-acid sequence of human glycoprotein IIb. The signal peptide is outlined. ⇒ shows the two sites at which the cleavage of GPIIb to give the α and β chains has been proposed to occur. ◆ shows the N-bound carbohydrate. The presumed transmembrane stretch is underlined.

Fibronectin receptor α-subunit

```
M G S R T P E S P L H A V Q L R W G P R R R P P L V P L L L L L V
P P P P R V G G F N L D A E A P A V L S G P P G S F F G F S V E F
Y R P G T D G V S V L V G A P K A N T S Q P G V L Q G G A V Y L C
P W G A S P T Q C T P I E F D S K G S R L L E S S L S S S E G E E
P V E Y K S L Q W F G A T V R A H G S S I L A C A P L Y S W R T E
K E P L S D P V G T C Y L S T D N F T R I L E Y A P C R S D F S W
A A G Q G Y C Q G G F S A E F T K T G R V V L G G P G S Y F W Q G
Q I L S A T Q E Q I A E S Y Y P E Y L I N L V Q G Q L Q T R Q A S
S I Y D D S Y L G Y S V A V G E F S G D D T E D F V A G V P K G N
L T Y G Y V T I L N G S D I R S L Y N F S G E Q M A S Y F G Y A V
A A T D V N G D G L D D L L V G A P L L M D R T P D G R P Q E V G
R V Y V Y L Q H P A G I E P T P T L T L T G H D E F G R F G S S L
T P L G D L D Q D G Y N D V A I G A P F G G E T Q Q G V V F V F P
G G P G G L G S K P S Q V L Q P L W A A S H T P D F F G S A L R G
G R D L D G N G Y P D L I V G S F G V D K A V V Y R G R P I V S A
S A S L T I F P A M F N P E E R S C S L E G N P V A C I N L S F C
L N A S G K H V A D S I G F T V E L Q L D W Q K Q K G G V R R A L
F L A S R Q A T L T Q T L L I Q N G A R E D C R E M K I Y L R N E
S E F R D K L S P I H I A L N F S L D P Q A P V D S H G L R P A L
H Y Q S K S R I E D K A Q I L L D C G E D N I C V P D L Q L E V F
G E Q N H V Y L G D K N A L N L T F H A Q N V G E G G A Y E A E L
R V T A P P E A E Y S G L V R H P G N F S S L S C D Y F A V N Q S
R L L V C D L G N P M K A G A S L W G G L R F T V P H L R D T K K
T I Q F D F Q I L S K N L N N S Q S D V V S F R L S V E A Q A Q V
T L N G V S K P E A V L F P V S D W H P R D Q P Q K E E D L G P A
V H H V Y E L I N Q G P S S I S Q G V L E L S C P Q A L E G Q Q L
L Y V T R V T G L N C T T N H P I N P K G L E L D P E G S L H H Q
Q K R E A P S R S S A S S G P Q I L K C P E A E C F R L R C E L G
P L H Q Q E S Q S L Q L H F R V W A K T F L Q R E H Q P F S L Q C
E A V Y K A L K M P Y R I L P R Q L P Q K E R Q V A T A V Q W T K
A E G S Y G V P L W I I I L A I L F G L L L L G L L I Y I L Y K L
G F F K R S L P Y G T A M E K A Q L K P P A T S D A
```

Fig. A69. Amino-acid sequence of the α subunit of the human fibronectin receptor. The signal peptide is outlined. ⇒ shows the proposed position of the bond that is cleaved when the heavy and light chains are formed. ◇ shows the potential attachment sites for N-bound carbohydrate. The presumed transmembrane stretch is underlined.

Fibronectin receptor β-subunit

M N L Q P I F W I G L I S S V C C V F A Q T D E N R C L K A N A K

S C G E C I Q A G P N C G W C T N S T F L Q E G M P T S A R C D D

L E A L K K K G C P P D D I E N P R G S K D I K K N K N V T N R S

K G T A E K L K P E D I H Q T Q P Q Q L V L R L R S G E P Q T F T

L K F K R A E D Y P I D L Y Y L M D L S Y S M K D D L E N V K S L

G T D L M N E M R R I T S D F R I G F G S F V E K T V M P Y I S T

T P A K L R N P C T S E Q N C T T P F S Y K N V L S L T N K G E V

F N E L V G K Q R I S G N L D S P E G G F D A I M Q V A V C G S L

I G W R N V T R L L V F S T D A G F H F A G D G K L G G I V L P N

D G Q C H L E N N M Y T M S H Y Y D Y P S I A H L V Q K L S E N N

I Q T I F A V T E E F Q P V Y K E L K N L I P K S A V G T L S A N

S S N V I Q L I I D A Y N S L S S E V I L E N G K L S E G V T I S

Y K S Y C K N G V N G T G E N G R K C S N I S I G D E V Q F E I S

I T S N K C P K K D S D S F K I R P L G F T E E V E V I L Q Y I C

E C E C Q S E G I P E S P K C H E G N G T F E C G A C R C N E G R

V G R H C E C S T D E V N S E D M D A Y C R K E N S S E I C S N N

G E C V C G Q C V C R K R D N T N E I Y S G K F C E C D N F N C D

R S N G L I C G G N G V C K C R V C E C N P N Y T G S A C D C S L

D T S T C E A S N G Q I C N G R G I C E C G V C K C T D P K F Q G

Q T C E M C Q T C L G V C A E H K E C V Q C R A F N K G E K K D T

C T Q E C S Y F N I T K V E S R D K L P Q P V Q P D P V S H C K E

K D V D D C W F Y F T Y S V N G N N E V M V H V V E N P E C P T G

P D I I P I V A G V V A G I V L I G L A L L L I W K L L M I I H D

R R E F A K F E K E K M N A K W D T G E N P I Y K S A V T T V V N

P K Y E G K

Fig. A70. Amino-acid sequence of the β subunit of the human fibronectin receptor. The presumed signal peptide is outlined. ◇ shows the potential attachment sites for N-bound carbohydrate. The presumed transmembrane stretch is underlined.

VLA-2 α-subunit

```
M G P E R T G A A P L P L L L V L A L S Q G I L N C C L A Y N V G
L P E A K I F S G P S S E Q F G Y A V Q Q F I N P K G N W L L V G
S P W S G F P E N R M G D V Y K C P V D L S T A T C E K L N L Q T
S T S I P N V T E M K T N M S L G L I L T R N M G T G G F L T C G
P L W A Q Q C G N Q Y Y T T G V C S D I S P D F Q L S A S F S P A
T Q P C P S L I D V V V V C D E S N S I Y P W D A V K N F L E K F
V Q G L D I G P T K T Q V G L I Q Y A N N P R V V F N L N T Y K T
K E E M I V A T S Q T S Q Y G G D L T N T F G A I Q Y A R K Y A Y
S A A S G G R R S A T K V M V V V T D G E S H D G S M L K A V I D
Q C N H D N I L R F G I A V L G Y L N R N A L D T K N L I K E I K
A I A S I P T E R Y F F N V S D E A A L L E K A G T L G E Q I F S
I E G T V Q G G D N F Q M E M S Q V G F S A D Y S S Q N D I L M L
G A V G A F G W S G T I V Q K T S H G H L I F P K Q A F D Q I L Q
D R N H S S Y L G Y S V A A I S T G E S T H F V A G A P R A N Y T
G Q I V L Y S V N E N G N I T V I Q A H R G D Q I G S Y F G S V L
C S V D V D K D T I T D V L L V G A P M Y M S D L K K E E G R V Y
L F T I K K G I L G Q H Q F L E G P E G I E N T R F G S A I A A L
S D I N M D G F N D V I V G S P L E N Q N S G A V Y I Y N G H Q G
T I R T K Y S Q K I L G S D G A F R S H L Q Y F G R S L D G Y G D
L N G D S I T D V S I G A F G Q V V Q L W S Q S I A D V A I E A S
F T P E K I T L V N K N A Q I I L K L C F S A K F R P T K Q N N Q
V A I V Y N I T L D A D G F S S R V T S R G L F K E N N E R C L Q
K N M V V N Q A Q S C P E H I I Y I Q E P S D V V N S L D L R V D
I S L E N P G T S P A L E A Y S E T A K V F S I P F H K D C G E D
G L C I S D L V L Q D V R I P A A Q E Q P F I V S N Q N K R L T F
S V T L K N K R E S A Y N T G I V V D F S E N L F F A S F S L P V
D G T E V T C Q V A A S Q K S V A C D V G Y P A L K R E Q Q V T F
T I N F D F N L Q N L Q N Q A S L S F Q A L S E S Q E E N K A D N
L V N L K I P L L Y D A E I H L T R S T N I N F Y E I S S D G N V
P S I V H S F E D V G P K F I F S L K V T T G S V P V S M A T V I
I H I P Q Y T K E K N P L M Y L T G V Q T D K A G D I S C N A D I
N P L K I G Q T S S S V S F K S E N F R H T K E L N C R T A S C S
N V T C W L K D V H M K G E Y F V N V T T R I W N G T F A S S T F
Q T V Q L T A A A E I N T Y N P E I Y V I E D N T V T I P L M I M
K P D E K A E V P T G V I I G S I I A G I L L L L A L V A I L W K
L G F F K R K Y E K M T K N P D E I D E T T E L S S
```

Fig. A71. Amino-acid sequence of the α subunit of human VLA-2. The signal peptide is outlined. ◇ shows the potential attachment sites for N-bound carbohydrate. The presumed transmembrane stretch is underlined. An R-G-D sequence is outlined with dashed lines.

LFA-1 α-subunit

M K D S C I T V M A M A L L S G F F F F A P A S S Y N L D V R G A
R S F S P P R A G R H F G Y R V L Q V G N G V I V G A P G E G N S
T G S L Y Q C Q S G T G H C L P V T L R G S N Y T S K Y L G M T L
A T D P T D G S I L A C D P G L S R T C D Q N T Y L S G L C Y L F
R Q N L Q G P M L Q G R P G F Q E C I K G N V D L V F L F D G S M
S L Q P D E F Q K I L D F M K D V M K K L S N T S Y Q F A A V Q F
S T S Y K T E F D F S D Y K V W K D P D A L L K H V K H M L L L T
N T F G A I N Y V A T E V F R E E L G A R P D A T K V L I I I T D
G E A T D S G N I D A A K D I I R Y I I G I G K H F Q T K E S Q E
T L H K F A S K P A S E F V K I L D T F E K L K D L F T E L Q K K
I Y V I E G T S K Q D L T S F N M E L S S S G I S A D L S R G H A
V V G A V G A K D W A G G F L D L K A D L Q D D T F I G N E P L T
P E V R A G Y L G Y T V T W L P S R Q K T S L L A S G A P R Y K H
M G R V L L F Q E P Q G G G H W S Q V Q T I H G T Q I G S Y F G G
E L C G V D V D Q D G E T E L L L I G A P L F Y G E Q R G G R V F
I Y Q R R Q L G F E E V S E L Q G D P G Y P L G R F G E A I T A L
T D I N G D G L V D V A V G A P L E E Q G A V Y I F N G R H G G L
S P Q P S Q R I E G T Q V L S G I Q W F G R S I H G V K D L E G D
G L A D V A V G A E S Q M I V L S S R P V V D M V T L M S F S P A
E I P V H E V E C S Y S T S N K M K E G V N I T I C F Q I K S L Y
P Q F Q G R L V A N L T Y T L Q L D G H R T R R R G L F P G G R H
E L R R N I A V T T S M S C T D F S F H F P V C V Q D L I S P I N
V S L N F S L W E E E G T P R D Q R A Q G K D I P P I L R P S L H
S E T W E I P F E K N C G E D K K C E A N L R V S F S P A R S R A
L R L T A F A S L S V E L S L S N L E E D A Y W V Q L D L H F P P
G L S F R K V E M L K P H S Q I P V S C E E L P E E S R L L S R A
L S C N V S S P I F K A G H S V A L Q M M F N T L V N S S W G D S
V E L H A N V T C N N E D S D L L E D N S A T T I I P I L Y P I N
I L I Q D Q E D S T L Y V S F T P K G P K I H Q V K H M Y Q V R I
Q P S I H D H N I P T L E A V V G V P Q P P S E G P I T H Q W S V
Q M E P P V P C H Y E D L E R L P D A A E P C L P G A L F R C P V
V F R Q E I L V Q V I G T L E L V G E I E A S S M F S L C S S L S
I S F N S S K H F H L Y G S N A S L A Q V V M K V D V V Y E K Q M
L Y L Y V L S G I G G L L L L L L I F I V L Y K V G F F K R N L K
E K M E A G R G V P N G I P A E D S E Q L A S G Q E A G D P G C L
K P L H E K D S E S G G G K D

Fig. A72. Amino-acid sequence of the α subunit of human LFA-1. The signal peptide is outlined. ◇ shows the potential attachment sites for N-bound carbohydrate. The presumed transmembrane stretch is underlined.

P150,95 α-chain

M T R T R A A L L L F T A L A T S L G F N L D T E E L T A F R V D
S A G F G D S V V Q Y A N S W V V V G A P Q K I T A A N Q T G G L
Y Q C G Y S T G A C E P I G L Q V P P E A V N M S L G L S L A S T
T S P S Q L L A C G P T V H H E C G R N M Y L T G L C F L L G P T
Q L T Q R L P V S R Q E C P R Q E Q D I V F L I D G S G S I S S R
N F A T M M N F V R A V I S Q F Q R P S T Q F S L M Q F S N K F Q
T H F T F E E F R R T S N P L S L L A S V H Q L Q G F T Y T A T A
I Q N V V H R L F H A S Y G A R R D A T K I L I V I T D G K K E G
D S L D Y K D V I P M A D A A G I I R Y A I G V G L A F Q N R N S
W K E L N D I A S K P S Q E H I F K V E D F D A L K D I Q N Q L K
E K I F A I E G T E T T S S S S F E L E M A Q E G F S A V F T P D
G P V L G A V G S F T W S G G A F L Y P P N M S P T F I N M S Q E
N V D M R D S Y L G Y S T E L A L W K G V Q S L V L G A P R Y Q H
T G K A V I F T Q V S R Q W R M K A E V T G T Q I G S Y F G A S L
C S V D V D T D G S T D L V L I G A P H Y Y E Q T R G G Q V S V C
P L P R G W R R W W C D A V L Y G E Q G H P W G R F G A A L T V L
G D V N G D K L T D V V I G A P G E E E N R G A V Y L F H G V L G
P S I S P S H S Q R I A G S Q L S S R L Q Y F G Q A L S G G Q D L
T Q D G L V D L A V G A R G Q V L L L R T R P V L W V G V S M Q F
I P A E I P R S A F E C R E Q V V S E Q T L V Q S N I C L Y I D K
R S K N L L G S R D L Q S S V T L D L A L D P G R L S P R A T F Q
E T K N R S L S R V R V L G L K A H C E N F N L L L P S C V E D S
V T P I T L R L N F T L V G K P L L A F R N L R P M L A A L A Q R
Y F T A S L P F E K N C G A D H I C Q D N L G I S F S F P G L K S
L L V G S N L E L N A E V M V W N D G E D S Y G T T I T F S H P A
G L S Y R Y V A E G Q K Q G Q L R S L H L T C D S A P V G S Q G T
W S T S C R I N H L I F R G G A Q I T F L A T F D V S P K A V L G
D R L L L T A N V S S E N N T P R T S K T T F Q L E L P V K Y A V
Y T V V S S H E Q F T K Y L N F S E S E E K E S H V A M H R Y Q V
N N L G Q R D L P V S I N F W V P V E L N Q E A V W M D V E V S H
P Q N P S L R C S S E K I A P P A S D F L A H I Q K N P V L D C S
I A G C L R F R C D V P S F S V Q E E L D F T L K G N L S F G W V
R Q I L Q K K V S V V S V A E I T F D T S V Y S Q L P G Q E A F M
R A Q T T T V L E K Y K V H N P T P L I V G S S I G G L L L L A L
I T A V L Y K V G F F K R Q Y K E M M E E A N G Q I A P E N G T Q
T P S P P S E K

Fig. A73. Amino-acid sequence of the α subunit of human p150,95. The signal peptide is outlined. ◇ shows the potential attachment sites for N-bound carbohydrate. The presumed transmembrane stretch is underlined.

Vitronectin receptor α-subunit

```
M A F P P R R R L R L G P R G L P L L L S G L L L P L C R A F N L
D V D S P A E Y S G P E G S Y F G F A V D F F V P S A S S R M F L
L V G A P K A N T T Q P G I V E G G Q V L K C D W S S T R R C Q P
I E F D A T G N R D Y A K D D P L E F K S H Q W F G A S V R S K Q
D K I L A C A P L Y H W R T E M K Q E R E P V G T C F L Q D G T K
T V E Y A P C R S Q D I D A D G Q G F C Q G G F S I D F T K A D R
V L L G G P G S F Y W Q G Q L I S D Q V A E I V S K Y D P N V Y S
I K Y N N Q L A T R T A Q A I F D D S Y L G Y S V A V G D F N G D
G I D D F V S G V P R A A R T L G M V Y I Y D G K N M S S L Y N F
T G E Q M A A Y F G F S V A A T D I N G D D Y A D V F I G A P L F
M D R G S D G K L Q E V G Q V S V S L Q R A S G D F Q T T K L N G
F E V F A R F G S A I A P L G D L D Q D G F N D I A I A A P Y G G
E D K K G I V Y I F N G R S T G L N A V P S Q I L E G Q W A A R S
M P P S F G Y S M K G A T D I D K N G Y P D L I V G A F G V D R A
I L Y R A R P V I T V N A G L E V Y P S I L N Q D N K T C S L P G
T A L K V S C F N V R F C L K A D G K G V L P R K L N F Q V E L L
L D K L K Q K G A I R R A L F L Y S R S P S H S K N M T I S R G G
L M Q C E E L I A Y L R D E S E F R D K L T P I T I F M E Y R L D
Y R T A A D T T G L Q P I L N Q F T P A N I S R Q A H I L L D C G
E D N V C K P K L E V S V D S D Q K K I Y I G D D N P L T L I V K
A Q N Q G E G A Y E A E L I V S I P L Q A D F I G V V R N N E A L
A R L S C A F K T E N Q T R Q V V C D L G N P M K A G T Q L L A G
L R F S V H Q Q S E M D T S V K F D L Q I Q S S N L F D K V S P V
V S H K V D L A V L A A V E I R G V S S P D H I F L P I P N W E H
K E N P E T E E D V G P V V Q H I Y E L R N N G P S S F S K A M L
H L Q W P Y K Y N N N T L L Y I L H Y D I D G P M N C T S D M E I
N P L R I K I S S L Q T T E K N D T V A G Q G E R D H L I T K R D
L A L S E G D I H T L G C G V A Q C L K I V C Q V G R L D R G K S
A I L Y V K S L L W T E T F M N K E N Q N H S Y S L K S S A S F N
V I E F P Y K N L P I E D I T N S T L V T T N V T W G I Q P A P M
P V P V W V I I L A V L A G L L L L A V L V F V M Y R M G F F K R
V R P P Q E E Q E R E Q L Q P H E N G E G N S E T
```

Fig. A74. Amino-acid sequence of the α subunit of the human vitronectin receptor. The signal peptide is outlined. ⇒ shows the proposed position of the bond that is cleaved when the heavy and light chains are formed. ◇ shows the potential attachment sites for N-bound carbohydrate. The presumed transmembrane stretch is underlined.

Heparin co-factor II

```
M K H S L N A L L I F L I I T S A W G G S K G P L D Q L E K G G E
T A Q S A D P Q W E Q L N N K N L S M P L L P A D F H K E N T V T
N D W I P E G E E D D D Y L D L E K I F S E D D D Y I D I V D S L
S V S P T D S D V S A G N I L Q L F H G K S R I Q R L N I L N A K
F A F N L Y R V L K D Q V N T F D N I F I A P V G I S T A M G M I
S L G L K G E T H E Q V H S I L H F K D F V N A S S K Y E I T T I
H N L F R K L T H R L F R R N F G Y T L R S V N D L Y I Q K Q F P
I L L D F R T K V R E Y Y F A E A Q I A D F S D P A F I S K T N N
H I M K L T K G L I K D A L E N I D P A T Q M M I L N C I Y F K G
S W V N K F P V E M T H N H N F R L N E R E V V K V S M M Q T K G
N F L A A N D Q E L D C D I L Q L E Y V G G I S M L I V V P H K M
S G M K T L E A Q L T P R V V E R W Q K S M T N R T R E V L L P K
F K L E K N Y N L V E S L K L M G I R M L F D K N G N M A G I S D
Q R I A I D L F K H Q G T I T V N E E G T Q A T T V T T V G F M P
L S T Q V R F T V D R P F L F L I Y E H R T S C L L F M G R V A N
P S R S
```

Fig. A75. Amino-acid sequence of human heparin co-factor II. The signal peptide is outlined. → shows the reactive site and ◇ the potential attachment sites for N-bound carbohydrate. Sulphated tyrosine residues are encircled. Position 218 (underlined) may be dimorphic (R/K).

α_1-proteinase inhibitor

```
M P S S V S W G I L L L A G L C C L V P V S L A E D P Q G D A A Q
K T D T S H H D Q D H P T F N K I T P N L A E F A F S L Y R Q L A
H Q S N S T N I F F S P V S I A T A F A M L S L G T K A D T H D E
I L E G L N F N L T E I P E A Q I H E G F Q E L L R T L N Q P D S
Q L Q L T T G N G L F L S E G L K L V D K F L E D V K K L Y H S E
A F T V N F G D T E E A K K Q I N D Y V E K G T Q G K I V D L V K
E L D R D T V F A L V N Y I F F K G K W E R P F E V K D T E E E D
F H V D Q V T T V K V P M M K R L G M F N I Q H C K K L S S W V L
L M K Y L G N A T A I F F L P D E G K L Q H L E N E L T H D I I T
K F L E N E D R R S A S L H L P K L S I T G T Y D L K S V L G Q L
G I T K V F S N G A D L S G V T E E A P L K L S K A V H K A V L T
I D E K G T E A A G A M F L E A I P M S I P P E V K F N K P F V F
L M I E Q N T K S P L F M G K V V N P T Q K
```

Fig. A76. Amino-acid sequence of human α_1-proteinase inhibitor. The signal peptide is outlined. → shows the reactive site and ◆ the N-bound carbohydrate.

α_1-antichymotrypsin

M E R M L P L L A L G L L A A G F C P A V L C H P N S P L D E E N

L T Q E N Q D R G T H V D L G L A S A N V D F A F S L Y K Q L V L

K A L D K N V I F S P L S I S T A L A F L S L G A H N T T L T E I

L K A S S S P H G D L L R Q K F T Q S F Q H L R A P S I S S S D E

L Q L S M G N A M F V K E Q L S L L D R F T E D A K R L Y G S E A

F A T D F Q D S A A A K K L I N D Y V K N G T R G K I T D L I K D

P D S Q T M M V L V N Y I F F K A K W E M P F D P Q D T H Q S R F

Y L S K K K W V M V P M M S L H H L T I P Y F R D E E L S C T V V

E L K Y T G N A S A L F I L P D Q D K M E E V E A M L L P E T L K

R W R D S L E F R E I G E L Y L P K F S I S R D Y N L N D I L L Q

L G I E E A F T S K A D L S G I T G A R N L A V S Q V V H K V V S

D V F E E G T E A S A A T A V K I T L L S A L V E T R T I V R F N

R P F L M I I V P T D T Q N I F F M S K V T N P S K P R A C I K Q

W G S Q

Fig. A77. Amino-acid sequence of human α_1-antichymotrypsin. The signal peptide is outlined. → shows the reactive site and ◇ the potential attachment sites for N-bound carbohydrate.

Angiotensinogen

M R K R A P Q S E M A P A G V S L R A T I L C L L A W A G L A A G

D R V Y I H P F H L V I H N E S T C E Q L A K A N A G K P K D P T

F I P A P I Q A K T S P V D E K A L Q D Q L V L V A A K L D T E D

K L R A A M V G M L A N F L G F R I Y G M H S E L W G V V H G A T

V L S P T A V F G T L A S L Y L G A L D H T A D R L Q A I L G V P

W K D K N C T S R L D A H K V L S A L Q A V Q G L L V A Q G R A D

S Q A Q L L L S T V V G V F T A P G L H L K Q P F V Q G L A L Y T

P V V L P R S L D F T E L D V A A E K I D R F M Q A V T G W K T G

C S L M G A S V D S T L A F N T Y V H F Q C K M K G F S L L A E P

Q E F W V D N S T S V S V P M L S G M G T F Q H W S D I Q D N F S

V T Q V P F T E S A C L L L I Q P H Y A S D L D K V E G L T F Q Q

N S L N W M K K L S P R T I H L T M P Q L V L Q G S Y D L Q D L L

A Q A G L P A I L H T E L N L Q K L S N D R I R V G E V L N S I F

F E L E A D E R E P T E S T Q Q L N K P E V L E V T L N R P F L F

A V Y D Q S A T A L H F L G R V A N P L S T A

Fig. A78. Amino-acid sequence of human angiotensinogen. The signal peptide is outlined. → shows the position of cleavage by renin which releases angiotensin I from angiotensinogen. ⇒ shows the position in angiotensin I which is cleaved by angiotensin-converting enzyme, leading to angiotensin II. ◆ shows the N-bound carbohydrate and ◇ the potential attachment sites for N-bound carbohydrate. Position 359 (underlined) may be dimorphic (L/M).

TBG

M S P F L Y L V L L V L G L H A T I H C A S P E G K V T A C H S S
Q P N A T L Y K M S S I N A D F A F N L Y R R F T V E T P D K N I
F F S P V S I S A A L V M L S F G A C C S T Q T E I V G T L G F N
L T D T P M V E I Q H G F Q H L I C S L N F P K K E L E L Q I G N
A L F I G K H L K P L A K F L N D V K T L Y E T E V F S T D F S N
I S A A K Q E I N S H V E M Q T K G K V V G L I Q D L K P N T I M
V L V N Y I H F K A Q W A N P F D P S K T E D S S S F L I D K T T
T V Q V P M M H Q M E Q Y Y H L V D M E L N C T V L Q M D Y S K N
A L A L F V L P K E G Q M E S V E A A M S S K T L K K W N R L L Q
K G W V D L F V P K F S I S A T Y D L G A T L L K M G I Q H A Y S
E N A D F S G L T E D N G L K L S N A A H K A V L H I G E K G T E
A A A V P E V E L S D Q P E N T F L H P I I Q I D R S F M L L I L
E R S T R S I L F L G K V V N P T E A

Fig. A79. Amino-acid sequence of human thyroxine-binding globulin (TBG). The signal peptide is outlined. ◇ shows the potential attachment sites for N-bound carbohydrate. Position 283 (underlined) is dimorphic and can be L or F.

CBG

M P L L L Y T C L L W L P T S E L W T V Q A M D P N A A Y V N M S
N H H R G L A S A N V D F A P S L Y K H L V A L S P K K N I F I S
P V S I S M A L A M L S L G T C G H T R A Q L L Q G L G F N L T E
R S E T E I H Q G F Q H L H Q L F A K S D T S L E M T M G N A L F
L D G S L E L L E S F S A D I K H Y Y E S E V L A M N F Q D W A T
A S R Q I N S Y V K N K T Q G K I V D L F S G L D S P A I L V L V
N Y I F F K G T W T Q P F D L A S T R E E N F Y V D E T T V V K V
P M M L Q S S T I S Y L H D S E L P C Q L V Q M N Y V G N G T V F
F I L P D K G K M N T V I A A L S R D T I N R W S A G L T S S Q V
D L Y I P K V T I S G V Y D L G D V L E E M G I A D L F T N Q A N
F S R I T Q D A Q L K S S K V V H K A V L Q L N E E G V D T A G S
T G V T L N L T S K P I I L R F N Q P F I I M I F D H F T W S S L
F L A R V M N P V

Fig. A80. Amino-acid sequence of human corticosteroid-binding globulin (CBG). The signal peptide is outlined. ◇ shows the potential attachment sites for N-bound carbohydrate.

Chicken ovalbumin

```
[M]G S I G A A S M E F C F D V F K E L K V H H A N E N I F Y C P I
A I M S A L A M V Y L G A K D S T R T Q I N K V V R F D K L P G F
G D(S)I E A Q C G T S V N V H S S L R D I L N Q I T K P N D V Y S
F S L A S R L Y A E E R Y P I L P E Y L Q C V K E L Y R G G L E P
I N F Q T A A D Q A R E L I N S W V E S Q T N G I I R N V L Q P S
S V D S Q T A M V L V N A I V F K G L W E K A F K D E D T Q A M P
F R V T E Q E S K P V Q M M Y Q I G L F R V A S M A S E K M K I L
E L P F A S G T M S M L V L L P D E V S G L E Q L E S I I N F E K
L T E W T S S N V M E E R K I K V Y L P R M K M E E K Y N L T S V
                            ◆
L M A M G I T D V F S S S A N L S G I S S A E S L K I S Q A V H A
A H A E I N E A G R E V V G(S)A E A G V D A A S V S E E F R A D H
P F L F C I K H I A T N A V L F F G R C V S P
```

Fig. A81. Amino-acid sequence of chicken ovalbumin. The protein is synthesized without a signal peptide. The methionine preceding Gly1 is outlined. ◆ shows the N-bound carbohydrate. The phosphorylated Ser residues are encircled.

REFERENCES

1. Davie, E.W. & Fujikawa, K. (1975). Basic mechanisms in blood coagulation. Ann. Rev. Biochem. 44, pp. 799–829.
2. Davie, E.W., Fujikawa, K., Kurachi, K. & Kisiel, W. (1979). The role of serine proteases in the blood coagulation cascade. Adv. Enzymol. 38, pp. 277–318.
3. Jackson, C.M. & Nemerson, Y. (1980). Blood coagulation. Ann. Rev. Biochem. 49, pp. 765–811.
4. Furie, B. & Furie, B.C. (1988). The molecular basis of blood coagulation. Cell 53, pp. 505–518.
5. Macfarlane, R.G. (1964). An enzyme cascade in the blood clotting mechanism, and its function as a biochemical amplifier. Nature 202, pp. 498–499.
6. Davie, E.W. & Ratnoff, O.D. (1964). Waterfall sequence for intrinsic blood clotting. Science 145, pp. 1310–1312.
7. Ratnoff, O.D. (1966). The biology and pathology of the initial stages of blood coagulation. Prog. Hematol. 5, pp. 204–245.
8. Ogston, D. (1981). Contact activation of coagulation. Recent Advances in Blood Coagulation Vol. III, pp. 109–123.
9. Cochrane, C.G. & Griffin, J.H. (1982). The biochemistry and pathophysiology of the contact system of plasma. Adv. Immunol. 33, pp. 241–306.
10. Colman, R.W. (1984). Surface-mediated defence reactions. J. Clin. Invest. 73, pp. 1249–1253.
11. Bouma, B.N. & Griffin, J.H. (1986). Initiation mechanisms: The contact activation system in plasma. In: New Comprehensive Biochemistry Vol. 13 (Zwaal, R.F.A. & Hemker, H.C., Eds.), pp. 103–128, Elsevier, Amsterdam.
12. Kaplan, A.P. & Silverberg, M. (1987). The coagulation-kinin pathway of human plasma. Blood 70, pp. 1–15.
13. Proud, D. & Kaplan, A.P. (1988). Kinin formation: Mechanism and role in inflammatory disorder. Ann. Rev. Immunol. 6, pp. 49–83.
14. Rosenthal, R.L., Dreskin, O.H. & Rosenthal, M. (1953). New hemophilia-like disease caused by deficiency of a third plasma thromboplastin factor. Proc. Soc. Exp. Biol. Med. 82, pp. 171–174.
15. Ratnoff, O.D. & Colopy, J.E. (1955). A familial hemorrhagic trait associated with deficiency of a clot-promoting fraction of plasma. J. Clin. Invest. 34, pp. 602–613.
16. Hathaway, W.E., Belhasen, L.P. & Hathaway, H.S. (1965). Evidence for a new plasma thromboplastin factor. Blood 26, pp. 521–532.
17. Wuepper, K.D. (1973). Prekallikrein deficiency in man. J. Exp. Med. 138, pp. 1345–1355.
18. Saito, H., Ratnoff, O.D., Waldmann, R. & Abraham, J.P. (1975). Fitzgerald trait. Deficiency of a hitherto unrecognized agent, Fitzgerald factor, participating in surface mediated reactions of clotting, fibrinolysis, generation of kinins, and the property of diluted plasma enhancing vascular permeability (PF/DIL). J. Clin. Invest. 55, pp. 1082–1089.

19. Colman, R.W., Bagdasarian, A., Talamo, R.C., Scott, C.F., Seavey, M., Guimaraes, J.A., Pierce, J.V. & Kaplan, A.P. (1975). Williams trait. Human kininogen deficiency with diminished levels of plasminogen proactivator and prekallikrein associated with abnormalities of the Hageman factor-dependent pathways. J. Clin. Invest. 56, pp. 1650–1662.
20. Wuepper, K.D., Miller, D.R. & Lacombe, M.J. (1975). Flaujeac trait. Deficiency of human plasma kininogen. J. Clin. Invest. 56, pp. 1663–1672.
21. Lutcher, C.L. (1976). Reid trait: A new expression of high molecular weight kininogen (HMW-kininogen) deficiency. Clinical Research 24, p. 47.
22. Ratnoff, O.D., Busse Jr., R.J. & Sheon, R.P. (1968). The demise of John Hageman. N. Eng. J. Med. 279, pp. 760–761.
23. Fujikawa, K., McMullen, B., Heimark, R.L., Kurachi, K. & Davie, E.W. (1980). The role of factor XII (Hageman factor) in blood coagulation and a partial amino acid sequence of human factor XII and its fragments. Protides Biol. Fluids 28, pp. 193–196.
24. Saito, H., Ratnoff, O.D. & Pensky, J. (1976). Radioimmunoassay of human Hageman factor (factor XII). J. Lab. Clin. Med. 88, pp. 506–514.
25. Fujikawa, K. & McMullen, B.A. (1983). Amino acid sequence of human β-factor XIIa. J. Biol. Chem. 258, pp. 10924–10934.
26. McMullen, B.A. & Fujikawa, K. (1985). Amino acid sequence of the heavy chain of human α-factor XIIa (activated Hageman factor). J. Biol. Chem. 260, pp. 5328–5341.
27. Cool, D.E., Edgell, C.-J., Louie, G.V., Zoller, M.J., Brayer, D.D. & MacGillivray, R.T.A. (1985). Characterization of human blood coagulation factor XII cDNA. J. Biol. Chem. 260, pp. 13666–13676.
28. Que, B.G. & Davie, E.W. (1986). Characterization of a cDNA coding for human factor XII (Hageman factor). Biochemistry 25, pp. 1525–1528.
29. Tripodi, M., Citarella, F., Guida, S., Galeffi, P., Fantoni, A. & Cortese, R. (1986). cDNA sequence coding for human coagulation factor XII (Hageman). Nucl. Acids Res. 14, p. 3146.
30. Cool, D.E. & MacGillivray, R.T.A. (1987). Characterization of the human blood coagulation factor XII gene. J. Biol. Chem. 262, pp. 13662–13673.
31. Royle, N.J., Nigli, M., Cool, D., MacGillivray, R.T.A. & Hamerton, J.L. (1988). Structural gene encoding human factor XII is located at 5q33–qter. Somat. Cell Mol. Genet. 14, pp. 217–221.
32. Petersen, T.E., Thøgersen, H.C., Skorstengaard, K., Vibe Pedersen, K., Sahl, P., Sottrup-Jensen, L. & Magnusson, S. (1983). Partial primary structure of bovine plasma fibronectin: Three types of internal homology. Proc. Natl. Acad. Sci. USA 80, pp. 137–141.
33. Savage Jr., C.R., Inagami, T. & Cohen, S. (1972). The primary structure of epidermal growth factor. J. Biol. Chem. 247, pp. 7612–7621.
34. Savage Jr., C.R., Hash, J.H. & Cohen, S. (1973). Epidermal growth factor. J. Biol. Chem. 248, pp. 7669–7672.
35. Magnusson, S., Petersen, T.E., Sottrup-Jensen, L. & Claeys, H. (1975). Complete primary structure of prothrombin: Isolation, structure and reactivity of ten carboxylated glutamic acid residues and regulation of prothrombin activation by thrombin. In: Proteases and Biological Control (Reich, E., Rifkin, D.D. & Shaw, E., Eds.), pp. 123–149, Cold Spring Harbor Laboratory, Cold Spring Harbor, New York.
36. Skorstengaard, K., Thøgersen, H.C. & Petersen, T.E. (1984). Complete primary structure of the collagen-binding domain of bovine fibronectin. Eur. J. Biochem. 140, pp. 235–243.
37. Miyata, T., Kawabata, S., Iwanaga, S., Takahashi, I., Alving, B. & Saito, H. (1989). Coagulation factor XII (Hageman factor) Washington D.C.: Inactive factor XIIa results from Cys-571 → Ser substitution. Proc. Natl. Acad. Sci. USA 86, pp. 8319–8322.
38. Bouma, B.N., Miles, L.A., Beretta, G. & Griffin, J.H. (1980). Human plasma prekallikrein. Studies of its activation by activated factor XII and of its inactivation by diisopropyl phosphofluoridate. Biochemistry 19, pp. 1151–1160.
39. Saito, H., Poon, M., Vicic, W., Goldsmith, G.H. & Menitove, J.E. (1978). Human plasma prekallikrein (Fletcher factor) clotting activity and antigen in health and disease. J. Lab. Clin. Med. 92, pp. 84–95.

40. Bouma, B.N., Kerbiriou, D.M., Vlooswijk, R.A.A. & Griffin, J.H. (1980). Immunological studies of prekallikrein, kallikrein, and high molecular-weight kininogen in normal and deficient plasmas and in normal plasma after cold-dependent activation. J. Lab. Clin. Med. 96, pp. 693–709.
41. Heimark, R.L. & Davie, E.W. (1981). Bovine and human plasma prekallikrein. Methods Enzymol. 80, pp. 157–172.
42. Hojima, Y., Pierce, J.V. & Pisano, J.J. (1985). Purification and characterization of multiple forms of human plasma prekallikrein. J. Biol. Chem. 260, pp. 400–406.
43. Chung, D.W., Fujikawa, K., MacMullen, B.A. & Davie, E.W. (1986). Human plasma prekallikrein, a zymogen to a serine protease that contains four tandem repeats. Biochemistry 25, pp. 2410–2417.
44. Heimark, R.L. & Davie, E.W. (1979). Isolation and characterization of bovine plasma prekallikrein. Biochemistry 18, pp. 5743–5750.
45. Mandel Jr., R. & Kaplan, A.P. (1977). Hageman factor substrates. J. Biol. Chem. 252, pp. 6097–6104.
46. Saito, H. & Goldsmith, G.H. (1977). Plasma thromboplastin antecedent (PTA, factor XI): A specific and sensitive radioimmunoassay. Blood 50, pp. 377–385.
47. Bouma, B.N., Vlooswijk, R.A.A. & Griffin, J.H. (1983). Immunologic studies of human coagulation factor XI and its complex with high molecular weight kininogen. Blood 62, pp. 1123–1131.
48. Bouma, B.N. & Griffin, J.H. (1977). Human blood coagulation factor XI. J. Biol. Chem. 252, pp. 6432–6437.
49. Kurachi, K. & Davie, E.W. (1977). Activation of human factor XI (plasma thromboplastin antecedent) by factor XIIa (activated Hageman factor). Biochemistry 16, pp. 5831–5839.
50. Fujikawa, K., Chung, D.W., Hendrickson, L.E. & Davie, E.W. (1986). Amino acid sequence of human factor XI, a blood coagulation factor with four tandem repeats that are highly homologous with plasma prekallikrein. Biochemistry 25, pp. 2417–2424.
51. Davie, E.W. (1989). Molecular genetics of the coagulation and fibrinolytic proteins. Thromb. Haemostas. 62, Abstract PL-11.
52. Asakai, R., Davie, E.W. & Chung, D.W. (1987). Organization of the gene for human factor XI. Biochemistry 26, pp. 7221–7228.
53. Asakai, R., Chung, D.W., Ratnoff, O.D. & Davie, E.W. (1989). Factor XI (plasma thromboplastin antecedent) deficiency in Ashkenazi Jews is a bleeding disorder that can result from three types of point mutations. Proc. Natl. Acad. Sci. USA 86, pp. 7667–7671.
54. Kato, K., Nagasawa, S. & Iwanaga, S. (1981). HMW and LMW kininogen. Methods Enzymol. 80, pp. 172–198.
55. Proud, D., Pierce, J.V. & Pisano, J.J. (1980). Radioimmunoassay of human high molecular weight kininogen in normal and deficient plasma. J. Lab. Clin. Med. 95, pp. 563–574.
56. Kerbiriou, D.M. & Griffin, J.H. (1979). Human high molecular weight kininogen. J. Biol. Chem. 254, pp. 12020–12027.
57. Schiffman, S., Mannhalter, C. & Tyner, K.D. (1980). Human high molecular weight kininogen. J. Biol. Chem. 255, pp. 6433–6438.
58. Lottspeich, F., Kellerman, J., Henschen, A., Foertsch, B. & Müller-Esterl, W. (1985). The amino acid sequence of the light chain of human high-molecular-mass kininogen. Eur. J. Biochem. 152, pp. 307–314.
59. Kellerman, J., Lottspeich, F., Henschen, A. & Müller-Esterl, W. (1986). Completion of the primary structure of human high-molecular-mass kininogen. Eur. J. Biochem. 154, pp. 471–478.
60. Takagaki, Y., Kitamura, N. & Nakanishi, S. (1985). Cloning and sequence analysis of cDNAs for human high molecular weight and low molecular weight prekininogens. J. Biol. Chem. 260, pp. 8601–8609.
61. Kitamura, N., Kitagawa, H., Fukushima, D., Takagaki, Y., Miyata, T. & Nakanishi, S. (1985). Structural organization of the human kininogen gene and a model for its evolution. J. Biol. Chem. 260, pp. 8610–8617.
62. Sasaguri, M., Ikeda, M., Ideishi, M. & Arakawa, K. (1988). Identification of

(hydroxyproline[3])-lysyl-bradykinin released from human plasma protein by kallikrein. Biochem. Biophys. Res. Commun. 150, pp. 511–516.
63. Maier, M., Reissert, G., Jerabek, I., Lottspeich, F. & Binder, B.R. (1988). Identification of (hydroxyproline[3])-lysyl-bradykinin released from human kininogens by human urinary kallikrein. FEBS Lett. 232, pp. 395–398.
64. Mindroiu, T., Carretero, O.A., Proud, D., Walz, D. & Scili, A.G. (1988). A new kinin moiety in human plasma kininogens. Biochem. Biophys. Res. Commun. 152, pp. 519–526.
65. Salvesen, G., Parkes, C., Abrahamson, M., Grubb, A. & Barrett, A.J. (1986). Human low-M_r kininogen contains three copies of a cystatin sequence that are divergent in structure and in inhibitory activity for cysteine proteinases. Biochem. J. 234, pp. 429–434.
66. Sueyoshi, T., Enjyoji, K., Shimada, T., Kato, H., Iwanaga, S., Bando, Y., Kominami, E. & Katanuma, N. (1985). A new function of kininogens as thiol-proteinase inhibitors: Inhibition of papain and cathepsins B, H and L by bovine, rat and human plasma kininogens. FEBS Lett. 182, pp. 193–195.
67. Müller-Esterl, W., Fritz, H., Marchleidt, W., Ritonja, A., Brzin, J., Kotnik, M., Turk, V., Kellermann, J. & Lottspeich, F. (1985). Human plasma kininogens are identical with α-cysteine proteinase inhibitor. FEBS Lett. 182, pp. 310–314.
68. Higashiyama, S., Ohkubo, I., Ishiguro, H., Kunimatsu, M., Sawaki, K. & Sasaki, M. (1986). Human high molecular weight kininogen as a thiol proteinase inhibitor: presence of the entire inhibition capacity in the native form of heavy chain. Biochemistry 25, pp. 1669–1675.
69. Kellerman, J., Thielen, C., Lottspeich, F., Henschen, A., Vogel, R. & Müller-Esterl, W. (1987). Arrangement of the disulphide bridges in human low-M_r kininogen. Biochem. J. 247, pp. 15–21.
70. Sueyoshi, T., Miyata, T., Hashimoto, N., Kato, H., Hayashida, H., Miyata, T. & Iwanaga, S. (1987). Bovine high molecular weight kininogen. J. Biol. Chem. 262, pp. 2768–2779.
71. Kato, H., Han, Y.N. & Iwanaga, S. (1977). Primary structure of bovine plasma low-molecular weight kininogen. J. Biochem. (Tokyo) 82, pp. 377–385.
72. Han, Y.N., Komiya, M., Iwanaga, S. & Suzuki, T. (1975). Studies on the primary structure of bovine high-molecular-weight kininogen. J. Biochem. (Tokyo) 77, pp. 55–68.
73. Han, Y.N., Kato, H., Iwanaga, S. & Suzuki, T. (1976). Primary structure of bovine plasma high-molecular-weight kininogen. J. Biochem. (Tokyo) 79, pp. 1201–1222.
74. Nawa, H., Kitamura, N., Hirose, T., Asai, M., Inayama, S. & Nakanishi, S. (1983). Primary structure of bovine liver low molecular weight kininogen precursors and their two mRNAs. Proc. Natl. Acad. Sci. USA 80, pp. 90–94.
75. Kitamura, N., Takagaki, Y., Furuto, S., Tanaka, T., Nawa, H. & Nakanishi, S. (1983). A single gene for bovine high molecular weight and low molecular weight kininogens. Nature 305, pp. 545–549.
76. Furuto-Kato, S., Matsumoto, A., Kitamura, N. & Nakanishi, S. (1985). Primary structures of the mRNAs encoding the rat precursors for bradykinin and T-kinin. J. Biol. Chem. 260, pp. 12054–12059.
77. Kitasawa, H., Kitamura, N., Hayashida, H., Miyata, T. & Nakanishi, S. (1987). Differing expression patterns and evolution of the rat kininogen gene family. J. Biol. Chem. 262, pp. 2190–2198.
78. Kageyama, R., Kitamura, N., Ohkubo, H. & Nakanishi, S. (1985). Differential expression of the multiple forms of rat prekininogen mRNAs after acute inflammation. J. Biol. Chem. 260, pp. 12060–12064.
79. Cole, T., Inglis, A.S., Roxburgh, C.M., Howlett, G.J. & Schreiber G. (1985). Major acute phase α_1-protein of the rat is homologous to bovine kininogen and contains the sequence for bradykinin: Its synthesis is regulated at the mRNA level. FEBS Lett. 182, pp. 57–61.
80. Anderson, K.P. & Heath, E.C. (1985). The relationship between rat major acute phase protein and the kininogens. J. Biol. Chem. 260, pp. 12065–12071.
81. Anderson, K.P., Croyle, M.L. & Lingrel, J.B. (1989). Primary structure of a gene

encoding rat T-kininogen. Gene 81, pp. 119–128.
82. Shafrir, E. & de Vries, A. (1955). Studies on the clot promoting activity of glass. J. Clin. Invest. 35, pp. 1183–1190.
83. Margolis, J. (1956). Glass surface and blood coagulation. Nature 178, pp. 805–806.
84. Margolis, J. (1957). Initiation of blood coagulation by glass and related surfaces. J. Physiol. (London) 137, pp. 95–109.
85. Margolis, J. (1958). The kaolin clotting time. J. Clin. Pathol. 11, pp. 406–409.
86. Ratnoff, O.D. & Crum, J.D. (1964). Activation of Hageman factor by solutions of ellagic acid. J. Lab. Clin. Med. 63, pp. 359–377.
87. Fujikawa, K., Heimark, R.L., Kurachi, K. & Davie, E.W. (1980). Activation of bovine factor XII (Hageman factor) by plasma kallikrein. Biochemistry 19, pp. 1322–1330.
88. Kluft, C. (1978). Determination of prekallikrein in human plasma: Optimal conditions for activating prekallikrein. J. Lab. Clin. Med. 91, pp. 83–95.
89. Griep, M.A., Fujikawa, K. & Nelsestuen, G.L. (1986). Possible basis for the apparent surface selectivity of the contact activation of human blood coagulation factors. Biochemistry 25, pp. 6688–6694.
90. Mandel, R.J., Colman, R.W. & Kaplan, A.P. (1976). Identification of prekallikrein and high-molecular-weight kininogen as a complex in human plasma. Proc. Natl. Acad. Sci. USA 73, pp. 4179–4183.
91. Thompson, R.E., Mandel, R. & Kaplan, A.P. (1977). Association of factor XI and high molecular weight kininogen in human plasma. J. Clin. Invest. 60, pp. 1376–1380.
92. Warn-Cramer, B.J. & Bajaj, S.P. (1985). Stoichiometry of binding of high molecular weight kininogen to factor XI/XIa. Biochem. Biophys. Res. Commun. 138, pp. 417–422.
93. Thompson, R.E., Mandel, R. & Kaplan, A.P. (1979). Studies of binding of prekallikrein and factor XI to high molecular weight kininogen and its light chain. Proc. Natl. Acad. Sci. USA 76, pp. 4862–4866.
94. Kerbiriou, D.M., Bouma, B.N. & Griffin, J.H. (1980). Immunochemical studies of human high molecular weight kininogen and of its complex with plasma prekallikrein and kallikrein. J. Biol. Chem. 255, pp. 3952–3958.
95. Bock, P.E. & Shore, J.D. (1983). Protein-protein interactions in contact activation of blood coagulation. J. Biol. Chem. 258, pp. 15079–15086.
96. Shimade, T., Kato, H., Maeda, H. & Iwanaga, S. (1985). Interaction of factor XII, high-molecular-weight (HMW) kininogen and prekallikrein with sulfatide: Analysis by fluorescence polarization. J. Biochem. (Tokyo) 97, pp. 1637–1644.
97. Bock, P.E., Shore, J.D., Tans, G. & Griffin, J.H. (1985). Protein-protein interactions in contact activation of blood coagulation. J. Biol. Chem. 260, pp. 12434–12443.
98. Tait, J.F. & Fujikawa, K. (1986). Identification of the binding site for plasma prekallikrein in human high molecular weight kininogen. J. Biol. Chem. 261, pp. 15396–15401.
99. Tait, J.F. & Fujikawa, K. (1987). Primary structure requirements for the binding of human high molecular weight kininogen to plasma prekallikrein and factor XI. J. Biol. Chem. 262, pp. 11651–11656.
100. Thompson, R.E., Mandel, R. & Kaplan, A.P. (1978). Characterization of human high molecular weight kininogen. J. Exp. Med. 147, pp. 488–499.
101. Ikari, N., Sugo, T., Fujii, S., Kato, H. & Iwanaga, S. (1981). The role of bovine high-molecular-weight (HMW) kininogen in contact mediated activation of bovine factor XII: Interaction of HMW kininogen with kaolin and plasma prekallikrein. J. Biochem. (Tokyo) 89, pp. 1699–1709.
102. van der Graaf, F., Tans, G., Bouma, B.N. & Griffin, J.H. (1982). Isolation and functional properties of the heavy and light chains of human plasma kallikrein. J. Biol. Chem. 257, pp. 14300–14305.
103. van der Graaf, F., Greengard, J.S., Bouma, B.N., Kerbiriou, D.M. & Griffin, J.H. (1983). Isolation and functional characterization of the active light chain of activated human blood coagulation factor XI. J. Biol. Chem. 258, pp. 9669–9675.
104. Revak, S.D. & Cochrane, C.G. (1976). The relationship of structure and function in human Hageman factor. J. Clin. Invest. 57, pp. 852–860.
105. Pixley, R.A., Stumpo, L.G., Birkmeyer, K., Silver, L. & Colman, R.W. (1987). A

monoclonal antibody recognizing an icosapeptide sequence in the heavy chain of human factor XII inhibits surface catalyzed activation. J. Biol. Chem. 262, pp. 10140–10145.

106. Clarke, B.J., Côté, H.C.F., Cool, D.E., Clark-Lewis, I., Saito, H., Pixley, R.A., Colman, R.W. & MacGillivray, R.T.A. (1989). Mapping of a putative surface-binding site of human coagulation factor XII. J. Biol. Chem. 264, pp. 11497–11502.
107. Heimark, R.L., Kurachi, K., Fujikawa, K. & Davie, E.W. (1980). Surface activation of blood coagulation, fibrinolysis and kinin formation. Nature 286, pp. 456–460.
108. Silverberg, M., Dunn, J.T., Garen, L. & Kaplan, A.P. (1980). Autoactivation of human Hageman factor. J. Biol. Chem. 255, pp. 7281–7286.
109. Tans, G., Rosing, J., Berrettini, M., Lämmle, B. & Griffin, J.H. (1987). Autoactivation of human plasma prekallikrein. J. Biol. Chem. 262, pp. 11308–11314.
110. Griffin, J.H. (1978). Role of surface in surface dependent activation of Hageman factor (blood coagulation factor XII). Proc. Natl. Acad. Sci. USA 75, pp. 1998–2002.
111. Nuijens, J.H., Huijbregts, C.C.M., Eerenberg-Belmer, A.J.M., Meijers, J.C.M., Bouma, B.N. & Hack, C.E. (1989). Activation of the contact system of coagulation by a monoclonal antibody directed against a neodeterminant in the heavy chain region of human coagulation factor XII (Hageman factor). J. Biol. Chem. 264, pp. 12941–12949.
112. Revak, S.D., Cochrane, C.G. & Griffin, J.H. (1977). The binding and cleavage characteristics of human Hageman factor during contact activation. J. Clin. Invest. 59, pp. 1157–1175.
113. Fujikawa, K., Kurachi, K. & Davie, E.W. (1977). Characterization of bovine factor XIIa (activated Hageman factor). Biochemistry 16, pp. 4182–4188.
114. Beretta, G. & Griffin, J.H. (1979). DFP studies on the mechanism of surface-dependent reactions of Hageman factor (HF, factor XII). Fed. Proc. 38, p. 811.
115. Scott, C.F., Silver, L.D., Schapira, M. & Colman, R.W. (1984). Cleavage of human high molecular weight kininogen markedly enhances its coagulant activity. J. Clin. Invest. 73, pp. 954–962.
116. Tans, G., Rosing, J. & Griffin, J.H. (1983). Sulfatide dependent activation of human blood coagulation factor XII (Hageman factor). J. Biol. Chem. 258, pp. 8215–8222.
117. Sugo, T., Kato, H., Iwanaga, S., Takada, K. & Sakakibara, S. (1985). Kinetic studies on surface mediated activation of bovine factor XII and prekallikrein. Eur. J. Biochem. 146, pp. 43–50.
118. Aggeler, P.M., White, S.G., Glendening, M.B., Page, E.W., Leake, T.B. & Bates, G. (1952). Plasma thromboplastin component (PTC) deficiency: A new disease resembling hemophilia. Proc. Soc. Exp. Biol. Med. 79, pp. 692–694.
119. Biggs, R., Douglas, A.S., Macfarlane, R.G., Dacie, J.V., Pitney, W.R., Merskey, C. & O'Brien, J.R. (1952). Christmas disease: A condition previously mistaken for hemophilia. Br. Med. J. 2, pp. 1378–1382.
120. Thompson, A.R. (1986). Structure, function and molecular defects of factor IX. Blood 67, pp. 565–572.
121. Thompson, A.R. (1977). Factor IX antigen by radioimmunoassay. Abnormal factor IX protein in patients on warfarin therapy and with hemophilia B. J. Clin. Invest. 59, pp. 900–910.
122. Österud, B., Bouma, B.N. & Griffin, J.H. (1978). Human blood coagulation factor IX. J. Biol. Chem. 253, pp. 5946–5951.
123. DiScipio, R.G., Kurachi, K. & Davie, E.W. (1978). Activation of human factor IX (Christmas factor). J. Clin. Invest. 61, pp. 1528–1538.
124. Katayama, K., Ericsson, L.H., Enfield, D.L., Walsh, K.A., Neurath, H., Davie, E.W. & Titani, K. (1979). Comparison of amino acid sequence of bovine coagulation factor IX (Christmas factor) with that of other vitamin K-dependent plasma proteins. Proc. Natl. Acad. Sci. USA 76, pp. 4990–4994.
125. Kurachi, K. & Davie, E.W. (1982). Isolation and characterization of a cDNA clone for human factor IX. Proc. Natl. Acad. Sci. USA 79, pp. 6461–6464.
126. Jaye, M., de la Salle, H., Schamber, F., Ballaud, A., Kohli, V., Findeli, A., Tolstoshev, P. & Lecocq, J.-P. (1983). Isolation of a human anti-haemophilic factor IX cDNA clone using a unique 52-base synthetic oligonucleotide probe deduced from the amino acid sequence of bovine factor IX. Nucl. Acids Res. 11, pp. 2325–2335.

127. Anson, D.S., Choo, K.H., Rees, D.J.G., Gianelli, F., Gould, K., Huddleston, J.A. & Brownlee, G.G. (1984). The gene structure of human anti-haemophilic factor IX. EMBO J. 3, pp. 1053–1060.
128. McGraw, R.A., Davis, L.M., Noyes, C.M., Lundblad, R.L., Roberts, H.R., Graham, J.B. & Stafford, D.W. (1985). Evidence for a prevalent dimorphism in the activation peptide of human coagulation factor IX. Proc. Natl. Acad. Sci. USA 82, pp. 2847–2851.
129. Evans, J.P., Watzke, H.H., Ware, J.L., Stafford, D.W. & High, K.A. (1989). Molecular cloning of a cDNA encoding canine factor IX. Blood 74, pp. 207–212.
130. Yoshitake, S., Schach, B.G., Foster, D.C., Davie, E.W. & Kurachi, K. (1985). Nucleotide sequence of the gene for human factor IX (antihemophilic factor B). Biochemistry 24, pp. 3736–3750.
131. Chance, P.F., Dyer, K.A., Kurachi, K., Yoshitake, S., Ropers, H.H., Wieacker, P. & Gartler, S.M. (1983). Regional localization of the human factor IX gene by molecular hybridization. Hum. Genet. 65, pp. 207–208.
132. Boyd, Y., Buckle, V.J., Munro, E.A., Choo, K.H., Migeon, B.R. & Craig, I.W. (1984). Assignment of the hemophilia B (factor IX) locus to the q26–qter region of the X chromosome. Ann. Hum. Genet. 48, pp. 145–152.
133. Camerino, G., Grzeschik, K.H., Jaye, M., de la Salle, H., Tolstoshev, P., Lecocq, J.-P., Heilig, R. & Mandel, J.L. (1984). Regional localization on the human X chromosome and polymorphism of the coagulation factor IX gene (hemophilia B locus). Proc. Natl. Acad. Sci. USA 81, pp. 498–502.
134. Buckle, V., Craig, I.W., Hunter, D. & Edwards, J.H. (1985). Fine assignment of the coagulation factor IX gene. Cytogenet. Cell Genet. 40, p. 593.
135. Jagadeeswaran, P., Lavelle, D.E., Kaul, R., Mohandas, T. & Warren, S.T. (1984). Isolation and characterization of human factor IX cDNA: Identification of Taq1 polymorphism and regional assignment. Somat. Cell Mol. Genet. 10, pp. 465–473.
136. Mattei, M.G., Baeteman, M.A., Heilig, R., Oberlé, I., Davies, K., Mandel, J.L. & Mattei, J.F. (1985). Localization by in situ hybridization of the coagulation factor IX gene and of two polymorphic DNA probes with respect to the fragile X site. Hum. Genet. 69, pp. 327–331.
137. Hase, S., Kawabata, S., Nishimura, H., Takeya, H., Sueyoshi, T., Miyata, T., Iwanaga, S., Takao, T., Shimonishi, Y. & Ikenaka, T. (1988). A new trisaccharide sugar chain linked to a serine residue in bovine blood coagulation factors VII and IX. J. Biochem. (Tokyo) 104, pp. 867–868.
138. Nishimura, H., Kawabata, S., Kisiel, W., Hase, S., Ikenaka, T., Takao, T., Shimonishi, Y. & Iwanaga, S. (1989). Identification of a disaccharide (Xyl-Glc) and a trisaccharide in the first epidermal growth factor-like domain of human factors VII and IX and protein Z and bovine protein Z. J. Biol. Chem. 264, pp. 20320–20335.
139. McMullen, B.A., Fujikawa, K. & Kisiel, W. (1983). The occurence of β-hydroxyaspartic acid in the vitamin K-dependent blood coagulation zymogens. Biochem. Biophys. Res. Commun. 115, pp. 8–14.
140. Sugo, T., Fernlund, P. & Stenflo, J. (1984). Erythro-β-hydroxyaspartic acid in bovine factor IX and factor X. FEBS Lett. 165, pp. 102–106.
141. Fernlund, P. & Stenflo, J. (1983). β-Hydroxyaspartic acid in vitamin K-dependent proteins. J. Biol. Chem. 258, pp. 12509–12512.
142. Bajaj, S.P., Rapaport, S.I. & Russell, W.A. (1983). Redetermination of the rate-limiting step in the activation of factor IX by factor XIa and by factor VIIa/tissue factor. Explanation for different electrophoretic radioactivity profiles obtained on activation of ^{3}H-and ^{125}I-labeled factor IX. Biochemistry 22, pp. 4047–4053.
143. McGraw, R.A., Davis, L.M., Lundblad, R.L., Stafford, D.W. & Roberts, H.R. (1985). Structure and function of factor IX: Defects in haemophilia B. Clin. Haematol. 14, pp. 359–383.
144. Brownlee, G.G. (1986). The molecular genetics of haemophilia A and B. J. Cell Sci. Suppl. 4, pp. 445–458.
145. Brownlee, G.G. (1987). The molecular pathology of haemophilia B. Biochem. Soc. Trans. 15, pp. 1–8.
146. Reitsma, P.H., Bertina, R.M., van Amstel, J.K.P., Riemens, A & Briët, E. (1988). The putative factor IX gene promotor in hemophilia B Leyden. Blood 72, pp. 1074–1076.

147. Reitsma, P.H., Mandalaki, T., Kasper, G.K., Bertina, R.M. & Briët, E. (1989). Two novel point mutations correlate with an altered developmental expression of blood coagulation factor IX (hemophilia B Leyden phenotype). Blood 73, pp. 743–746.
148. Koeberl, D.D., Bottema, C.D.K., Buerstedde, J.M. & Sommer, S.S. (1989). Functionally important regions of the factor IX gene have a low rate of polymorphism and a high rate of mutation in the dinucleotide CpG. Am. J. Hum. Genet. 45, pp. 448–457.
149. Green, P.M., Bentley, D.R., Mibashan, R.S., Nilsson, I.M. & Gianelli, F. (1989). Molecular pathology of haemophilia B. EMBO J. 8, pp. 1067–1072.
150. Bentley, A.K., Rees, D.J.G., Rizza, C. & Brownlee, G.G. (1986). Defective propeptide processing of blood clotting factor IX caused by mutation of arginine to glutamine at position –4. Cell 45, pp. 343–348.
151. Ware, J., Diuguid, D.L., Liebman, H.A., Rabiet, M.-J., Kasper, C.K., Furie, B.C., Furie, B. & Stafford, D.W. (1989). Factor IX San Dimas. J. Biol. Chem. 264, pp. 11401–11406.
152. Liddell, M.B., Lillicrap, D.P., Peake, I.R. & Bloom, A.L. (1989). Defective propeptide processing and abnormal activation underlie the molecular pathology of factor IX Troed-Y-Rhiw. Br. J. Haematol. 72, pp. 208–215.
153. Sugimoto, M., Miyata, T., Kawabata, S., Yoshioka, A., Fukui, H. & Iwanaga, S. (1989). Factor IX Kawachinagano: Impaired function of the Gla-domain caused by attached propeptide region due to the substitution of arginine by glutamine at position –4. Br. J. Haematol. 72, pp. 216–221.
154. Diguid, D.L., Rabiet, H.J., Furie, B.C., Liebman, H.A. & Furie, B. (1986). Molecular basis of hemophilia B: A defective enzyme due to an unprocessed propeptide is caused by a point mutation in the factor IX precursor. Proc. Natl. Acad. Sci. USA 83, pp. 5803–5807.
155. Winship, P.R. (1989). Characterization of the molecular defect in haemophilia B patients using the polymerase chain reaction procedure. Thromb. Haemostas. 62, Abstract SY-XXI-4.
156. Chen, S.-H., Thompson, A.R., Zhang, M. & Scott, C.R. (1989). Three point mutations in the factor IX genes of five hemophilia B patients. J. Clin. Invest. 84, pp. 113–118.
157. Davis, L.M., MacGraw, R.M., Ware, J.L., Ruberts, H.R. & Stafford, D.W. (1987). Factor $IX_{Alabama}$: A point mutation in a clotting protein results in hemophilia B. Blood 69, pp. 140–143.
158. Lozier, J.N., Stanfield-Oakley, S.A. & High, K.A. (1989). Factor $IX_{New\ London}$: A point mutation causing hemophilia B. Thromb. Haemostas. 62, Abstract 483.
159. Spitzer, S., Katzman, D., Kasper, C. & Bajaj, S.P. (1989). Factor $IX_{Hollywood}$: Substitution of 55 Pro to Ala in the first EGF domain. Thromb. Haemostas. 62, Abstract 617.
160. Denton, P.H., Fowlkes, D.M., Lord, S.T. & Reisner, H.M. (1988). Hemophilia B_{Durham}: A mutation in the first EGF-like domain of factor IX that is characterized by polymerase chain reaction. Blood 72, pp. 1407–1411.
161. Noyes, C.M., Griffith, M.J., Roberts, H.R. & Lundblad, R.L. (1983). Identification of the molecular defect in factor $IX_{Chapel\ Hill}$: Substitution of histidine for arginine at position 145. Proc. Natl. Acad. Sci. USA 80, pp. 4200–4202.
162. Green, P.M., Montandon, A.J., Bentley, D.R., Ljung, R., Nilsson, I.M. & Gianelli, F. (1990). The incidence and distribution of CpG→TpG transitions in the coagulation Factor IX gene. A fresh look at CpG mutational hotspots. *Nucl. Acids Res.* 18, pp. 3227–3231.
163. Diguid, D.L., Rabiet, M.-J., Furie, B.C. & Furie, B. (1989). Molecular defects of factor $IX_{Chicago-2}$ (Arg145 → His) and $prothrombin_{Madrid}$ (Arg271 → Cys): Arginine mutations that preclude zymogen activation. Blood 74, pp. 193–200.
164. Toomey, J.R., Stafford, D. & Smith, K. (1989). Factor IX Albuquerque (arginine 145 to cysteine) is cleaved slowly by factor XIa and has reduced coagulant activity. Blood 72, p. 312a, Abstract 1158.
165. Liddell, M.B., Peaker, I.R., Taylor, S.A.M., Lillicrap, D.P., Giddings, J.C. & Bloom, A.L. (1989). Factor IX Cardiff: A variant factor IX protein that shows abnormal activation is caused by an arginine to cysteine substitution at position 145. Br. J. Haematol. 72, pp. 556–560.
166. Huang, M.-N., Kasper, C.K., Roberts, H.R., Stafford, D.W. & High, K.A. (1989).

Molecular defect in factor IX_{Hilo}, a hemophilia B_m variant: Arg → Gln at the carboxyterminal cleavage site of the activation peptide. Blood 73, pp. 718–721.

167. Bertina, R.M., van der Linden, I.K., Mannucci, P.M., Reinalda-Poot, H.H., Cupers, R., Poort, S.R. & Reitsma, P.H. (1989). Mutations in haemophilia B_m occur at the 180Arg-Val activation site or in the catalytic domain. Thromb. Haemostas. 62, Abstract 618.
168. Suehiro, K., Kawabata, S., Miyata, T., Takamatsu, J., Ogata, K., Kamiya, T., Saito, H., Niho, Y. & Iwanaga, S. (1989). Blood coagulation factor IX B_m Nagoya: Substitution of arginine 180 by trypotophan and its activation by chymotrypsin. Thromb. Haemostas. 62, Abstract 482.
169. Sakai, T., Yoshioka, A., Yamamoto, K., Niiomi, K., Fujimura, Y., Fukui, H., Miyata, T. & Iwanaga, S. (1989). Blood clotting factor IX Kashihara: Amino acid substitution of valine-182 by phenylalanine. J. Biochem. (Tokyo) 105, pp. 756–759.
170. Siguret, V., Anselem, S., Vidaud, M., Assouline, Z., Kerbiriou-Nabias, D., Piétu, G., Goosens, M., Larrien, M.J., Bahnak, B., Meyer, D. & Lavergne, J.M. (1988). Identification of a CpG mutation in the coagulation factor-IX gene by analysis of amplified DNA sequences. Br. J. Haematol. 70, pp. 411–416.
171. Chen, S.-H., Scott, C.R., Schoof, J., Lovrien, E.E. & Kurachi, K. (1989). Factor $IX_{Portland}$: A nonsense mutation (C̲GA to A) resulting in hemophilia. Am. J. Hum. Genet. 44, pp. 567–569.
172. Tsang, T.C., Bentley, D.R., Mibashan, R.S. & Gianelli, F. (1988). A factor IX mutation, verified by direct genomic sequencing, causes haemophilia B by a novel mechanism. EMBO J. 7, pp. 3009–3015.
173. Ludwig, M., Schwaab, R., Eigel, A., Horst, J., Egli, H., Brackmann, H.-H. & Olek, K. (1989). Identification of a single nucleotide C-to-T transition and five different deletions in patients with severe hemophilia B. Am. J. Hum. Genet. 45, pp. 115–122.
174. Driscoll, M.C., Bouhassira, E. & Aledort, L.M. (1989). A codon 338 nonsense mutation in the factor IX gene in unrelated hemophilia B patients. Factor $IX^{338}_{New\ York}$. Blood 74, pp. 737–742.
175. Spitzer, S.G., Pendurthi, U.R., Kasper, C.K. & Bajaj, S.P. (1988). Molecular defect in factor $IX_{Bm\ Lake\ Elsinore}$. J. Biol. Chem. 263, pp. 10545–10548.
176. Sugimoto, M., Miyata, T., Kawabata, S., Yoshioka, A., Fukui, H., Takahashi, H. & Iwanaga, S. (1988). Blood clotting factor IX Niigata: Substitution of alanine-390 by valine in the catalytic domain. J. Biochem. (Tokyo) 104, pp. 878–880.
177. Attree, O., Vidaud, D., Vidaud, M., Anselem, S., Lavergne, J.-M. & Goossens, H. (1989). Mutations in the catalytic domain of human coagulation factor IX: Rapid characterization by direct genomic sequencing of DNA fragments displaying an altered melting behavior. Genomics 4, pp. 266–272.
178. Geddes, V.A., Le Bonniec, B.F., Louie, G.V., Brayer, G.D., Thompson, A.R. & MacGillivray, R.T.A. (1989). A moderate form of hemophilia B is caused by a novel mutation in the protease domain of factor $IX_{Vancouver}$. J. Biol. Chem. 264, pp. 4689–4697.
179. Ware, J., Davis, L., Frazier, D., Bajaj, S.P. & Stafford, D.W. (1988). Genetic defect responsible for the dysfunctional protein: Factor $IX_{Long\ Beach}$. Blood 72, pp. 820–822.
180. Rees, D.J.G., Rizza, C.R. & Brownlee, G.G. (1985). Haemophilia B caused by a point mutation in a donor splice junction of the human factor IX gene. Nature 316, pp. 643–645.
181. Hassan, H.J., Leonardi, A., Guerriero, R., Chelucci, C., Cianetti, L., Ciavarella, N., Ranieri, P., Picolli, D. & Peschle, C. (1985). Hemophilia B with inhibitor, molecular analysis of the subtotal deletion of the factor IX gene. Blood 66, pp. 728–730.
182. Chen, S.-H., Yoshitake, S., Chance, P.F., Bray, G.L., Thompson, A.R., Scott, C.R. & Kurachi, K. (1985). An intragenic deletion of the factor IX gene in a family with hemophilia B. J. Clin. Invest. 76, pp. 2161–2164.
183. Vidaud, M., Chabret, C., Gazengel, C., Grunenbaum, L. Cazenave, J.P. & Goossens, M. (1986). A de novo intragenic deletion of the potential EGF domain of the factor IX gene in a family with severe hemophilia B. Blood 68, pp. 961–963.
184. Matthews, R.J., Anson, D.S., Peake, I.R. & Bloom, A.L. (1987). Heterogeneity of factor IX locus in 9 hemophilia B patients. J. Clin. Invest. 79, pp. 746–753.
185. Mikami, S., Nishino, M., Nishimura, T. & Fukki, H. (1987). RFLPs of factor IX gene in

japanese haemophilia B families and gene deletion in two high-responder-inhibitor patients. Japan. J. Human. Genet. 32, pp. 21–31.
186. Ludwig, M., Schwaab, R., Brackmann, H.H., Egli, H. & Olek, K. (1987). Molecular defects of the factor IX gene causing severe haemophilia B. Thromb. Haemostas. 58, Abstract 1277.
187. Schach, B.G., Yoshitake, S. & Davie, E.W. (1987). Hemophilia B (Factor $IX_{Seattle\ 2}$) due to a single nucleotide deletion in the gene for factor IX. J. Clin. Invest. 80, pp. 1023–1028.
188. Tanimoto, M., Kojima, T., Kamiya, T., Takamatsu, J., Ogata, K., Obata, Y., Inagaki, M., Iizuka, A., Nagao, T., Kurachi, K. & Saito, H. (1988). DNA analysis of seven patients with hemophilia B who have anti-factor IX antibodies: Relationship to clinical manifestations and evidence that the abnormal gene was inherited. J. Lab. Clin. Med. 112, pp. 307–313.
189. Green, P.M., Bentley, D.R., Mibashan, R.S. & Gianelli, F. (1988). Partial deletion by illegitimate recombination of the factor IX gene in haemophilia B family with two inhibitor patients. Mol. Biol. Med. 5, pp. 95–106.
190. Wadelius, C., Blombäck, M. & Petterson, U. (1988). Molecular studies of haemophilia B in Sweden. Hum. Genet. 81, pp. 13–17.
191. Chen, S.-H., Scott, C.R., Edson, J.R. & Kurachi, K. (1988). An insertion within the factor IX gene: Hemophilia $B_{El\ Salvador}$. Am. J. Hum. Genet. 25, pp. 581–584.
192. Veltkamp, J.J., Meilof, J., Remmelts, H.G., van der Klerk, D. & Loeliger, E.A. (1970). Another genetic variant of haemophilia B: Haemophilia B Leyden. Scand. J. Haemat. 7, pp. 82–90.
193. Briët, E., Bertina, R.M., Van Tilburg, N.H. & Veltkamp, J.J. (1982). Hemophilia B Leyden: A sex-linked hereditary disorder that improves after puberty. N. Eng. J. Med. 306, pp. 788–790.
194. Briët, E., Van Tilburg, N.H. & Veltkamp, J.J. (1978). Oral contraception and the detection of carriers in haemophilia B. Thromb. Res. 13, pp. 379–388.
195. Briët, E., Wijnands, M.C. & Veltkamp, J.J. (1985). The prophylactic treatment of haemophilia B Leyden with anabolic steroids. Ann. Intern. Med. 103, pp. 225–226.
196. Bray, G.L. & Thompson, A.R. (1986). Partial factor IX protein in pedigree with hemophilia B to a partial gene deletion. J. Clin. Invest. 77, pp. 1194–1200.
197. Evans, J.P., Brinkhouse, K.M., Brayer, G.D., Reisner, H.M. & High, K.A. (1989). Canine hemophilia B resulting from a point mutation with unusual consequences. Proc. Natl. Acad. Sci. USA 86, pp. 10095–10099.
198. Bajaj, S.P. (1982). Cooperative Ca^{2+} binding to human factor IX. J. Biol. Chem. 257, pp. 4127–4132.
199. Walsh, P.N., Bradford, H., Sinha, D., Piperno, J.R. & Tuszynski, G.P. (1984). Kinetics of the factor XIa catalyzed activation of human blood coagulation factor IX. J. Clin. Invest. 73, pp. 1392–1399.
200. Sinha, D., Koshy, A., Seaman, F.S. & Walsh, P.N. (1985). Functional characterization of human blood coagulation factor XIa using hybridoma antibodies. J. Biol. Chem. 260, pp. 10714–10719.
201. Sinha, D., Seaman, F.S. & Walsh, P.N. (1987). Role of calcium ions and the heavy chain of factor XIa in the activation of human coagulation factor IX. Biochemistry 26, pp. 3738–3775.
202. Baglia, F.A., Sinha, D. & Walsh, P.N. (1989). Functional domains in the heavy-chain region of factor XI: A high molecular weight kininogen-binding site and a substrate-binding site for factor IX. Blood 74, pp. 244–251.
203. Liebman, H.A., Furie, B.C. & Furie, B. (1987). The factor IX phospholipid-binding site is required for calcium-dependent activation of factor IX by factor XIa. J. Biol. Chem. 262, pp. 7605–7612.
204. Amphlett, G.W., Kisiel, W. & Castellino, F.J. (1981). The interaction of Ca^{2+} with human factor IX. Arch. Biochem. Biophys. 208, pp. 576–585.
205. Amphlett, G.W., Byrne, R. & Castellino, F.J. (1978). The binding of metal ions to bovine factor IX. J. Biol. Chem. 253, pp. 6774–6779.
206. Morita, T., Isaacs, B.S., Esmon, C.T. & Johnson, A.E. (1984). Derivatives of blood coagulation factor IX contain a high affinity Ca^{2+} binding site that lacks γ-

carboxyglutamic acid. J. Biol. Chem. 259, pp. 5698–5704.
207. Morita, T., Isaacs, B.S., Esmon, C.T. & Johnson, A.E.(1985). J. Biol. Chem. 260, p. 2583. (Correction to reference (206)).
208. Morita, T. & Kisiel, W. (1985). Calcium binding to human factor IXa derivative lacking γ–carboxyglutamic acid: Evidence for two high-affinity sites that do not involve β-hydroxyaspartic acid. Biochem. Biophys. Res. Commun. 130, pp. 841–847.
209. Huang, L.H., Ke, X.-H., Sweeney, W. & Tam, J.P. (1989). Calcium binding and putative activity of the epidermal growth factor domain of blood coagulation factor IX. Biochem. Biophys. Res. Commun. 160, pp. 133–139.
210. Rees, D.J.G., Jones, I.M., Handford, P.A., Walter, S.J., Esnouf, M.P., Smith, K.J. & Brownlee, G.G. (1988). The role of β-hydroxyaspartate and adjacent carboxylate residues in the first EGF domain of human factor IX. EMBO J. 7, pp. 2053–2061.
211. Stenflo, J., Holme, E., Lindstedt, S., Chandramouli, N., Huang, L.H.T., Tam, J.P. & Merrifield, R.B. (1989). Hydroxylation of aspartic acid in domains homologous to the epidermal growth factor precursor is catalyzed by a 2-oxoglutarate-dependent dioxygenase. Proc. Natl. Acad. Sci. USA 86, pp. 444–447.
212. Derian, C.K., VanDusen, W., Przysiecki, C.T., Walsh, P.N., Berkner, K.L., Kaufman, R.J. & Friedman, P.A. (1989). Inhibitors of 2-ketoglutarate-dependent dioxygenases block aspartyl β-hydroxylation of recombinant human factor IX in several mammalian expression systems. J. Biol. Chem. 264, pp. 6615–6618.
213. Gronke, R.S., VanDusen, W.J., Garsky, V.M., Jacobs, J.W., Sardana, M.K., Stern, A.M. & Friedman, P.A. (1989). Aspartyl β-hydroxylase: *In vitro* hydroxylation of a synthetic peptide based on the structure of the first growth factor-like domain of human factor IX. Proc. Natl. Acad. Sci. USA 86, pp. 3609–3613.
214. Fowler, S.A., Paulson, D., Owen, B.A. & Owen, W.G. (1986). Binding of iron by factor IX. J. Biol. Chem 261, pp. 4371–4372.
215. Mannhalter, C., Shiffman, S. & Deutsch, E. (1984). Phospholipids accelerate factor IX activation by surface bound factor XIa. Br. J. Haematol. 56, pp. 261–271.
216. Mannhalter, C. (1987). Biochemical and functional properties of factor XI and prekallikrein. Sem. Thromb. Haemostas. 13, pp. 25–35.
217. Warn-Cramer, B.J. & Bajaj, S.P. (1986). Intrinsic versus extrinsic coagulation. Biochem. J. 239, pp. 757–762.
218. Soons, H. (1987). Studies on the inhibition of blood coagulation factor XIa. PhD thesis, University of Limburg, Maastricht, the Netherlands, pp. 49–61.
219. Op den Kamp, J.A.F. (1979). Lipid asymmetry in membranes. Ann. Rev. Biochem. 48, pp. 47–71.
220. Chap, H.J., Zwaal, R.F.A. & van Deenen, L.L.M. (1977). Action of highly purified phospholipase on blood platelets. Evidence for an asymmetric distribution of phospholipids in the surface membrane. Biochem. Biophys. Acta 467, pp. 146–164.
221. Zimmerman, T.S., Ruggeri, Z.M. & Fulcher, C.A. (1983). Factor VIII/von Willebrand factor. Progress in Hematology 13, pp. 273–309.
222. Lawn, R.M. (1985). The molecular genetics of hemophilia: Blood clotting factors VIII and IX. Cell 42, pp. 405–406.
223. Eaton, D.L. & Vehar, G.A. (1986). Factor VIII structure and proteolytic processing. In: Progress in Hemostasis and Thrombosis Vol. 11 (Spaet, T.D., Ed.), pp. 47–70, Grune & Stratton, Orlando.
224. Fay, P., Chavin, S.I., Meyer, D. & Marder, V.J. (1986). Nonenzymatic cofactors: Factor VIII. In: New Comprehensive Biochemistry Vol 13 (Zwaal, R.F.A. & Hemker, H.C., Eds.), pp. 35–57, Elsevier, Amsterdam.
225. Kane, W.H. & Davie, E.W. (1988). Blood coagulation factors V and VIII: Structural and functional similarities and their relationship to hemorrhagic and thrombotic disorders. Blood 71, pp. 539–555.
226. Addis, T. (1911). Pathogenesis of hereditary haemophilia. J. Pathol. Bacteriol. 15, pp. 427–452.
227. Hoyer, L.W. (1981). The factor VIII complex: Structure and function. Blood 58, pp. 1–13.
228. Weiss, H.J., Sussman, I.I. & Hoyer, L.W. (1977). Stabilization of factor VIII in plasma by the von Willebrand factor. J. Clin. Invest. 60, pp. 390–404.

229. Vehar, G.A. & Davie, E.W. (1980). Preparation and properties of bovine factor VIII (antihemophilic factor). Biochemistry 19, pp. 401–410.
230. Fass, D.N., Knutzon, G.J. & Katzmann, J.A. (1982). Monoclonal antibodies to porcine factor VIII coagulant and their use in the isolation of active coagulant protein. Blood 59, pp. 594–600.
231. Knutzon, G.J. & Fass, D.N. (1982). Porcine factor VIII:C prepared by affinity interaction with von Willebrand factor and heterologous antibodies: Sodium dodecyl sulfate polyacrylamide gel analysis. Blood 59, pp. 615–624.
232. Fulcher, C.A. & Zimmerman, T.S. (1982). Characterization of the human factor VIII procoagulant protein with a heterologous precipitating antibody. Proc. Natl. Acad. Sci. USA 79, pp. 1648–1652.
233. Fay, P.J., Chavin, S.I., Schroeder, D., Young, F.E. & Marder, V.J. (1982). Purification and characterization of highly purified human factor VIII consisting of a single type of polypeptide chain. Proc. Natl. Acad. Sci. USA 79, pp. 7200–7204.
234. Fulcher, C.A., Roberts, J.R. & Zimmerman, T.S. (1983). Thrombin proteolysis of purified factor VIII procoagulant protein: Correlation of activation with generation of a specific polypeptide. Blood 61, pp. 807–811.
235. Gitschier, J., Wood, W.I., Goralka, T.M., Wion, K.L., Chen, E.Y., Eaton, D.H., Vehar, G.A., Capon, D.J. & Lawn, R.M. (1984). Characterization of the human factor VIII gene. Nature 312, pp. 326–330.
236. Wood, W.I., Capon, D.J., Simonsen, C.C., Eaton, D.H., Gitschier, J.J., Keyt, B., Seeburg, P.H., Smith, D.H., Hollingshead, P., Wion, K.L., Delwart, E., Tuddenham, E.G.D., Vehar, G.A. & Lawn, R.M. (1984). Expression of active human factor VIII from recombinant DNA clones. Nature 312, pp. 330–337.
237. Vehar, G.A., Keyt, B., Eaton, D., Rodriguez, H., O'Brien, D.P., Rotblat, F., Opperman, M., Keck, R., Wood, W.I., Harkins, R.N., Tuddenham, E.G.D., Lawn, R.M. & Capon, D.J. (1984). Structure of human factor VIII. Nature 312, pp. 337–342.
238. Toole, J.J., Knopf, J.L., Wozney, J.M., Sultzman, L.A., Buecker, J.L., Pittman, D.D., Kaufman, R.J., Brown, E., Shoemaker, C., Orr, E.C., Amphlett, G.W., Foster, W.B., Coe, M.L., Knutson, G.C., Fass, D.N. & Hewick, R.M. (1984). Molecular cloning of a cDNA encoding human antihaemophilic factor. Nature 312, pp. 342–347.
239. Truett, M.A., Blacher, R., Burke, R.L., Caput, D., Chu, C., Dina, D., Hartog, K., Kuo, C.H., Masiarz, F.R., Merryweather, J.P., Najarian, R., Pachl, C., Potter, S.J., Puma, J., Quiroga, M., Rall, L.B., Randolph, A., Urdea, M.S., Valenzuela, P., Dahl, H.H., Favalaro, J., Hansen, J., Nordfang, O. & Ezban, M. (1985). Characterization of the polypeptide composition of human factor VIII:C and the nucleotide sequence and expression of the human kidney cDNA. DNA 4, pp. 333–346.
240. Purello, M., Alhadeff, B., Esposito, D., Szabo, P., Rocchi, M., Truett, M., Masiarz., F. & Siniscalo, M. (1985). The human genes for hemophilia A and hemophilia B flank the X chromosome fragile site at Xq27.3. EMBO J. 4, pp. 725–729.
241. Church, W.R., Jernigan, R.L., Toole, J., Hewick, R.M., Knopf, J., Knutson, G.J., Nesheim, M.E., Mann, K.G. & Fass, D.N. (1984). Coagulation factors V and VIII and ceruloplasmin constitute a family of structurally related proteins. Proc. Natl. Acad. Sci. USA 81, pp. 6934–6937.
242. Fass, D.N., Hewick, R.M., Knutson, G.J., Nesheim, M.E. & Mann, K.G. (1985). Internal duplication and sequence homology in factors V and VIII. Proc. Natl. Acad. Sci. USA 82, pp. 1688–1691.
243. Takahashi, N., Ortel, T.L. & Putnam, F.W. (1984). Single-chain structure of human ceruloplasmin: The complete amino acid sequence of the whole molecule. Proc. Natl. Acad. Sci. USA 81, pp. 390–394.
244. Yang, F., Naylor, S.L., Lum, J.B., Cutshaw, S., McCombs, J.L., Naberhaus, K.H., McGill, J.R., Adrian, G.S., Moore, C.M., Barnett, D.R. & Bowman, B.H. (1986). Characterization, mapping, and expression of the human ceruloplasmin gene. Proc. Natl. Acad. Sci. USA 83, pp. 3257–3261.
245. Koschinshy, M.L., Funk, W.D., van Oost, B.A. & MacGillivray, R.T.A. (1986). Complete cDNA sequence of human proceruloplasmin. Proc. Natl. Acad. Sci. USA 83, pp. 5086–5090.
246. Royle, N.J., Irwin, D.M., Koschinsky, M.L., MacGillivray, R.T.A. & Hamerton, J.L.

(1987). Human genes encoding prothrombin and ceruloplasmin map to 11p11–q12 and 3q21–24, respectively. Somat. Cell Mol. Genet. 13, pp. 285–292.

247. Takahashi, N., Bauman, R.A., Ortel, T.L., Dwulet, F.E., Wang, C.-C. & Putnam, F. (1983). Internal triplication in the structure of human ceruloplasmin. Proc. Natl. Acad. Sci. USA 80, pp. 115–119.
248. Eaton, D., Rodriguez, H. & Vehar, G.A. (1986). Proteolytic processing of human factor VIII. Correlation of specific cleavages by thrombin, factor Xa, and activated protein C with activation and inactivation of factor VIII coagulant activity. Biochemistry 25, pp. 505–512.
249. Rotblat, F., O'Brien, D.P., O'Brien, F.J., Goodall, A.H. & Tuddenham, E.G.D. (1985). Purification of human factor VIII:C and its characterization by western blotting using monoclonal antibodies. Biochemistry 24, pp. 4294–4300.
250. Anderson, L.-O., Forsman, N., Huang, K., Larsen, K., Lundin, A., Paulu, B., Sandberg, H., Sewerin, K. & Smart, J. (1986). Isolation and characterization of human factor VIII: Molecular forms in commercial factor VIII concentrate, cryoprecipitate, and plasma. Proc. Natl. Acad. Sci. USA 83, pp. 2979–2983.
251. Fay, P.J., Anderson, M.T., Chavin, S.I. & Marder, V.J. (1986). The size of human factor VIII heterodimers and the effects produced by thrombin. Biochim. Biophys. Acta 871, pp. 268–278.
252. Hamer, R.J., Koedam, J.A., Beeser-Visser, N.H. & Sixma, J.J. (1986). Human factor VIII: Purification from commercial factor VIII concentrate, characterization, identification and radiolabeling. Biochim. Biophys. Acta 873, pp. 356–366.
253. Kaufman, R.J., Wasley, L.C. & Dorner, A.J. (1988). Synthesis, processing, and secretion of recombinant human factor VIII expressed in mammalian cells. J. Biol. Chem. 263, pp. 6352–6362.
254. Mikkelsen, J., Bayne, S. & Ezban, M. (1989). Characterization of N-linked carbohydrates on recombinant and plasma derived FVIII. Thromb. Haemostas. 62, abs 595.
255. Pittman, D.D., Wasley, L.C., Murray, B.L., Wang, J.H. & Kaufman, R.J. (1987). Analysis of structural requirements for factor VIII function using site directed mutagenesis. Thromb. Haemostas. 58, Abstract 1245.
256. Mikkelsen, J., Thomsen, J., Kongerslev, L., Christensen, M. & Ezban, M. (1989). Heterogeneity in the tyrosine sulfation of chinese hamster ovary cell produced recombinant FVIII. Thromb. Haemostas. 62, abs 594.
257. Fay, P.J. (1988). Reconstitution of human factor VIII from isolated subunits. Arch. Biochem. Biophys. 262, pp. 525–531.
258. Rapaport, S.I., Schiffman, S., Patch, M.J. & Ames, S.B. (1963). The importance of activation of antihemophilic globulin and proaccelerin by traces of thrombin in the generation of intrinsic prothrombinase activity. Blood 21, pp. 221–235.
259. Eaton, D.L., Hass, P.E., Riddle, L., Mather, J., Wieber, M., Gregory, T. & Vehar, G. (1987). Characterization of recombinant human factor VIII. J. Biol. Chem. 262, pp. 3285–3290.
260. Fay, P.J. (1987). Subunit structure of thrombin activated human factor VIIIa. Biochim. Biophys. Acta 952, pp. 181–190.
261. Harmon, J.T., Jamieson, G.A. & Rock, G.A. (1982). The functional molecular weights of factor VIII activities in whole plasma as determined by electron irradiation. J. Biol. Chem. 257, pp. 14245–14249.
262. Pittman, D.D. & Kaufman, R.J. (1988). Proteolytic requirements for thrombin activation of anti-hemophilic factor (factor VIII). Proc. Natl. Acad. Sci. USA 85, pp. 2429–2433.
263. Toole, J.J., Pittman, D.D., Orr, E.C., Murtha, P., Wasley, L.C. & Kaufman, R.J. (1986). A large region ($\approx$95 kDa) of human factor VIII is dispensable for *in vitro* procoagulant activity. Proc. Natl. Acad. Sci. USA 83, pp. 5939–5942.
264. Burke, R.L., Pachl, C., Quiroga, M., Rosenberg, S., Haigwood, N., Nordfang, O. & Ezban, M. (1986). The functional domains of coagulation factor VIII:C. J. Biol. Chem. 261, pp. 12574–12578.
265. Eaton, D.L., Wood, W.I., Eaton, D., Hass, P.E., Hollingshead, P., Wion, K., Mather, J., Lawn, R.M., Vehar, G.A. & Gorman, C. (1986). Construction and characterization of an active factor VIII variant lacking the central one-third of the molecule. Biochemistry 25, pp. 8343–8347.

266. Pavirani, A., Meulien, P., Harrer, H., Dott, K., Mischler, F., Wiesel, M.L., Mazurier, C., Cazenave, J.-P. & Lecocq, J.-P. (1987). Two independent domains of factor VIII coexpressed using recombinant vaccinia viruses have procoagulant activity. Biochem. Biophys. Res. Commun. 145, pp. 234–240.
267. Meulin, P., Faure, T., Mischler, F., Harrer, H., Ulrich, P., Bouderbata, B., Dott, K., Marie, M.S., Mazurier, C., Wiesel, M.-L., Van de Pol, H., Cazenave, J.-P., Courtney, M. & Pavirani, A. (1988). A new recombinant procoagulant protein derived from the cDNA encoding human factor VIII. Protein Engineering 2, pp. 301–306.
268. Dorner, A.J., Bole, D.G. & Kaufman, R.J. (1987). The relationship of N-linked glycosylation and heavy chain-binding protein association with the secretion of glycoproteins. J. Cell Biol. 105, pp. 2665–2674.
269. Hamer, R.J., Koedam, J.A., Beeser-Visser, N.H., Bertina, R.M., van Mourik, J.A. & Sixma, J.J. (1987). Factor VIII binds to von Willebrand factor via its M_r-80,000 light chain. Eur. J. Biochem. 166, pp. 37–43.
270. Foster, P.A., Fulcher, C.A., Houghten, R.A. & Zimmerman, T.S. (1988). An immunogenic region within residues Val^{1670}-Glu^{1684} of the factor VIII light chain induces antibodies which inhibit binding of factor VIII to von Willebrand factor. J. Biol. Chem. 263, pp. 5230–5234.
271. Leyte, A., Verbeet, M.P., Brooniewicz-Proba, T., van Mourik, J.A. & Mertens, K. (1989). The interaction between human blood coagulation factor VIII and von Willebrand factor. Biochem. J. 257, pp. 679–683.
272. Hamer, R.J., Koedam, J.A., Beeser-Visser, N.H. & Sixma, J.J. (1987). The effect of thrombin on the complex between factor VIII and von Willebrand factor. Eur. J. Biochem. 167, pp. 253–259.
273. Lollar, P., Knutson, G.J. & Fass, D.N. (1985). Activation of porcine factor VIII:C by thrombin and factor Xa. Biochemistry 24, pp. 8056–8064.
274. Lollar, P. & Parker, C.G. (1989). Subunit structure of thrombin-activated porcine factor VIII. Biochemistry 28, pp. 666–674.
275. Lollar, P., Hill-Eubanks, D.C. & Parker, C.G. (1988). Association of the factor VIII light chain with von Willebrand factor. J. Biol. Chem. 263, pp. 10451–10455.
276. Antonarakis, S.E., Youssoufian, H. & Kazazian Jr., H.H. (1987). Molecular genetics of hemophilia A in man (factor VIII deficiency). Mol. Biol. Med. 4, pp. 81–94.
277. Sadler, J.E. & Davie, E.W. (1987). Hemophilia A, hemophilia B, and von Willebrand disease. In: The Molecular Basis of Blood Diseases (Stamatoyannopoulos, G., Nienhuis, A.W., Leder, P. & Majerus, P.W., Eds.), pp. 575–630, W.B. Saunders, Philadelphia.
278. White II, G.C. & Shoemaker, C.B. (1989). Factor VIII gene and hemophilia A. Blood 73, pp. 1–12.
279. Antonarakis, S.E. (1988). The molecular genetics of hemophilia A and B in man. Adv. Hum. Genet. 17, pp. 27–59 & 201–203.
280. Hoyer, L.W. & Breckenridge, R.T. (1968). Immunologic studies of antihemophilic factor (AHF, factor VIII): Cross-reacting material in a genetic variant of hemophilia A. Blood 32, pp. 962–971.
281. Denson, K.W.E., Biggs, R., Haddon, M.E., Borrett, R. & Cobb, K. (1969). Two types of haemophilia (A^+ and A^-): A study of 48 cases. Br. J. Haematol. 17, pp. 163–171.
282. Youssoufian, H., Wong, C., Aronis, S., Platokoukis, H., Kazazian Jr., H.H. & Antonarakis, S.E. (1988). Moderately severe hemophilia A resulting from Glu → Gly substitution in exon 7 of the factor VIII gene. Am. J. Hum. Genet. 42, pp. 867–871.
283. Antonarakis, S.E., Waber, P.G., Kittur, S.D., Patel, A.S., Kazazian Jr., H.H., Mellis, M.A., Counts, R.B., Stamatoyannopoulos, G., Bowie, E.J.W., Fass, D.N., Pittman, D.D., Wozney, J.M. & Toole, J.J. (1985). Hemophilia A: Detection of molecular defects and of carriers by DNA analysis. N. Eng. J. Med. 313, pp. 842–848.
284. Higuchi, M., Kochman, L., Schwaab, R., Egli, H., Brackman, H.-H., Horst, J. & Olek, K. (1989). Molecular defects in hemophilia A: Identification and characterization of mutations in the factor VIII gene and family analysis. Blood 74, pp. 1045–1051.
285. Youssoufian, H., Kazazian Jr., H.H., Phillips, D.G., Aronis, S., Tsiftis, G., Brown, V.A. & Antonarakis, S.E. (1986). Recurrent mutation in haemophilia A give evidence for CpG mutation hot-spots. Nature 324, pp. 380–382.
286. Levinson, B., Janco, R., Phillips III, J. & Gitschier, J. (1987). A novel missense mutation

in the factor VIII gene identified by analysis of amplified hemophilia DNA sequences. Nucl. Acids Res. 15, pp. 9797–9805.

287. Youssoufian, H., Antonarakis, S.E., Bell, W., Griffin, A.M. & Kazazian Jr., H.H. (1988). Nonsense and missense mutations in hemophilia A: Estimate of the relative mutation rate at CG dinucleotides. Am. J. Hum. Genet 42, pp. 718–725.
288. Gitschier, J., Wood, W.I., Tuddenham, E.G.D., Shuman, M.A., Goralka, T.M., Chen, E.Y. & Lawn, R.M. (1985). Detection and sequence of mutations in the factor VIII gene of haemophiliacs. Nature 315, pp. 427–430.
289. Marchetti, G., Patracchini, P., Gegnani, C., Rodrigio, G., De Rosa, V. & Bernardi, F. (1989). Detection of missense mutations in exon 24 of factor VIII gene by polymerase chain reaction amplification. Thromb. Haemostas. 62, Abstract 612.
290. Bernardi, F., Legnani, C., Volinia, S., Patrascchini, P., Rodorigo, G., DeRosa, V. & Marchetti, G. (1988). A HindIII rflp and a gene lesion in the coagulation factor VIII gene. Hum. Genet. 78, pp. 359–362.
291. Gitschier, J., Wood, W.I., Shuman, M.A. & Lawn, R.M. (1986). Identification of a missense mutation in the factor VIII gene of a mild hemophilia. Science 232, pp. 1415–1416.
292. Inaba, H., Fujimaki, M., Kazazian Jr., H.H. & Antonarakis, S.E. (1989). Mild hemophilia A resulting from Arg-to-Leu substitution in exon 26 of the factor VIII gene. Hum. Genet. 81, pp. 335–338.
293. Youssoufian, H., Kazazian Jr., H.H., Patel, A., Pronis, S., Tsiftis, G., Hoyer, L.W. & Antonarakis, S.E. (1988). Mild hemophilia associated with cryptic donor splice site mutation in intron 4 of the factor VIII gene. Genomics 2, pp. 32–36.
294. Gitschier, J., Kogan, S., Levinson, G. & Tuddenham, E.G.D. (1988). Mutations of factor VIII cleavage sites in hemophilia A. Blood 72, pp. 1022–1028.
295. Shima, M., Ware, J., Yoshioka, A., Fukui, H. & Fulcher, C. (1989). An arginine to cysteine amino acid substitution at a critical thrombin cleavage site in a dysfunctional factor VIII molecule. Blood 74, pp. 1612–1617.
296. Arai, M., Inaga, H., Higuchi, M., Antonarakis, S.E., Kazazian Jr., H.H., Fujimaki, M. & Hoyer, L.W. (1989). Direct characterization of factor VIII in plasma: Detection of a mutation altering a thrombin cleavage site (Arginine-372 → histidine). Proc. Natl. Acad. Sci. USA 86, pp. 4277–4281.
297. Higuchi, M., Traystman, M., Wong, C., Olek, K., Kazazian Jr., H.H. & Antonarakis, S.E. (1989). Detection of point mutations in hemophilia A using PCR amplification of selected regions of the factor VIII gene. Thromb. Haemostas. 62, abs 610.
298. Arai, M., Higuchi, M., Antonarakis, S.E., Kazazian Jr., H.H., Fujimaki, M. & Hoyer, L.W. (1989). Characterization of the molecular and functional defects in CRM positive hemophilia A: Identification of thrombin cleavage site mutations in three patients. Thromb. Haemostas. 62, Abstract 480.
299. Youssoufian, H., Antonarakis, S.E., Aronis, S., Tsiftis, G., Phillips, D.G. & Kazazian Jr., H.H. (1987). Characterization of five partial deletions of the factor VIII gene. Proc. Natl. Acad. Sci. USA 84, pp. 3772–3776.
300. Bardoni, B., Sampietro, M., Romano, M., Crapanzano, M., Mannucci, P.M. & Camerino, G. (1988). Characterization of a partial deletion of the factor VIII gene in a hemophiliac with inhibitor. Hum. Genet. 79, pp. 86–88.
301. Casarino, L., Pecorara, M., Mori, P.G., Morfini, M., Mancuso, G., Scrivano, L., Molinari, A.C., Lanza, T., Giavarella, G., Loi, A., Perseu, L., Cao, A. & Pirastu, M. (1986). Molecular basis for hemophilia A in Italians. Res. Clin. Lab. 16, p. 227.
302. Din, N., Schwartz, M., Kruse, T., Vestergaard, S.R., Ahrens, P., Scheibel, E., Nordfang, O. & Ezban, M. (1986). Factor VIII gene specific probes used to study heritage and molecular defects in hemophilia A. Res. Clin. Lab. 16, p. 182.
303. Lillicrap, D., Taylor, S.A.M., Grover, H., Teitel, J., Giles, A.R., Holden, J.J.A. & White, B.N. (1986). Genetic analysis in hemophilia A. Identification of a large F.VIII gene deletion in a patient with high titre antibodies to human and porcine F.VIII. Blood 68, 337a.
304. Youssoufian, H., Patel, A., Phillips, D., Kazazian Jr., H.H. & Antonarakis, S.E. (1987). Hemophilia A: A recurrent mutation and an unusual deletion. Pediatr. Res. 21, 296a.
305. Barker, D., Schafer, M. & White, R. (1984). Restriction sites containing CpG show a higher frequency of polymorphism in human DNA. Cell 36, pp. 131–138.

306. Cooper, D.N. & Youssoufian, H. (1988). The CpG dinucleotide and human genetic disease. Hum. Genet. 78, pp. 151–155.
307. Kazazian Jr., H.H., Wong, C., Youssoufian, H., Scott, A.F., Phillips, D.G. & Antonarakis, S.E. (1988). Haemophilia A resulting from de novo insertion of L1 sequences represents a novel mechanism for mutation in man. Nature 332, pp. 164–166.
308. DiScipio, R.G., Hermodson, M.A., Yates, S.G. & Davie, E.W. (1977). A comparison of human prothrombin, factor IX (Christmas factor), factor X (Stuart factor), and protein S. Biochemistry 16, pp. 698–706.
309. Telfer, T.P., Denson, K.W. & Wright, D.R. (1956). A 'new' coagulation defect. Br. J. Haematol. 2, pp. 308–316.
310. Hougie, C., Barrow, E.M. & Graham, J.B. (1957). Stuart clotting defect. I. Segregation of an hereditary hemorrhagic state from the heterogeneous group heretofore called 'stable factor' (SPCA, proconvertin, factor VII) deficiency. J. Clin. Invest. 36, pp. 485–496.
311. Enfield, D.L., Ericsson, L.H., Walsh, K.A., Neurath, H. & Titani, K. (1975). Bovine factor X_1 (Stuart factor). Primary structure of the light chain. Proc. Natl. Acad. Sci. USA 72, pp. 16–19.
312. Enfield, D.L., Ericsson, L.H., Fujikawa, K., Walsh, K.A., Neurath, H. & Titani, K. (1980). Amino acid sequence of the light chain of bovine factor X_1 (Stuart factor). Biochemistry 19, pp. 659–667.
313. Titani, K., Fujikawa, K., Enfield, D.L., Ericsson, L.H., Walsh, K.A. & Neurath, H. (1975). Bovine factor X_1 (Stuart factor): Amino acid sequence of heavy chain. Proc. Natl. Acad. Sci. USA 72, pp. 3082–3086.
314. McMullen, B.A., Fujikawa, K., Kisiel, W., Sasagawa, T., Howald, W.N., Kwa, E.Y. & Weinstein, B. (1983). Complete amino acid sequence of the light chain of human coagulation factor X: Evidence for identification of residue 63 as β-hydroxyaspartic acid. Biochemistry 22, pp. 2875–2884.
315. Leytus, S.P., Chung, D.W., Kisiel, W., Kurachi, K. & Davie, E.W. (1984). Characterization of a cDNA coding for human factor X. Proc. Natl. Acad. Sci. USA 81, pp. 3699–3702.
316. Fung, M.R., Hay, C.W. & MacGillivray, R.T.A. (1985). Characterization of an almost full-length cDNA coding for human blood coagulation factor X. Proc. Natl. Acad. Sci. USA 82, pp. 3591–3595.
317. Leytus, S.P., Foster, D.C., Kurachi, K. & Davie, E.W. (1986). Gene for human factor X: A blood coagulation factor whose gene organization is essentially identical with that of factor IX and protein C. Biochemistry 25, pp. 5098–5102.
318. Fung, M.R., Campbell, R.M. & MacGillivray, R.T.A. (1984). Blood coagulation factor X mRNA encodes a single polypeptide chain containing a prepro leader sequence. Nucleic Acids Res. 12, pp. 4481–4492.
319. Kaul, R.K., Hildebrand, B., Roberts, S. & Jagadeeswaran, P. (1986). Isolation and characterization of human blood-coagulation factor X cDNA. Gene 41, pp. 311–314.
320. Jagadeeswaran, P., Rao, K.J. & Zhou, Z.-Q. (1989). Characterization of factor X deficiency mutation by use of inexpensive PCR device. Thromb. Haemostas. 62, Abstract 947.
321. Royle, N.J., Fung, M.R., MacGillivray, R.T.A. & Hamerton, J.L. (1986). The gene for clotting factor 10 is mapped to 13q32 $\rightarrow$ qter. Cytogenet. Cell Genet. 41, pp. 185–188.
322. Scrabler, P.J. & Williamson, R. (1985). The structural gene for human coagulation factor X is located on chromosome 13q34. Cytogenet. Cell Genet. 39, pp. 231–233.
323. Thøgersen, H.C., Petersen, T.E., Sottrup-Jensen, L., Magnusson, S. & Morris, H.R. (1978). The N-terminal sequences of blood coagulation factor X_1 and X_2 light chains. Biochem. J. 175, pp. 613–627.
324. Morita, T. & Jackson, C.M. (1986). Localization of the structural differences between coagulation factors X_1 and X_2 to tyrosine 18 in the activation peptide. J. Biol. Chem. 261, pp. 4008–4014.
325. Højrup, P. & Magnusson, S. (1987). Disulphide bridges of bovine factor X. Biochem. J. 245, pp. 887–892.
326. Girolami, A., de Marco, L., Dal Bo Zanon, R., Patrassi, G. & Cappellato, M.G. (1985). Rarer quantitative and qualitative abnormalities of coagulation. Clin. Haematol. 14, pp. 385–411.

327. Fujiwara, Y., Takai, K. & Sanada, M. (1988). A case of factor X anomaly (Prower defect). Blood & Vessel 19, pp. 143–149.
328. Reddy, S.V., Zhou, Z.-Q., Rao, K.J., Scott, J.P., Watzke, H., High, K.A. & Jagadeeswaran, P. (1989). Molecular characterization of human factor $X_{San\ Antonio}$. Blood 74, pp. 1486–1490.
329. Bernardi, F., Marchetti, G., Patracchini, P., Volinia, S., Gemmati, D., Simoni, P. & Girolami, A. (1989). Partial gene deletion in a family with factor X deficiency. Blood 73, pp. 2123–2127.
330. Straight, D.L., Sherrill, G.B., Noyes, C.M., Trapp, H.G., Wright, S.F., Roberts, H.R., Hiskey, R.G. & Griffith, M.J. (1985). Structural and functional characteristics of activated human factor IX after chemical modification of γ-carboxyglutamic acid residues. J. Biol. Chem. 260, pp. 2890–2893.
331. Jones, M.E., Griffith, M.J., Monroe, D.M., Roberts, H.R. & Lentz, B.R. (1985). Comparison of lipid binding and kinetic properties of normal, variant and γ-carboxyglutamic acid modified human factor IX and factor IXa. Biochemistry 24, pp. 8064–8069.
332. Bajaj, S.P., Rapaport, S.I. & Maki, S.L. (1985). A monoclonal antibody to factor IX that inhibits the factor VIIICa potentiation of factor X activation. J. Biol. Chem. 260, pp. 11574–11580.
333. Lin, S.-W., Smith, K.J., Welsch, D. & Stafford, D.W. (1990). Expression and characterization of human factor IX and factor IX-factor X chimeras in mouse C127 cells. J. Biol. Chem. 265, pp. 144–150.
334. Cited as personal communication in reference (294).
335. Bertina, R.M., Cupers, R. & van Wijngaarden, A. (1984). Factor IXa protects activated factor VIII against inactivation by activated protein C. Biochem. Biophys. Res. Commun. 125, pp. 177–183.
336. Walker, F.J., Chavin, S.I. & Fay, P.J. (1987). Inactivation of factor VIII by activated protein C and protein S. Arch. Biochem. Biophys. 252, pp. 322–328.
337. Ware, J., Toomey, J.R. & Stafford, D.W. (1988). Localization of a factor VIII-inhibiting antibody epitope to a region between residues 338 and 362 of factor VIII heavy chain. Proc. Natl. Acad. Sci. USA 85, pp. 3165–3169.
338. Foster, P.A., Fulcher, C.A., Houghten, R.A., Mahouey, S.d.G. & Zimmerman, T.S. (1988). Localization of the binding regions of a murine monoclonal anti-factor VIII antibody and a human anti-factor VIII alloantibody, both of which inhibit factor VIII procoagulant activity, to amino acid residues threonine351-serine365 of the factor VIII heavy chain. J. Clin. Invest. 82, pp. 123–128.
339. Kaufman, R.J., Pittman, D.D., Marquette, K.A., Murray, B.L. & Wang, J.H. (1989). Structural requirements for factor VIII function. Thromb. Haemostas. 62, Abstract SY-XV-1.
340. Anderson, L.-O. & Brown, J.E. (1981). Interaction of factor VIII-von Willebrand factor with phospholipid vesicles. Biochem. J. 200, pp. 161–167.
341. Lajmanovich, A., Hudry-Clergeon, G., Freyssinet, J.-M. & Marguerie, G. (1981). Human factor VIII procoagulant activity and phospholipid interaction. Biochim. Biophys. Acta 678, pp. 132–136.
342. Bloom, J.W. (1987). The interaction of rDNA factor VIII, factor $VIII_{des\text{-}797\text{-}1562}$ and factor $VIII_{des\text{-}797\text{-}1562}$-derived peptides with phospholipid. Thromb. Res. 48, pp. 439–448.
343. Kemball-Cook, G., Edwards, S.J., Sewerin, K., Anderson, L.-O. & Barrowcliffe, T.W. (1988). Factor VIII procoagulant protein interacts with phospholipid vesicles via its 80 kDa light chain. Thromb. Haemostas. 60, pp. 442–446.
344. Foster, P.A., Fulcher, C.A., Houghten, R.A. & Zimmerman, T.S. (1989). Synthetic factor VIII peptides with amino acid sequences contained within the C2 domain of factor VIII inhibit factor VIII binding to phosphatidylserine. Thromb. Haemostas. 62, Abstract 69.
345. Arai, M., Scandella, D. & Hoyer, L.W. (1989). Molecular basis of factor VIII inhibition by human antibodies. J. Clin. Invest. 83, pp. 1978–1984.
346. Barteles, J.R., Galvin, N.J. & Fraizer, W.A. (1982). Discoidin I-membrane interactions: Discoidin I binds to and agglutinates negatively charged phospholipid vesicles. Biochim. Biophys. Acta 687, pp. 129–136.

347. van Dieijen, G., Tans, G., Rosing, J. & Hemker, H.C. (1981). The role of phospholipid and factor VIIIa in the activation of bovine factor X. J. Biol. Chem. 256, pp. 3433–3442.
348. Link, R.P. & Castellino, F.J. (1983). Kinetic comparison of bovine blood coagulation factors IXaα and IXaβ toward bovine factor X. Biochemistry 22, pp. 4033–4041.
349. van Dieijen, G., van Rijn, J.L.M.L., Govers-Riemslag, J.W.P., Hemker, H.C. & Rosing, J. (1985). Assembly of the intrinsic factor X activating complex – interaction between factor IXa, factor VIIIa and phospholipid. Thromb. Haemostas. 53, pp. 396–400.
350. Beals, J.M., Chibber, B.A.K. & Castellino, F.J. (1989). The kinetic assembly of the intrinsic bovine factor X activation system. Arch. Biochem. Biophys. 268, pp. 485–501.
351. Griffith, M.J., Reisner, H.M., Lundblad, R.L. & Roberts, H.R. (1982). Measurement of human factor IXa activity in an isolated factor X activation system. Thromb. Res. 27, pp. 289–301.
352. Hultin, M.B. (1982). Role of human factor VIII in factor X activation. J. Clin. Invest. 69, pp. 950–958.
353. Mertens, K. & Bertina, R.M. (1984). The contribution of Ca^{2+} and phospholipids to the activation of human blood coagulation factor X by activated factor IX. Biochem. J. 223, pp. 607–615.
354. Mertens, K., Van Wijngaarden, A. & Bertina, R.M. (1985). The role of factor VIII in the activation of human blood coagulation factor X by activated factor IX. Thromb. Haemostas. 54, pp. 654–660.
355. Neuenschwander, P. & Jesty, J. (1988). A comparison of phospholipid and platelets in the activation of human factor VIII by thrombin and factor Xa, and in the activation of factor X. Blood 72, pp. 1761–1770.
356. Chattopadhyay, A. & Fair, D.S. (1989). Molecular recognition in the activation of human blood coagulation factor X. J. Biol. Chem. 264, pp. 11035–11043.
357. Mann, K.G., Nesheim, M.E. & Tracy, P.B. (1986). Nonenzymatic cofactors: Factor V. In: New Comprehensive Biochemistry Vol. 13 (Zwaal, R.F.A. & Hemker, H.C., Eds.), pp. 15–34, Elsevier, Amsterdam.
358. Owren, P.A. (1947). Parahæmophilia. Lancet 1, pp. 446–448.
359. Esmon, C.T. (1979). Thrombin-catalyzed activation of single chain bovine factor V. J. Biol. Chem. 254, pp. 964–973.
360. Nesheim, M.E., Myrmel, K.H., Hibbard, L. & Mann, K.G. (1979). Isolation and characterization of single chain bovine factor V. J. Biol. Chem. 254, pp. 508–517.
361. Kane, W.H. & Majerus, P.W. (1981). Purification and characterization of human coagulation factor V. J. Biol. Chem. 256, pp. 1002–1007.
362. Dahlbäck, B. (1980). Human coagulation factor V, purification and thrombin-catalyzed activation. J. Clin. Invest. 66, pp. 583–591.
363. Katzmann, J.A., Nesheim, M.E., Hibbard, L.S. & Mann, K.G. (1981). Isolation of functional human coagulation factor V by using a hybridoma antibody. Proc. Natl. Acad. Sci. USA 78, pp. 162–166.
364. Mann, K.G., Nesheim, M.E. & Tracy, P.B. (1981). Molecular weight of undegraded plasma factor V. Biochemistry 20, pp. 28–33.
365. Tracy, P.B., Eidi, L.L., Bowie, E.J.W. & Mann, K.G. (1982). Radioimmunoassay of factor V in human plasma and platelets. Blood 60, pp. 59–63.
366. Jenny, R.J., Pittman, D.D., Toole, J.J., Kriz, R.W., Aldape, R.A., Hewick, R.M., Kaufman, R.J. & Mann, K.G. (1987). Complete cDNA and derived amino acid sequence of human factor V. Proc. Natl. Acad. Sci. USA 84, pp. 4846–4850.
367. Kane, W.H. & Davie, E.W. (1986). Cloning of a cDNA coding for human factor V, a blood coagulation factor homologous to factor VIII and ceruloplasmin. Proc. Natl. Acad. Sci. USA 83, pp. 6800–6804.
368. Kane, W.H., Ichinose, A., Hagen, F.S. & Davie, E.W. (1987). Cloning of cDNAs coding for the heavy chain region and connecting region of human factor V, a blood coagulation factor with four types of internal repeats. Biochemistry 26, pp. 6508–6514.
369. Wang, H., Riddell, D.C., Guinto, E.R., MacGillivray, R.T.A. & Hamerton, J.L. (1988). Localization of the gene encoding human factor V to chromosome 1q21–25. Geneomics 2, pp. 324–328.
370. Dahlbäck, B., Hanson, C., Islam, M.Q., Szpirer, C., Lundwall, Å. & Levan, G. (1988). Assignment of gene for coagulation factor V to chromosome 1 in man and chromsome

13 in rat. Somat. Cell Mol. Genet. 14, pp. 509–514.
371. Guinto, E.R., Odegaard, B., Mann, K.G. & MacGillivray, R.T.A. (1989). Characterization of the cDNA for bovine factor V: Prediction of the amino acid sequence of bovine factor V and comparison with human factor V. Thromb. Haemostas. 62, Abstract 228.
372. Dahlbäck, B. (1985). Ultrastructure of human coagulation factor V. J. Biol. Chem. 260, pp. 1347–1349.
373. Lampe, P.D., Pusey, M.L., Wei, G.H. & Nelsestuen, G.L. (1984). Electron microscopy and hydrodynamic properties of blood clotting factor V and activation fragments of factor V with phospholipid vesicles. J. Biol. Chem. 259, pp. 9959–9964.
374. Mosesson, M.W., Nesheim, M.E., DiOrio, J., Hainfield, J.F., Walls, J.S. & Mann, K.G. (1985). Studies on the structure of bovine factor V by scanning transmission electron microscopy. Blood 65, pp. 1158–1162.
375. Dahlbäck, B. (1986). Bovine coagulation factor V visualized with electron microscopy. J. Biol. Chem. 261, pp. 9495–9501.
376. Mann, K.G., Lawler, C.M., Vehar, G.A. & Church, W.R. (1984). Coagulation factor V contains copper ion. J. Biol. Chem. 259, pp. 12949–12951.
377. Nesheim, M.E. & Mann, K.G. (1979). Thrombin-catalyzed activation of single chain bovine factor V. J. Biol. Chem. 254, pp. 1326–1334.
378. Suzuki, K., Dahlbäck, B. & Stenflo, J. (1982). Thrombin-catalyzed activation of human coagulation factor V. J. Biol. Chem. 257, pp. 6556–6564.
379. Nesheim, M.E., Foster, W.B., Hewick, R., & Mann, K.G. (1984). Characterization of factor V activation intermediates. J. Biol. Chem. 259, pp. 3187–3196.
380. Foster, W.B., Nesheim, M.E. & Mann, K.G. (1983). Factor Xa activation of factor V. J. Biol. Chem. 258, pp. 13970–13977.
381. Tracy, P.B., Nesheim, M. & Mann, K.G. (1983). Proteolytic alteration of factor Va bound to platelets. J. Biol. Chem. 258, pp. 662–669.
382. Odegaard, B. & Mann, K.G. (1987). Proteolysis of factor Va by factor Xa and activated protein C. J. Biol. Chem. 262, pp. 11233–11236.
383. Seeler, R.A. (1972). Parahemophilia. Med. Clin. North. Am. 56, pp. 119–125.
384. Tracy, P.B. & Mann, K.G. (1987). Abnormal formation of the prothrombinase complex: Factor V deficiency and related disorders. Hum. Pathol. 18, pp. 162–169.
385. Waltz, D.A., Hewett-Emmett, D. & Seegers, W.H. (1977). Amino acid sequence of human prothrombin fragments 1 and 2. Proc. Natl. Acad. Sci. USA 74, pp. 1969–1972.
386. Butkowski, R.J., Elton, J., Dowing, M.R. & Mann, K.G. (1977). Primary structure of human prethrombin 2 and α-thrombin. J. Biol. Chem. 252, pp. 4942–4957.
387. MacGillivray, R.T.A. & Davie, E.W. (1984). Characterization of bovine prothrombin mRNA and its translation product. Biochemistry 23, pp. 1626–1634.
388. Degen, S.J.F., MacGillivray, R.T.A. & Davie, E.W. (1983). Characterization of the complementary deoxyribonucleic acid and gene coding for human prothrombin. Biochemistry 22, pp. 2087–2097.
389. Degen, S.J.F. & Davie, E.W. (1987). Nucleotide sequence of the gene for human prothrombin. Biochemistry 26, pp. 6165–6177.
390. Irwin, D.M., Ahern, K.G., Pearson, G.D. & MacGillivray, R.T.A. (1985). Characterization of the bovine prothrombin gene. Biochemistry 24, pp. 6854–6861.
391. Irwin, D.M., Robertson, K.A. & MacGillivray, R.T.A. (1988). Structure and evolution of the prothrombin gene. J. Mol. Biol. 200, pp. 31–45.
392. Tulinsky, A., Park, C.H. & Skrzypczak, E. (1988). Structure of prothrombin fragment 1 refined at 2.8 Å resolution. J. Mol. Biol. 202, pp. 885–901.
393. Soriano-Garcia, M., Park, C.H., Tulinski, A., Ravichandran, K.G. & Skrzypczak-Jankun, E. (1989). Structure of Ca^{2+} prothrombin fragment 1 including the conformation of the Gla domain. Biochemistry 28, pp. 6805–6810.
394. Bode, W., Mayr, I., Bauman, U., Huber, R., Stone, S. & Hofsteenge, J. (1989). The refined 1.9 Å crystal structure of human a-thrombin: Interaction with D-Phe-Pro-Arg chloromethylketone and significance of the Tyr-Pro-Pro-Trp insertion segment. EMBO J. 8, pp. 3467–3475.
395. Inimoto, T., Shirakami, A., Kawauchi, S., Shigekiyo, T., Saito, S., Miyoshi, K., Morita, T. & Iwanaga, S. (1987). Prothrombin Tokushima: Characterization of dysfunctional thrombin derived from a variant of human prothrombin. Blood 69, pp. 565–569.

396. Rabiet, M.-J., Furie, B.C. & Furie, B. (1986). Molecular defect of prothrombin Barcelona. J. Biol. Chem. 261, pp. 15045–15048.
397. Miyata, T., Morita, T., Inomoto, T., Kawauchi, S., Shirakami, A. & Iwanaga, S. (1987). Prothrombin Tokushima, a replacement of arginine-418 by tryptophan that impairs the fibrinogen clotting activity of derived thrombin Tokushima. Biochemistry 26, pp. 1117–1122.
398. Henriksen, R.A. & Mann, K.G. (1988). Identification of the primary structural defect in the dysthrombin thrombin Quick I: Substitution of cysteine for arginine-382. Biochemistry 27, pp. 9160–9165.
399. Henriksen, R.A. & Mann, K.G. (1989). Substitution of valine for glycine-558 in the congenital dysthrombin thrombin Quick II alters primary substrate specificity. Biochemistry 28, pp. 2078–2082.
400. Mann, K.G. (1987). The assembly of blood clotting complexes on membranes. Trends Biochem. 12, pp. 229–233.
401. Mann, K.G., Jenny, R.J. & Krishnaswamy, S. (1988). Cofactor proteins in the assembly and expression of blood clotting enzyme complexes. Ann. Rev. Biochem. 57, pp. 915–956.
402. Skogen, W.F., Esmon, C.T. & Cox, A.C. (1984). Comparison of coagulation factor Xa and des-(1–44)factor Xa in the assembly of prothrombinase. J. Biol. Chem. 259, pp. 2306–2310.
403. Morita, T. & Jackson, C.M. (1986). Preparation and properties of derivatives of bovine factor X and factor Xa from which the γ-carboxyglutamic acid containing residue has been removed. J. Biol. Chem. 261, pp. 4015–4023.
404. Sugo, T., Björk, I., Holmgren, A. & Stenflo, J. (1984). Calcium binding properties of bovine factor X lacking the γ-carboxyglutamic acid containing region. J. Biol. Chem. 259, pp. 5705–5710.
405. Persson, E., Selander, M., Linse, S., Drakenberg, T., Öhlin, A.-K. & Stenflo, J. (1989). Calcium binding to the isolated β-hydroxyaspartic acid-containing epidermal growth factor-like domain of bovine factor X. J. Biol. Chem. 264, pp. 16897–16904.
406. Annamalai, A.E., Rao, A.K., Chiu, H.C., Wang, D., Dutta-Roy, A., Walsh, P.N. & Colman, R.W. (1987). Epitope mapping of functional domains of human factor Va with human and murine monoclonal antibodies. Evidence for the interaction of heavy chain with factor Xa and calcium. Blood 70, pp. 139–146.
407. Tucker, M.M., Foster, W.B., Katzman, J.A. & Mann, K.G. (1983). A monoclonal antibody which inhibits the factor Va–factor Xa interaction. J. Biol. Chem. 258, pp. 1210–1214.
408. Guinto, E.R. & Esmon, C.T. (1984). Loss of prothrombin and factor Xa-factor Va interactions upon inactivation of factor Va by activated protein C. J. Biol. Chem. 259, pp. 13986–13992.
409. Suzuki, K., Stenflo, J., Dahlbäck, B. & Teodorsson, B. (1983). Inactivation of human coagulation factor V by activated protein C. J. Biol. Chem. 258, pp. 1914–1920.
410. Walker, F.J., Sexton, P.W. & Esmon, C.T. (1979). The inhibition of blood coagulation by activated protein C through the selective inactivation of activated factor V. Biochem. Biophys. Acta 571, pp. 333–342.
411. Pusey, M.L. & Nelsestuen, G.L. (1984). Membrane binding properties of blood coagulation factor V and derived peptides. Biochemistry 23, pp. 6202–6210.
412. Bloom, J.W., Nesheim, M.E. & Mann, K.G. (1979). Phospholipid binding properties of bovine factor V and factor Va. Biochemistry 18, pp. 4419–4425.
413. Pusey, M.L., Mayer, L.D., Wei, J., Bloomfield, V.A. & Nelsestuen, G.L. (1982). Kinetic and hydrodynamic analysis of blood clotting factor V-membrane binding. Biochemistry 21, pp. 5262–5269.
414. van de Waart, P., Bruls, H., Hemker, H.C. & Lindhout, T. (1983). Interaction of bovine blood clotting factor Va and its subunits with phospholipid vesicles. Biochemistry 22, pp. 2427–2432.
415. Higgins, D.L. & Mann, K.G. (1983). The interaction of bovine factor V and factor V-derived peptides with phospholipid vesicles. J. Biol. Chem. 258, pp. 6503–6508.
416. Isaacs, B.S., Husten, E.J., Esmon, C.T. & Johnson, A.E. (1986). A domain of membrane-bound blood coagulation factor Va is located far from the phospholipid surface. A fluorescent energy transfer measurement. Biochemistry 25, pp. 4958–4969.

417. Krishnaswamy, S. & Mann, K.G. (1988). The binding of factor Va to phospholipid vesicles. J. Biol. Chem. 263, pp. 5714–5723.
418. Krieg, U.C., Isaacs, B.S. Yemul, S.S., Esmon, C.T., Bayley, H. & Johnson, A.E. (1987). Interaction of blood coagulation factor Va with phospholipid vesicles examined by using lipophilic photoreagents. Biochemistry 26, pp. 103–109.
419. Lecompte, M.F., Krishnaswamy, S., Mann, K.G., Nesheim, M.E. & Gitler, C. (1987). Membrane penetration of bovine factor V and Va detected by labeling with 5-iodonaphthalene-1-azide. J. Biol. Chem. 262, pp. 1935–1937.
420. Krishnaswamy, S., Jones, K.C. & Mann, K.G. (1988). Prothrombinase complex assembly. J. Biol. Chem. 263, pp. 3823–3834.
421. Nesheim, M.E., Kettner, C., Shaw, E. & Mann, K.G. (1981). Cofactor dependence of factor Xa incorporation into the prothrombinase complex. J. Biol. Chem. 256, pp. 6537–6540.
422. Husten, E.J., Esmon, C.T. & Johnson, A.E. (1987). The active site of blood coagulation factor Xa. J. Biol. Chem. 262, pp. 12953–12961.
423. Gitel, S.N., Owen, W.G., Esmon, C.T. & Jackson, C.M. (1973). A polypeptide region of bovine prothrombin specific for binding to phospholipid. Proc. Natl. Acad. Sci. USA 70, pp. 1344–1348.
424. Dombrose, F.A., Gitel, S.N., Zawalich, K. & Jackson, C.M. (1979). The association of bovine prothrombin fragment 1 with phospholipid. J. Biol. Chem. 254, pp. 5027–5040.
425. Malhotra, O.P., Nesheim, M.E. & Mann, K.G. (1985). The kinetics of activation of normal and γ-carboxyglutamic acid-deficient prothrombins. J. Biol. Chem. 260, pp. 279–287.
426. Weber, D.J., Pollock, J.S., Pedersen, L.G. & Hiskey, R.G. (1988). The determination of a calcium-dependent binding constant of the bovine prothrombin Gla domain (residues 1–45) to phospholipid vesicles. Biochem. Biophys. Res. Commun. 155, pp. 230–235.
427. Pollock, J.S., Shepard, A.J., Weber, D.J., Olson, D.L., Klapper, D.G., Pedersen, L.G. & Hiskey, R.G. (1988). Phospholipid binding properties of bovine prothrombin peptide residues 1–45. J. Biol. Chem. 263, pp. 14216–14223.
428. Luckow, E.A., Lyons, D.A., Ridgeway, T.M., Esmon, C.T. & Laue, T.M. (1989). Interaction of clotting factor V heavy chain with prothrombin and prethrombin 1 and role of activated protein C in regulating this interaction: Analysis by analytical ultracentrifugation. Biochemistry 28, pp. 2348–2354.
429. Esmon, C.T. & Jackson, C.M. (1974). The conversion of prothrombin to thrombin. J. Biol. Chem. 249, pp. 7791–7797.
430. Bajaj, S.P., Butowski, R.J. & Mann, K.G. (1975). Prothrombin fragments. J. Biol. Chem. 250, pp. 2150–2156.
431. Rosing, J., Zwaal, R.F. A. & Tans, G. (1986). Formation of meizothrombin as intermediate in factor Xa-catalyzed prothrombin activation. J. Biol. Chem. 261, pp. 4224–4228.
432. Krishnaswamy, S., Mann, K.G. & Nesheim, M.E. (1986). The prothrombinase-catalyzed activation of prothrombin proceeds through the intermediate meizothrombin in an ordered sequential reaction. J. Biol. Chem. 261, pp. 8977–8984.
433. Stenn, K.S. & Blout, E.R. (1972). Mechanism of bovine prothrombin activation by an insoluble preparation of bovine factor Xa (thrombokinase). Biochemistry 11, pp. 4502–4515.
434. Esmon, C.T. & Jackson, C.M. (1974). The conversion of prothrombin to thrombin. J. Biol. Chem. 249, pp. 7782–7790.
435. Esmon, C.T., Owen, W.G. & Jackson, C.M. (1974). The conversion of prothrombin to thrombin. J. Biol. Chem. 249, pp. 7798–7807.
436. Esmon, C.T., Owen, W.G. & Jackson, C.M. (1974). A plausible mechanism for prothrombin activation by factor Xa, factor Va, phospholipid, and calcium ions. J. Biol. Chem. 249, pp. 8045–8047.
437. Krishnaswamy, S., Church, W.R., Nesheim, M.E. & Mann, K.G. (1987). Activation of human prothrombin by human prothrombinase. J. Biol. Chem. 262, pp. 3291–3299.
438. Rabiet, M.J., Blashill, A., Furie, B. & Furie, B.C. (1986). Prothrombin fragment 1.2.3, a major product of prothrombin activation in human plasma. J. Biol. Chem. 261, pp. 13210–13215.

439. Rosing, J., Tans, G., Govers-Riemslag, J.W.P., Zwaal, R.F.A. & Hemker, H.C. (1980). The role of phospholipids and factor Va in the prothrombinase complex. J. Biol. Chem. 255, pp. 274–283.
440. Nesheim, M.E., Taswell, J.B. & Mann, K.G. (1979). The contribution of bovine factor V and factor Va to the activity of prothrombinase. J. Biol. Chem. 254, pp. 10952–10962.
441. Tracy, P.B., Nesheim, M.E. & Mann, K.G. (1981). Coordinate binding of factor Va and factor Xa to the unstimulated platelet. J. Biol. Chem. 256, pp. 743–751.
442. Kane, W.H., Lindhout, M.J., Jackson, C.M. & Majerus, P.W. (1980). Factor Va-dependent binding of factor Xa to human platelets. J. Biol. Chem. 255, pp. 1170–1174.
443. Nesheim, M.E. & Mann, K.G. (1983). The kinetics and cofactor dependence of the two cleavages involved in prothrombin activation. J. Biol. Chem. 258, pp. 5386–5391.
444. Nesheim, M.E., Tracy, R.P. & Mann, K.G. (1984). 'Clotspeed', a mathematical simulation of the functional properties of prothrombinase. J. Biol. Chem. 259, pp. 1447–1453.
445. Nelsestuen, G.L. (1978). Interactions of vitamin K-dependent proteins with calcium ions and phospholipid membranes. Fed. Proc. 37, pp. 2621–2625.
446. Nelsestuen, G.L. & Broderius, M. (1977). Interaction of prothrombin and blood-clotting factor X with membranes of varying composition. Biochemistry 16, pp. 4172–4176.
447. Pusey, M.L. & Nelsestuen, G.L. (1983). The physical significance of K_M in the prothrombinase reaction. Biochem. Biophys. Res. Commun. 114, pp. 526–533.
448. van Rijn, J.L.M.L., Govers-Riemslag, J.W.P., Zwaal, R.F.A. & Rosing, J. (1984). Kinetic studies of prothrombin activation: Effect of factor Va and phospholipid on the formation of the enzyme-substrate complex. Biochemistry 23, pp. 4557–4564.
449. Österud, B. (1986). Initiation mechanism: Activation induced by thromboplastin. In: New Comprehensive Biochemistry Vol. 13 (Zwaal, R.F.A. & Hemker, H.C., Eds.), pp. 129–139, Elsevier, Amsterdam.
450. Broze Jr., G.J. & Majerus, P.W. (1980). Purification and properties of human coagulation factor VII. J. Biol. Chem. 255, pp. 1242–1247.
451. Bajaj, S.P., Rapaport, S.I. & Brown, S.F. (1981). Isolation and characterization of human factor VII. J. Biol. Chem. 256, pp. 253–259.
452. Fair, D.S. (1983). Quantitation of factor VII in the plasma of normal and warfarin-treated individuals by radioimmunoassay. Blood 62, pp. 784–791.
453. Hagen, F.S., Gray, C.L., O'Hara, P., Grant, F.J., Saari, G.C., Woodbury, R.G., Insley, M., Kisiel, W., Kurachi, K. & Davie, E.W. (1986). Characterization of a cDNA coding for human factor VII. Proc. Natl. Acad. Sci. USA 83, pp. 2412–2416.
454. Thim, L., Bjørn, S., Christensen, M., Nicolaisen, E.M., Lund-Hansen, T., Pedersen, A.H. & Hedner, U. (1988). Amino acid sequence and posttranslational modifications of human factor VIIa from plasma and transfected baby hamster kidney cells. Biochemistry 27, pp. 7785–7793.
455. Takeya, H., Kawagata, S., Nakagawa, K., Yamamichi, Y., Miyata, T., Iwanaga, S., Takao, T., & Shimonishi, Y. (1988). Bovine factor VII. J. Biol. Chem. 263, pp. 14868–14877.
456. O'Hara, P., Grant, F.J., Haldeman, B.A., Gray, C.L., Insley, M.Y., Hagen, F.S. & Murray, M.J. (1987). Nucleotide sequence of the gene coding for human factor VII, a vitamin K-dependent protein participating in blood coagulation. Proc. Natl. Acad. Sci. USA 84, pp. 5158–5162.
457. Pfeiffer, R.A., Ott, R., Gilgenkrantz, S. & Alexandre, P. (1982). Deficiency of coagulation factors VII and X associated with deletion of a chromosome 13(q34). Evidence from two cases with 46,XY,t(13;Y)(q11;q34). Hum. Genet. 62, pp. 358–360.
458. de Grouchy, J., Dautzenberg, M.-D., Turleau, C., Beguin, S. & Chavin-Colin, F. (1984). Regional mapping of clotting factor VII and X to 13q34. Expression of factor VII through chromosome 8. Hum. Genet. 66, pp. 230–233.
459. Gilgenkranz, S., Briquel, M.-E., André, E., Alexandre, P., Jalbert, P., Lemarec, B., Pouzol, P. & Pommereuil, M. (1986). Structural genes of coagulation factors VII and X located on 13q34. Ann. Génét. 29, pp. 32–35.
460. Kisiel, W. & McMullen, B.A. (1981). Isolation and characterization of human factor VIIa. Thromb. Res. 22, pp. 375–380.

461. Radcliffe, R. & Nemerson, Y. (1975). Activation and control of factor VII by activated factor X and thrombin. J. Biol. Chem. 250, pp. 388–395.
462. Zur, M. & Nemerson, Y. (1978). The esterase activity of coagulation factor VII. J. Biol. Chem. 253, pp. 2203–2209.
463. Zur, M., Radcliffe, R.D., Oberdick, J. & Nemerson, Y. (1982). The dual role of factor VII in blood coagulation. J. Biol. Chem. 257, pp. 5623–5631.
464. Alexander, B., Goldstein, R., Landwehr, G. & Cook, C.D. (1951). Congenital SPCA deficiency: A hitherto unrecognized coagulation defect with hemorrhage rectified by serum and serum fractions. J. Clin. Invest. 30, pp. 596–608.
465. Astrup, T. (1965). Assay and content of tissue thromboplastin in different organs. Thromb. Diath. Haemorrh. 14, pp. 401–406.
466. Pitlick, F.A. & Nemerson, Y. (1970). Binding of the protein component of tissue factor to phospholipid. Biochemistry 9, pp. 5105–5113.
467. Hvatum, M. & Prydz, H. (1969). Studies on tissue thromboplastin – its splitting into two separable parts. Thromb. Diath. Haemorrh. 21, pp. 217–222.
468. Zwaal, R.F.A. (1978). Membrane and lipid involvement in blood coagulation. Biochim. Biophys. Acta 515, pp. 163–205.
469. Nemerson, Y. & Pitlick, F.A. (1970). Purification and characterization of the protein component of tissue factor. Biochemistry 9, pp. 5100–5105.
470. Bach, R., Nemerson, Y. & Konigsberg, W. (1981). Purification and characterization of bovine tissue factor. J. Biol. Chem. 256, pp. 8324–8331.
471. Broze Jr., G.J., Leykam, J.E., Schwartz, B.D. & Miletich, J.P. (1985). Purification of human brain tissue factor. J. Biol. Chem. 260, pp. 10917–10920.
472. Bom, V.J.J., Ram, I.E., Alderkamp, G.H.J., Reinalda-Poot, H.H. & Bertina, R. (1986). Application of factor VII-Sepharose affinity chromatography in the purification of human tissue factor apoprotein. Thromb. Res. 42, pp. 635–643.
473. Tanaka, H., Janssen, B., Preissner, K.T. & Müller-Berghaus, G. (1985). Purification of glycosylated apoprotein of tissue factor from human brain and inhibition of its procoagulant activity by a specific antibody. Thromb. Res. 40, pp. 745–756.
474. Guha, A., Bach, R., Konigsberg, W. & Nemerson, Y. (1986). Affinity purification of human tissue factor: Interaction of factor VII and tissue factor in detergent micelles. Proc. Natl. Acad. Sci. USA 83, pp. 299–302.
475. Morrissey, J.H., Fakhrai, H. & Edgington, T.S. (1987). Molecular cloning of the cDNA for tissue factor, the cellular receptor for the initiation of the coagulation protease cascade. Cell 50, pp. 129–135.
476. Scarpati, E.M., Wen, D., Broze Jr., G.J., Miletich, J.P., Flandermeyer, R.R., Siegel, N.R. & Sadler, J.E. (1987). Human tissue factor: cDNA sequence and chromosome localization of the gene. Biochemistry 26, pp. 5234–5238.
477. Spicer, E.K., Horton, R., Bloem, L., Bach, R., Williams, K.R., Guha, A., Kraus, J., Lin, T.-C., Nemerson, Y. & Konigsberg, W.H. (1987). Isolation of cDNA clones coding for human tissue factor: Primary structure of the protein and cDNA. Proc. Natl. Acad. Sci. USA 84, pp. 5148–5152.
478. Fisher, K.L., Gorman, C.M., Vehar, G.A., O'Brien, D.P. & Lawn, R.M. (1987). Cloning and expression of human tissue factor cDNA. Thromb. Res. 48, pp. 89–99.
479. Mackman, N., Morrisey, J.H., Fowler, B. & Edgington, T.S. (1989). Complete sequence of the human tissue factor gene, a highly regulated cellular receptor that initiates the coagulation protease cascade. Biochemistry 28, pp. 1755–1762.
480. Carson, S.D., Henry, W.M. & Shows, T.B. (1985). Tissue factor gene localized to human chromosome 1 (1pter → 1p21). Science 229, pp. 991–993.
481. Kao, F.-T., Hariz, J., Horton, R., Nemerson, Y. & Carson, S.D. (1988). Regional assignment of human tissue factor gene (F3) to chromosome 1p21–p22. Somat. Cell Mol. Genet. 14, pp. 407–410.
482. Bach, R., Konigsberg, W.H. & Nemerson, Y. (1988). Human tissue factor contains thioester-linked palmitate and stearate on the cytoplasmic half-cysteine. Biochemistry 27, pp. 4227–4231.
483. Carson, S.D. (1987). Continuous chromogenic tissue factor assay: Comparison to clot-based assays and sensitivity established using pure tissue factor. Thromb. Res. 47, pp. 379–387.

484. Carson, S.D., Ross, S.E. & Gramzinski, R.A. (1988). Protein co-isolated with human tissue factor impairs recovery of activity. Blood 71, pp. 520–523.
485. Morrissey, J.H., Revak, D., Tejada, P., Fair, D.S. & Edgington, T.S. (1988). Resolution of monomeric and heterodimeric forms of tissue factor, the high affinity cellular receptor for factor VII. Thromb. Res. 50, pp. 481–493.
486. Nemerson, Y. (1988). Tissue factor and hemostasis. Blood 71, pp. 1–8.
487. Bach, R., Gentry, R. & Nemerson, Y. (1986). Factor VII binding in reconstituted phospholipid vesicles: Induction of cooperativity by phosphatidylserine. Biochemistry 25, pp. 4007–4020.
488. Silverberg, S.A., Nemerson, Y. & Zur, M. (1977). Kinetics of the activation of bovine coagulation factor X by components of the extrinsic pathway. J. Biol. Chem. 252, pp. 8481–8488.
489. Forman, S.D. & Nemerson, Y. (1986). Membrane dependent coagulation reaction is independent of the concentration of phospholipid-bound substrate: Fluid-phase factor X regulates the extrinsic system. Proc. Natl. Acad. Sci. USA 83, pp. 4675–4679.
490. Nemerson, Y. & Gentry, R. (1986). An ordered addition, essential activation model of the tissue factor pathway of coagulation: Evidence for a conformational cage. Biochemistry 25, pp. 4020–4033.
491. Fujikawa, K., Coan, M.H., Legaz, M.E. & Davie, E.W. (1974). The mechanism of activation of bovine factor X (Stuart factor) by intrinsic and extrinsic pathways. Biochemistry 13, pp. 5290–5299.
492. Österud, B. & Rapaport, S.I. (1977). Activation of factor IX by the reaction product of tissue factor and factor VII: Additional pathway for initiating blood coagulation. Proc. Natl. Acad. Sci. USA 74, pp. 5260–5264.
493. Rao, L.V.M., Rapaport, S.I. & Bajaj, S.P. (1986). Activation of human factor VII in the initiation of tissue factor-dependent coagulation. Blood 68, pp. 685–691.
494. Rao, L.V.M. & Rapaport, S.I. (1988). Activation of factor VII bound to tissue factor: A key early step in the tissue factor pathway of blood coagulation. Proc. Natl. Acad. Sci. USA 85, pp. 6687–6691.
495. Williams, E.B., Krishnaswamy, S. & Mann, K.G. (1989). Zymogen/enzyme discrimination using peptide chloromethyl ketones. J. Biol. Chem. 264, pp. 7536–7545.
496. Seligsohn, U., Kasper, C.K., Österud, B. & Rapaport, S.I. (1978). Activated factor VII: Presence in factor IX concentrates and persistence in the circulation after infusion. Blood 53, pp. 828–833.
497. Bauer, K.A., Weiss, L.M., Sparrow, D., Vokonas, P.S. & Rosenberg, R.D. (1987). Aging-associated changes in indices of thrombin generation and protein C activation in humans. J. Clin. Invest. 80, pp. 1527–1534.
498. Bauer, K.A., Kass, B.L., ten Cate, H., Bednarck, M.A., Hawiger, J.J. & Rosenberg, R.R. (1989). Detection of factor X activation in humans. Blood 74, pp. 2007–2015.
499. Miller, B.C., Hultin, M.B. & Jesty, J. (1985). Altered factor VII activity in hemophilia. Blood 65, pp. 845–849.
500. Radcliffe, R. & Nemerson, Y. (1976). Mechanism of activation of bovine factor VII. J. Biol. Chem. 251, pp. 4797–4802.
501. Nemerson, Y. & Repk, D. (1985). Tissue factor accelerates the activation of coagulation factor VII: The role of a bifunctional coagulation cofactor. Thromb. Res. 40, pp. 351–358.
502. Masys, D.R., Bajaj, S.P. & Rapaport, S.I. (1982). Activation of human factor VII by activated factors IX and X. Blood 60, pp. 1143–1150.
503. Henschen, A. & McDonagh, J. (1986). Fibrinogen, fibrin and factor XIII. In: New Comprehensive Biochemistry Vol. 13 (Zwaal, R.F.A. & Hemker, H.C., Eds.), pp. 171–241, Elsevier, Amsterdam.
504. Shafer, J.A. & Higgins, D.L. (1988). Human fibrinogen. CRC Crit. Rev. Clin. Lab. Sci. 26, pp. 1–41.
505. Iwanaga, S., Blombäck, B., Grøndahl, N.J., Hessel, B. & Wallen, P. (1968). Amino acid sequence of the N-terminal part of γ-chain in human fibrinogen. Biochim. Biophys. Acta 160, pp. 280–288.
506. Töpfer-Petersen, E., Lottspeich, F. & Henschen, A. (1976). Carbohydrate linkage site in the γ-chain of human fibrin. Hoppe-Seyler's Z. Physiol. Chem. 357, pp. 1509–1513.

507. Watt, K.W.K., Cottrell, B.A., Strong, D.D. & Doolittle, R.F. (1979). Amino acid sequence studies on the α-chain of human fibrinogen. Overlapping sequences providing the complete sequence. Biochemistry 18, pp. 5410–5416.
508. Henschen, A., Lottspeich, F. & Hessel, B. (1979). Amino acid sequence of human fibrin. Hoppe-Seyler's Z. Physiol. Chem. 360, pp. 1951–1956.
509. Blombäck, B., Hessel, B., Iwanaga, S., Reuterby, J. & Blombäck, M. (1972). Primary structure of human fibrinogen and fibrin. J. Biol. Chem. 247, pp. 1496–1512.
510. Gårdlund, B. (1977). Human fibrinogen – amino acid sequence of fragment E and of adjacent structures in the Aα- and Bβ-chain. Thromb. Res. 10, pp. 689–702.
511. Doolittle, R.F., Cassman, K.G., Cottrell, B.A., Friezner, S.J., Hucko, J.T. & Takagi, T. (1977). Amino acid sequence studies on the α chain of human fibrinogen. Characterization of 11 cyanogen bromide fragments. Biochemistry 16, pp. 1703–1709.
512. Doolittle, R.F., Cassman, K.G., Cottrell, B.A., Friezner, S.J. & Takagi, T. (1977). Amino acid sequence studies on the α chain of human fibrinogen. Covalent structure of the α-chain portion of fragment D. Biochemistry 16, pp. 1710–1715.
513. Takagi, T. & Doolittle, R.F. (1975). Amino acid sequence studies on the α chain of human fibrinogen. Location of four plasmin attack points and a covalent cross-linking site. Biochemistry 14, pp. 5149–5156.
514. Lottspeich, F. & Henschen, A. (1978). Amino acid sequence of human fibrin. Hoppe-Seyler's Z. Physiol. Chem. 359, pp. 1451–1455.
515. Henschen, A., Lottspeich, F. & Hessel, B. (1978). Amino acid sequence of human fibrin. Hoppe-Seyler's Z. Physiol. Chem. 359, pp. 1607–1610.
516. Lottspeich, F. & Henschen, A. (1978). Amino acid sequence of human fibrin. Hoppe-Seyler's Z. Physiol. Chem. 359, pp. 1611–1616.
517. Watt, K.W.K., Takagi, T. & Doolittle, R.F. (1979). Amino acid sequence of the β-chain of human fibrinogen. Biochemistry 18, pp. 68–76.
518. Henschen, A. & Lottspeich, F. (1977). Amino acid sequence of human fibrin. Hoppe-Seyler's Z. Physiol. Chem. 358, pp. 1643–1646.
519. Blombäck, B., Hogg, D.H., Gårdlund, B., Hessel, B. & Kudryk, B. (1976). Fibrinogen and fibrin formation. Thromb. Res. 8 Supp. II, pp. 329–346.
520. Hessel, B., Makino, M., Iwanaga, S. & Blombäck, B. (1979). Primary structure of human fibrinogen and fibrin. Eur. J. Biochem. 98, pp. 521–534.
521. Lottspeich, F. & Henschen, A. (1977). Amino acid sequence of human fibrin. Hoppe-Seyler's Z. Physiol. Chem. 358, pp. 1521–1524.
522. Lottspeich, F. & Henschen, A. (1977). Amino acid sequence of human fibrin. Hoppe-Seyler's Z. Physiol. Chem. 358, pp. 1639–1642.
523. Watt, K.W.K., Takagi, T. & Doolittle, R.F. (1978). Amino acid sequence of the β chain of human fibrinogen: Homology with the γ chain. Proc. Natl. Acad. Sci. USA 75, pp. 1731–1735.
524. Henschen, A. & Lottspeich, F. (1977). Sequence homology between γ-chain and β-chain in human fibrin. Thromb. Res. 11, pp. 869–880.
525. Blombäck, B., Grøndahl, N.J., Hessel, B., Iwanaga, S. & Wallén, P. (1973). Primary structure of human fibrinogen and fibrin. J. Biol. Chem. 248, pp. 5806–5820.
526. Collen, D., Kudryk, B., Hessel, B. & Blombäck, B. (1975). Primary structure of human fibrinogen and fibrin. J. Biol. Chem. 250, pp. 5808–5817.
527. Takagi, T. & Doolittle, R.F. (1975). Amino acid sequence of the carboxy-terminal cyanogen bromide peptide of the human fibrinogen β-chain: Homology with the corresponding γ-chain peptide and presence in fragment D. Biochim. Biophys. Acta 386, pp. 617–622.
528. Takagi, T. & Doolittle, R.F. (1975). Amino acid sequence studies on plasmin-derived fragments of human fibrinogen: Amino-terminal sequences of intermediate and terminal fragments. Biochemistry 14, pp. 940–946.
529. Henschen, A. & Warbinek, R. (1975). Amino acid sequence of human fibrin. Hoppe-Seyler's Z. Physiol. Chem. 356, pp. 1981–1984.
530. Henschen, A. & Lottspeich, F. (1975). Amino acid sequence of human fibrin. Hoppe-Seyler's Z. Physiol. Chem. 356, pp. 1985–1988.
531. Henschen, A., Lottspeich, F., Sekita, T. & Warbinek, R. (1976). Amino acid sequence of human fibrin. Hoppe-Seyler's Z. Physiol. Chem. 357, pp. 605–608.

532. Lottspeich, F. & Henschen, A. (1977). Amino acid sequence of human fibrin. Hoppe-Seyler's Z. Physiol. Chem. 358, pp. 703–707.
533. Lottspeich, F. & Henschen, A. (1977). Amino acid sequence of human fibrin. Hoppe-Seyler's Z. Physiol. Chem. 358, pp. 935–938.
534. Henschen, A., Lottspeich, F., Southan, C. & Töpfer-Petersen, E. (1980). Human fibrinogen: Sequence, sulfur bridges, glycosylation and some structural variants. Protides Biol. Fluids 28, pp. 51–56.
535. Kant, J.A., Lord, S.T. & Crabtree, G.R. (1983). Partial mRNA sequences for human Aα, Bβ, and γ-fibrinogen chains: Evolution and functional implications. Proc. Natl. Acad. Sci. USA 80, pp. 3953–3957.
536. Rixon, M.W., Chan, W.-Y., Davie, E.W. & Chung, D.W. (1983). Characterization of a complementary deoxyribonucleic acid coding for the α chain of human fibrinogen. Biochemistry 22, pp. 3237–3244.
537. Chung, D.W., Que, B.G., Rixon, M.W., Mace Jr., M. & Davie, E.W. (1983). Characterization of complementary deoxyribonucleic acid and genomic deoxyribonucleic acid for the β chain of human fibrinogen. Biochemistry 22, pp. 3244–3250.
538. Chung, D.W., Chan, W.-Y. & Davie, E.W. (1983). Characterization of a complementary deoxyribonucleic acid coding for the γ chain of human fibrinogen. Biochemistry 22, pp. 3250–3256.
539. Chung, D.W. & Davie, E.W. (1984). γ And γ' chains of human fibrinogen are produced by alternative mRNA processing. Biochemistry 23, pp. 4232–4236.
540. Fornace Jr., A.J., Cummings, D.E., Comeau, C.M., Kant, J.A. & Crabtree, G.R. (1984). Structure of human γ-fibrinogen gene. J. Biol. Chem. 259, pp. 12826–12830.
541. Rixon, M.W., Chung, D.W. & Davie, E.W. (1985). Nucleotide sequence of the gene for the γ-chain of human fibrinogen. Biochemistry 24, pp. 2077–2086.
542. McKee, P.A., Rogers, L.A., Marler, E. & Hill, R.L. (1966). The subunit polypeptides of human fibrinogen. Arch. Biochem. Biophys. 116, pp. 271–279.
543. Doolittle, R.F. (1984). Fibrinogen and fibrin. Ann. Rev. Biochem. 53, pp. 195–229.
544. Kaiser, C., Seydewitz, H.H. & Witt, I. (1984). Studies on the primary structure of the Aα-chain of human fibrinogen: Clarification of hitherto uncertain amino acid residues. Thromb. Res. 33, pp. 543–548.
545. Blombäck, B., Hessel, B. & Hogg, D. (1976). Disulfide bridges in the NH_2-terminal part of human fibrinogen. Thromb. Res. 8, pp. 639–658.
546. Blombäck, B., Blombäck, M., Edman, P. & Hessel, B. (1966). Human fibrinopeptides: Isolation, characterization and structure. Biochem. Biophys. Acta 115, pp. 371–396.
547. Seydewitz, H.H., Kaiser, C., Rothweiler, H. & Witt, I. (1984). The location of a second *in vivo* phosphorylation site in the Aα-chain of human fibrinogen. Thromb. Res. 33, pp. 487–498.
548. Kudryk, B., Okada, M., Redman, C.M. & Blombäck, B. (1982). Biosynthesis of dog fibrinogen. Eur. J. Biochem. 125, pp. 673–682.
549. Seydewitz, H.H. & Witt, I. (1985). Increased phosphorylation of human fibrinopeptide under acute phase conditions. Thromb. Res. 40, pp. 29–39.
550. Cited as personal communication in reference (553).
551. Crabtree, G.R., Comeau, C.M., Fowlkes, D.M., Fornace Jr, A.J., Malley, J.D. & Kant, J.A. (1985). Evolution and structures of fibrinogen genes. J. Mol. Biol. 185, pp. 1–19.
552. Kriegelstein, K. & Henschen, A. (1989). The primary structure of bovine fibrinogen and its functional implications. Thromb. Haemostas. 62, Abstract 192.
553. Wang, Y.Z., Patterson, J., Gray, J.E., Yu, C., Cottrell, B.A., Shimizu, A., Graham, D., Riley, M. & Doolittle, R.F. (1989). Complete sequence of the lamprey fibrinogen α chain. Biochemistry 28, pp. 9801–9806.
554. McKee, P.A., Mattock, P. & Hill, R.L. (1970). Subunit structure of human fibrinogen, soluble fibrin, and cross-linked insoluble fibrin. Proc. Natl. Acad. Sci. USA 66, pp. 738–744.
555. Doolittle, R.F. (1980). The evolution of vertebrate fibrinogen. Protides Biol. Fluids 28, pp. 41–46.
556. Doolittle, R.F. & Blombäck, B. (1964). Amino-acid sequence investigations of fibrinopeptides from various mammals: Evolutionary implications. Nature 202, pp. 147–152.

557. Martinelli, R.A., Inglis, A.S., Rubira, M.R., Hageman, T.C., Hurrell, J.G.R., Leach, S.J. & Scheraga, H.A. (1979). Amino acid sequence of portions of the α and β chains of bovine fibrinogen. Arch. Biochem. Biophys. 192, pp. 27–32.
558. Chung, D.W., Rixon, M.W., MacGillivray, R.T.A. & Davie, E.W. (1981). Characterization of a cDNA clone coding for the β chain of bovine fibrinogen. Proc. Natl. Acad. Sci. USA 78, pp. 1466–1470.
559. Cottrell, B.A. & Doolittle, R.F. (1976). Amino acid sequences of lamprey fibrinopeptides A and B and characterization of the junctions split by lamprey and mammalian thrombins. Biochim. Biophys. Acta 453, pp. 426–438.
560. Bohonus, V.L., Doolittle, R.F., Pontes, M. & Strong, D.D. (1986). Complementary DNA sequence of lamprey fibrinogen β chain. Biochemistry 25, pp. 6512–6516.
561. Wolfenstein-Todel, C. & Mosesson, M.W. (1981). Carboxy-terminal amino acid sequence of a human fibrinogen γ-chain variant (γ'). Biochemistry 21, pp. 6146–6149.
562. Francis, C.W., Müller, E., Henschen, A., Simpson, P.J. & Marder, V.J. (1988). Carboxyterminal amino acid sequence of two variant forms of the γ chain of human plasma fibrinogen. Proc. Natl. Acad. Sci. USA 85, pp. 3358–3362.
563. Francis, C.W., Keele, E.M. & Marder, V.J. (1984). Purification of three γ-chains with different molecular weights from normal plasma fibrinogen. Biochim. Biophys. Acta 797, pp. 328–335.
564. Crabtree, G.R. & Kant, J.A. (1982). Organization of the rat γ-fibrinogen gene: Alternative mRNA splice patterns produce the γA and γB(γ') chains of fibrinogen. Cell 31, pp. 159–166.
565. Strong, D.D., Moore, M., Cottrell, B.A., Bohonus, V.L., Pontes, M., Evans, B., Riley, M. & Doolittle, R.F. (1985). Lamprey fibrinogen γ chain: Cloning, cDNA sequencing, and general characterization. Biochemistry 24, pp. 92–101.
566. Morgan, J.C., Holbrook, N.J. & Crabtree, G.R. (1987). Nucleotide sequence of the gamma chain gene of rat fibrinogen: Conserved intronic sequences. Nucl. Acids Res. 15, pp. 2774–2776.
567. Jevons, F.R. (1963). Tyrosine-O-sulfate in fibrinogen and fibrin. Biochem. J. 89, pp. 621–624.
568. Henry, I., Uzan, G., Weil, D., Nicolas, H., Kaplan, J.C., Marguerie, C., Kahn, A. & Junien, C. (1984). The genes coding for Aα-, Bβ- and γ-chains of fibrinogen maps to 4q2. Am. J. Hum. Genet. 36, pp. 760–768.
569. Humphries, S.E., Imam, A.M.A., Robbins, T.P., Cook, M., Carrit, B., Ingle, C. & Williamson, R. (1984). The identification of a DNA polymorphism of the a fibrinogen gene, and the regional assignment of the human fibrinogen genes to 4q26–qter. Hum. Genet 68, pp. 148–158.
570. Kant, J.A., Fornace Jr., A.J., Saxe, D., Simon, M.I., McBride, O.W. & Crabtree, G.R. (1985). Evolution and organization of the fibrinogen locus on chromosome 4: Gene duplication accompanied by transposition and inversion. Proc. Natl. Acad. Sci. USA 82, pp. 2344–2348.
571. Crabtree, G.R. & Kant, J.A. (1982). Coordinate accumulation of the mRNAs for the α, β, and γ chains of rat fibrinogen following defibrination. J. Biol. Chem. 257, pp. 7277–7279.
572. Yu, S., Sher, B., Kudryk, B. & Redman, C.M. (1984). Fibrinogen precursors. J. Biol. Chem. 259, pp. 10574–10581.
573. Hoeprich Jr., P.D. & Doolittle, R.F. (1983). Dimeric half molecules of human fibrinogen are joined through disulfide bonds in an antiparallel orientation. Biochemistry 22, pp. 2049–2055.
574. Henschen, A. (1978). Disulfide bridges in the middle part of human fibrinogen. Hoppe-Seyler's Z. Physiol. Chem. 359, pp. 1757–1770.
575. Bouma, H., Takagi, T. & Doolittle, R.F. (1978). The arrangement of disulfide bonds in fragment D from human fibrinogen. Thromb. Res. 13, pp. 557–562.
576. Nieuwenhuizen, W., von Ruijven-Vermeer, I.A.M., Nooijen, W.J., Vermond, A., Haverkate, F. & Hermans, J. (1981). Recalculation of calcium binding properties of human and rat fibrin(ogen) and their degradation products. Thromb. Res. 22, pp. 653–657.
577. Marguerie, G., Chagniel, G. & Suscillon, M. (1977). The binding of calcium to bovine fibrinogen. Biochim. Biophys. Acta 490, pp. 94–103.

578. Váradi, A. & Scheraga, H.A. (1986). Localization of segments essential for polymerization and for calcium binding in the γ-chains of human fibrinogen. Biochemistry 25, pp. 519–528.
579. Dang, C.V., Ebert, R.F. & Bell, W.R. (1985). Localization of a fibrinogen calcium binding site between γ-subunit position 311 and 336 by terbium fluorescence. J. Biol. Chem. 260, pp. 9713–9719.
580. Nieuwenhuizen, W., Vermond, A. & Hermans, J. (1983). Evidence for the localization of a calcium-binding site in the amino-terminal disulfide knot of fibrin(ogen). Thromb. Res. 31, pp. 81–86.
581. Hall, C.E. & Slayter, H.S. (1959). The fibrinogen molecule: Its size, shape, and mode of polymerization. J. Biophys. Biochem. Cytol. 5, pp. 11–15.
582. Price, T.M., Strong, D.D., Rudee, M.L. & Doolittle, R.F. (1981). Shadow cast electron microscopy of fibrinogen with antibody fragments bound to specific regions. Proc. Natl. Acad. Sci. USA 78, pp. 200–204.
583. Norton, P.A. & Slayter, H.S. (1981). Immune labeling of the D and E regions of human fibrinogen by electron microscopy. Proc. Natl. Acad. Sci. USA 78, pp. 1661–1665.
584. Telford, J.N., Nagy, J.A., Hatcher, P.A. & Scheraga, H.A. (1980). Localization of peptide fragments in the fibrinogen molecule by immunoelectron microscopy. Proc. Natl. Acad. Sci. USA 77, pp. 2372–2376.
585. Mosesson, M.W., Hainfeld, J., Wall, J. & Haschenmeyer, R. (1981). Identification and mass analysis of human fibrinogen molecules and their domains by scanning transmission electron microscopy. J. Mol. Biol. 153, pp. 695–718.
586. Fowler, W.E., Fretto, L.J., Erickson, H.P. & McKee, P.A. (1980). Electron microscopy of plasmic fragments of human fibrinogen as related to trinodular structure of the intact molecule. J. Clin. Invest. 66, pp. 50–56.
587. Crabtree, G.R. (1987). The molecular biology of fibrinogen. In: The Molecular Biology of Blood Diseases (Stamatoyannopoulos, G., Nienhuis, A.W., Leder, P. & Majerus, P.W., Eds.), pp. 631–661, W.B. Saunders, Philadelphia.
588. Morris, S., Denninger, M.-H., Finlayson, J.S. & Menache, D. (1981). Fibrinogen Lille: $A\alpha 7Asp \rightarrow Asn$. Thromb. Haemostas. 46, Abstract 315.
589. Henschen, A., Lottspeich, F., Kehl, M. & Southan, C. (1983). Covalent structure of fibrinogen. Ann. N. Y. Acad. Sci. 408, pp. 28–43.
590. Higgins, D.L. & Shafer, J.A. (1981). Fibrinogen Petoskey, a dysfibrinogenemia characterized by replacement of Arg-$A\alpha 16$ by a histidyl residue: Evidence for thrombin-catalyzed hydrolysis at a histidyl residue. J. Biol. Chem. 256, pp. 12013–12017.
591. Southan, C., Kehl, M., Henschen, A. & Lane, D.A. (1983). Fibrinogen Manchester: Identification of an abnormal fibrinopeptide A with a C-terminal arginine to histidine substitution. Br. J. Haematol. 54, pp. 143–151.
592. Southan, C., Lane, D.A., Bode, W. & Henschen, A. (1985). Thrombin-induced fibrinopeptide release from a fibrinogen variant (fibrinogen Sydney I) with an $A\alpha$ Arg-16 → His substitution. Eur. J. Biochem. 147, pp. 593–600.
593. Galanakis, D.K., Henschen, A., Keeling, M., Kehl, M., Dismore, R. & Peerschke, E.I. (1983). Fibrinogen Louisville: A 16 $A\alpha$ Arg-His defect that forms no hybrid molecule in heterozygous individuals and inhibits aggregation of normal fibrin monomers. Ann. N. Y. Acad. Sci. 408, pp. 644–648.
594. Henschen, A. & Alving, B. (1986). Fibrinogen Giessen I, a classical dysfibrinogen: The molecular defect. In: Fibrinogen and its Derivatives (Berghaus, G.M., Borchel, U.S., Selmayr, E. & Henschen, A., Eds.), pp. 23–24, Elsevier, Amsterdam.
595. Soria, J., Soria, C., Bertyrand, O., Roussel, B., Dieval, J., Vendrely, C. & Delobel, J. (1987). Molecular basis for the A alpha chain substitutions in dysfibrinogenemia with a description of two new cases of the disorder: Fibrinogen Amiens I and fibrinogen Amiens II. In: Fibrinogen 2. Biochemistry, Physiology and Clinical Relevance (Lowe, G. D. O., Douglas, J. T., Forbes, C. D. & Henschen, A., Eds.), pp. 57–61, Elsevier, Amsterdam.
596. Lane, D.A., Thompson, E. & Southan, C. (1987). Rapid screening method of fibrinopeptide analysis applied to congenital dysfibrinogenaemia: Summary of 23 cases. In: Fibrinogen 2. Biochemistry, Physiology and Clinical Relevance (Lowe, G.D.O., Douglas, J.T., Forbes, C.D. & Henschen, A., Eds.), pp. 71–74, Elsevier, Amsterdam.
597. Henschen, A., Kehl, M. & Deutch, E. (1983). Novel structure elucidation strategy for

genetically abnormal fibrinogens with incomplex fibrinopeptide release as applied to fibrinogen Schwarzach. Hoppe-Seyler's Z. Physiol. Chem. 364, pp. 1747–1751.

598. Henschen, A., Southan, C., Kehl, M. & Lottspeich, F. (1981). The structural error and its relationship to the malfunction in some abnormal fibrinogens. Thromb. Haemostas. 46, ads. 559.

599. Reber, P., Furlan, M., Beck, E.A., Finazzi, G., Buelli, M. & Barbui, T. (1985). Fibrinogen Bergamo I (Aα16Arg → Cys): Susceptibility towards thrombin following aminoethylation, methylation or carboxamidomethylation of cysteine residues. Thromb. Haemostas. 54, pp. 390–393.

600. Miyata, T., Terukina, S., Matsuda, M., Kasamatsu, A., Takeda, Y., Murakami, T. & Iwanaga, S. (1987). Fibrinogen Kawaguchi and Osaka: An amino acid substitution of Aα arginine-16 to cysteine which forms an extra interchain disulfide bridge between the two Aα chains. J. Biol. Chem. (Tokyo) 102, pp. 93–101.

601. Galanakis, D.K., Henschen, A., Peerschke, E.I.G. & Kehl, M. (1989). Fibrinogen Stony Brook, a heterozygous Aα16Arg → Cys dysfibrinogenemia. J. Clin. Invest. 84, pp. 295–304.

602. Matsuda, M. (1989). Molecular abnormalities of fibrinogen. Thromb. Haemostas. 62, Abstract SA-11.

603. Blombäck, M., Blombäck, B., Mammen, E.F. & Prasad, A.S. (1968). Fibrinogen Detroit – a molecular defect in the N-terminal disulphide knot of human fibrinogen? Nature 218, pp. 134–137.

604. Pirkle, H., Kaudewitz, H., Henschen, A., Theodor, I. & Simmons, G. (1987). Substitution of Bβ14 arginine by cyst(e)ine in fibrinogen Seattle I. In: Fibrinogen 2. Biochemistry, Physiology and Clinical Relevance (Lowe, G. D. O., Douglas, J. T., Forbes, C. D. & Henschen, A., Eds.), pp. 49–52, Elsevier, Amsterdam.

605. Kaudewitz, H., Henschen, A., Soria, C., Soria, J., Bertrand, O. & Heaton, D. (1986). The molecular defect of the genetically abnormal fibrinogen Christchurch II. In: Fibrinogen and its Derivatives (Berghaus, G.M., Borchel, U.S., Selmayr, E. & Henschen, A., Eds.), pp. 31–36, Elsevier, Amsterdam.

606. Liu, C.Y., Koehn, J.A. & Morgan, F.J. (1985). Characterization of fibrinogen New York I. J. Biol. Chem. 260, pp. 4390–4396.

607. Kaudewitz, H., Henschen, A., Soria, J. & Soria, C. (1986). Fibrinogen Pentoise – a genetically abnormal fibrinogen with defective fibrin polymerization but normal fibrinopeptide release. In: Fibrinogen, Fibrin Formation, and Fibrinolysis Vol. 4 (Lane, D.A., Henschen, A. & Jasani, M.K., Eds.), pp. 91–96, Walter de Gruyter, Berlin.

608. Reber, P., Furlan, M., Henschen, A., Kaudewitz, H., Barbui, T., Hilgard, P., Nenci, G.G., Berrettini, M. & Bech, E.A. (1986). Three abnormal fibrinogen variants with the same amino acid substitution (γ275 Arg → His): Fibrinogen Bergamo II, Essen and Perugia. Thromb. Haemostas. 56, pp. 401–406.

609. Yoshida, N., Ota, K., Moroi, M. & Matsuda, M. (1988). An apparently higher molecular weight γ-chain variant in a new congenital abnormal fibrinogen Tochigi characterized by the replacement of γ arginine-275 by cysteine. Blood 71, pp. 480–487.

610. Terukina, S., Matsuda, M., Hirata, H., Miyata, T., Takao, T. & Shimonishi, Y. (1988). Substitution of γ Arg-275 by Cys in an abnormal fibrinogen, 'fibrinogen Osaka II'. J. Biol. Chem. 263, pp. 13579–13587.

611. Yoshida, N., Terukina, S., Okuma, M., Moroi, M., Aoki, N. & Matsuda, M. (1988). Characterization of an apparently lower molecular weight γ-chain variant in fibrinogen Kyoto I. J. Biol. Chem. 263, pp. 13848–13858.

612. Yamazumi, K., Shimura, K., Terukina, S., Takahashi, N. & Matsuda, M. (1989). A γ methionine-310 to threonine substitution and consequent N-glycosylation at γ asparagine-308 identified in a congenital dysfibrinogenemia associated with posttraumatic bleeding, fibrinogen Asahi. J. Clin. Invest. 83, pp. 1590–1597.

613. Koopman, J., Haverkate, F., Lord, S.T., Caekebeke-Peerlinck, K., Brommer, E. & Briët, E. (1989). A six base deletion in the γ-chain gene of dysfibrinogen Vlissingen, coding for Asn319-Asp320, resulting in defective interaction with Ca^{2+}. Thromb. Haemostas. 62, abs 470.

614. Miyata, T., Furukawa, K., Iwanaga, S., Takamatsu, J. & Saito, H. (1989). Fibrinogen Nagoya, a replacement of glutamine-329 by arginine in the γ-chain that impairs the polymerization of fibrin monomer. J. Biochem. (Tokyo) 105, pp. 10–14.

615. Reber, P., Furlan, M., Rupp, C., Kehl, M., Henschen, A., Mannucci, P.M. & Beck, E.A. (1986). Characterization of fibrinogen Milano I: Amino acid exchange γ330Asp → Val impairs fibrin polymerization. Blood 67, pp. 1751–1756.
616. Terukina, S., Yamazumi, K., Okamoto, K., Akasaka, K., Yamashita, H., Ito, Y. & Matsuda, M. (1989). Fibrinogen Kyoto III: A congenital dysfibrinogenemia with a replacement of γ Asp330 by Tyr. Thromb. Haemostas. 62, Abstract 472.
617. Schwartz, M.L., Pizzo, S.V., Hill, R.L. & McKee, P.A. (1973). Human factor XIII from plasma and platelets. J. Biol. Chem. 248, pp. 1395–1407.
618. Yorifuji, H., Anderson, K., Lynch, G.W., Van De Water, L. & McDonagh, J. (1988). B protein of human factor XIII: Differentiation between free B and complexed B. Blood 62, pp. 1645–1650.
619. Takahashi, N., Takahashi, Y. & Putnam, F.W. (1986). Primary structure of blood coagulation factor XIIIa (fibrinoligase, transglutaminase) from human placenta. Proc. Natl. Acad. Sci. USA 83, pp. 8019–8023.
620. Grundmann, U., Amann, E., Zettlmeissl, G. & Küpper, H.A. (1986). Characterization of cDNA coding for human factor XIIIa. Proc. Natl. Acad. Sci. USA 83, pp. 8024–8028.
621. Ichinose, A., Hendrickson, L.E., Fujikawa, K. & Davie, E.W. (1986). Amino acid sequence of the a-subunit of human factor XIII. Biochemistry 25, pp. 6900–6906.
622. Ichinose, A., McMullen, B.A., Fujikawa, K. & Davie, E.W. (1986). Amino acid sequence of the b subunit of human factor XIII, a protein composed of ten repetitive segments. Biochemistry 25, pp. 4633–4638.
623. Takagi, J., Kasahara, K., Sekiya, F., Inada, Y. & Saito, Y. (1988). Subunit B of factor XIII is present in bovine platelets. Thromb. Res. 50, pp. 767–774.
624. Ichinose, A. & Davie, E.W. (1988). Characterization of the gene for the a subunit of human factor XIII (plasma transglutaminase), a blood coagulation factor. Proc. Natl. Acad. Sci. USA 85, pp. 5829–5833.
625. Board, P.G., Reid, M. & Serjeantson, S. (1984). The gene for coagulation factor XIII a subunit (F13A) is distal to HLA on chromosome 6. Hum. Genet. 67, pp. 406–408.
626. Olaisen, B., Gedde-Dahl Jr., T., Teisberg, P., Thorsby, E., Siverts, A., Jonassen, R. & Wilhelmy, M.C. (1985). A structural locus for the coagulation factor XIIIa (F13A) i located distal to the HLA region on chromosome 6 in man. Am. J. Hum. Genet. 37, pp. 215–220.
627. Weisberg, L.J., Shiu, D.T., Greenberg, C.S., Kan, Y.W. & Shuman, M.A. (1987). Localization of the gene for coagulation factor XIII a-chain to chromosome 6 and identification of sites of synthesis. J. Clin. Invest. 79, pp. 649–652.
628. Board, P.G., Webb, G.C., McKee, J. & Ichinose, A. (1988). Localization of the coagulation factor XIIIa subunit gene (F13A) to chromosome bands 6p24 → p25. Cytogenet. Cell Genet. 48, pp. 25–27.
629. Nagy, J.A., Kradin, R.L. & McDonagh, J. (1988). Biosynthesis of factor XIII A and B subunits. Adv. Exp. Med. Biol. 231, pp. 29–49.
630. Ikura, K., Nasu, T., Yokota, H., Tsuchiya, Y., Sasaki, R. & Chiba, H. (1988). Amino acid sequence of guinea pig liver transglutaminase from its cDNA sequence. Biochemistry 27, pp. 2898–2905.
631. Webb, G.C., Coggan, M., Ichinose, A. & Board, P.G. (1989). Localization of the coagulation factor XIII B subunit gene (F13B) to chromosome bands 1q31–32.1 and restriction fragment length polymorphisms at the locus. Hum. Genet. 81, pp. 157–160.
632. Lozier, J., Takahashi, N. & Putnam, F.W. (1984). Complete amino acid sequence of human plasma β_2-glycoprotein I. Proc. Natl. Acad. Sci. USA 81, pp. 3640–3644.
633. Mole, J.E., Anderson, J.K., Davison, E.A. & Woods, D.E. (1984). Complete primary structure for the zymogen of human complement factor B. J. Biol. Chem. 259, pp. 3407–3412.
634. Morley, B.J. & Campbell, R.D. (1984). Internal homologies of the Ba fragment from human complement component factor B, a class III MHC antigen. EMBO J. 3, pp. 153–157.
635. Chung, L.P., Bentley, D.R. & Reid, K.B.M. (1985). Molecular cloning and characterization of the cDNA coding for C4b-binding protein, a regulatory protein of the classical pathway of the human complement system. Biochem. J. 230, pp. 133–141.
636. Kristensen, T., Ogata, R.T., Chung, L.P., Reid, K.B.M. & Tack, B.F. (1987). cDNA

structure of murine C4b-binding protein, a regulatory component of the serum complement system. Biochemistry 26, pp. 4668–4674.
637. Bentley, D.R. (1986). Primary structure of human complement component C2. Biochem. J. 239, pp. 339–345.
638. Ripoche, J., Day, A.J., Harris, T.J.R. & Sim, R.B. (1988). The complete amino acid sequence of human complement factor H. Biochem. J. 249, pp. 593–602.
639. Kristensen, T. & Tack, B.F. (1986). Murine protein H is comprised of twenty repeating units, sixty-one amino acids in length. Proc. Natl. Acad. Sci. USA 83, pp. 3963–3967.
640. Klickstein, L.B., Wong, W.W., Smith, J.A., Weis, J.H., Wilson, J.G. & Fearon, D.T. (1987). Human C3b/C4b receptor (CR1). J. Exp. Med. 165, pp. 1095–1112.
641. Klickstein, L.B., Bartow, T.J., Miletic, V., Rabson, L.D., Smith, J.A. & Fearon, D.T. (1988). Identification of distinct C3b and C4b recognition sites in the human C3b/C4b receptor (CR1,CD35) by deletion mutagenisis. J. Exp. Med. 168, pp. 1699–1717.
642. Paul, M.S., Aegerter, M., O'Brien, S.E., Kurtz, C.B. & Weis, J.H. (1989). The murine complement receptor gene family. Analysis of mCRY gene products and their homology to human CR1. J. Immunol. 142, pp. 582–589.
643. Moore, M.D., Cooper, N.R., Tack, B.F. & Nemerow, G.R. (1987). Molecular cloning of the cDNA encoding the Epstein-Barr virus/C3d receptor (complement receptor type 2) of human B lymphocytes. Proc. Natl. Acad. Sci. USA 84, pp. 9194–9198.
644. Weis, J.J., Toothaker, L.E., Smith, J.A., Weis, J.H. & Fearon, D.T. (1988). Structure of the human B lymphocyte receptor for C3d and the Epstein-Barr virus and relatedness to other members of the family of C3/C4 binding proteins. J. Exp. Med. 167, pp. 1047–1066.
645. Fingerroth, J.D., Benedict, M.A., Levy, D.N. & Strominger, J.L. (1989). Identification of murine complement receptor type 2. Proc. Natl. Acad. Sci. USA 86, pp. 242–246.
646. Leytus, S.P., Kurachi, K., Sakariassen, K.S. & Davie, E.W. (1986). Nucleotide sequence of the cDNA coding for human complement C1r. Biochemistry 25, pp. 4855–4863.
647. Journet, A. & Tosi, M. (1986). Cloning and sequencing of full-length cDNA encoding the precursor of human complement component C1r. Biochem. J. 240, pp. 783–787.
648. Arlaud, G.J., Willis, A.C. & Gagnon, J. (1987). Complete amino acid sequence of the A chain of human complement-classic-pathway enzyme C1r. Biochem. J. 241, pp. 711–720.
649. MacKinnon, C.M., Carter, P.E., Smyth, S.J., Dunbar, B. & Fothergill, J.E. (1987). Molecular cloning of cDNA for human complement component C1s. Eur. J. Biochem. 169, pp. 547–553.
650. Tosi, M., Duponchel, C., Meo, T. & Julier, C. (1987). Complete cDNA sequence of human complement C1s and close physical linkage of the homologous genes for C1s and C1r. Biochemistry 26, pp. 8516–8524.
651. Caras, I.W., Davitz, M.A., Rhee, L., Weddell, G., Martin Jr., D.W. & Nussenzweig, V. (1987). Cloning of decay-accelerating factor suggests novel use of splicing to generate two proteins. Nature 325, pp. 545–549.
652. Medof, M.E., Lublin, D.M., Holers, V.M., Ayers, D.J., Getty, R.R., Leykam, J.F., Atkinson, J.P. & Tykocinski, M.L. (1987). Cloning and characterization of cDNAs encoding the complete sequence of decay-accelerating factor of human complement. Proc. Natl. Acad. Sci. USA 84, pp. 2007–2011.
653. Lublin, D.M., Liszewski, M.K., Post, T.W., Arce, M.A., LeBeau, M.M., Rebentisch, M.B., Lemons, R.S., Seya, T. & Atkinson, J.P. (1988). Molecular cloning and chromosomal localization of human membrane cofactor protein (MCP). J. Exp. Med. 168, pp. 181–194.
654. Goldberg, G., Burns, G.A.P., Rits, M., Edge, M.D. & Kwiatkowski, D.J. (1987). Human complement factor I: Analysis of cDNA-derived primary structure and assignment of its gene to chromosome 4. J. Biol. Chem. 262, pp. 10065–10071.
655. DiScipio, R.G. & Hugli, T.E. (1989). The molecular architecture of human complement component C6. J. Biol. Chem. 264, pp. 16197–16206.
656. Haefliger, J.-A., Tschopp, J., Vial, N. & Jenne, D. (1989). Complete primary structure and functional characterization of the sixth component of the human complement system. J. Biol. Chem. 264, pp. 18041–18051.

657. DiScipio, R.G., Chakravarti, D.N., Müller-Eberhard, H.J. & Fey, G.H. (1988). The structure of human complement component C7 and the C5b–7 complex. J. Biol. Chem. 263, pp. 549–560.
658. Leonard, W.J., Depper, J.M., Kanehisa, M., Krönke, M., Peffer, N.J., Svetlik, P.B., Sullivan, M. & Greene, W.C. (1985). Structure of the human interleukin-2 gene. Science 230, pp. 633–639.
659. Kotwal, G.J. & Moss, B. (1988). Vaccinia virus encodes a secretory polypeptide structurally related to complement control proteins. Nature 335, pp. 176–178.
660. Carrell, N.A., Erickson, H.P. & McDonagh, J. (1989). Electron microscopy and hydrodynamic properties of factor XIII subunits. J. Biol. Chem. 264, pp. 551–556.
661. Hageman, T.C. & Scheraga, H.A. (1974). Mechanism of action of thrombin on fibrinogen. Arch. Biochem. Biophys. 164, pp. 707–715.
662. Van Nispen, J.W., Hageman, T.C. & Scheraga, H.A. (1977). Mechanism of action of thrombin on fibrinogen. Arch. Biochem. Biophys. 182, pp. 227–243.
663. Marsh, H.C., Meinwald, Y.C., Lee, S. & Scheraga, H.A. (1982). Mechanism of action of thrombin on fibrinogen. Direct evidence for the involvement of phenylalanine at position P_9. Biochemistry 21, pp. 6167–6171.
664. Marsh, H.C., Meinwald, Y.C., Tannhauser, T.W. & Scheraga, H.A. (1983). Mechanism of action of thrombin on fibrinogen. Kinetic evidence for involvement of aspartic acid at position P_{10}. Biochemistry 22, pp. 4170–4174.
665. Ni, F., Konishi, Y., Fraizer, R.B., Scheraga, H.A. & Lord, S.T. (1989). High-resolution NMR studies of fibrinogen-like peptides in solution: Interaction of thrombin with residue 1–23 of the Aα chain of human fibrinogen. Biochemistry 28, pp. 3082–3094.
666. Ni, F., Meinwald, Y.S., Vásquez, M. & Scheraga, H.A. (1989). High-resolution NMR studies of fibrinogen-like peptides in solution: Structure of a thrombin-bound peptide corresponding to residues 7–16 of the Aα-chain of human fibrinogen. Biochemistry 28, pp. 3094–3105.
667. Ni, F., Konishi, Y., Bullock, L.D., Rivetna, M.N. & Scheraga, H.A. (1989). High-resolution NMR studies on fibrinogen-like peptides in solution: Structural basis for the bleeding disorder caused by a single mutation of Gly(12) to Val(12) in the Aα chain of human fibrinogen Rouen. Biochemistry 28, pp. 3106–3119.
668. Váli, Z. & Scheraga, H.A. (1988). Localization of the binding site on fibrin for the secondary binding site of thrombin. Biochemistry 27, pp. 1956–1963.
669. Kaczmarek, E. & McDonagh, J. (1988). Thrombin binding to the Aα, Bβ, and γ-chains of fibrinogen and to their remnants contained in fragment E. J. Biol. Chem. 263, pp. 13896–13900.
670. Hageman, T.C. & Scheraga, H.A. (1977). Mechanism of thrombin action on fibrinogen. Arch. Biochem. Biophys. 179, pp. 506–517.
671. Higgins, D.L., Lewis, S.D. & Schafer, J.A. (1983). Steady state kinetic parameters for the thrombin-catalyzed conversion of human fibrinogen to fibrin. J. Biol. Chem. 258, pp. 9276–9282.
672. Kudryk, B.J., Collen, D., Woods, K.R. & Blombäck, B. (1974). Evidence for localization of polymerization sites in fibrinogen. J. Biol. Chem. 249, pp. 3322–3325.
673. Hantgan, R.R. & Hermans, J. (1979). Assembly of fibrin: A light scattering study. J. Biol. Chem. 254, pp. 11272–11281.
674. Fowler, W.E., Hantgan, R.R., Hermans, J. & Erickson, H.P. (1981). Structure of the fibrin protofibril. Proc. Natl. Acad. Sci. USA 78, pp. 4872–4876.
675. Landis, W.J. & Waugh, D.F. (1975). Interaction of bovine fibrinogen and thrombin. Arch. Biochem. Biophys. 168, pp. 498–511.
676. Jammey, P.A., Erdile, L., Bale, M.D. & Ferry, J.D. (1983). Kinetics of fibrin oligomer formation observed by electron microscopy. Biochemistry 22, pp. 4336–4340.
677. Dietler, G., Känzig, W., Haeberli, A. & Straub, P.W. (1986). Experimental tests of a geometrical abstraction of fibrin polymerization. Biopolymers 25, pp. 905–929.
678. Laudano, A.P. & Doolittle, R.F. (1978). Synthetic peptide derivatives that bind to fibrinogen and prevent the polymerization of fibrin monomers. Proc. Natl. Acad. Sci. USA 75, pp. 3085–3089.
679. Laudano, A.P. & Doolittle, R.F. (1980). Studies on synthetic peptides that bind to fibrinogen and prevent fibrin polymerization. Structural requirements, number of

binding sites, and species differences. Biochemistry 19, pp. 1013–1019.
680. Olexa, S.A. & Budzynski, A.Z. (1981). Localization of a fibrin polymerization site. J. Biol. Chem. 256, pp. 3544–3549.
681. Horwitz, B.H., Váradi, A. & Scheraga, H.A. (1984). Localization of a fibrin γ-chain polymerization site within segment Thr 374 to Glu 396 of human fibrinogen. Proc. Natl. Acad. Sci. USA 81, pp. 5980–5984.
682. Southan, C., Thompson, E., Panico, M., Etienne, T., Morris, H.R. & Lane, D.A. (1985). Characterization of peptides cleaved by plasmin from the C-terminal polymerization domain of human fibrinogen. J. Biol. Chem. 260, pp. 13095–13101.
683. Blombäck, B., Hessel, B., Hogg, D. & Therkildsen, L. (1978). A two-step fibrinogen-fibrin transition in blood coagulation. Nature 275, pp. 501–505.
684. Olexa, S.A. & Budzynski, A.Z. (1980). Evidence for four different polymerization sites involved in human fibrin formation. Proc. Natl. Acad. Sci. USA 77, pp. 1374–1378.
685. Shimizu, A., Saito, Y. & Inada, Y. (1986). Distinctive role of histidine-16 of the B-chain of fibrinogen in the end-to-end association of fibrin. Proc. Natl. Acad. Sci. USA 83, pp. 591–593.
686. Shen, L.L., Hermans, J., McDonagh, J. & McDonagh, R.P. (1987). Role of fibrinopeptide B release: Comparison of fibrins produced by thrombin and ancrod. Am. J. Physiol. 232, pp. 629–633.
687. Mikuni, Y., Iwanaga, S. & Konishi, K. (1973). A peptide released from plasma fibrin stabilizing factor in the conversion to the active enzyme by thrombin. Biochem. Biophys. Res. Commun. 54, pp. 1393–1402.
688. Takagi, T. & Doolittle, R.F. (1974). Amino acid sequence studies on factor XIII and the peptide released during its activation by thrombin. Biochemistry 13, pp. 750–756.
689. Lorand, L. & Konoshi, K. (1964). Activation of fibrin stabilizing factor by thrombin. Arch. Biochem. Biophys. 105, pp. 58–67.
690. Curtis, C.G., Stenberg, P., Chou, C.-H., Gray, A., Brown, K.L. & Lorand, L. (1973). Titration and subunit localization of active center cysteine in fibrinoligase (thrombin-activated fibrin stabilizing factor). Biochem. Biophys. Res. Commun. 52, pp. 51–56.
691. Lorand, L., Gray, A., Brown, K.L., Credo, R.B., Curtis, C.G., Domanik, R.A. & Stenberg, P. (1974). Dissociation of the subunit structure of fibrin stabilizing factor during activation of the zymogen. Biochem. Biophys. Res. Commun. 56, pp. 914–922.
692. Cooke, R.D. & Holbrook, J.J. (1974). The calcium-induced dissociation of human plasma clotting factor XIII. Biochem. J. 141, pp. 79–84.
693. Cooke, R.D., Pestell, T.C. & Holbrook, J.J. (1974). Calcium and thiol reactivity of human plasma clotting factor XIII. Biochem. J. 141, pp. 675–682.
694. Cooke, R.D. (1974). Calcium-induced dissociation of human plasma factor XIII and the appearance of catalytic activity. Biochem. J. 141, pp. 683–691.
695. Credo, R.B., Curtis, C.G. & Lorand, L. (1978). Ca^{2+}-related regulatory function of fibrinogen. Proc. Natl. Acad. Sci. USA 75, pp. 4234–4237.
696. Credo, R.B., Curtis, C.G. & Lorand, L. (1981). α-Chain domain of fibrinogen controls generation of fibrinoligase (coagulation factor XIIIa). Calcium ion regulatory aspect. Biochemistry 20, pp. 3770–3778.
697. Greenberg, C.S. & Shuman, M.A. (1982). The zymogen forms of blood coagulation factor XIII bind specifically to fibrinogen. J. Biol. Chem. 257, pp. 6096–6101.
698. Janus, T.J., Lewis, S.D., Lorand, L. & Shafer, J.A. (1983). Promotion of thrombin-catalyzed activation of factor XIII by fibrinogen. Biochemistry 22, pp. 6269–6272.
699. Greenberg, C.S., Miraglia, C.C., Rickless, F.R. & Shuman, M. (1985). Cleavage of blood coagulation factor XIII and fibrinogen by thrombin during *in vitro* clotting. J. Clin. Invest. 75, pp. 1463–1470.
700. Greenberg, C.S. & Miraglia, C.C. (1985). The effect of fibrin polymers on thrombin-catalyzed plasma factor XIIIa formation. Blood 66, pp. 466–469.
701. Greenberg, C.S., Dobson, J.V. & Miraglia, C.C. (1985). Regulation of plasma factor XIII binding to fibrin *in vitro*. Blood 66, pp. 1028–1034.
702. Lewis, S.D., Janus, T.J., Lorand, L. & Shafer, J.A. (1985). Regulation of formation of factor XIIIa by its fibrin substrates. Biochemistry 24, pp. 6772–6777.
703. Greenberg, C.S., Achyuthan, K.E. & Fenton II, J.W. (1987). Factor XIIIa formation promoted by complexation of a-thrombin, fibrin, and plasma factor XIIIa. Blood 69, pp. 867–871.

704. Greenberg, C.S., Achyuthan, K.E., Rajagopalan, S. & Pizzo, S.V. (1988). Characterization of the fibrin polymer structure that accelerates thrombin cleavage of plasma factor XIII. Arch. Biochem. Biophys. 262, pp. 142–148.
705. Greenberg, C.S., Enghild, J.J., Mary, A., Dobson, J.V. & Achyuthan, K.E. (1988). Isolation of a fibrin-binding fragment from blood coagulation factor XIII capable of cross-linking fibrin(ogen). Biochem. J. 256, pp. 1013–1019.
706. Mary, A., Achyuthan, K.E. & Greenberg, C.S. (1987). Factor XIII binds to the Aα-and Bβ-chains in the D-domain of fibrinogen: An immunoblotting study. Biochem. Biophys. Res. Commun. 147, pp. 608–614.
707. Matacic, S. & Loewy, A.G. (1968). The identification of isopeptide crosslinks in insoluble fibrin. Biochem. Biophys. Res. Commun. 30, pp. 356–362.
708. Pisano, J.J., Finlayson, J.S. & Peyton, M.P. (1968). Crosslink in fibrin polymerized by factor XIII: (ϵ-(γ-glutamyl)-lysine. Science 160, pp. 892–893.
709. Lorand, L., Downey, J., Gotoh, T., Jacobsen, A. & Tokura, S. (1968). The transpeptidase system which crosslinks fibrin by γ-glutamyl-ϵ-lysine bonds. Biochem. Biophys. Res. Commun. 31, pp. 222–230.
710. Doolittle, R.F., Chen, R. & Lau, F. (1971). Hybrid fibrin: Proof of the intermolecular nature of γ-γ-crosslinking units. Biochem. Biophys. Res. Commun. 44, pp. 94–100.
711. Schwartz, M.L., Pizzo, S.V., Hill, R.L. & McKee, P.A. (1971). The effect of fibrin stabilizing factor on the subunit structure of human fibrin. J. Clin. Invest. 50, pp. 1506–1513.
712. Chen, R. & Doolittle, R.F. (1970). Isolation, characterization, and location of a donor-acceptor unit from cross-linked fibrin. Proc. Natl. Acad. Sci. USA 66, pp. 472–479.
713. Chen, R. & Doolittle, R.F. (1971). γ-γ Cross-linking sites in human and bovine fibrin. Biochemistry 10, pp. 4486–4491.
714. Purves, L., Purves, M. & Brandt, W. (1987). Cleavage of fibrin-derived D-dimer into monomers by endopeptidase from puff adder venom (Bitis arietans) acting at cross-linked sites of the γ-chain. Sequence of carboxy-terminal cyanogen bromide γ-chain fragments. Biochemistry 26, pp. 4640–4646.
715. Cottrell, B.A., Strong, D.D., Watt, K.W.K. & Doolittle, R.F. (1979). Amino acid sequence studies on the α chain of human fibrinogen. Exact location of cross-linking acceptor sites. Biochemistry 18, pp. 5405–5410.
716. Corcoran, D.H., Ferguson, E.W., Fretto, L.J. & McKee, P. A. (1980). Localization of a cross-link donor site in the α-chain of human fibrin. Thromb. Res. 19, pp. 883–888.
717. Mosesson, M.W., Siebenlist, K.R., Amrani, D.L. & DiOrio, J.P. (1989). Identification of covalently linked trimeric and tetrameric D domains in crosslinked fibrin. Proc. Natl. Acad. Sci. USA 86, pp. 1113–1116.
718. Erickson, H.P. & Fowler, W.E. (1983). Electron microscopy of fibrinogen, its plasmic fragments and small polymers. Ann. N. Y. Acad. Sci. 408, pp. 146–163.
719. Selmayr, E., Thiel, W. & Müller-Berghaus, G. (1985). Crosslinking of fibrinogen to immobilized desAA-fibrin. Thromb. Res. 39, pp. 459–465.
720. Selmayr, E., Deffner, M., Bachmann, L. & Müller-Berghaus, G. (1988). Chromatography and electron microscopy of cross-linked fibrin polymers – a new model describing the cross-linking at the DD-trans contact of the fibrin molecules. Biopolymers 27, pp. 1733–1748.
721. Shen, L. & Lorand, L. (1983). Contribution of fibrin stabilization to clot strength. J. Clin. Invest. 71, pp. 1336–1341.
722. Lorand, L., Losowsky, M.S. & Miloszewski, K.J.M. (1980). Human factor XIII: Fibrin stabilizing factor. Progress in Hemostasis and Thrombosis 5, pp. 245–290.
723. Sakata, Y. & Aoki, N. (1980). Cross-linking of α_2-plasmin inhibitor to fibrin by fibrin-stabilizing factor. J. Clin. Invest. 65, pp. 290–297.
724. Tamaki, T. & Aoki, N. (1981). Cross-linking of α_2-plasmin inhibitor and fibronectin to fibrin by fibrin-stabilizing factor. Biochim. Biophys. Acta 661, pp. 280–286.
725. Sakata, Y. & Aoki, N. (1982). Significance of cross-linking of α_2-plasmin inhibitor to fibrin in inhibition of fibrinolysis and in hemostasis. J. Clin. Invest. 69, pp. 536–542.
726. Tamaki, T. & Aoki, N. (1982). Cross-linking of α_2-plasmin inhibitor to fibrin catalyzed by activated fibrin-stabilizing factor. J. Biol. Chem. 257, pp. 14767–14772.
727. Kimura, S. & Aoki, N. (1986). Cross-linking site in fibrinogen for α_2-plasmin inhibitor. J. Biol. Chem. 261, pp. 15591–15595.

728. Mosher, D.F. (1975). Cross-linking of cold-insoluble globulin by fibrin-stabilizing factor. J. Biol. Chem. 250, pp. 6614–6621.
729. Mosher, D.F. (1976). Action of fibrin-stabilizing factor on cold-insoluble globulin and α_2-macroglobulin in clotting plasma. J. Biol. Chem. 251, pp. 1639–1645.
730. McDonagh, R.P., McDonagh, J., Petersen, T.E., Thøgersen, H.C., Skorstengaard, K., Sottrup-Jensen, L. & Magnusson, S. (1981). Amino acid sequence of the factor XIIIa acceptor site in bovine plasma fibronectin. FEBS Lett. 127, pp. 174–178.
731. Okada, M., Blombäck, B., Chang, M.-D. & Horowitz, B. (1985). Fibronectin and fibrin gel structure. J. Biol. Chem. 260, pp. 1811–1820.
732. Carr, M.E., Gabriel, D.A. & McDonagh, J. (1987). Influence of factor XIII and fibronectin on fiber size and density in thrombin-induced fibrin-gels. J. Lab. Clin. Med. 110, pp. 747–752.
733. Mosher, D.F., Schad, P.E. & Kleinman, H.K. (1979). Cross-linking of fibronectin to collagen by blood coagulation factor XIIIa. J. Clin. Invest. 64, pp. 781–787.
734. Mosher, D.F., Schad, P.E. & Vann, J.M. (1980). Cross-linking of collagen and fibronectin by factor XIIIa. J. Biol. Chem. 255, pp. 1181–1188.
735. Hada, M., Kaminski, M., Bockenstedt, P. & McDonagh, J. (1986). Covalent crosslinking of von Willebrand factor to fibrin. Blood 68, pp. 95–101.
736. Bale, M.D., Westrick, L.G. & Mosher, D.F. (1985). Incorporation of thrombospondin into fibrin clots. J. Biol. Chem. 260, pp. 7502–7508.
737. Bale, M.D. & Mosher, D.F. (1986). Thrombospondin is a substrate for blood coagulation factor XIIIa. Biochemistry 25, pp. 5667–5673.
738. Bockenstedt, P., McDonagh, J. & Handin, R.I. (1986). Binding and covalent cross-linking of purified von Willebrand factor to native monomeric collagen. J. Clin. Invest. 78, pp. 551–556.
739. Francis, R.T., McDonagh, J. & Mann, K.G. (1986). Factor V is a substrate for the transamidase factor XIIIa. J. Biol. Chem. 261, pp. 9787–9792.
740. Huh, M.M., Schick, B.P., Schick, P.K. & Colman, R.W. (1988). Covalent crosslinking of human coagulation factor V by activated factor XIII from guinea pig megakaryocytes and human plasma. Blood 71, pp. 1693–1702.
741. Sane, D.C., Moser, T.L., Pippen, A.M.M., Parker, C.J., Achyuthan, K.E. & Greenberg, C.S. (1988). Vitronectin is a substrate for transglutaminases. Biochem. Biophys. Res. Commun. 157, pp. 115–120.
742. Asijee, G.M., Muszbek, L., Kappelmayer, J., Polgár, J., Horváth, A. & Sturk, A. (1988). Platelet vinculin: A substrate for activated factor XIII. Biochim. Biophys. Acta 954, pp. 303–308.
743. Halkier, T., Vestergaard, A.B., Andersen, H.F. & Magnusson, S. (1990). Histidine-rich glycoprotein inhibits contact activation of blood coagulation and functions as a substrate for plasma factor XIIIa. Submitted.
744. Travis, J. & Salvesen, G.S. (1983). Human plasma proteinase inhibitors. Ann. Rev. Biochem. 52, pp. 655–709.
745. Carrell, R.W., Jeppsson, J.-O., Laurell, C.-B., Brennan, S.O., Owen, M.C., Vaughan, L. & Boswell, D.R. (1982). Structure and variation of human α_1-antitrypsin. Nature 298, pp. 329–334.
746. Long, G.L., Chandra, T., Woo, S.L.C., Davie, E.W. & Kurachi, K. (1984). Complete sequence of the cDNA for human α_1-antitrypsin and the gene for the S variant. Biochemistry 23, pp. 4828–4837.
747. Petersen, T.E., Dudek-Wojchiechowska, G., Sottrup-Jensen, L. & Magnusson, S. (1979). Primary structure of antithrombin III (heparin cofactor). Partial homology between α_1-antitrypsin and antithrombin III. In: The Physiological Inhibitors of Blood Coagulation and Fibrinolysis (Collen, D., Wiman, B. & Verstraete, M. Eds.), pp. 43–54, Elsevier, Amsterdam.
748. Bock, S.C., Wion, K.L., Vehar, G.A. & Lawn, R.M. (1982). Cloning and expression of the cDNA for human antithrombin III. Nucl. Acids Res. 10, pp. 8113–8125.
749. Bock, S.C., Skriver, K., Nielsen, E., Thøgersen, H.C., Wiman, B., Donaldson, V.H., Eddy, R.L., Marrinan, J., Radziejewska, E., Huber, R., Shows, T.B. & Magnusson, S. (1986). Human $C\bar{1}$-inhibitor: Primary structure, cDNA cloning, and chromosomal localization. Biochemistry 25, pp. 4292–4301.

750. Pixley, R.A., Schapira, M. & Colman, R.W. (1985). The regulation of human factor XIIa by plasma proteinase inhibitors. J. Biol. Chem. 260, pp. 1723–1729.
751. Harpel, P.C., Lewin, M.F. & Kaplan, A.P. (1985). Distribution of plasma kallikrein between C1 inactivator and α_2-macroglobulin in plasma utilizing a new assay for α_2-macroglobulin-kallikrein complexes. J. Biol. Chem. 260, pp. 4257–4263.
752. Scott, C.F., Schapira, M., James, H.L., Cohen, A.B. & Colman, R.W. (1982). Inactivation of factor XIa by plasma proteinase inhibitors. J. Clin. Invest. 69, pp. 844–852.
753. Soons, H., Janssen-Claessen, T., Tans, G. & Hemker, H.C. (1987). Inhibition of factor XIa by antithrombin III. Biochemistry 26, pp. 4624–4629.
754. Meijers, J.C.M., Vlooswijk, R.A.A. & Bouma, B.N. (1988). Inhibition of human blood coagulation factor XIa by $C\bar{1}$-inhibitor. Biochemistry 27, pp. 959–963.
755. Abildgaard, U. (1968). Highly purified antithrombin III with heparin cofactor activity prepared by disc electrophoresis. Scand. J. Clin. Lab. Invest. 21, pp. 89–91.
756. Rosenberg, R.D. & Damus, P.S. (1973). The purification and mechanism of action of human antithrombin-heparin cofactor. J. Biol. Chem. 248, pp. 6490–6505.
757. Österud, B., Miller-Andersson, M., Abildgaard, U. & Prydz, H. (1976). The effect of antithrombin III on the activity of the coagulation factors VII, IX and X. Thromb. Haemostas. 35, pp. 295–304.
758. Jesty, J. (1978). The inhibition of activated bovine coagulation factors X and VII by antithrombin III. Arch. Biochem. Biophys. 185, pp. 165–173.
759. Kurachi, K., Schmer, G., Hermodson, M.A., Teller, D.C. & Davie, E.W. (1976). Inhibition of bovine factor IXa and factor Xa_β by antithrombin III. Biochemistry 15, pp. 373–377.
760. Nordenman, B., Nystrøm, C. & Bjørk, I. (1979). The size and shape of human and bovine antithrombin III. Eur. J. Biochem. 78, pp. 195–204.
761. Murano, G., Williams, L., Miller-Anderson, M., Aronson, D.L. & King, C. (1980). Some properties of antithrombin III and its concentration in human plasma. Thromb. Res. 18, pp. 259–262.
762. Chandra, T., Stackhouse, R., Kidd, V.J. & Woo, S.L.C. (1983). Isolation and sequence characterization of a cDNA clone of human antithrombin III. Proc. Natl. Acad. Sci. USA 80, pp. 1845–1848.
763. Prochownik, E.V., Markman, A.F. & Orkin, S.H. (1983). Isolation of a cDNA clone for human antithrombin III. J. Biol. Chem. 258, pp. 8389–8394.
764. Prochownik, E.V., Bock, S.C. & Orkin, S.H. (1985). Intron structure of the human antithrombin III gene differs from that of other members of the serine protease inhibitor superfamily. J. Biol. Chem. 260, pp. 9608–9612.
765. Kao, F.T., Morse, H.G., Law, M.L., Lidsky, A., Chandra, T. & Woo, S.L.C. (1984). Genetic mapping of the structural gene for antithrombin III to human chromosome 1. Hum. Genet. 67, pp. 34–36.
766. Bock, S.C., Harris, J.F., Balazs, I. & Trent, J.M. (1985). Assignment of the human antithrombin III structural gene to chromosome 1q23–25. Cytogenet. Cell Genet. 39, pp. 67–69.
767. Björk, I., Danielsson, Å., Fenton, J.W. & Jörnvall, H. (1981). The site in human antithrombin for functional proteolytic cleavage by human thrombin. FEBS Lett. 126, pp. 257–260.
768. Björk, I., Jackson, C.M., Jörnvall, H., Lavine, K.K., Nordling, K. & Salsgiver, W.J. (1982). The active site of antithrombin. J. Biol. Chem. 257, pp. 2406–2411.
769. Rosenberg, R. (1987). Regulation of the hemostatic mechanism. In: The Molecular Basis of Blood Diseases (Stamatoyannopoulos, G., Nienhuis, A.W., Leder, P. & Majerus, P.W., Eds.), pp.575–630, W.B. Saunders, Philadelphia.
770. Oscarson, L.-G., Pejler, G. & Lindahl, U. (1989). Location of the antithrombin-binding sequence in the heparin chain. J. Biol. Chem. 264, pp. 296–304.
771. Chang, J.-Y. & Tran, T.H. (1986). Antithrombin III_{Basel}. J. Biol. Chem. 261, pp. 1174–1176.
772. Brunel, F., Duchange, N., Fischer, A.M., Cohen, G.U. & Zakin, M.M. (1987). Antithrombin III Alger: A new case of Arg47 to Cys mutation. Am. J. Hematol. 25, pp. 223–224.

773. Duchange, N., Chasse, J.-F., Cohen, G.N. & Zakin, M.M. (1986). Antithrombin III Tours gene: Identification of a point mutation leading to an arginine-cysteine replacement in a silent deficiency. Nucl. Acids Res. 14, p. 2408.
774. Koide, T., Odani, S., Takahashi, K., Ono, T. & Sukuragawa, N. (1984). Antithrombin III Toyama: Replacement of arginine-47 by cysteine in hereditary abnormal antithrombin III that lacks heparin-binding ability. Proc. Natl. Acad. Sci. USA 81, pp. 289–293.
775. Owen, M.C., Borg, J.Y., Soria, C., Soria, J., Caen, J. & Carrell, R.W. (1987). Heparin binding defect in a new antithrombin III variant: Rouen, 47 Arg to His. Blood 69, pp. 1275–1279.
776. Borg, J.Y., Owen, M.C., Soria, C., Soria, J., Caen, J. & Carrell,R.W. (1988). Proposed heparin binding site in antithrombin based on arginine 47. J. Clin. Invest. 81, pp. 1292–1296.
777. Blackburn, M.N., Smith, R.L., Carson, J. & Sibley, C.C. (1984). The heparin-binding site of antithrombin III. J. Biol. Chem. 259, pp. 939–941.
778. Chang, J.-Y. (1989). Binding of heparin to human antithrombin III activates selective chemical modification at lysine 236. J. Biol. Chem. 264, pp. 3111–3115.
779. Liu, C.-S. & Chang, J.-Y. (1987). The heparin binding site of human antithrombin III. J. Biol. Chem. 262, pp. 17356-17361.
780. Peterson, C.B., Noyes, C.M., Pecon, J.M., Church, F.C. & Blackburn, M.N. (1987). Identification of a lysyl residue in antithrombin which is essential for heparin binding. J. Biol. Chem. 262, pp. 8061–8065.
781. Brennan, S.O., George, P.M. & Jordan, R.E. (1987). Physiological variant of antithrombin III lacks carbohydrate side chain at Asn135. FEBS Lett. 219, pp. 431–436.
782. Brennan, S.O., Borg, J.-Y., George, P.M., Soria, C., Soria, J., Caen, J. & Carrell, R.W. (1988). New carbohydrate site in mutant antithrombin (7Ile → Asn) with decreased heparin affinity. FEBS Lett. 237, pp. 118–122.
783. Molho-Sabatier, P., Aiach, M., Gaillard, I., Fiessinger, J.-N., Fisher, A.-M., Chadeuf, G. & Clauser, E. (1989). Molecular characterization of antithrombin III (ATIII) variants using polymerase chain reaction. J. Clin. Invest. 84, pp. 1236–1242.
784. Devraj-Kizuk, R., Chui, D.H.K., Prochownik, E.V., Carter, C.J., Ofuso, F.A. & Blajchman, M.A. (1988). Antithrombin-III-Hamilton: A gene with a point mutation (guanine to adenine in codon 382) causing impaired serine protease reactivity. Blood 72, pp. 1518–1523.
785. Sørensen, L. & Petersen, T.E. (1989). Personal communication.
786. Perry, D.J., Harper, P.L., Fairham, S., Daly, M. & Carrell, R.W. (1989). Antithrombin Cambridge, 384 Ala to Pro: A new variant identified using the polymerase chain reaction. FEBS Lett. 254, pp. 174–176.
787. Erdjument, H., Lane, D.A., Panico, M., Di Marzo, V. & Morris, H.R. (1988). Single amino acid substitutions in the reactive site of antithrombin leading to thrombosis. J. Biol. Chem. 263, pp. 5589–5593.
788. Thein, S.L. & Lane, D.A. (1988). Use of synthetic oligonucleotides in the characterization of antithrombin III Northwick Park (393CGT → TGT) and antithrombin III Glasgow (393CGT → CAT). Blood 72, pp. 1817–1821.
789. Lane, D.A., Erdjument, H., Flynn, A., Di Marzo, V., Panico, M., Morris, H.R., Greaves, M., Dolan, G. & Preston, F.E. (1989). Antithrombin Sheffield: Amino acid substitution at the reactive site (Arg393 to His) causing thrombosis. Br. J. Haematol. 71, pp. 91–96.
790. Erdjument, H., Lane, D.A., Panico, M., Di Marzo, V., Morris, H.R., Bauer, K.A. & Rosenberg, R.D. (1989). Antithrombin Chicago, amino acid substitution of arginine 393 to histidine. Thromb. Res. 54, pp. 613–619.
791. Erdjument, H., Lane, D.A., Ireland, H., Di Marzo, V., Panico, M., Morris, H.R., Tripodi, A. & Manucci, P.M. (1988). Antithrombin Milano, single amino acid substitution at the reactive site, Arg393 to Cys. Thromb. Haemostas. 60, pp. 471–475.
792. Lane, D.A., Erdjument, H., Thompson, E., Panico, M., Di Marzo, V., Morris, H.R., Leone, G., De Stefano, V. & Thein, S.L. (1989). A novel amino acid substitution in the reactive site of a congenital variant antithrombin. J. Biol. Chem. 264, pp. 10200–10204.
793. Stephens, A.W., Thalley, B.S. & Hirs, C.H.W. (1987). Antithrombin-III Denver, a reactive site variant. J. Biol. Chem. 262, pp. 1044–1048.

794. Bock, S.C., Silberman, J.A., Wikoff, W., Abildgaard, U. & Hultin, M.B. (1989). Identification of a threonine for alanine substitution at residue 404 of antithrombin III Oslo suggests integrity of the 404–407 region is important for maintaining normal plasma inhibitor levels. Thromb. Haemostas. 62, Abstract 1550.
795. Bock, S.C., Marrinan, J.A. & Radziejewska, E. (1988). Antithrombin III Utah: Proline-407 to leucine mutation in a highly conserved region near the inhibitor reactive site. Biochemistry 27, pp. 6171–6178.
796. Rapaport, S.I. (1989). Inhibition of factor VIIa/tissue factor-induced blood coagulation, with particular emphasis upon a factor Xa-dependent inhibitory mechanism. Blood 73, pp. 359–365.
797. Hjorth, P.F (1957). Intermediate reactions in the coagulation of blood with tissue thromboplastin. Scan. J. Clin. Lab. Invest. 9, Suppl. 27, pp. 1–183.
798. Sanders, N.L., Bajaj, S.P., Zivelin, A. & Rapaport, S.I. (1985). Inhibition of tissue factor/factor VIIa activity in plasma requires factor X and an additional plasma component. Blood 66, pp. 204–212.
799. Rao, L.V.M. & Rapaport, S.I. (1987). Studies of a mechanism inhibiting the initiation of the extrinsic pathway of coagulation. Blood 69, pp. 645–651.
800. Hubbard, A.R. & Jennings, C.A. (1986). Inhibition of tissue thromboplastin-mediated blood coagulation. Thromb. Res. 42, pp. 489–498.
801. Broze Jr., G.J. & Miletich, J.P. (1987). Characterization of the inhibition of tissue factor in serum. Blood 69, pp. 150–155.
802. Sandset, P.M., Abildgaard, U.A. & Pettersen, M. (1987). A sensitive assay of extrinsic coagulation pathway inhibitor (EPI) in plasma and plasma fractions. Thromb. Res. 47, pp. 389–400.
803. Broze Jr., G.J. & Miletich, J.P. (1987). Isolation of the tissue factor inhibitor produced by HepG2 heptoma cells. Proc. Natl. Acad. Sci. USA 84, pp. 1886–1890.
804. Warn-Cramer, B.J., Rao, L.V.H., Maki, S.L. & Rapaport, S.I. (1988). Modification of extrinsic pathway inhibitor (EPI) and factor Xa that affects their ability to interact and to inhibit factor VIIa/tissue factor: Evidence for a two-step model of inhibition. Thromb. Haemostas. 60, pp. 453–456.
805. Broze Jr., G.J., Warren, L.A., Novotny, W.F., Higuchi, D.A., Girard, J.J. & Miletich, J.P. (1987). The lipoprotein-associated coagulation inhibitor that inhibits the factor VII-tissue factor complex also inhibits factor Xa: Insight into its possible mechanism of action. Blood 71, pp. 335–343.
806. Warn-Cramer, B.J., Maki, S.L., Zivelin, A. & Rapaport, S.I. (1987). Partial purification and characterization of extrinsic pathway inhibitor (the factor Xa-dependent inhibitor of factor VIIa/tissue factor). Thromb. Res. 48, pp. 11–22.
807. Hubbard, A.R. & Jennings, C.A. (1987). Inhibition of the tissue factor factor VII complex: Involvement of factor Xa and lipoproteins. Thromb. Res. 46, pp. 527–537.
808. Novotny, W.F., Girard, T.J., Miletich, J.P. & Broze Jr., G.J. (1989). Purification and characterization of the lipoprotein-associated coagulation inhibitor from human plasma. J. Biol. Chem. 264, pp. 18832–18837.
809. Wun, T.-C, Kretzmer, K.K., Girard, T.J., Miletich, J.P. & Broze Jr., G.J. (1988). Cloning and characterization of a cDNA coding for the lipoprotein-associated coagulation inhibitor shows that it consists of three tandem Kunitz-type inhibitory domains. J. Biol. Chem. 263, pp. 6001–6004.
810. Girard, T.J., Warren, L.A., Novotny, W.F., Likert, K.M., Brown, S.G., Miletich, J.P. & Broze Jr., G.J. (1989). Functional significance of the Kunitz-type inhibitory domains of lipoprotein-associated coagulation inhibitor. Nature 338, pp. 518–520.
811. Esmon, C.T. (1987). The regulation of natural anticoagulant pathways. Science 235, pp. 1348–1352.
812. Esmon, C.T. (1989). The roles of protein C and thrombomodulin in the regulation of blood coagulation. J. Biol. Chem. 264, pp. 4743–4746.
813. Dahlbäck, B., Fernlund, P. & Stenflo, J. (1986). Inhibitors: Protein C. In: New Comprehensive Biochemistry Vol. 13 (Zwaal, R.F.A. & Hemker, H.C., Eds.), pp. 285–306, Elsevier, Amsterdam.
814. Kisiel, W. (1979). Human plasma protein C. J. Clin. Invest. 64, pp. 761–769.
815. Heeb, M.J., Schwartz, H.P., White, T., Lämmle, B., Berrettini, M. & Griffin, J.H. (1988). Immunoblotting studies of the molecular forms of protein C in plasma.

Thromb. Res. 52, pp. 33–43.

816. Bauer, K.A., Kass, B.L., Beeler, D.L. & Rosenberg, R.D. (1984). Detection of protein C activation in humans. J. Clin. Invest. 74, pp. 2033–2041.
817. Stenflo, J. (1976). A new vitamin K-dependent protein. J. Biol. Chem. 251, pp. 355–363.
818. Seegers, W.H., Novoa, E., Henry, R.L. & Hassouna, H.I. (1976). Relatioship of 'new' vitamin K-dependent protein C and 'old' autoprothrombin IIa. Thromb. Res. 8, pp. 543–552.
819. Mammen, E.F., Thomas, W.R. & Seegers, W.H. (1960). Activation of purified prothrombin to autoprothrombin I or autoprothrombin II (platelet cofactor II) or autoprothrombin IIa. Thromb. Diath. Haemorrh. 5., pp. 218–249.
820. Fernlund, P. & Stenflo, J. (1982). Amino acid sequence of the light chain of bovine protein C. J. Biol. Chem. 257, pp. 12170–12179.
821. Stenflo, J. & Fernlund, P. (1982). Amino acid sequence of the heavy chain of bovine protein C. J. Biol. Chem. 257, pp. 12180–12190.
822. Long, G.L., Belagaje, R.M. & MacGillivray, R.T.A. (1984). Cloning and sequencing of liver cDNA coding for bovine protein C. Proc. Natl. Acad. Sci. USA 81, pp. 5653–5656.
823. Beckmann, R.J., Schmidt, R.J., Santerre, R.F., Plutzky, J., Crabtree, G.R. & Long, G.L. (1985). The structure and evolution of a 461 amino acid human protein C precursor and its messenger RNA, based upon the DNA sequence of cloned human liver cDNAs. Nucl. Acids Res. 13, pp. 5233–5247.
824. Foster, D.C., Yoshitake, S. & Davie, E.W. (1985). The nucleotide sequence of the gene for human protein C. Proc. Natl. Acad. Sci. USA 82, pp. 4673–4677.
825. Plutzky, J., Hoskins, J.A., Long, G.L. & Crabtree, G.R. (1986). Evolution and organization of human protein C gene. Proc. Natl. Acad. Sci. USA 83, pp. 546–550.
826. Rocchi, M., Roncuzzi, L., Santamaria, R., Archidiacono, N., Dente, L. & Romeo, G. (1986). Mapping through somatic cell hybrids and cDNA probes of protein C to chromosome 2, factor X to chromosome 13 and α1-acid glycoprotein to chromosome 9. Hum. Genet. 74, pp. 30–33.
827. Long, G.L., Marshall, A., Gardner, J.C. & Naylor, S.L. (1988). Genes for human vitamin K-dependent plasma proteins C and S are located on chromosome 2 and 3, respectively. Somat. Cell Mol. Genet. 14, pp. 93–98.
828. Kato, A., Miura, O., Sumi, Y. & Aoki, N. (1988). Assignment of the human protein C gene (PROC) to chromosome region 2q14–q21 by in situ hybridization. Cytogenet. Cell Genet. 47, pp. 46–47.
829. Patracchini, P., Aiello, V., Palazzi, P., Calzolari, E. & Bernardi, F. (1989). Sublocalization of the human protein C gene on chromosome 2q13–14. Hum. Genet. 81, pp. 191–192.
830. Drakenberg, T., Fernlund, P., Roepstorff, P. & Stenflo, J. (1983). β-Hydroxyaspartic acid in vitamin K-dependent protein C. Proc. Natl. Acad. Sci. USA 80, pp. 1802–1806.
831. Griffin, J.H., Evatt, B., Zimmerman, T.S., Kleiss, A.J. & Wideman, C. (1981). Deficiency of protein C in congenital thrombotic research. J. Clin. Invest. 68, pp. 1370–1373.
832. Bertina, R.M., Broekmans, A.W., van der Linden, I.K. & Mertens, K. (1982). Protein C deficiency in a Dutch family with thrombotic disease. Thromb. Haemostas. 48, pp. 1–5.
833. Seligsohn, U., Berger, A., Abend, M., Rubin, L., Attias, D., Zivelin, A. & Rapaport, S.I. (1984). Homozygous protein C deficiency manifested by massive venous thrombosis in the newborn. N. Eng. J. Med. 310, pp. 559–562.
834. Benson, H., Katz, J., Marble, R. & Griffin, J.H. (1983). Inherited protein C deficiency and a coumarin-responsive chronic relapsing purpura fulminans in a neonatate. Lancet 2, pp. 1156–1168.
835. Marciniak, E., Wilson, H.D. & Marlar, R.A. (1985). Neonatal purpura fulminans: A genetic disorder related to the absence of protein C in blood. Blood 65, pp. 15–20.
836. Romeo, G., Hassan, H.J., Staempfli, S., Roncuzzi, L., Cianetti, L., Leonardi, A., Vicente, V., Manucci, P.M., Bertina, R., Peschle, C. & Cortese, R. (1987). Hereditary thrombophilia: Identification of nonsense and missense mutations in the protein C gene. Proc. Natl. Acad. Sci. USA 84, pp. 2829–2832.
837. Matsuda, M., Sugo, T., Sakata, Y., Murayama, H., Mimuro, J., Tanabe, S. & Yoshitake, S. (1988). A thrombotic state due to an abnormal protein C. N. Eng. J. Med. 319, pp. 1265–1268.

838. Esmon, C.T. & Owen, W.G. (1981). Identification of an endothelial cell cofactor for thrombin catalyzed activation of protein C. Proc. Natl. Acad. Sci. USA 78, pp. 2249–2252.
839. Owen, W.G. & Esmon, C.T. (1981). Functional properties of an endothelial cell cofactor for thrombin-catalyzed activation of protein C. J. Biol. Chem. 256, pp. 5532–5535.
840. Esmon, N.L., Owen, W.G. & Esmon, C.T. (1982). Isolation of a membrane-bound cofactor for thrombin-catalyzed activation of protein C. J. Biol. Chem. 257, pp. 859–864.
841. Maruyama, I. (1987). Protein C-thrombomodulin system. Blood & Vessel 18, pp. 289–298.1–1228.
842. Esmon, C.T., Esmon, N.L. & Harris, K.W. (1982). Complex formation between thrombin and thrombomodulin inhibits both thrombin catalyzed fibrin formation and factor V activation. J. Biol. Chem. 257, pp. 7944–7947.
843. Jakubowski, H.V., Cline, M.D. & Owen, W.G. (1986). The effect of bovine thrombomodulin on the specificity of bovine thrombin. J. Biol. Chem. 261, pp. 3876–3882.
844. Polgár, J., Léránt, I., Muszbek, L. & Machovich, R. (1987). Thrombomodulin inhibits the activation of factor XIII by thrombin. Thromb. Haemostas. 58, Abstract 506.
845. Esmon, N.L., Caroll, R.C. & Esmon, C.T. (1983). Thrombomodulin blocks the ability of thrombin to activate platelets. J. Biol. Chem. 258, pp. 12238–12242.
846. Musci, G., Berliner, L.J. & Esmon, C.T. (1988). Evidence for multiple conformational changes in the active center of thrombin induced by complex formation with thrombomodulin: An analysis employing nitroxide spin-labels. Biochemistry 27, pp. 769–773.
847. Salem, H.H., Maruyama, I., Ishii, H. & Majerus, P.W. (1984). Isolation and characterization of thrombomodulin from human placenta. J. Biol. Chem. 259, pp. 12246–12251.
848. Jackman, R.W., Beeler, D.L., VanDeWater, L. & Rosenberg, R.D. (1986). Characterization of a thrombomodulin cDNA reveals structural similarity to the low density lipoprotein receptor. Proc. Natl. Acad. Sci. USA 83, pp. 8834–8838.
849. Dittman, W.A., Kumada, T., Sadler, J.E. & Majerus, P.W. (1988). The structure and function of mouse thrombomodulin. J. Biol. Chem. 263, pp. 15815–15822.
850. Suzuki, K., Kusumoto, H., Deyashiki, Y., Nishioka, J., Maruyama, I., Zushi, M., Kawahara, S., Honda, G., Yamamoto, S. & Horiguchi, S. (1987). Structure and expression of human thrombomodulin, a thrombin receptor on endothelium acting as a cofactor for protein C activation. EMBO J. 6, pp. 1891–1897.
851. Wen, D., Dittman, W.A., Ye, R.D., Deaven, L.L., Majerus, P.W. & Sadler, J. E. (1987) Human thrombomodulin: Complete cDNA sequence and chromosome localization of the gene. Biochemistry 26, pp. 4350–4357.
852. Jackman, R.W., Beeler, D.L., Fritze, L., Soff, G. & Rosenberg, R.D. (1987). Human thrombomodulin gene is intron depleted: Nucleic acid sequences of the cDNA and gene predict protein structure and suggest sites of regulatory control. Proc. Natl. Acad. Sci. USA 84, pp. 6425–6429.
853. Shirai, T., Shiojiri, S., Ito, H., Yamamoto, S., Kusomoto, H., Deyashiki, Y., Muruyama, I. & Suzuki, K. (1988). Gene structure of human thrombomodulin, a cofactor for thrombin-catalyzed activation of protein C. J. Biochem. (Tokyo) 103, pp. 281–285.
854. Maruyama, I. & Majerus, P.W. (1985). The turnover of thrombin-thrombomodulin complex in cultured human umbilical vein endothelial cells and A549 lung cancer cells. J. Biol. Chem. 260, pp. 15432–15438.
855. Maruyama, I. & Majerus, P.W. (1987). Protein C inhibits endocytosis of thrombin-thrombomodulin complexes in A549 lung cancer cells and human umbilical vein endothelial cells. Blood 69, pp. 1481–1484.
856. Hofsteenge, J., Taguchi, H. & Stone, S.R. (1986). Effect of thrombomodulin on the kinetics of the interaction of thrombin with substrates and inhibitors. Biochem. J. 237, pp. 243–251.
857. Preissner, K.T., Delvos, U. & Müller-Berghaus, G. (1987). Binding of thrombin to thrombomodulin accelerates inhibition of the enzyme by antithrombin III. Evidence

for a heparin-independent mechanism. Biochemistry 26, pp. 2521–2528.
858. Lollar, P. & Owen, W.G. (1980). Clearance of thrombin from circulation in rabbits by high-affinity binding sites on endothelium. J. Clin. Invest. 66, pp. 1222–1230.
859. Suzuki, K., Kusumoto, H. & Hashimoto, S. (1986). Isolation and characterization of thrombomodulin from bovine lung. Biochim. Biophys. Acta 882, pp. 343–352.
860. Bourin, M.-C., Boffa, M.-C., Björk, I. & Lindahl, U. (1986). Functional domains of rabbit thrombomodulin. Proc. Natl. Acad. Sci. USA 83, pp. 5924–5928.
861. Bourin, M.-C., Öhlin, A.-K., Lane, D.A., Stenflo, J. & Lindahl, U. (1988). Relationship between anticoagulant activities and polyanionic properties of rabbit thrombomodulin. J. Biol. Chem. 263, pp. 8044–8052.
862. Hofstenge, J. & Stone, S. (1987). The effect of thrombomodulin on the cleavage of fibrinogen and fibrinogen fragments by thrombin. Eur. J. Biochem. 168, pp. 49–56.
863. Freyssinet, J.-M., Gauchy, J. & Cazenave, J.-P. (1986). The effect of phospholipids on the activation of protein C by the human thrombin-thrombomodulin complex. Biochem. J. 238, pp. 151–157.
864. Johnson, A.E., Esmon, N.L., Laue, T.M. & Esmon, C.T. (1983). Structural changes required for activation of protein C are induced by Ca^{2+} binding to a high affinity site that does not contain γ-carboxyglutamic acid. J. Biol. Chem. 258, pp. 5554–5560.
865. Esmon, N.L., DeBault, L.E. & Esmon, C.T. (1983). Proteolytic formation and properties of γ-carboxyglutamic acid-domain-less protein C. J. Biol. Chem. 258, pp. 5548–5553.
866. Öhlin, A.-K. & Stenflo, J. (1987). Calcium-dependent interaction between the epidermal growth factor precursor-like region of human protein C and a monoclonal antibody. J. Biol. Chem. 262, pp. 13798–13804.
867. Öhlin, A.-K., Linse, S. & Stenflo, J. (1988). Calcium binding to the epidermal growth factor homology region of bovine protein C. J. Biol. Chem. 263, pp. 7411–7417.
868. Öhlin, A.-K., Landes, G., Bourdon, P., Oppenheimer, C., Wydro, R. & Stenflo, J. (1988). β-Hydroxyaspartic acid in the first epidermal growth factor-like domain of protein C. J. Biol. Chem. 263, pp. 19240–19248.
869. Stearns, D.J., Kurosawa, S., Sims, P.J., Esmon, N.L. & Esmon, C.T. (1988). The interaction of a Ca^{2+}-dependent monoclonal antibody with the protein C activation peptide region. J. Biol. Chem. 263, pp. 826–832.
870. Kurosawa, S., Stearns, D.J., Jackson, K.W. & Esmon, C.T. (1988). A 10-kDa cyanogen bromide fragment from the epidermal growth factor homology domain of rabbit thrombomodulin contains the primary thrombin binding site. J. Biol. Chem. 263, pp. 5993–5996.
871. Stearns, D.J., Kurosawa, S. & Esmon, C.T (1989). Microthrombomodulin. J. Biol. Chem. 264, pp. 3352–3356.
872. Suzuki, K., Hayashi, T., Nishioka, J., Kosaka, Y., Zushi, M., Honda, G. & Yamamoto, S. (1989). A domain composed of epidermal growth factor–like structures of human thrombomodulin is essential for thrombin binding and for protein C activation. J. Biol. Chem. 264, pp. 4872–4876.
873. Kimura, S., Nagoya, S. & Aoki, N. (1989). Monoclonal antibodies to human thrombomodulin whose binding is calcium dependent. J. Biochem. (Tokyo) 105, pp. 478–483.
874. Stenflo, J., Öhlin, A.-K., Owen, W.G. & Schneider, W.J. (1988). β-Hydroxyaspartic acid or β-hydroxyasparagine in bovine low density lipoprotein receptor and in bovine thrombomodulin. J. Biol. Chem. 263, pp. 21–24.
875. DiScipio, R.G. & Davie, E.W. (1979). Characterization of protein S, a γ-carboxyglutamic acid containing protein from bovine and human plasma. Biochemistry 18, pp. 899–904.
876. Dahlbäck, B. (1983). Purification of human vitamin K-dependent protein S and its limited proteolysis by thrombin. Biochem. J. 209, pp. 837–846.
877. Dahlbäck, B. & Stenflo, J. (1981). High molecular weight complex in human plasma between vitamin K-dependent protein S and complement component C4b-binding protein. Proc. Natl. Acad. Sci. USA 78, 2512–2516.
878. Dahlbäck, B. (1983). Purification of human C4b-binding protein and formation of its complex with vitamin K-dependent protein S. Biochem. J. 209, pp. 847–856.

879. Dahlbäck, B., Lundwall, Å. & Stenflo, J. (1986). Primary structure of bovine vitamin K-dependent protein S. Proc. Natl. Acad. Sci. USA 83, pp. 4199–4203.
880. Lundwall, Å., Dackowski, W., Cohen, E., Shaffer, M., Mahr, A., Dahlbäck, B., Stenflo, J. & Wydro, R. (1986). Isolation and sequence of the cDNA for human protein S, a regulator of blood coagulation. Proc. Natl. Acad. Sci. USA 83, pp. 6716–6720.
881. Hoskins, J., Norman, D.K., Beckmann, R.J. & Long, G.L. (1987). Cloning and characterization of human liver cDNA encoding a protein S precursor. Proc. Natl. Acad. Sci. USA 84, pp. 349–353.
882. Ploos van Amstel, H.K., van der Zanden, A.L., Reitsma, P.H. & Bertina, R.M. (1987). Human protein S cDNA encodes Phe(–16) and Tyr222 in consensus sequences for post-translational processing. FEBS Lett. 222, pp. 186–190.
883. Bertina, R.M., Ploos van Amstel, H.K., van Wijngaarden, A., Coenen, J., Deutz-Terlouw, P.P., Reitsma, P.H. & van der Linden, I.K. (1989). The Heerlen polymorphism of protein S, an immunologic polymorphism due to dimorphism of residue 460. Thromb. Haemostas. 62, Abstract 975.
884. Edenbrandt, C.-M., Gershagen, S., Fernlund, P., Wydro, R., Stenflo, J. & Lundwall, Å. (1987). Gene structure of vitamin K-dependent protein S; a region homologous to sex hormone binding globulin (SHBG) replaces the serine protease region of factors IX, X and protein C. Thromb. Haemostas. 58, Abstract 1841.
885. Watkins, P.C., Eddy, R., Fukushima, Y., Byers, M.G., Cohen, E.H., Dackowski, W.R., Wydro, R.M. & Shows, T.B. (1988). The gene for protein S maps near the centromere of human chromosome 3. Blood 71, pp. 238–241.
886. Ploos van Amstel, H.K., van der Zanden, A.L., Bakker, E., Reitsma, P.H. & Bertina, R.M. (1987). Two genes homologous with human protein S cDNA are located on chromosome 3. Thromb. Haemostas. 58, pp. 982–987.
887. Ploos van Amstel, H.K., Reitsma, P.H. & Bertina, R.M. (1989). Identification of the protein Sb gene as a pseudogene. Thromb. Haemostas. 62, Abstract 873.
888. Dahlbäck, B., Lundwall, Å. & Stenflo, J. (1986). Localization of thrombin cleavage sites in the amino-terminal region of bovine protein S. J. Biol. Chem. 261, pp. 5111–5115.
889. Stenflo, J., Lundwall, Å. & Dahlbäck, B. (1987). β-Hydroxyasparagine in domains homologous to the epidermal growth factor precursor in vitamin K-dependent protein S. Proc. Natl. Acad. Sci. USA 84, pp. 368–372.
890. Baker, M.E., French, F.S. & Joseph, D.R. (1987). Vitamin K-dependent protein S is similar to rat androgen-binding protein. Biochem. J. 243, pp. 293–296.
891. Joseph, D.R., Hall, S.H. & French, F.S. (1987). Rat androgen-binding protein: Evidence for identical subunits and amino acid sequence homology with human sex hormone-binding globulin. Proc. Natl. Acad. Sci. USA 84, pp. 339–343.
892. Walsh, K.A., Titani, K., Takio, K., Kumar, S., Hayes, R. & Petra, P.H. (1986). Amino acid sequence of the sex steroid binding protein of human blood plasma. Biochemistry 25, pp. 7584–7590.
893. Gershagen, S., Fernlund, P. & Lundwall, Å. (1987). A cDNA coding for human sex hormone binding globulin. FEBS Lett. 220, pp. 129–135.
894. Walker, F.J. (1989). Characterization of a synthetic peptide that inhibits the interaction between protein S and C4b–binding protein. J. Biol. Chem. 264, pp. 17645–17648.
895. Comp, P.C., Nixon, R.R., Cooper, M.R. & Esmon, C.T. (1984). Familial protein S deficiency is associated with recurrent thrombosis. J. Clin. Invest. 74, pp. 2082–2088.
896. Schwartz, H.P., Fisher, M., Hopmeier, P., Batard, M.A. & Griffin, J.H. (1984). Plasma protein S deficiency in familial thrombotic disease. Blood 64, pp. 1297–1300.
897. Kamiya, T., Sugihara, T., Ogata, K., Saito, H., Suzuki, K., Nishioka, J., Hashimoto, S. & Yamagata, K. (1986). Inherited deficiency of protein S in a Japanese family with recurrent venous thrombosis: A study of three generations. Blood 67, pp. 406–410.
898. Ploos van Amstel, H.K., Huisman, M.V., Reitsma, P.H., ten Cate, J.W. & Bertina, R.M. (1989). Partial protein S gene deletion in a family with hereditary thrombophilia. Blood 73, pp. 479–483.
899. Suzuki, K., Stenflo, J., Dahlbäck, B. & Teodorsson, B. (1983). Inactivation of human coagulation factor V by activated protein C. J. Biol. Chem. 258, pp. 1914–1920.
900. Walker, F.J., Chavin, S.I. & Fay, P.J. (1987). Inactivation of factor VIII by activated

protein C and protein S. Arch. Biochem. Biophys. 252, pp. 322–328.
901. Nelsestuen, G.L., Kisiel, W. & Discipio, R.G. (1978). Interaction of vitamin K dependent proteins with membranes. Biochemistry 17, pp. 2134–2138.
902. Walker, F.J. (1980). Regulation of activated protein C by a new protein. J. Biol. Chem. 256, pp. 5521–5524.
903. Walker, F.J. (1981). Regulation of activated protein C by protein S. J. Biol. Chem. 256, pp. 11128–11131.
904. Suzuki, K., Nishioka, J. & Hashimoto, S. (1983). Regulation of activated protein C by thrombin modified protein S. J. Biochem. (Tokyo) 94, pp. 699–705.
905. Walker, F.J. (1984). Regulation of vitamin K dependent protein S. J. Biol. Chem. 259, pp. 10335–10339.
906. Mitchell, C.A., Hau, L. & Salem, H.H. (1986). Control of thrombin mediated cleavage of protein S. Thromb. Haemostas. 56, pp. 151–154.
907. Sugo, T., Dahlbäck, B., Holmgren, A. & Stenflo, J. (1986). Calcium binding of bovine protein S. J. Biol. Chem. 261, pp. 5116–5120.
908. Walker, F.J. (1986). Properties of chemically modified protein S: Effect of conversion of γ-carboxyglutamic acid to γ-methylene-glutamic acid on functional properties. Biochemistry 25, pp. 6305–6311.
909. Bertina, R.M., Cupers, R. & van Wijngaarden, A. (1984). Factor IXa protects activated factor VIII against inactivation by activated protein C. Biochem. Biophys. Res. Commun. 125, pp. 177–183.
910. Nesheim, M.E., Canfield, W.M., Kisiel, W. & Mann, K.G. (1982). Studies on the capacity of factor Xa to protect factor Va from inactivation by activated protein C. J. Biol. Chem. 257, pp. 1443–1447.
911. Solymoss, S., Tucker, M.M. & Tracy, P.B. (1988). Kinetics of inactivation of membrane-bound factor Va by activated protein C. J. Biol. Chem. 263, pp. 14884–14890.
912. Koedam, J.A., Meijers, J.C.M., Sixma, J.J. & Bouma, B.N. (1988). Inactivation of human factor VIII by activated protein C. J. Clin. Invest. 82, pp. 1236–1243.
913. Krishnaswamy, S., Williams, E.B. & Mann, K.G. (1986). The binding of activated protein C to factors V and Va. J. Biol. Chem. 261, pp. 9684–9693.
914. Fay, P.J. & Walker, F.J. (1989). Inactivation of human factor VIII by activated protein C: Evidence that the factor VIII light chain contains the activated protein C binding site. Biochim. Biophys. Acta 994, pp. 142–148.
915. Dahlbäck, B. (1986). Inhibition of protein Ca cofactor function of human and bovine protein S by C4b-binding protein. J. Biol. Chem. 261, pp. 12022–12027.
916. Walker, F.J. (1986). Identification of a new protein involved in the regulation of the anticoagulant activity of activated protein C. J. Biol. Chem. 261, pp. 10941–10944.
917. Suzuki, K., Nishioka, J. & Hashimoto, S. (1983). Protein C inhibitor. J. Biol. Chem. 258, pp. 163–168.
918. Suzuki, K., Nishioka, J., Kusumoto, H. & Hashimoto, S. (1984). Mechanism of inhibition of activated protein C inhibitor. J. Biochem. (Tokyo) 95, pp. 187–195.
919. Suzuki, K., Deyashiki, Y., Nishioka, J., Kurachi, K., Akira, M., Yamamoto, S. & Hashimoto, S. (1987). Characterization of a cDNA for human protein C inhibitor. J. Biol. Chem. 262, pp. 611–616.
920. Heeb, M.J., España, F., Geiger, M., Collen, D., Stump, D.C. & Griffin, J.H. (1987). Immunological identity of heparin-dependent plasma and urinary protein C inhibitor and plasminogen activator inhibitor-3. J. Biol. Chem. 262, pp. 15813–15816.
921. Heeb, M.J. & Griffin, J.H. (1988). Physiological inhibition of human activated protein C by α_1-antitrypsin. J. Biol. Chem. 263, pp. 11613–11616.
922. Meijers, J.C.M., Vlooswijk, R.A.A., Kanters, D.H.A.J., Hessing, M. & Bouma, B.N. (1988). Identification of monoclonal antibodies that inhibit the function of protein C inhibitor. Evidence for heparin-independent inhibition of activated protein C in plasma. Blood 72, pp. 1401–1403.
923. Heeb, M.J., España, F. & Griffin, J.H. (1989). Inhibition and complexation of activated protein C by two major inhibitors in plasma. Blood, pp. 446–454.
924. Schousboe, I. (1985). β_2-Glycoprotein I: A plasma inhibitor of the contact activation of the intrinsic blood coagulation pathway. Blood 66, pp. 1086–1091.

925. Nimpf, J., Bevers, E.M., Bomans, P.H.H., Till, U., Wurm, H., Kostner, G.M. & Zwaal, R.F.A. (1986). Prothrombinase activity of human platelets is inhibited by β_2-glycoprotein I. Biochim. Biophys. Acta 884, pp. 142–149.
926. Halkier, T. & Magnusson, S. (1988). Contact activation of blood coagulation is inhibited by plasma factor XIII b-chain. Thromb. Res. 51, pp. 313–324.
927. Funakoshi, T., Heimark, R.L., Hendrickson, L.E., McMullen, B.A. & Fujikawa, K. (1987). Human placental anticoagulant protein. Isolation and characterization. Biochemistry 26, pp. 5572–5578.
928. Reutlingsperger, C.P.M., Hornstra, G. & Hemker, H.C. (1985). Isolation and partial purification of a novel anticoagulant from arteries of human umbilical cord. Eur. J. Biochem. 151, pp. 625–629.
929. Reutlingsperger, C.P.M., Koip, J.M.M., Hornstra, G. & Hemker, H.C. (1988). Purification and characterization of a novel protein from bovine aorta that inhibits coagulation. Eur. J. Biochem. 175, pp. 171–175.
930. Maki, M., Murata, M. & Shidra, Y. (1984). Inhibitors of platelet aggregation and blood coagulation isolated from the human placenta. Eur. J. Obstet. Gynecol. Reprod. Biol. 17, pp. 149–154.
931. Shidara, Y. (1984). Isolation and purification of placental coagulation inhibitor. Acta Obst. Gynaec. Jpn. 36, pp. 2583–2592.
932. Funakoshi, T., Hendrickson, L.E., McMullen, B.A. & Fujikawa, K. (1987). Primary structure of human placental anticoagulant protein. Biochemistry 26, pp. 8087–8092.
933. Iwasaki, A., Suda, M., Nakao, H., Nagoya, Y., Saino, Y., Arai, K., Mizoguchi, T., Sato, F., Yoshizaki, H., Hirata, M., Miyata, T., Shidra, Y., Murata, M. & Maki, M. (1987). Structure and expression of cDNA for an inhibitor of blood coagulation isolated from human placenta: A new lipocortin-like protein. J. Biochem. (Tokyo) 102, pp. 1261–1273.
934. Grundmann, U., Abel, K.-J., Bohn, H., Löbermann, H., Lottspeich, F. & Küpper, H. (1988). Characterization of cDNA encoding human placental anticoagulant protein (pp4): Homology with the lipocortin family. Proc. Natl. Acad. Sci. USA 85, pp. 3708–3712.
935. Mauer-Fogy, I., Reutlingsperger, C.P., Pieters, J., Bodo, G., Stratowa, C. & Hauptmann, R. (1988). Cloning and expression of cDNA for human vascular anticoagulant, a Ca^{2+}-dependent phospholipid-binding protein. Eur. J. Biochem. 174, pp. 585–592.
936. Schlaepfer, D.D., Mehlman, T., Burgess, W.H. & Haigler, H.T. (1987). Structural and functional characterization of endonexin II, a calcium-and phospholipid binding protein. Proc. Natl. Acad. Sci. USA 84, pp. 6078–6082.
937. Kapalan, R., Jaye, M., Burgess, W.H., Schlaepfer, D.D. & Haigler, H.T. (1988). Cloning and expression of cDNA for human endonexin II, a Ca^{2+} and phospholipid binding protein. J. Biol. Chem. 263, pp. 8037–8043.
938. Wallner, B.P., Mattaliano, R.J., Hession, C., Cate, R.L., Tizard, R., Sinclair, L.K., Foeller, C., Chow, E.P., Browning, J.L., Ramachandran, K.L. & Pepinsky, R.B. (1986). Cloning and expression of human lipocortin, a phospholipase A2 inhibitor with potential anti-inflammatory activity. Nature 320, pp. 77–81.
939. Huang, K.-S., Wallner, B.P., Mattaliano, R.J., Tizard, R., Burne, C., Frey, A., Hession, C., McGraw, P., Sinclair, L.K., Chow, E.P., Browning, J.L., Ramachandran, K.L., Tang, J., Smart, J.E. & Pepinsky, R.B. (1986). Two human 35 kd inhibitors of phospholipase A2 are related to substrates of $pp60^{v\text{-}src}$ and of epidermal growth factor receptor/kinase. Cell 46, pp. 191–199.
940. Saris, C.J.M., Tack, B.F., Kristensen, T., Glenney, J.R. & Hunter, T. (1986). The cDNA sequence for the protein-tyrosine kinase substrate p36 (calpactin I heavy chain) reveals a multi-domain protein with internal repeats. Cell 46, pp. 201–212.
941. Kristensen, T., Saris, C.J.M., Hunter, T., Hicks, L.J., Noonan, D.J., Glenney, J.R. & Tack, B.F. (1986). Primary structure of bovine calpactin I heavy chain (p36), a major cellular substrate for retroviral protein-tyrosine kinases: Homology with the human phospholipase A2 inhibitor lipocortin. Biochemistry 25, pp. 4497–4503.
942. Tait, J.F., Sakata, M., McMullen, B.A., Miao, C.H., Funakoshi, T., Hendrickson, L.E. & Fujikawa, K. (1988). Placental anticoagulant proteins: Isolation and comparative

characterization of four members of the lipocortin family. Biochemistry 27, pp. 6268–6272.
943. Rabiner, S.F., Goldfine, D., Hart, A., Summaria, L. & Robbins, K.C. (1969). Radioimmunoassay of human plasminogen and plasmin. J. Lab. Clin. Med. 74, pp. 265–273.
944. Sottrup-Jensen, L., Claeys, H., Zajdel, M., Petersen, T.E. & Magnusson, S. (1978). The primary structure of human plasminogen. Isolation of two lysine-binding fragments and one 'mini'-plasminogen (M.W. 38,000) by elastase-catalyzed-specific limited proteolysis. In: Progress in Chemical Fibrinolysis and Thrombolysis (Davidson, J.F., Rowan, R.M., Samama, M.M. & Desnoyers, P.C., Eds.) Vol. 3, pp. 191–209, Raven Press, New York.
945. Sottrup-Jensen, L., Petersen, T.E. & Magnusson, S. (1978). In: Atlas of Protein Sequence and Structure (Dayhoff, M.O., Ed.) Vol. 5, Suppl. 3, p. 91, National Biomedical Research Foundation, Silver Spring, MD.
946. Wiman, B. (1977). Primary structure of the B-chain of human plasmin. Eur. J. Biochem. 76, pp. 129–137.
947. Malinowski, D.P., Sadler, J.E. & Davie, E.W. (1984). Characterization of a complemetary deoxyribonucleic acid coding for human and bovine plasminogen. Biochemistry 23, pp. 4243–4250.
948. Forsgren, M., Råden, B., Israelson, M., Larsson, K. & Hedén, L.-O. (1987). Molecular cloning and characterization of a full length cDNA for human plasminogen. FEBS Lett. 213, pp. 254–260.
949. Ichinose, A., Espling, F.S., Takamatsu, J., Saito, H., Synmyozu, K., Murayama, I., Martzen, M.R., Petersen, T.E. & Davie, E.W. (1989). Characterization and diagnosis of the abnormal genes for human plasminogen. Thromb. Haemostas. 62, Abstract 1552.
950. Murray, J.C., Buetow, K.H., Donovan, M., Hornung, S., Motulsky, A.G., Disteche, C., Dyer, K., Swissheim, K., Anderson, J., Giblett, E., Sadler, E., Eddy, R. & Shows, T.B. (1987). Linkage disequilibrium of plasminogen polymorphisms and assignment of the gene to human chromosome 6q26–6q27. Am. J. Hum. Genet. 40, pp. 338–350.
951. Frank, S.L., Klisak, I., Sparkes, R.S., Mohandas, T., Tomlinson, J.E., McLean, J.W., Lawn, R.M. & Lusis, A.J. (1988). The apolipoprotein (a) gene resides on the human chromosome 6q26–27, in close proximity to the homologous gene for plasminogen. Hum. Genet. 79, pp. 352–356.
952. Frank, S.L., Klisak, I., Sparkes, R.S. & Lusis, A.J. (1989). A gene homologous to plasminogen located on human chromosome 2q11–p11. Genomics 4, pp. 449–451.
953. Schaller, J., Moser, P.W., Dannegger-Müller, G.A.K., Rösselet, S.J., Kämpfer, U. & Rickli, E.E. (1985). Complete amino acid sequence of bovine plasminogen. Eur. J. Biochem. 149, pp. 267–278.
954. Marti, T., Schaller, J. & Rickli, E.E. (1985). Determination of the complete amino-acid sequence of porcine miniplasminogen. Eur. J. Biochem. 149, pp. 279–285.
955. Schaller, J., Marti, T., Rösselet, S.J., Kämpfer, U. & Rickli, E.E. (1987). Amino acid sequence of the heavy chain of porcine plasmin. Fibrinolysis 1, pp. 91–102.
956. Geynes, M. & Patthy, L. (1985). The kringle 4 domain of chicken plasminogen. Biochim. Biophys. Acta 832, pp. 326–330.
957. Schaller, J. & Rickli, E.E. (1988). Structural aspects of the plasminogen of various species. Enzyme 40, pp. 63–69.
958. Tomlinson, J.E., McLean, J.W. & Lawn, R.M. (1989). Rhesus monkey apolipoprotein (a). J. Biol. Chem. 264, pp. 5957–5965.
959. Brockway, W.J. & Castellino, F.J. (1972). Measurement of the binding of antifibrinolytic amino acids to various plasminogens. Arch. Biochem. Biophys. 151, pp. 194–199.
960. Hayes, M.L. & Castellino, F.J. (1979). Carbohydrate of the human plasminogen variants. J. Biol. Chem. 254, pp. 8768–8771.
961. Hayes, M.L. & Castellino, F.J. (1979). Carbohydrate of the human plasminogen variants. J. Biol. Chem. 254, pp. 8772–8776.
962. Hayes, M.L. & Castellino, F.J. (1979). Carbohydrate of the human plasminogen variants. J. Biol. Chem. 254, pp. 8777–8780.
963. Wiman, B. & Wallén, P. (1977). The specific interaction between plasminogen and

fibrin. A physiological role of the lysine binding site in plasminogen. Thromb. Res. 10, pp. 213–222.
964. Wiman, B., Lijnen, H.R. & Collen, D. (1979). On the specific interaction between the lysine-binding sites in plasmin and complemetary sites in α_2-antiplasmin and in fibrinogen. Biochem. Biophys. Acta 579, pp. 142–154.
965. Thorsen, S., Clemmensen, I., Sottrup-Jensen, L. & Magnusson, S. (1981) Adsorption to fibrin of native fragments of known primary structure from human plasminogen. Biochem. Biophys. Acta 668, pp. 377–387.
966. Markus, G., DePasquale, J.L. & Wissler, F.C. (1978). Quantitative determination of the binding of ϵ-aminocaproic acid to native plasminogen. J. Biol. Chem. 253, pp. 727–732.
967. Markus, G., Evers, J.L. & Hobika, G.H. (1978). Comparison of some properties of native (Glu) and modified (Lys) human plasminogen. J. Biol. Chem. 253, pp. 733–739.
968. Markus, G., Priore, R.L. & Wissler, F.C. (1979). The binding of tranexamic acid to native (Glu) and modified (Lys) human plasminogen and its effect on conformation. J. Biol. Chem. 254, pp. 1211–1216.
969. Lerch, P.G., Rickli, E.E., Lergier, W. & Gillessen, D. (1980). Localization of individual lysine-binding regions in human plasminogen and investigations of their complex-forming properties. Eur. J. Biochem. 107, pp. 7–13.
970. Váli, Z. & Patthy, L. (1982). Location of the intermediate and high affinity ω-aminocarboxylic acid-binding site in human plasminogen. J. Biol. Chem. 257, pp. 2104–2110.
971. Váli, Z. & Patthy, L. (1980). Essential carboxyl- and guanidino-groups in the lysine-binding site of human plasminogen. Biochem. Biophys. Res. Commun. 96, pp. 1804–1811.
972. Trexler, M., Váli, Z. & Patthy, L. (1982). Structure of the ω-aminocarboxylic acid binding site of human plasminogen. J. Biol. Chem. 257, pp. 7401–7406.
973. Hochschwender, S.M. & Laursen, R.A. (1981). The lysine binding sites of human plasminogen. J. Biol. Chem. 256, pp. 11172–11176.
974. De Marco, A., Laursen, R.A. & Llinas, M. (1986). ^{1}H-NMR spectroscopic manifestations of ligand binding to kringle 4 domain of human plasminogen. Arch. Biochem. Biophys. 244, pp. 727–741.
975. Váli, Z. & Patthy, L. (1984). The fibrin-binding site of human plasminogen. J. Biol. Chem. 259, pp. 13690–13694.
976. Lucas, M.A., Fretto, L.J. & McKee, P.A. (1983). The binding of human plasminogen to fibrin and fibrinogen. J. Biol. Chem. 258, pp. 4249–4256.
977. Christensen, U. (1984). The AH-site of plasminogen and two C-terminal fragments. Biochem. J. 223, pp. 413–421.
978. Váradi, A. & Patthy, L. (1983). Location of plasminogen-binding sites in human fibrin(ogen). Biochemistry 22, pp. 2440–2446.
979. Miyata, T.S., Iwanaga, S., Sakata, Y. & Aoki, N. (1982). Plasminogen Tochigi: Inactive plasmin resulting from replacement of alanine-600 by threonine in the active site. Proc. Natl. Acad. Sci. USA 79, pp. 6132–6136.
980. Miyata, T.S., Iwanaga, S., Sakata, Y., Aoki, N., Takamatsu, J. & Kamiya, T. (1984). Plasminogens Tochigi II and Nagoya: Two additional molecular defects with Ala-600-Thr replacement found in the plasmin light chain variants. J. Biochem. (Tokyo) 96, pp. 277–287.
981. Scanu, A.M. (1988). Lipoprotein(a): A genetically determined lipoprotein containing a glycoprotein of the plasminogen family. Sem. Thromb. Hemostas. 14, pp. 266–270.
982. Kostner, G.M., Avogaro, P., Cazzolato, G., Marth, E., Bittolo-Bon, G. & Qunici, G.B. (1981). Lipoprotein Lp(a) and the risk for myocardial infarction. Atherosclerosis 38, pp. 51–61.
983. Költringer, P. & Jürgens, G. (1985). A dominant role of lipoprotein(a) in the investigation and evaluation of parameters indicating the development of cervical atherosclerosis. Atherosclerosis 58, pp. 187–198.
984. Zenker, G., Költringer, P., Boné, G., Niederkorn, K., Pfeiffer, K. & Jürgens, G. (1986). Lipoprotein(a) as a strong indicator for cerebrovascular disease. Stroke 17, pp. 942–945.

985. Rhoads, G.G., Dahlen, G., Berg, K., Morton, N.E. & Danneberg, A.L. (1986). Lp(a) lipoprotein as a risk factor for myocardial infarction. J. Am. Med. Ass. 256, pp. 2540–2544.
986. Dahlen, G.H., Guyton, J.R., Attar, M., Farmer, J.A., Kautz, J.A. & Gotto Jr., A.M. (1986). Association of levels of lipoprotein Lp(a), plasma lipids, and other lipoproteins with coronary artery disease documented by angiography. Circulation 74, pp. 758–765.
987. Murai, A., Miyahara, T., Fujimoto, N., Matsuda, M. & Kameyama, M. (1986). Lp(a) lipoprotein as a risk factor for coronary heart disease and cerebral infarction. Atherosclerosis 59, pp. 199–204.
988. Armstrong, V.W., Cremer, P., Eberle, E., Manke, A., Schulze, F., Wieland, H., Kreuzer, H. & Seidel, D. (1986). The association between serum Lp(a) concentrations and angiographically assessed coronary atherosclerosis. Atherosclerosis 62, pp. 249–257.
989. Durrington, P.N., Ishola, M., Hunt, L., Arrol, S. & Bhatnagar, D. (1988). Apolipoprotein(a), AI, and B and parental history in men with early onset ischaemic heart disease. Lancet i, pp. 1070–1073.
990. Hoff, H.F., Beck, G.J., Skibinski, C.I., Jürgens, G., O'Niel, J., Kramer, J. & Lytle, B. (1988). Serum Lp(a) levels as a predictor of vein graft stenosis after coronary artery bypass surgery in patients. Circulation 77, pp. 1238–1244.
991. Uterman, G., Menzel, H.J., Kraft, H.G., Duba, H.C., Kemmler, H.G. & Seitz, C. (1987). Lp(a) glycoprotein phenotypes. J. Clin. Invest. 80, pp. 458–465.
992. Eaton, D.L., Fless, G.M., Kohr, W.J., McLean, J.W., Xu, Q.-T., Miller, C.G., Lawn, R.M. & Scanu, A.M. (1987). Partial amino acid sequence of apolipoprotein(a) shows that it is homologous to plasminogen. Proc. Natl. Acad. Sci. USA 84, pp. 3224–3228.
993. Kratzin, H., Armstrong, V.W., Niehaus, M., Hilschmann, N. & Seidel, D. (1987). Structural relationship of an apolipoprotein(a) phenotype (570 kDa) to plasminogen: Homologous kringle domains are linked by carbohydrate-rich regions. Biol. Chem. Hoppe-Seyler. 368, pp. 1533–1544.
994. McLean, J.W., Tomlinson, J.E., Kuang, W.-J., Eaton, D.L., Chen, E.Y., Fless, G.M., Scanu, A.M. & Lawn, R.M. (1987). cDNA sequence of human apolipoprotein(a) is homologous to plasminogen. Nature 330, pp. 132–137.
995. Weitkamp, L.R., Guttormsen, S.A. & Schultz, J.S. (1988). Linkage between the loci for the Lp(a) lipoprotein (LP) and plasmingen (PLG). Hum. Genet. 79, pp. 80–82.
996. Lindahl, G., Gersdorf, E., Menzel, H.J., Duba, C., Cleve, H., Humphries, S. & Uterman, G. (1989). The gene for the Lp(a)-specific glycoprotein is closely linked to the gene for plasminogen on chromosome 6. Hum. Genet. 81, pp. 149–152.
997. Salonen, E.-M., Jauhiainen, M., Zardi, L., Vaheri, A. & Ehnholm, M.C. (1989). Lipoprotein(a) binds to fibronectin and has serine proteinase activity capable of cleaving it. EMBO J. 8, pp. 4035–4040.
998. Gonzales-Gronow, M., Edelberg, J.M. & Pizzo, S.V. (1989). Further characterization of the cellular plasminogen binding site: Evidence that plasminogen 2 and lipoprotein (a) compete for the same site. Biochemistry 28, pp. 2374–2377.
999. Harpel, P.C., Gordon, B.R. & Parker, T.S. (1989). Plasmin catalyzes binding of lipoprotein(a) to immobilized fibrinogen and fibrin. Proc. Natl. Acad. Sci. USA 86, pp. 3847–3851.
1000. Miles, L.A., Fless, G.M., Levin, E.G., Scanu, A.M. & Plow, E.F. (1989). A potential basis for the thrombotic risks associated with lipoprotein(a). Nature 339, pp. 301–303.
1001. Hajjar, K.A., Gavish, D., Breslow, J.L. & Nachman, R.L. (1989). Lipoprotein(a) modulation of endothelial cell surface fibrinolysis and its potential role in atherosclerosis. Nature 339, pp. 303–305.
1002. Blasi, F., Vassalli, J.-D. & Danø, K. (1987). Urokinase-type plasminogen activator: Proenzyme, receptor and inhibitors. J. Cell Biol. 104, pp. 801–804.
1003. Eaton, D.L., Scott, R.W. & Baker, J.B. (1984). Purification of human fibroblast urokinase proenzyme and analysis of its regulation by proteases and protease nexin. J. Biol. Chem. 259, pp. 6241–6247.
1004. Nielsen, L.S., Hansen, J.G., Skriver, L., Wilson, E.L., Kaltoft, K., Zeuthen, J. & Danø, K. (1982). Purification of zymogen to plasminogen activator from human glioblastoma cells by affinity chromatography with monoclonal antibody. Biochemistry 21, pp. 6410–6415.

1005. Wun, T.-C., Ossowski, L. & Reich, E. (1982). A proenzyme form of human urokinase. J. Biol. Chem. 257, pp. 7262–7268.
1006. Wun, T.-C., Schleuning, W.-D. & Reich, E. (1982). Isolation and characterization of urokinase from human plasma. J. Biol. Chem. 257 pp. 3276–3283.
1007. Kasai, S., Arimura, H., Nishida, M. & Suyama, T. (1985). Proteolytic cleavage of single-chain pro-urokinase induces conformational change which follows activation of the zymogen and reduction of its high affinity for fibrin. J. Biol. Chem. 260, pp. 12377–12381.
1008. Stump, D.C., Lijnen, H.R. & Collen, D. (1986). Purification and characterization of single-chain urokinase-type plasminogen activator from human cell cultures. J. Biol. Chem. 261, pp. 1274–1278.
1009. Günzler, W.A., Steffens, G.J., Ötting, F., Buse, G. & Flohé, L. (1982). Structural relationship between human high and low molecular mass urokinase. Hoppe-Seyler's Z. Physiol. Chem. 363, pp. 133–141.
1010. Steffens, G.J., Günzler, W.A., Ötting, F., Frankus, E. & Flohé, L. (1982). The complete amino acid sequence of low molecular mass urokinase from human urine. Hoppe-Seyler's Z. Physiol. Chem. 363, pp. 1043–1058.
1011. Günzler, W.A., Steffens, G.J., Ötting, F., Kim, S.-M.A., Frankus, E. & Flohé, L. (1982). The primary structure of high molecular mass urokinase from human urine. Hoppe-Seyler's Z. Physiol. Chem. 363, pp. 1155–1165.
1012. Kasai, S., Arimura, H., Nishida, M. & Suyama, T. (1985). Primary structure of single chain pro-urokinase. J. Biol. Chem. 260, pp. 12382–12389.
1013. Verde, P., Stoppelli, M.P., Galeffi, P., Di Nocera, P. & Blasi, F. (1984). Identification and primary sequence of an unspliced human urokinase poly(A)+ RNA. Proc. Natl. Acad. Sci. USA 81, pp. 4727–4731.
1014. Holmes, W.E., Pennica, D., Blaber, M., Rey, M.W., Guenzler, W.A., Steffens, G.J. & Heyneker, H.L. (1985). Cloning and expression of the gene for pro-urokinase in *Escherichia coli*. Bio/Technology 3, pp. 923–929.
1015. Nagamine, Y., Pearson, D., Altus, M.S. & Reich, E. (1984). cDNA and gene nucleotide sequences of porcine plasminogen activator. Nucl. Acids Res. 12, pp. 9525–9541.
1016. Belin, D., Vassalli, J.-D, Combépine, C., Godeau, F., Nagamine, Y., Reich, E., Kocher, H.P. & Duvoisin, R.M. (1985). Cloning, nucleotide sequencing and expression of cDNAs encoding mouse urokinase-type plasminogen activator. Eur. J. Biochem. 148, pp. 225–232.
1017. Riccio, A., Grimaldi, G., Verde, P., Sebastio, G., Boast, S. & Blasi, F. (1985). The human urokinase-plasminogen activator gene and its promotor. Nucl. Acids Res. 13, pp. 2759–2771.
1018. Degen, S.J.F., Heckel, J.L., Reich, E. & Degen, J.L. (1987). The murine urokinase-type plasminogen activator gene. Biochemistry 26, pp. 8270–8279.
1019. Triputti, P., Blasi, F., Verde, P., Cannizzaro, L.A., Emanuel, B.S. & Croce, C.M. (1985). Human urokinase gene is located on the long arm of chromosome 10. Proc. Natl. Acad. Sci. USA 82, pp. 4448–4452.
1020. Rajput, B., Degen, S.F., Reich, E., Waller, E.K., Axelrod, J., Eddy, R.L. & Shows, T.B. (1985). Chromosomal locations of human tissue plasminogen activator and urokinase genes. Science 230, pp. 672–674.
1021. Gerard, R.D., Chien, K.R. & Meidell, R.S. (1986). Molecular biology of tissue plasminogen activator and endogenous inhibitors. Mol. Biol. Med. 3, pp. 449–457.
1022. Rijken, D.C. (1988). Relationships between structure and function of tissue-type plasminogen activator. Klin. Wochenschr. 66 (Suppl. XII), pp. 33–39.
1023. Rijken, D.C., Juhan-Vague, I., De Cock, F. & Collen, D. (1983). Measurement of human tissue-type plasminogen activator by a two-site immunoradiometric assay. J. Lab. Clin. Med. 101, pp. 274–284.
1024. Verstraete, M. & Collen, D. (1986). Thrombolytic therapy in the eighties. Blood 67, pp. 1529–1541.
1025. Pennica, D., Holmes, W.E., Kohr, W.J., Harkins, R.N., Vehar, G.A., Ward, C.A., Bennett, W.F., Yelverton, E., Seeburg, P.H., Heyneker, H.L., Goeddel, D.V. & Collen D. (1983). Cloning and expression of human tissue-type plasminogen activator cDNA in *E. coli*. Nature 301, pp. 214–221.
1026. Harris, T.J.R., Patel, T., Marston, F.A.O., Little, S., Emtage, J.S., Opdenakker, G.,

Volekaert, G., Rombauts, W., Billiau, A. & De Somer, P. (1986). Cloning of cDNA for human tissue-type plasminogen activator and its expression in *Escherichia coli*. Mol. Biol. Med. 3, pp. 279–292.

1027. Wallén, P., Pohl, G., Bergsdorf, N., Rånby, M., Ny, T. & Jörnvall, H. (1983). Purification and characterization of a melanoma cell plasminogen activator. Eur. J. Biochem. 132, pp. 681–686.
1028. Pohl, G., Kallström, M., Bergsdorf, N., Wallén, P. & Jörnvall, H. (1984). Tissue plasminogen activator: Peptide analyses confirm an indirectly derived amino acid sequence, identify the active site serine residue, establish glycosylation sites, and localize variant differences. Biochemistry 23, pp. 3701–3707.
1029. Jörnvall, H., Pohl, G., Bergsdorf, N. & Wallén, P. (1983). Differential proteolysis and evidence for a residue exchange in tissue plasminogen activator suggest possible association between two types of protein microheterogeneity. FEBS Lett. 156, pp. 47–50.
1030. Ny, T., Elgh, F. & Lund, B. (1984). The structure of the human tissue-type plasminogen activator gene: Correlation of intron and exon structures to functional domains. Proc. Natl. Acad. Sci. USA 81, pp. 5355–5359.
1031. Ficher, R., Waller, E.K., Grossi, G., Thompson, D., Tizard, R. & Schleuning, W.-D. (1985). Isolation and characterization of the human tissue-type plasminogen activator structural gene including its 5' flanking region. J. Biol. Chem. 260, pp. 11223–11230.
1032. Degen, S.J.F., Rajput, B. & Reich, E. (1986). The human tissue plasminogen activator gene. J. Biol. Chem. 261, pp. 6972–6985.
1033. Verheijen, J.H., Visse, R., Wijnen, J.T., Chang, G.T.G., Kluft, C. & Khan, P. M. (1986). Assignment of the human tissue-type plasminogen activator gene (PLAT) to chromosome 8. Hum. Genet. 72, pp. 153–156.
1034. Gardell, S.J., Duong, L.T., Diehl, R.E., York, J.D., Hare, T.R., Register, R.B., Jacobs, J.W., Dixon, R.A.F. & Friedman, P.A. (1989). Isolation, characterization, and cDNA cloning of a vampire bat salivary plasminogen activator. J. Biol. Chem. 264, pp. 17947–17952.
1035. Saito, H. (1988). α_2-Plasmin inhibitor and its deficiency states. J. Lab. Clin. Med. 112, pp. 671–678.
1036. Collen, D. (1976). Identification and some properties of a new fast acting plasmin inhibitor in plasma. Eur. J. Biochem. 69, pp. 209–216.
1037. Moroi, M. & Aoki, N. (1976). Isolation and characterization of α_2-plasmin inhibitor from human plasma. J. Biol. Chem. 251, pp. 5956–5965.
1038. Müllertz, S. & Clemmensen, I. (1976). The primary inhibitor of plasmin in human plasma. Biochem. J. 159, pp. 545–553.
1039. Lijnen, H.R., Holmes, W.E., van Hoff, B., Wiman, B., Rodriguez, H. & Collen, D. (1987). Amino acid sequence of human α_2-antiplasmin. Eur. J. Biochem. 166, pp. 565–574.
1040. Holmes, W.E., Nelles, L., Lijnen, H.R. & Collen, D. (1987). Primary structure of human α_2-antiplasmin, a serine proteinase inhibitor (Serpin). J. Biol. Chem. 262, pp. 1659–1664.
1041. Sumi, Y., Nakamura, Y., Aoki, N., Sakai, M. & Maramatsu, M. (1986). Structure of the carboxyterminal half of the human α_2-plasmin inhibitor deduced from that of the cDNA. J. Biochem. (Tokyo) 100, pp. 1399–1402.
1042. Tone, M., Kikuno, R., Kume-Iwaki, A. & Hashimoto-Gotoh, T. (1987). Structure of human α_2-plasmin inhibitor deduced from the cDNA sequence. J. Biochem. (Tokyo) 102, pp. 1033–1041.
1043. Kato, A., Nakamura, Y., Miura, O., Hirosawa, S., Sumi, Y. & Aoki, N. (1988). Assignment of the human α_2-plasmin inhibitor gene (PLI) to chromosome region 18p11.1 → q11.2 by in situ hybridization. Cytogenet. Cell Genet. 47, pp. 209–211.
1044. Hirosawa, S., Nakamura, Y., Miura, O., Sumi, Y. & Aoki, N. (1988). Organization of the human α_2-plasmin inhibitor gene. Proc. Natl. Acad. Sci. USA 85, pp. 6836–6840.
1045. Hortin, G., Fok, K.F., Toren, P.C. & Strauss, A.W. (1987). Sulfation of a tyrosine residue in the plasmin-binding domain of α_2-antiplasmin. J. Biol. Chem. 262, pp. 3082–3085.
1046. Shieh, B.-H & Travis, J. (1987). The reactive site of human α_2-antiplasmin. J. Biol. Chem. 262, pp. 6055–6059.

1047. Holmes, W.E., Lijnen, H.R., Nelles, L., Kluft, C., Nieuwenhuis, K.H., Rijken, D.C. & Collen, D. (1987). α_2-Antiplasmin Enschede: Alanine insertion and abolition of plasmin inhibitory activity. Science 238, pp. 209–211.
1048. Rijken, D.C., Groeneveld, E., Kluft, C. & Nieuwenhuis, K.H. (1988). α_2-antiplasmin Enschede is not an inhibitor, but a substrate, of plasmin. Biochem. J. 255, pp. 609–615.
1049. Miura, O., Sugahara, Y. & Aoki, N. (1989). Hereditary α_2-plasmin inhibitor deficiency caused by a transport-deficient mutation (α_2-PI-Okinawa). J. Biol. Chem. 264, pp. 18213–18219.
1050. Miura, O., Hirosawa, S., Kato, A. & Aoki, N. (1989). Molecular basis for congenital defficiency of α_2-plasmin inhibitor. J. Clin. Invest. 83, pp. 1598–1604.
1051. Sprengers, E.D. & Kluft, C. (1987). Plasminogen activator inhibitors. Blood 69, pp. 381–387.
1052. Chmielewska, J., Rånby, M. & Wiman, B. (1983). Evidence for a rapid inhibitor to tissue plasminogen activator in plasma. Thromb. Res. 31, pp. 427–436.
1053. Kruithof, E.K.O., Tran-Thang, C., Ransijn, A. & Bachman, F. (1984). Demonstration of a fast-acting inhibitor of plasminogen activators in human plasma. Blood 64, pp. 907–913.
1054. van Mourik, J.A., Lawrence, D.A. & Loskutoff, D.J. (1984). Purification of an inhibitor of plasminogen activator (antiactivator) synthesized by endothelial cells. J. Biol. Chem. 259, pp. 14914–14921.
1055. Åstedt, B., Lecander, I., Brodin, T., Lundblad, A. & Low, K. (1985). Purification of a specific placental plasminogen activator inhibitor by monoclonal antibody and its complex formation with plasminogen activator. Thromb. Haemostas. 53, pp. 122–125.
1056. Stump, D.C., Thienpont, M. & Collen, D. (1986). Purification and characterization of a novel inhibitor of urokinase from human urine. J. Biol. Chem. 261, pp. 12759–12766.
1057. Heeb, M.J., España, F., Geiger, M., Collen, D., Stump, D.C. & Griffin, J.H. (1987). Immunological identity of heparin-dependent plasma and urinary protein C inhibitor and plasminogen activator inhibitor-3. J. Biol. Chem. 262, pp. 15813–15816.
1058. Scott, R.W., Bergman, B.L., Bajpai, A. & Hersh, R.T., (1987). Comparative thrombolytic properties of single-chain forms of urokinase-type plasminogen activator. Blood 69, pp. 592–596.
1059. Ny, T., Sawdey, M., Lawrence, D., Millan, J.L. & Loskutoff, D.J. (1986). Cloning and sequence of a cDNA coding for the human β-migrating endothelial-cell-type plasminogen activator inhibitor. Proc. Natl. Acad. Sci. USA 83, pp. 6776–6780.
1060. Pannekoek, H., Veerman, H., Lamders, H., Diergaarde, P., Verweij, C.L., van Zonneveld, A.J. & van Mourik, J.A. (1986). Endothelial plasminogen activator inhibitor (PAI): A new member of the Serpin gene family. EMBO J. 5, pp. 2539–2544.
1061. Ginsburg, D., Zeheb, R., Yang, A.Y., Rafferty, U.M., Andreasen, P.A., Nielsen, L., Danø, K., Lebo, R.V. & Gelehrter, T.D. (1986). cDNA cloning of human plasminogen activator-inhibitor from endothelial cells. J. Clin. Invest. 78, pp. 1673–1680.
1062. Wun, T.-Z. & Kretzmer, K.K. (1987). cDNA cloning and expression in *E. coli* of a plasminogen activator inhibitor (PAI) related to a PAI produced by HepG2 heptoma cell. FEBS Lett. 210, pp. 11–16.
1063. Zaheb, R. & Gelehrter, T.D. (1988). Cloning and sequencing of cDNA for the rat plasminogen activator inhibitor-1. Gene 73, pp. 459–468.
1064. Mimuro, J., Sawdey, M., Hattori, M. & Loskutoff, D.J. (1989). cDNA for bovine type 1 plasminogen activator inhibitor (PAI-1). Nucl. Acids Res. 17, p. 8872.
1065. Loskutoff, D.J., Linders, M., Keijer, J., Veerman, H., van Heerikhuizen, H. & Pannekoek, H. (1987). Structure of the human plasminogen activator inhibitor 1 gene: Nonrandom distribution of introns. Biochemistry 26, pp. 3763–3768.
1066. Bosma, P.J., van den Berg, E.A., Kooistra, T., Siemieniak, D.R. & Slightom, J.L. (1988). Human plasminogen activator inhibitor-1 gene. J. Biol. Chem. 263, pp. 9129–9141.
1067. Strandberg, L., Lawrence, D. & Ny, T. (1988). The organization of the human plasminogen-activator-inhibitor-1 gene. Eur. J. Biochem. 176, pp. 609–619.
1068. Klinger, K.W., Winquist, R., Riccio, A., Andreasen, P.A., Sartorion, R., Nielsen, L.S., Stuart, N., Stanislovitis, P., Watkins, P., Douglas, R., Grzeschik, K.-H., Alitalo, K., Blasi, F. & Danø, K. (1987). Plasminogen activator inhibitor type 1 gene is located at region q21.3–q22 of chromosome 7 and genetically linked with cystic fibrosis. Proc.

Natl. Acad. Sci. USA 84, pp. 8548–8552.
1069. Andreasen, P.A., Riccio, A., Welinder, K.G., Douglas, R., Sartorio, R., Nielsen, L.S., Oppenheim, C., Blasi, F. & Danø, K. (1986). Plasminogen activator inhibitor type-1: Reactive center and amino terminal heterogeneity determined by protein and cDNA sequencing. FEBS Lett. 209, pp. 213–218.
1070. Kruithof, E.K.O., Nicolosa, G. & Bachman, F. (1987). Plasminogen activator inhibitor 1: Development of a radioimmunoassay and observation on its plasma concentration during venous occlusion and after platelet aggregation. Blood 70, pp. 1645–1653.
1071. Delerck, P.J., De Mol, M., Alessi, M.-C., Baudner, S., Pâques, E.-P., Preissner, K.T., Müller-Berghaus, G. & Collen, D. (1988). Purification and characterization of plasminogen activator inhibitor 1 binding protein from human plasma. J. Biol. Chem. 263, pp. 15454–15461.
1072. Wiman, B., Almquist, Å., Sigurdardottir, O. & Lindahl, T. (1988). The plasminogen activator inhibitor 1 (PAI) is bound to vitronectin in plasma. FEBS Lett. 242, pp. 125–128.
1073. Mimuro, J. & Loskutoff, D.J. (1989). Purification of a protein from bovine plasma that binds to type 1 plasminogen activator inhibitor and prevents its interaction with extracellular matrix. J. Biol. Chem. 264, pp. 936–939.
1074. Salonen, E.-M., Vaheri, A., Pöllänen, J., Stephens, R., Andreasen, P., Mayer, M., Danø, K., Gailit, J. & Ruoslahti, E. (1989). Interaction of plasminogen activator inhibitor (PAI-1) with vitronectin. J. Biol. Chem. 264, pp. 6339–6343.
1075. Åstedt, B. (1989). The placental type plasminogen activator inhibitor, PAI-2. Blood & Vessel 20, pp. 121–130.
1076. Kopitar, M., Rozman, B., Babnik, J., Turk, V., Mullins, D.E. & Wun, T.-Z. (1985). Human leukocyte urokinase inhibitor – purification, characterization and comparative studies against different plasminogen activators. Thromb. Haemostasis 54, pp. 750–755.
1077. Kruithof, E.K.O., Vassalli, J.D., Schleuning, W.-D., Mattaliano, R.J. & Bachman, F. (1986). Purification and characterization of a plasminogen activator inhibitor from the histiocytic lymphoma cell line U-937. J. Biol. Chem. 261, pp. 11207–11213.
1078. Lecander, I. & Åstedt, B. (1986). Isolation of a new specific plasminogen activator inhibitor from pregnancy plasma. Br. J. Haematol. 62, pp. 221–228.
1079. Ye, R.D., Wun, T.-Z. & Sadler, J.E. (1987). cDNA cloning and expression in *Escherichia coli* of a plasminogen activator inhibitor from human placenta. J. Biol. Chem. 262, pp. 3718–3725.
1080. Webb, A.C., Collins, K.L., Snyder, S.E., Alexander, S.J., Rosenwasser, L.J., Eddy, R.L., Shows, T.B. & Auron, P.E. (1987). Human monocyte Arg-serpin cDNA. J. Exp. Med. 166, pp. 77–94.
1081. Schleuning, W.-D., Medcalf, R.L., Hession, C., Rothenbühler, R., Shaw, A. & Kruithof, E.K.O. (1987). Plasminogen activator inhibitor 2: Regulation of gene transcription during phorbol ester-mediated differentiation of U-937 human histiocytic lymphoma cells. Mol. Cell. Biol. 7, pp. 4564–4567.
1082. Antalis, T.M., Clark, M.A., Barnes, T., Lehrbach, P.R., Devine, P.L., Schevzov, G., Goss, N.H., Stephens, R.W. & Tolstoshev, P. (1988). Cloning and expression of a cDNA coding for human monocyte-derived plasminogen activator inhibitor. Proc. Natl. Acad. Sci. USA 85, pp. 985–989.
1083. Ye, R.D., Ahern, S.M., Le Beau, M.M., Lebo, R.V. & Sadler, J.E. (1989). Structure of the gene for human plasminogen activator inhibitor-2. J. Biol. Chem. 264, pp. 5495–5502.
1084. Ye, R.D., Wun, T.-C. & Sadler, J.E. (1988). Mammalian protein secretion without signal peptide removal. J. Biol. Chem. 263, pp. 4869–4875.
1085. Lecander, I. & Åstedt, B. (1987). A specific plasminogen activator inhibitor of placental type occurring in amniotic fluid and cord blood. J. Lab. Clin. Med. 110, pp. 602–605.
1086. Kiso, U., Kaudewitz, H., Henschen, A., Åstedt, B., Kruithof, E.K.O. & Bachman, F. (1988). Determination of intermediate products and cleavage site in the reaction between plasminogen activator inhibitor type-2 and urokinase. FEBS Lett. 230, pp. 51–56.
1087. McGrogan, M., Kennedy, J., Li, M.P., Hsu, C., Scott, R.W., Simonsen, C.C. & Baker,

J.B. (1988). Molecular cloning and expression of the two forms of human protease nexin I. Bio/Technology 6, pp. 172–177.
1088. Gloor, S., Odink, K., Guenther, J., Nick, H. & Monard, D. (1986). A glia-derived neurite promoting factor with protease inhibitory activity belongs to the protease nexins. Cell 47, pp. 687–693.
1089. Sommer, J., Gloor, S.M., Rovelli, G.F., Hofstenge, J., Nick, H., Meier, R. & Monard, D. (1987). cDNA sequence coding for rat glia derived nexin and its homology to members of the serpin superfamily. Biochemistry 26, pp. 6407–6410.
1090. Francis, C.W. & Marder, V.J. (1986). Concepts of clot lysis. Ann. Rev. Med. 37, pp. 187–204.
1091. Ichinose, A., Fujikawa, K. & Suyama, T. (1986). The activation of pro-urokinase by plasma kallikrein and its inactivation by thrombin. J. Biol. Chem. 261, pp. 3486–3489.
1092. Lijnen, H.R., Zamarron, C., Blaber, M., Winkler, M. & Collen, D. (1986). Activation of plasminogen by pro-urokinase. J. Biol. Chem. 261, pp. 1253–1258.
1093. Stump, D.C., Thienpont, M. & Collen, D. (1986). Urokinase-related proteins in human urine. J. Biol. Chem. 261, pp. 1267–1275.
1094. Pannell, R. & Gurewich, V. (1987). Activation of plasminogen by single-chain urokinase or by two-chain urokinase – a demonstration that single-chain urokinase has low catalytic activity. Blood 69, pp. 22–26.
1095. Ellis, V., Scully, M.F. & Kakkar, V.V. (1987). Plasminogen activation by single-chain urokinase in functional isolation. J. Biol. Chem. 262, pp. 14998–15003.
1096. Urano, T., de Serrano, V.S., Gaffney, P.J. & Castellino, F.J. (1988). The activation of human (Glu[1])plasminogen by human single-chain urokinase. Arch. Biochem. Biophys. 264, pp. 222–230.
1097. Petersen, L.C., Lund, L.R., Nielsen, L.S., Danø, K. & Skriver, L. (1988). One-chain urokinase type plasminogen activator from human sarcoma cells is a proenzyme with little or no intrinsic activity. J. Biol. Chem. 263, pp. 11189–11195.
1098. Nelles, L., Lijnen, H.R., Collen, D. & Holmes, W. (1987). Characterization of recombinant single chain urokinase-type plasminogen activator mutants produced by site-specific mutagenesis of lysine-158. J. Biol. Chem. 262, pp. 5682–5689.
1099. Gurewich, V., Pannell, R., Broeze, R.J. & Mao, J. (1988). Characterization of the intrinsic fibrinolytic properties of pro-urokinase through a study of plasmin-resistant mutant forms produced by site-specific mutagenesis of lysine[158]. J. Clin. Invest. 82, pp. 1956–1962.
1100. Danø, K., Andreasen, P.A., Grøndahl-Hansen, J., Kristensen, P., Nielsen, L.S. & Skriver, L. (1985). Plasminogen activators, tissue degradation and cancer. Adv. Cancer Res. 44, pp. 139–266.
1101. Sumi, H., Maruyama, M., Matsuo, O., Mihara, H. & Toki, N. (1982). Higher fibrin-binding and thrombolytic properties of single-polypeptide-chain, high-molecular-weight urokinase. Thromb. Haemostas. 47, p. 297.
1102. Gurewich, V., Pannell, R., Louie, S., Kelly, P., Suddith, R.L. & Greenlee, R. (1984). Effective and fibrin-specific clot lysis by a zymogen precursor form of urokinase (pro-urokinase). J. Clin. Invest. 73, pp. 1731–1739.
1103. Stump, D.C., Stassen, J.M., Demarsin, E. & Collen, D. (1987). Comparative thrombolytic properties of single-chain forms of urokinase-type plasminogen activator. Blood 69, pp. 592–596.
1104. Pannell, R. & Gurewich, V. (1986). Pro-urokinase – a study of its stability in plasma and of a mechanism for its selective fibrinolytic effect. Blood 67, pp. 1215–1223.
1105. Vassalli, J.-D., Baccino, D. & Belin, D. (1985). A cellular binding site for the Mr 55,000 form of the human plasminogen activator, urokinase. J. Cell Biol. 100, pp. 86–92.
1106. Cubellis, M.V., Nolli, M.L., Cassani, G. & Blasi, F. (1986). Binding of singlechain prourokinase to the urokinase receptor of human U937 cells. J. Biol. Chem. 261, pp. 15819–15822.
1107. Nielsen, L.S., Kellerman, G.M., Behrendt, N., Picone, R., Danø, K. & Blasi, F. (1988). A 55,000–60,000 M_r receptor protein for urokinase-type plasminogen activator. J. Biol. Chem. 263, pp. 2358–2363.
1108. Estreicher, A., Wohlwend, A., Belin, D., Schleuning, W.-D. & Vassalli, J.-D. (1989). Characterization of the cellular binding site for the human urokinase-type plasminogen

activator. J. Biol. Chem. 264, pp. 1180–1189.
1109. Stoppelli, M.P., Corti, A., Soffientini, A., Cassani, G., Blasi, F. & Associan, R. K. (1985). Differentiation-enhanced binding of the amino-terminal fragment of human urokinase plasminogen activator to a specific receptor on U937 monocytes. Proc. Natl. Acad. Sci. USA 82, pp. 4939–4943.
1110. Appella, E., Robinson, L.A., Ullrich, S.J., Stoppelli, M.P., Corti, A., Cassani, G. & Blasi, F. (1987). The receptor-binding sequence of urokinase. J. Biol. Chem. 262, pp. 4437–4440.
1111. Komoriya, A., Hortsch, M., Meyers, C., Smith, M., Kanety, H. & Schlessinger, J. (1984). Biologically active synthetic fragments of epidermal growth factor: Localization of a major receptor-binding region. Proc. Natl. Acad. Sci. USA 81, pp. 1351–1355.
1112. Heath, W.F. & Merrifield, R.B. (1986). A synthetic approach to structure-function relationships in murine epidermal growth factor molecule. Proc. Natl. Acad. Sci. USA 83, pp. 6367–6371.
1113. Thorsen, S., Glas-Greenwalt, P. & Astrup, T. (1972). Differences in the binding to fibrin of urokinase and tissue plasminogen activator. Thrombos. Diathes. Haemorrh. 28, pp. 65–74.
1114. Rijken, D.C., Hoylaerts, H. & Collen, D. (1982). Fibrinolytic properties of one-chain and two-chain human extrinsic (tissue-type) plasminogen activator. J. Biol. Chem. 257, pp. 2920–2925.
1115. Tate, K.M., Higgins, D.L., Holmes, W.E., Winkler, M.E., Heyneker, H.L. & Vehar, G.A. (1987). Functional role of proteolytic cleavage at arginine-275 of human tissue plasminogen activator as assessed by site-directed mutagenesis. Biochemistry 26, pp. 338–343.
1116. Higgins, D.L. & Vehar, G.A. (1987). Interaction of one-chain and two-chain tissue plasminogen activator with intact and plasmin-degraded fibrin. Biochemistry 26, pp. 7786–7791.
1117. Husain, S.S., Hasan, A.A.K. & Budzynski, A.Z. (1989). Differences between binding of one-chain and two-chain tissue plasminogen activator to non-cross-linked and cross-linked fibrin clots. Blood 74, pp. 999–1006.
1118. Petersen, L.C., Johannessen, M., Foster, D., Kumar, A. & Mulvihill, E. (1988). The effect of polymerised fibrin on the catalytic activities of one chain tissue-type plasminogen activator as revealed by an analogue resistant to plasmin cleavage. Biochim. Biophys. Acta 952, pp. 245–254.
1119. Boose, J.A., Kuismanen, E., Gerard, R., Sambrook, J. & Gethling, M.-J. (1989). The single-chain form of tissue-type plasminogen activator has catalytic activity: Studies with a mutant that lacks the cleavage site. Biochemistry 28, pp. 635–643.
1120. Urano, S., Metzger, A.R. & Castellino, F.J. (1989). Plasmin-mediated fibrinolysis by variant recombinant tissue plasminogen activators. Proc. Natl. Acad. Sci. USA 86, pp. 2568–2571.
1121. MacDonald, M.E., van Zonnenveld, A.J. & Pannekoek, H. (1986). Functional analysis of the human tissue-type plasminogen protein: The light chain. Gene 42, pp. 59–67.
1122. Ichinose, A., Takio, K. & Fujikawa, K. (1986). Localization of the binding site of tissue-type plasminogen activator to fibrin. J. Clin. Invest. 78, pp. 163–169.
1123. van Zonnenveld, A.J., Veerman, H. & Pannekoek, H. (1986). Autonomous functions of structural domains in human tissue-type plasminogen activator. Proc. Natl. Acad. Sci. USA 83, pp. 4670–4674.
1124. Verheijen, J.H., Caspers, M.P.M., Chang, G.T.C., de Munk, G.A.W., Pouwels, P.H. & Enger-Valk, B.E. (1986). Involvement of finger domain and kringle 2 domain of tissue-type plasminogen activator in fibrin binding and stimulation of activity by fibrin. EMBO J. 5, pp. 3525–3530.
1125. van Zonnenveld, A.J., Veerman, H. & Pannekoek, H. (1986). On the interaction of the finger and the kringle-2 domain of tissue-type plasminogen activator. J. Biol. Chem. 261, pp. 14214–14218.
1126. Ehrlich, H.J., Bang, N.U., Little, S.P., Jaskunas, S.R., Weigel, B.J., Mattler, L.E. & Harms, C.S. (1987). Biological properties of a kringleless tissue plasminogen activator (t-PA) mutant. Fibrinolysis 1, pp. 75–81.
1127. de Vries, C., Veerman, H. & Pannekoek, H. (1989). Identification of domains of tissue-

type plasminogen activator involved in the augmented binding to fibrin after limited digestion with plasmin. J. Biol. Chem. 264, pp. 12604–12610.
1128. Gething, M.-J., Adler, B., Boose, J.-A., Gerard, R.D., Madison, E.L., McGookey, D., Meidell, R.S., Roman, L.M. & Sambrook, J. (1988). Variants of human tissue-type plasminogen activator that lack specific structural domains of the heavy chain. EMBO J. 7, pp. 2731–2740.
1129. Beebe, D.P. (1987). Binding of tissue plaminogen activator to human umbilical vein endothelial cells. Thromb. Res. 46, pp. 241–254.
1130. Haijar, K.A., Hamel, N.M., Harpel, P.C. & Nachman, R.L. (1987). Binding of tissue plasminogen activator to cultured human endothelial cells. J. Clin. Invest. 80, pp. 1712–1719.
1131. Barnathan, E.S., Kuo, A., Van der Keyl, H., McCrae, K.R., Larsen, G.R. & Cines, D.B. (1988). Tissue-type plasminogen activator binding to human endothelial cells. J. Biol. Chem. 263, pp. 7792–7799.
1132. Beebe, D.P., Miles, L.A. & Plow, E.F. (1989). A linear amino acid sequence involved in the interaction of t-PA with its endothelial cell receptor. Blood 74, pp. 2034–2037.
1133. Nelles, L., Lijnen, H.R., Collen, D. & Holmes, W.E. (1987). Characterization of a fusion protein consisting of amino acids 1 to 263 of tissue-type plasmingen activator and amino acids 144 to 411 of urokinase-type plasminogen activator. J. Biol. Chem. 262, pp. 10855–10862.
1134. Piérard, L., Jacobs, P., Gheysen, D., Hoylaerts, M., André, B., Topisirovic, L., Cravador, A., de Foresta, F., Herzog, A., Collen, D., De Wilde, M. & Bollen, A. (1987). Mutant and chimeric recombinant plasminogen activators. J. Biol. Chem. 262, pp. 11771–11775.
1135. Gheysen, D., Lijnen, H.R., Piérard, L., de Foresta, F., Demarsin, E., Jacobs, P., De Wilde, M., Bollen, A. & Collen, D. (1987). Characterization of a recombinant fusion protein of the finger domain of tissue-type plasminogen activator with a truncated single chain urokinase plasminogen activator. J. Biol. Chem. 262, pp. 11779–11784.
1136. Lee, S.G., Kalyan, N., Wilhelm, J., Hum, W.-T., Rappaport, R., Cheng, S.-M., Dheer, S., Urbano, C., Hartzell, R.W., Ronchetti-Blume, M., Leuner, M. & Hung, P.P. (1988). Construction and expression of hybrid plasminogen activators prepared from tissue-type plasminogen activator and urokinase plasminogen activator genes. J. Biol. Chem. 263, pp. 2917–2924.
1137. Piérard, L., Quintana, L.G., Reff, M.E. & Bollen, A. (1989). Production in eukaryotic cells and characterization of four hybrids of tissue-type and urokinase-type plasminogen activators. DNA 8, pp. 321–328.
1138. Robbins, K.C. & Tanaka, Y. (1986). Covalent molecular weight ≈92,000 hybrid plasminogen activator derived from human plasmin amino-terminal and urokinase carboxy-terminal domains. Biochemistry 25, pp. 3603–3611.
1139. Robbins, K.C. & Boreisha, I.G. (1987). A covalent molecular weight ≈92,000 hybrid plasminogen activator derived from human plasmin fibrin-binding and tissue plasminogen activator catalytic domains. Biochemistry 26, pp. 4661–4667.
1140. Nieuwenhuizen, W., Verheijen, J., Vermond, A. & Chang, G.T.G. (1983). Plasminogen activation by tissue activator is accelerated in the presence of firin(ogen) cyanogen bromide fragment FCB-2. Biochim. Biophys. Acta 755, pp. 531–533.
1141. Voskuilen, M., Vermond, A., Veeneman, G.H., van Boom, J.H., Klasen, E.A., Zegers, N.D. & Nieuwenhuizen, W. (1987). Fibrinogen lysine residue Aα157 plays a crucial role in the fibrin-induced acceleration of plasminogen activation, catalyzed by tissue-type plasminogen activator. J. Biol. Chem. 262, pp. 5944–5946.
1142. Bosma, P.J., Rijken, D.C. & Nieuwenhuizen, W. (1988). Binding of tissue-type plasminogen activator to fibrinogen fragments. Eur. J. Biochem. 172, pp. 399–404.
1143. Schielen, W.J.G., Voskuilen, M., Tesser, G.I. & Nieuwenhuizen, W. (1989). The sequence Aα(148–160) in fibrin, but not in fibrinogen, is accessible to monoclonal antibodies. Proc. Natl. Acad. Sci. USA 86, pp. 8951–8954.
1144. Tran-Tang, C., Kruithof, E.K.O. & Bachman, F. (1986). High affinity binding site for human Glu-plasminogen unveiled by limited plasmic degradation. Eur. J. Biochem. 160, pp. 599–604.
1145. Pannell, R., Black, J. & Gurewich, V. (1988). Complementary modes of action of tissue-

type plasminogen activator and pro-urokinase by which their synergistic effect on clot lysis may be explained. J. Clin. Invest. 81, pp. 853–859.

1146. Pizzo, S.V., Schwartz, M.L., Hill, R.L. & McKee, P.A. (1973). The effect of plasmin on the subunit structure of human fibrin. J. Biol. Chem. 248, pp. 4574–4583.

1147. Francis, C.W., Marder, V.J. & Barlow, G.H. (1980). Plasmic degradation of crosslinked fibrin. J. Clin. Invest. 66, pp. 1033–1043.

1148. Andreasen, P.A., Nielsen. L.S., Kristensen. P., Grøndahl-Hansen, J., Skriver, L. & Danø, K. (1986). Plasminogen activator inhibitor from human fibrosarcoma cells binds urokinase-type plasminogen activator, but not its proenzyme. J. Biol. Chem. 261, pp. 7644–7651.

1149. Vassalli, J.-D., Dayer, J.-M., Wohlwend, A. & Belin, D. (1984). Concomitant secretion of prourokinase and of a plasminogen activator-specific inhibitor by cultured human monocytes-macrophages. J. Exp. Med. 159, pp. 1653–1668.

1150. Hekman, C.M. & Loskutoff, D.J. (1985). Endothelial cells produce a latent inhibitor of plasminogen activators that can be activated by denaturants. J. Biol. Chem. 260, pp. 11581–11587.

1151. Lambers, J.W.J., Cammenca, M., König, B.W., Mertens, K., Pannekoek, H. & van Mourik, J.A. (1987). Activation of human endothelial cell-type plasmingen activator inhibitor (PAI-1) by negatively charged phospholipids. J. Biol. Chem. 262, pp. 17492–17496.

1152. Declerck, P.J., Alessi, M.C., Verstreken, M., Kruithof, E.K.O., Juhan-Vague, I. & Collen, D. (1988). Measurement of plasminogen activator inhibitor 1 in biologic fluids with a murine monoclonal antibody-based enzyme-linked immunosorbent assay. Blood 71, pp. 220–225.

1153. Urden, G., Hamsten, A. & Wiman, B. (1987). Comparison of plasminogen activator inhibitor activity and antigen in plasma samples. Clin. Chim. Acta 169, pp. 189–196.

1154. Murayama, H. & Bang, N.U. (1987). Incorporation of plasminogen activator inhibitor into fibrin, an alternative regulatory pathway of fibrinolysis. Thromb. Haemostas. 58, Abstract 1643.

1155. Wagner, O.F., de Vries, C., Hohmann, C., Veerman, H. & Pannekoek, H. (1989). Interaction between plasminogen activator inhibitor type 1 (PAI-1) bound to fibrin and either tissue-type plasminogen activator (t-PA) or urokinase-type plasminogen activator (u-PA). J. Clin. Invest 84, pp. 647–655.

1156. Madison, E.L., Goldsmith, E.J., Gerard, R.D., Gething, M.-J.H. & Sambrook, J.F. (1989). Serpin-resistant mutant of human tissue-type plasminogen activator. Nature 339, pp. 721–724.

1157. Clemmensen, I. (1979). Different molecular forms of α_2-antiplasmin. In: The Physiological Inhibitors of Coagulation and Fibrinolysis (Collen, D., Wiman, B. & Verstraete, M., Eds.), pp. 131–136, Elsevier, Amsterdam.

1158. Kluft, C., Los, P., Jie, A.F.H., Van Hinsbergh, V.W.M., Vellenga, E., Jespersen, J. & Henny, C.H.P. (1986). The mutual relationship between the two molecular forms of the major fibrinolysis inhibitor α_2-antiplasmin in blood. Blood 67, pp. 616–622.

1159. Sasaki, T., Morita, T. & Iwanaga, S. (1986). Identification of the plasminogen binding site of human α_2-plasmin inhibitor. J. Biochem. (Tokyo) 99, pp. 1699–1705.

1160. Hortin, G.L., Gibson, B.L. & Fok, K.F. (1988). α_2-Antiplasmin's carboxy-terminal lysine residue is a major site of interaction with plasmin. Biochem. Biophys. Res. Commun. 155, pp. 591–596.

1161. Hortin, G.L., Trimpe, B.L. & Fok, K.F. (1989). Plasmin's peptide-binding specificity: Characterization of ligand sites in α_2-antiplasmin. Thromb. Res. 54, pp. 621–632.

1162. Kluft, C., Los, P. & Jie, A.F.H. (1984). The molecular form of α_2-antiplasmin with affinity for plasminogen is selectively bound to fibrin by factor XIII. Thromb. Res. 33, pp. 419–425.

1163. Christensen, U. & Clemmensen, I. (1977). Kinetic properties of the primary inhibitor of plasmin from human plasma. Biochem. J. 163, pp. 389–391.

1164. Wiman, B. & Collen, D. (1978). On the kinetics of the reaction between human antiplasmin and plasmin. Eur. J. Biochem. 84, pp. 573–578.

1165. Tran-Thang, C., Kruithof, E.K.O. & Bachman, F. (1984). Tissue-type plasminogen activator increases the binding of Glu-plasminogen to clots. J. Clin. Invest. 74, pp. 2009–2016.

1166. Suenson, E., Lützen, O. & Thorsen, S. (1984). Initial plasmin-degradation of fibrin as the basis of a positive feed-back mechanism in fibrinolysis. Eur. J. Biochem. 140, pp. 513–522.
1167. Harpel, P.C., Chang, T.-S. & Verderber, E. (1985). Tissue plasminogen activator and urokinase mediate the binding of Glu-plasminogen to plasma fibrin I. J. Biol. Chem. 260, pp. 4432–4440.
1168. Koide, T., Foster, D., Yoshitake, S. & Davie, E.W. (1986). Amino acid sequence of human histidine-rich glycoprotein derived from the nucleotide sequence of its cDNA Biochemistry 25, pp. 2220 2225.
1169. Lijnen, H.R., Hoylaerts, M. & Collen, D. (1980). Isolation and characterization of a human plasma protein with affinity for the lysine binding sites in plasminogen. J. Biol. Chem. 255, pp. 10214–10222.
1170. Ichinose, A., Mimuro, J., Koide, T. & Aoki, N. (1984). Histidine-rich glycoprotein and α_2-plasmin inhibitor in inhibition of plasminogen binding to fibrin. Thromb. Res. 33, pp. 401–407.
1171. Silverstein, R.L., Nachman, R.L., Leung, L.L.K. & Harpel, P.C. (1985). Activation of immobilized plasminogen by tissue activator. J. Biol. Chem. 260, pp. 10346–10352.
1172. Silverstein, R.L., Leung, L.L.K., Harpel, P.C. & Nachman, R.L. (1984). Complex formation of platelet thrombospondin with plasminogen. J. Clin Invest. 74, pp. 1625–1633.
1173. DePolli, P., Bacon-Baguley, T., Kendra-Franczak, S., Caderholm, M.T. & Walz, D.A. (1989). Thrombospondin interaction with plasminogen. Evidence for binding to a specific region of the kringle structure of plasminogen. Blood 73, pp. 976–982.
1174. Leung, L.L.K., Nachman, R.L. & Harpel, P.C. (1984). Complex formation of platelet thrombospondin with histidine-rich glycoprotein. J. Clin. Invest. 73, pp. 5–12.
1175. Leung, L.L.K. (1986). Interaction of histidine-rich glycoprotein with fibrinogen and fibrin. J. Clin. Invest. 77, pp. 1305–1311.
1176. Porter, R.R. & Reid, K.B.M. (1978). The biochemistry of complement. Nature 275, pp. 699–704.
1177. Porter, R.R. & Reid, K.B.M. (1979). Activation of the complement system by antibody-antigen complexes: The classical pathway. Adv. Protein Chem. 33, pp. 1–71.
1178. Müller-Eberhard, H.J. & Schreiber, R.D. (1980). Molecular biology and chemistry of the alternative pathway of complement. Adv. Immunol. 29, pp. 1–53.
1179. Reid, K.B.M. & Porter, R.R. (1981). The proteolytic activation systems of complement. Ann. Rev. Biochem. 50, pp. 433–464.
1180. Ziccardi, R.J. (1983). The first component of human complement C1: Activation and control. Springer Semin. Immunopathol. 6, pp. 213–230.
1181. Reid, K.B.M. (1983). Proteins involved in the activation and control of human complement. Biochem. Soc. Trans. 11, pp. 1–12.
1182. Fearon, D.T. & Wong, W.W. (1983). Complement ligand-receptor interactions that mediate biological responses. Ann. Rev. Immunol. 1, pp. 243–271.
1183. Ziccardi, R.J. (1984). The first component of human complement (C1): Activation and control. Springer Semin. Immunopathol. 7, pp. 33–50.
1184. Müller-Eberhard, H.J. (1984). The membrane attack complex. Springer Semin. Immunopath. 7, pp. 93–141.
1185. Podack, E. & Tschopp, J. (1984). Membrane attack by complement. Mol. Immunol. 21, pp. 589–603.
1186. Pangburn, M.K. & Müller-Eberhard, H.J. (1984). The alternative pathway of complement. Springer Semin. Immunopath. 7, pp. 163–192.
1187. Cooper, N.R. (1985). The classical complement pathway: Activation and regulation of the first complement component. Adv. Immunol. 37, pp. 151–216.
1188. Reid, K.B.M. (1985). Application of molecular cloning to studies on the complement system. Immunology 55, pp. 185–196.
1189. Holers, V.M., Cole, J.L., Lublin, D.M., Seya, T. & Atkinson, J.P. (1985). Human C3b- and C4b-regulatory proteins: A new multi-gene family. Immunol. Today 6, pp. 188–192.
1190. Ross, G.D. & Medof, M.E. (1985). Membrane complement receptors specific for bound fragments of C3. Adv. Immunol. 37, pp. 217–267.
1191. Reid, K.B.M. (1986). Activation and control of the complement system. Essays in

Biochemistry 22, pp. 27–68.
1192. Campbell, R.D., Carroll, M.C. & Porter, R.R. (1986). The molecular genetics of components of complement. Adv. Immunol. 38, pp. 203–244.
1193. Müller-Eberhard, H.J. (1986). The membrane attack complex of complement. Ann. Rev. Immunol. 4, pp. 503–528.
1194. Reid, K.B.M., Bentley, D.R., Campbell, R.D., Chung, L.P., Sim, R.B., Kristensen, T. & Tack, B.F. (1986). Complement system proteins which interact with C3b or C4b. Immunol. Today 7, pp. 230–234.
1195. Schur, P.H. (1986). Inherited complement component abnormalities. Ann. Rev. Med. 37, pp. 333–346.
1196. Schumaker, V.N., Zavodszky, P. & Poon, P.H. (1987). Activation of the first component of complement. Ann. Rev. Immunol. 5, pp. 21–42.
1197. Chakravarti, D.N. & Kristensen, T. (1987). Recent advances in the study of the molecular structure of the complement proteins. Pathol. Immunopathol. Res. 5, pp. 317–351.
1198. Kristensen, T., D'Eustachio, P., Ogata, R.T., Chung, L.P., Reid, K.B.M. & Tack, B.F. (1987). The superfamily of C3b/C4b-binding proteins. Fed. Proc. 46, pp. 2463–2469.
1199. Müller-Eberhard, H.J. (1988). Molecular organization and function of the complement system. Ann. Rev. Biochem. 57, pp. 321–347.
1200. Campbell, R.D., Law, S.K.A., Reid, K.B.M. & Sim, R.B. (1988). Structure, organization, and regulation of the complement genes. Ann. Rev. Immunol. 6, pp. 161–195.
1201. Bentley, D.R. (1988). Structural superfamilies of the complement system. Expl. Clin. Immunogenet. 5, pp. 69–80.
1202. Davis III, A.E. (1988). C1 inhibitor and hereditary angioneurotic edema. Ann. Rev. Immunol. 6, pp. 595–628.
1203. Mollnes, T.E. & Lachmann, P.J. (1988). Regulation of complement. Scand. J. Immunol. 27, pp. 127–142.
1204. Cooper, N.R., Moore, M.D. & Nemerow, G.R. (1988). Immunobiology of CR2, the B lymphocyte receptor for Epstein-Barr virus and the C3d complement fragment. Ann. Rev. Immunol. 6, pp. 85–113.
1205. Hourcade, D., Holers, V.M. & Atkinson, J.P. (1989). The regulators of complement activation (RCA) gene cluster. Adv. Immunol. 45, pp. 381–416.
1206. Morgan, B.P. (1989). Complement membrane attack on nucleated cells: Resistance, recovery and non-lethal effects. Biochem. J. 264, pp. 1–14.
1207. Reid, K.B.M. (1974). A collagen-like amino acid sequence in a polypeptide chain of human C1q (a subcomponent of the first component of complement). Biochem. J. 141, pp. 189–203.
1208. Reid, K.B.M. (1977). Amino acid sequence of the N-terminal forty-two amino acid residues of the C chain of subcomponent C1q of the first component of human complement. Biochem. J. 161, pp. 247–251.
1209. Reid, K.B.M. & Thompson, E.O.P. (1978). Amino acid sequence of the N-terminal 108 amino acid residues of the B chain of subcomponent C1q of the first component of human complement. Biochem. J. 173, pp. 863–868.
1210. Reid, K.B.M. (1979). Complete amino acid sequence of the three collagen-like regions present in subcomponent C1q of the first component of human complement. Biochem. J. 179, pp. 367–371.
1211. Reid, K.B.M. (1982). Completion of the amino acid sequences of the A and B chains of subcomponent C1q of the first component of human complement. Biochem. J. 203, pp. 559–569.
1212. Reid, K.B.M. (1985). Molecular cloning and characterization of the complementary DNA and gene coding for the B-chain of subcomponent C1q of the human complement system. Biochem. J. 231, pp. 729–735.
1213. Solomon, E., Skok, J., Griffin, J. & Reid, K.B.M. (1985). Human C1q B chain (C1qB) is on chromosome 1p. Cytogenet. Cell Genet. 40, p. 749.
1214. Sellar, G.C., Goundis, D., McAdam, R.A., Solomon, E. & Reid, K.B.M. (1987). Cloning and chromosomal localisation of human C1q A-chain. Identification of a molecular defect in a C1q deficient patient. Complement 4, Abstract 267.

1215. Wood, L., Pulaski, S. & Vogeli, G. (1988). cDNA clones coding for the complete murine B chain of complement C1q: Nucleotide and derived amino acid sequences. Immunol. Lett. 17, pp. 59–62.
1216. Reid, K.B.M., Lowe, D.M. & Porter, R.R. (1972). Isolation and characterization of C1q, a subcomponent of the first component of complement, from human and rabbit sera. Biochem. J. 130, pp. 749–763.
1217. Calcott, M.A. & Müller-Eberhard, H.J. (1972). C1q protein of human complement. Biochemistry 11, pp. 3443–3450.
1218. Knobel, H.R., Villiger, W. & Isliker, H. (1975). Chemical analysis and electron microscopy studies of human C1q prepared by different methods. Eur. J. Immunol. 5, pp. 78–82.
1219. Brodsky-Doyle, B., Leonard, K.R. & Reid, K.B.M. (1976). Circular dichroism and electron microscopy studies of human subcomponent C1q before and after limited proteolysis by pepsin. Biochem. J. 159, pp. 279–286.
1220. Perkins, S.J. (1985). Molecular modelling of human complement subcomponent C1q and its complex with $C1r_2C1s_2$ derived from neutron-scattering curves and hydrodynamic properties. Biochem. J. 228, pp. 13–26.
1221. Kilchherr, E., Hofman, H., Steigemann, W. & Engel, J. (1985). Structural model of the collagen-like region of C1q comprising the kink region and the fiber-like packing of six triple helices. J. Mol. Biol. 186, pp. 403–415.
1222. McAdam, R.A., Goundis, D. & Reid, K.B.M. (1988). A homozygous point mutation results in a stop codon in the C1q B-chain of a C1q-deficient individual. Immunogenetics 27, pp. 259–264.
1223. Sim, R.B., Porter, R.R., Reid, K.B.M. & Gigli, I. (1977). The structure and enzymic activities of the C1r and C1s subcomponent of C1, the first component of human serum complement. Biochem. J. 163, pp. 219–227.
1224. Tschopp, J., Villiger, W., Fuchs, H., Kilchherr, E. & Engel, J. (1980). Assembly of subcomponents C1r and C1s of the first component of complement: Electron microscopy and ultracentrifugal studies. Proc. Natl. Acad. Sci. USA 77, pp. 7014–7018.
1225. Ziccardi, R.J. & Cooper, N.R. (1977). The subunit composition and sedimentation properties of human C1. J. Immunol. 118, pp. 2047–2052.
1226. Arlaud, G.J. & Gagnon, J. (1983). Complete amino acid sequence of the catalytic chain of human complement component C1r. Biochemistry 22, pp. 1758–1764.
1227. Arlaud, G.J. & Gagnon, J. (1985). Identification of the peptide bond cleaved during activation of human C1r. FEBS Lett. 180, pp. 234–238.
1228. Van Cong, N., Tosi, M., Gross, M.S., Cohen-Haguenauaer, O., Jegou-Foubert, C., de Tand, M.F., Meo, T. & Frézal, J. (1988). Assignment of the complement serine protease genes C1r and C1s to chromosome 12 region 12p13. Hum. Genet. 78, pp. 363–368.
1229. Arlaud, G.J., van Dorsselaer, A., Bell, A., Mancini, M., Aude, C. & Gagnon, J. (1987). Identification of erythro-β-hydroxyasparagine in the EGF-like domain of human C1r. FEBS Lett. 222, pp. 129–134.
1230. Carter, P.E., Dunbar, B. & Fothergill, J.E. (1983). The serine proteinase chain of human complement component C1s. Biochem. J. 215, pp. 565–571.
1231. Carter, P.E., Dunbar, B. & Fothergill, J.E. (1984). Structure and activity of C1r and C1s. Phil. Trans. R. Soc. Lond. B 306, pp. 293–299.
1232. Spycher, S.E., Nick, H. & Rickli, E.E. (1986). Human complement component C1s. Eur. J. Biochem. 156, pp. 49–57.
1233. Kusimoto, H., Hirosawa, S., Salier, J.P., Hagen, F.S. & Kurachi, K. (1988). Human genes for complement components C1r and C1s in a close tail-to-tail arrangement. Proc. Natl. Acad. Sci. USA 85, pp. 7307–7311.
1234. Tosi, M., Duponchel, C., Meo, T. & Couture-Tosi, E. (1989). Complement genes C1r and C1s feature an intronless serine protease domain closely related to haptoglobin. J. Mol. Biol. 208, pp. 709–714.
1235. Prysiecki, C.T., Staggers, J.E., Ramjit, H.G., Musson, D.G., Stern, A.M., Bennett, C.D. & Friedman, P.A. (1987). Occurrence of β-hydroxylated asparagine residues in non-vitamin K-dependent proteins containing epidermal growth factor-like domains. Proc. Natl. Acd. Sci. USA 84, pp. 7856–7860.
1236. Assimeh, S.N., Chapuis, R.M. & Isliker, H. (1978). Studies on the precursor form of the

first component of complement. Immunochemistry 15, pp. 13–17.
1237. Okamura, K. & Fujii, S. (1978). Isolation and characterization of different forms of C1r, a subcomponent of the first component of human complement. Biochim. Biophys. Acta 534, pp. 258–266.
1238. Arlaud, G.J., Villiers, C.L., Chesne, S. & Colomb, M.G. (1980). Purified proenzyme C1r. Some characteristics of its activation and subsequent proteolytic cleavage. Biochim. Biophys. Acta 616, pp. 116–129.
1239. Arlaud, G.J., Gagnon, J., Villiers, C.L. & Colomb, M.G. (1986). Molecular characterization of the catalytic domains of human complement serine protease C1r. Biochemistry 25, pp. 5177–5182.
1240. Valet, G. & Cooper, N.R. (1974). Isolation and characterization of the proenzyme form of the C1s subunit of the first complement component. J. Immunol. 112, pp. 339–350.
1241. Valet, G. & Cooper, N.R. (1974). Isolation and characterization of the proenzyme form of the C1r subunit of the first complement component. J. Immunol. 112, pp. 1667–1673.
1242. Villiers, C.L., Arlaud, G.J., Painter, R.H. & Colomb, M.G. (1980). Calcium binding properties of the C1 subcomponents C1q, C1r and C1s. FEBS Lett. 117, pp. 289–294.
1243. Villiers, C.L., Arlaud, G.J. & Colomb, M.G. (1985). Domain structure and associated functions of subcomponents C1r and C1s of the first component of human complement. Proc. Natl. Acad. Sci. USA 82, pp. 4477–4481.
1244. Busby, T.F. & Ingham, K.C. (1988). Domain structure, stability, and interaction of human complement C1s: Characterization of a derivative lacking most of the B chain. Biochemistry 27, pp. 6127–6135.
1245. Weiss, V., Fauser, C. & Engel, J. (1986). Functional model of subcomponent C1 of human complement. J. Mol. Biol. 189, pp. 573–581.
1246. Perkins, S.J. & Nealis, A.S. (1989). The quaternary structure in solution of human complement subcomponent $C1r_2C1s_2$. Biochem. J. 263, pp. 463–469.
1247. Ziccardi, R.J. (1982). Spontaneous activation of the first component of human complement (C1) by an intramolecular catalytic activity. J. Immunol. 128, pp. 2500–2504.
1248. Strang, C.J., Seigel, R.C., Phillips, M.L., Poon, P.H. & Schumaker, V.N. (1982). Ultrastructure of the first component of human complement: Electron microscopy of the crosslinked complex. Proc. Natl. Acad. Sci. USA 79, pp. 586–590.
1249. Poon, P.H., Schumaker, V.N., Phillips, M.L. & Strang, C.J. (1983). Conformation and restricted segmental flexibility of C1, the first component of human complement. J. Mol. Biol. 168, pp. 563–577.
1250. Schumaker, V.N., Hanson, D.C., Kilchherr, E., Phillips, M.L. & Poon, P.H. (1986). A molecular mechanism for the activation of the first component of complement by immune complexes. Mol. Immunol. 23, pp. 557–565.
1251. Arlaud, G.J., Colomb, M.G. & Gagnon, J. (1987). A functional model of the human C1 complex. Immunology Today 8, pp. 106–111.
1252. Müller-Eberhard, H.J. (1969). Complement. Ann. Rev. Biochem. 38, pp. 389–414.
1253. Schumaker, V.N., Calcott, M.A., Spiegelberg, H.L. & Müller-Eberhard, H.J. (1976). Ultracentrifuge studies of the binding of IgG of different subclasses to the C1q subunit of the first component of complement. Biochemistry 15, pp. 5175–5181.
1254. Augener, W., Grey, H.M., Cooper, N.R. & Müller-Eberhard, H.J. (1971). The reaction of monomeric and aggregated immunoglobulins with C1. Immunochemistry 8, pp. 1011–1020.
1255. Müller-Eberhard, H.J. & Calcott, M.A. (1966). Interaction between C1q and γG-globulin. Immunochemistry 3, p. 500.
1256. Ishizaka, T., Ishizaka, K., Salomon, S. & Fudenberg, H. (1967). Biologic activities of aggregated γ-globulin. J. Immunol. 99, pp. 82–91.
1257. Müller-Eberhard, H.J. (1975). Complement. Ann. Rev. Biochem. 44, pp. 697–724.
1258. Lepow, I.H., Naff, G.B., Todd, E.W., Pensky, J. & Hinz, C.F. (1963). Chromatographic resolution of the first component of human complement into three activities. J. Exp. Med. 117, pp. 983–1008.
1259. Müller-Eberhard, H.J., Nilsson, U.R., Dalmasso, A.P., Polley, M.J. & Calcott, M.A. (1966). A molecular concept of immune cytolysis. Arch. Pathol. 82, pp. 205–217.
1260. Yasmeen, D., Ellerson, J.R., Dorrington, K.J. & Painter, R.H. (1976). The structure and function of immunoglobulin domains. J. Immunol. 116, pp. 518–526.

1261. Kehoe, J.M. & Fougereau, M. (1969). Immunoglobulin peptide with complement fixing activity. Nature 224, pp. 1212–1213.
1262. Colomb, M. & Porter, R.R. (1975). Characterization of a plasmin-digest fragment of rabbit immunoglobulin gamma that binds antigen and complement. Biochem. J. 145, pp. 177–183.
1263. Arlaud, G.J., Meyer, C.M. & Colomb, M.G. (1976). Use of an IgG fragment prepared with particulate plasmin to study the C1 binding and activation. FEBS Lett. 66, pp. 132–136.
1264. Utsumi, S., Okada, M., Udaka, K. & Amano, T. (1985). Preparation and biological characterization of fragments containing dimeric and monomeric Cγ2 domain of rabbit IgG. Mol. Immunol. 22, pp. 811–819.
1265. Okada, M. & Utsumi, S. (1989). Role for the third constant domain of the IgG H chain in activation of complement in the presence of C1 inhibitor. J. Immunol. 142, pp. 195–201.
1266. Burton, D.R. (1985). Immunoglobulin G: Functional sites. Mol. Immunol. 22, pp. 161–206.
1267. Duncan, A.R. & Winter, G. (1988). The binding site for C1q on IgG. Nature 332, pp. 738–740.
1268. Hurst, M.M., Volanakis, J.E., Hester, R.B., Stroud, R.M. & Bennett, J.C. (1974). The structural basis for binding of complement by immunoglobulin M. J. Exp. Med. 140, pp. 1117–1121.
1269. Plaut, A.G., Cohen, S. & Tomasi, T.B. (1972). Immunoglobulin M: Fixation of human complement by the Fc fragment. Science 176, pp. 55–56.
1270. Wright, J.F., Shulman, M.J., Isenman, D.E. & Painter, R.H. (1988). C1 binding by murine IgM. J. Biol. Chem. 263, pp. 11221–11226.
1271. Hughes-Jones, N.C. & Gardener, B. (1979). Reaction between the isolated globular subunits of the complement component C1q and IgG-complexes. Mol. Immunol. 16, pp. 697–701.
1272. Pâques, E.P., Huber, R., Preiss, A. & Wright, J.K. (1979). Isolation of the globular region of the subcomponent q of the C1 component of complement. Hoppe-Seyler's Z. Physiol. Chem. 360, pp. 177–183.
1273. Tschopp, J., Villiger, W., Lustig, A., Jaton, J.-C. & Engel, J. (1980). Antigen-independent binding of IgG dimers to C1q as studied by sedimentation equilibrium, complement fixation and electron microscopy. Eur. J. Immunol. 10, pp. 529–535.
1274. Jaton, J.C., Huser, H., Riesen, W.F., Schlessinger, J. & Givol, D. (1976). The binding of complement by complexes formed between a rabbit antibody and oligosaccharides of increasing size. J. Immunol. 116, pp. 1363–1366.
1275. Liberti, P.A., Bausch, P.M. & Schoenberg, L.M. (1982). On the mechanism of C1q binding to antibody. Mol. Immunol. 19, pp. 143–149.
1276. Ziccardi, R.J. & Cooper, N.R. (1976). Activation of C1r by proteolytic cleavage. J. Immunol. 116, pp. 504–509. p3
1277. Dodds, A.W., Sim, R.B., Porter, R.R. & Kerr, M.A. (1978). Activation of the first component of human complement (C1) by antibody-antigen aggregates. Biochem. J. 175, pp. 383–390.
1278. Naff, G.B. & Ratnoff, O.D. (1968). The enzymatic nature of C1r. J. Exp. Med. 128, pp. 571–593.
1279. Sakai, K. & Stroud, R.M. (1974). The activation of C1s with purified C1r. Immunochemistry 11, pp. 191–196.
1280. Kerr, M.A. & Porter, R.R. (1978). Purification and properties of the second component of human complement. Biochem. J. 171, pp. 99–107.
1281. Tomana, M., Nieman, M., Garner, C. & Volanakis, J.E. (1985). Carbohydrate composition of the second, third and fifth components and factor B and D of human complement. Mol. Immunol. 22, pp. 107–111.
1282. Kerr, M.A. (1979). Limited proteolysis of complement components C2 and factor B. Biochem. J. 183, pp. 615–622.
1283. Parkes, C., Gagnon, J. & Kerr, M.A. (1983). The reaction of iodine and thiol-blocking reagents with human complement components C2 and factor B. Biochem. J. 213, pp. 201–209.
1284. Gagnon, J. (1984). Structure and activation of C2 and factor B. Phil. Trans. R. Soc.

Lond. B 306, pp. 301–309.
1285. Horiuchi, T., Macon, K.J., Kidd, V.J. & Volanakis, J.E. (1989). cDNA cloning and expression of human complement component C2. J. Immunol. 142, pp. 2105–2111.
1286. Falus, A., Wakeland, E.K., McConnell, T.J., Gitlin, J., Whitehead, A.S. & Colten, H.R. (1987). DNA polymorphism of MHC III genes in inbred and wild mouse strains. Immunogenetics 25, pp. 290–298.
1287. Carroll, M.C., Campbell, R.D., Bentley, D.R. & Porter, R.R. (1984). A molecular map of the human major histocompatibility complex class III region linking complement genes C4, C2 and factor B. Nature 307, pp. 237–241.
1288. Bentley, D.R., Campbell, R.D. & Cross, S.J. (1985). DNA polymorphism of the C2 locus. Immunogenetics 21, pp. 377–390.
1289. Polley, M.J. & Müller-Eberhard, H.J. (1967). Enhancement of the hemolytic activity of the second component of complement by oxidation. J. Exp. Med. 126, pp. 1013–1025.
1290. Tack, B.F. & Prahl, J.W. (1976). Third component of human complement: Purification from plasma and physicochemical characterization. Biochemistry 15, pp. 4513–4521.
1291. Tack, B.F., Morris, S.C. & Prahl, J.W. (1979). Third component of human complement: Structural analysis of the peptide chains of C3 and C3b. Biochemistry 18, pp. 1497–1503.
1292. Hugli, T.E. (1975). Human anaphylatoxin (C3a) from the third component of complement: Primary structure. J. Biol. Chem. 250, pp. 8293–8301.
1293. Davis, A.E., Harrison, R.A. & Lachmann, P.J. (1984). Physiologic inactivation of fluid phase C3b: Isolation and structural characterization of C3c, C3d,g (α2D), and C3g. J. Immunol. 132, pp. 1960–1966.
1294. Eggertsen, G., Hellman, U., Lundwall, Å., Folkersen, J. & Sjöquist, J. (1985). Characterization of tryptic fragments of human complement C3. Mol. Immunol. 22, pp. 833–841.
1295. Hellman, U., Eggertsen, G., Engström, A. & Sjöquist, J. (1985). Amino acid sequence of the trypsin-generated C3d fragment from human complement factor C3. Biochem. J. 230, pp. 353–361.
1296. Lundwall, Å., Hellman, U., Eggertsen, G. & Sjöquist, J. (1984). Chemical characterization of cyanogen bromide fragments from the β-chain of human complement factor C3. FEBS Lett. 169, pp. 57–62.
1297. Welinder, K.G. & Svendsen, A. (1986). Amino acid sequence analysis of the glycopeptides from human complement component C3. FEBS Lett. 202, pp. 59–62.
1298. de Bruijn, M.H.L. & Fey, G.H. (1985). Human complement component C3: cDNA coding sequence and derived primary structure. Proc. Natl. Acad. Sci. USA 82, pp. 708–712.
1299. Lundwall, Å., Wetsel, R.A., Domdey, H., Tack, B.F. & Fey, G.H. (1984). Structure of murine complement component C3. J. Biol. Chem. 259, pp. 13851–13856.
1300. Wetsel, R.A., Lundwall, Å., Davidson, F., Gibson, T., Tack, B.F. & Fey, G.H. (1984). Structure of murine complement component C3. J. Biol. Chem. 259, pp. 13857–13862.
1301. Kusano, M., Choi, N.-H., Tomita, M., Yamamoto, K., Migata, S., Sekeiya, T. & Nishimura, S. (1986). Nucleotide sequence of DNA and derived amino acid sequence of rabbit complement component C3 α-chain. Immunol. Invest. 15, pp. 365–378.
1302. Grossberger, D., Marcuz, A., Du Pasquier, L. & Lambris, J.D. (1989). Conservation of structural and functional domains in complement component C3 of Xenopus and mammals. Proc. Natl. Acad. Sci. USA 86, pp. 1323–1327.
1303. Whitehead, A.S., Solomon, E., Chambers, S., Bodmer, W.F., Povey, S. & Fey, G.H. (1982). Assignment of the structural gene for the third component of complement to chromosome 19. Proc. Natl. Acad. Sci. USA 79, pp. 5021–5025.
1304. Pericak-Vance, M.A., Yamaoka, L., Assinder, R., Bartlett, R.J., Ross, D.A., Fey, G., Humphries, S., Williamson, R., Conneally, P.M. & Roses, A.D. (1985). Tight linkage of APOC2 with myotonic dystrophy on chromosome 19. Cytogenet. Cell Genet. 40, pp. 721–722.
1305. Barnum, S.R., Amiguet, P., Amiguet-Barras, F., Fey, G. & Tack, B.F. (1989). Complete intron/exon organization of DNA encoding the α' chain of human C3. J. Biol. Chem. 264, pp. 8471–8474.
1306. Huber, R., Scholze, H., Paques, E.P. & Deisenhofer, J. (1980). Crystal structure

analysis and molecular model of human C3a anaphylatoxin. Hoppe-Seyler's Z. Physiol. Chem. 361, pp. 1389–1399.
1307. Matsuda, T., Seya, T. & Nagasawa, S. (1985). Location of the interchain disulfide bonds of the third component of human complement. Biochem. Biophys. Res. Commun. 127, pp. 264–269.
1308. Janatova, J. (1986). Detection of disulphide bonds and location of interchain linkages in the third (C3) and fourth (C4) components of human complement. Biochem. J. 233, pp. 819–825.
1309. Tack, B.F., Harrison, R.A., Janatova, J., Thomas, M.L. & Prahl, J.W. (1980). Evidence for presence of an internal thiol-ester bond in third component of complement. Proc. Natl. Acad. Sci. USA 77, pp. 5764–5768.
1310. Thomas, M.L., Janatova, J., Gray, W.R. & Tack, B.F. (1982). Third component of human complement: Localization of the internal thiolester bond. Proc. Natl. Acad. Sci. USA 79, pp. 1054–1058.
1311. Thomas, M.L., Davidson, F.F. & Tack, B.F. (1983). Reduction of the β-Cys-γ-Glu thiol ester bond of human C3 with sodium borohydride. J. Biol. Chem. 258, pp. 13580–13586.
1312. Sottrup-Jensen, L., Petersen, T.E. & Magnusson, S. (1980). A thiolester in α_2-macroglobulin cleaved during proteinase complex formation. FEBS Lett. 121, pp. 275–279.
1313. Sottrup-Jensen, L., Stepanik, T.M., Kristensen, T., Lønblad, P.B., Jones, C.M., Wierzbicki, D.M., Magnusson, S., Domdey, H., Wetsel, R.A., Lundwall, Å., Tack, B.F. & Fey, G.H. (1985). Common evolutionary origin of α_2-macroglobulin and complement components C3 and C4. Proc. Natl. Acad. Sci. USA 82, pp. 9–13.
1314. Martin, S.C. (1989). Phosphorylation of complement factor C3 *in vivo*. Biochem. J. 261, pp. 1051–1054.
1315. Botto, M., Fong, K.Y., So, A.K., Koch, C. & Walport, M.J. (1990). Molecular basis of polymorphisms of human complement component C3. J. Exp. Med. 172, pp. 1011–1017.
1316. Gigli, I., von Zabern, I. & Porter, R.R. (1977). The isolation and structure of C4, the fourth component of human complement. Biochem. J. 165, pp. 439–446.
1317. Bolotin, C., Morris, S.C., Tack, B.F. & Prahl, J.W. (1977). Purification and structural analysis of the fourth component of complement. Biochemistry 16, pp. 2008–2015.
1318. Moon, K.E., Gorski, J.P. & Hugli, T.E. (1981). Complete primary structure of human C4a anaphylatoxin. J. Biol. Chem. 256, pp. 8685–8692.
1319. Press, E.M. & Gagnon, J. (1981). Human complement component C4. Biochem. J. 199, pp. 351–357.
1320. Campbell, R.D., Gagnon, J. & Porter, R.R. (1981). Amino acid sequence around the thiol and reactive acyl groups of human complement component C4. Biochem. J. 199, pp. 359–370.
1321. Chakravati, D.N., Campbell, R.D. & Gagnon, J. (1983). Amino acid sequence of a polymorphic segment from fragment C4d of human complement component C4. FEBS Lett. 154, pp. 387–390.
1322. Lundwall, Å., Hellman, U., Eggertsen, G. & Sjöquist, J. (1982). Isolation of tryptic fragments of human C4 expressing Chido and Rodgers antigens. Mol. Immunol. 19, pp. 1655–1665.
1323. Hellman, U., Eggertsen, G., Lundwall, Å., Engström, A. & Sjöquist, J. (1984). Primary sequence differences between Chido and Rodgers variants of tryptic C4d of the human complement system. FEBS Lett. 170, pp. 254–258.
1324. Law, S.K.A. & Gagnon, J. (1985). The primary structure of the fourth component of human complement (C4)-C-terminal peptides. Biosci. Rep. 5, pp. 913–921.
1325. Chakravarti, D.N., Campbell, D.N. & Porter, R.R. (1987). The chemical structure of the C4d fragment of the human complement component C4. Mol. Immunol. 24, pp. 1187–1197.
1326. Belt, K.T., Carroll, M.C. & Porter, R.R. (1984). The structural basis of the multiple forms of human complement component C4. Cell 36, pp. 907–914.
1327. Belt, K.T., Yu, C.-Y., Carroll, M.C. & Porter, R.R. (1985). Polymorphism in human complement C4. Immunogenetics 21, pp. 173–180.

1328. Yu, C.Y., Belt, K.T., Giles, C.M., Campbell, R.D. & Porter, R.R. (1986). Structural basis of the polymorphism ofhuman complement component C4A and C4B: Gene size, reactivity and antigenicity. EMBO J. 5, pp. 2873–2881.

1329. Carroll, M.C., Palsdottir, A., Belt, K.T. & Yu, C.Y. (1986). Molecular genetics of the fourth component of human complement. Biochem. Soc. Symp. 51, pp. 29–36.

1330. Carroll, M.C., Campbell, R.D. & Porter, R.R. (1985). Mapping of steroid 21-hydroxylase genes adjacent to complement component C4 genes in HLA, the major histocompatability complex in man. Proc. Natl. Acad. Sci. USA 82, pp. 521–525.

1331. White, P.C., Grossberger, D., Onufer, B.J., Chaplin, D.D., New, M.I., Dupont, B. & Strominger, J.L. (1985). Two genes encoding steroid 21-hydroxylase are located near the genes encoding the fourth component of complement in man. Proc. Natl. Acad. Sci. USA 82, pp. 1089–1093.

1332. Hortin, G., Chan, A.C., Fok, K.F., Strauss, A.W. & Atkinson, J.P. (1986). Sequence analysis of the COOH terminus of the α-chain of the fourth component of complement. J. Biol. Chem. 261, pp. 9065–9069.

1333. Karp, D.R. (1983). Post-translational modification of the fourth component of complement. J. Biol. Chem. 258, pp. 12745–12748.

1334. Hortin, G., Sims, H. & Strauss, A.W. (1986). Identification of the site of sulfatation of the fourth component of human complement. J. Biol. Chem. 261, pp. 1786–1793.

1335. Harrison, R.A., Thomas, M.L. & Tack, B.F. (1981). Sequence determination of the thiolester site of the fourth component of human complement. Proc. Natl. Acad. Sci. USA 78, pp. 7388–7392.

1336. Sepich, D.S., Noonan, D.J. & Ogata, R.T. (1985). Complete nucleotide sequence of the fourth component of murine complement. Proc. Natl. Acad. Sci. USA 82, pp. 5895–5899.

1337. Nonaka, M., Nakayama, K., Yeul, Y.D. & Takahashi, M. (1985). Complete nucleotide and derived amino acid sequences of the fourth component of mouse complement (C4). J. Biol. Chem. 260, pp. 10936–10943.

1338. Ogata, R.T. & Sepich, D.C. (1985). Murine sex-limited protein: Complete cDNA sequence and comparison with murine fourth complement component. J. Immunol. 135, pp. 4239–4244.

1339. Nonaka, M., Nakayana, K., Yeul, Y.D. & Takahashi, M. (1986). Complete nucleotide and derived amino acid sequence of sex-limited protein (Slp). Nonfunctional isotype of the fourth component of mouse complement (C4). J. Immunol. 136, pp. 2989–2993.

1340. Karp, D.R., Capra, J.D., Atkinson, J.P. & Shreffler, D.C. (1982). Structural and functional characterization of an incompletely processed form of murine C4 and Slp. J. Immunol. 128, pp. 2336–2341.

1341. Ogata, R.T., Cooper, N.R., Bradt, B.M., Mathias, P. & Picchi, M.A. (1989). Murine complement component C4 and sex-limited protein: Identification of amino acid residues essential for C4 function. Proc. Natl. Acad. Sci. USA 86, pp. 5575–5579.

1342. Ogata, R.T., Rosa, P.A. & Zepf, N.E. (1989). Sequence of the gene for murine complement component C4. J. Biol. Chem. 264, pp. 16565–16572.

1343. Lepow, I.H., Ratnoff, O.D., Rosen, F.S. & Pillemer, L. (1956). Observation on a pro-esterase associated with partially purified first component of human complement (C′1). Proc. Soc. Exp. Biol. Med. 92, pp. 32–37.

1344. Stroud, R.M., Austen, K.F. & Mayer, M.M. (1965). Catalysis of C′2 fixation by C′1a. Immunochemistry 2, pp. 219–234.

1345. Stroud, R.M., Mayer, M.M., Miller, J.A. & McKenzie, A.T. (1966). C′2a^{d}, an inactive derivate of C′2 released during decay of EAC′4,2a. Immunochemistry 3, pp. 163–176.

1346. Polley, M.J. & Müller-Eberhard, H.J. (1968). The second component of human complement: Its isolation, fragmentation by C′1 esterase, and incorporation into C′3 convertase. J. Exp. Med. 128, pp. 533–551.

1347. Müller-Eberhard, H.J. & Lepow, I.H. (1965). C′1 esterase effect on activity and physicochemical properties of the fourth component of complement. J. Exp. Med. 121, pp. 819–833.

1348. Patrick, R.A., Taubman, S.B. & Lepow, I.H. (1970). Cleavage of the fourth component of human complement (C4) by activated C1s. Imunochemistry 7, pp. 217–225.

1349. Hortin, G.L., Farries, T.C., Graham, J.P. & Atkinson, J.P. (1989). Sulfation of tyrosine residues increases activity of the fourth component of complement. Proc. Natl. Acad.

Sci. USA 86, pp. 1338–1342.
1350. Matsumoto, M., Nagaki, K., Kitamura, H., Kuramitsu, S., Nagasawa, S. & Seya, T. (1989). Probing the C4-binding site on C1s with monoclonal antibodies. J. Immunol. 142, pp. 2743–2750.
1351. Law, S.K.A., Lichtenberg, N.A., Holcombe, F.H. & Levine, R.P. (1980). Interaction between the labile binding sites of the fourth (C4) and fifth (C5) human complement proteins and eryhrocyte cell membranes. J. Immunol. 125, pp. 634–639.
1352. Law, S.K.A., Lichtenberg, N.A. & Levine, R.P. (1980). Covalent binding and hemolytic activity of complement proteins. Proc. Natl. Acad. Sci. USA 77, pp. 7194–7198.
1353. Law, S.K.A., Minich, T.M. & Levine, R.P. (1984). Covalent binding efficiency of the third and fourth complement proteins in relation to pH, nucleophilicity and availability of hydroxyl groups. Biochemistry 23, pp. 3267–3272.
1354. Isenman, D.E. & Young, J.R. (1984). The molecular basis for the difference in immune hemolysis activity of the Chido and Rodgers isotypes of human complement component C4. J. Immunol. 132, pp. 3019–3027.
1355. Law, S.K.A., Dodds, A.W. & Porter, R.R. (1984). A comparison of the properties of two classes, C4A and C4B, of human complement component C4. EMBO J. 3, pp. 1819–1823.
1356. Isenman, D.E. & Young, J.R. (1986). Covalent binding properties of the C4A and C4B isotypes of the fourth component of human complement on several C1-bearing cell surfaces. J. Immunol. 136, pp. 2542–2550.
1357. Dodds, A.W. & Law, S.K.A. (1988). Structural basis of the binding specificity of the thioester-containing proteins, C4, C3 and alpha-2-macroglobulin. Complement 5, pp. 89–97.
1358. Müller-Eberhard, H.J., Polley, M.J. & Calcott, M.A. (1967). Formation and functional significance of a molecular complex derived from the second and the fourth component of human complement. J. Exp. Med. 125, pp. 359–380.
1359. Kerr, M.A. (1980). The human complement system: Assembly of the classical pathway C3 convertase. Biochem. J. 189, pp. 173–181.
1360. Nagasawa, S. & Stroud, R.M. (1977). Cleavage of C2 by $C\bar{1}$ into the antigenically distinct fragments C2a and C2b: Demonstration of binding of C2b to C4. Proc. Natl. Acad. Sci. USA 74, pp. 2998–3001.
1361. Oglesby, T.J., Accavitti, M.A. & Volanakis, J.E. (1988). Evidence for a C4b binding site on the C2b domain of C2. J. Immunol. 141, pp. 926–931.
1362. Gigli, I. & Austen, K.F. (1969). Fluid phase destruction of $C2^{hu}$ by $C1^{hu}$. J. Exp. Med. 130, pp. 833–846.
1363. Strunk, R. & Colten, H. (1974). The first components of human complement (C1): Kinetics of reaction with its natural substrates. J. Immunol. 112, pp. 905–910.
1364. Thielens, N.M., Villiers, M.B., Reboul, M.B., Villiers, C.L. & Colomb, M.G. (1982). Human complement subcomponent C2: Purification and proteolytic cleavage in fluid phase by C1s, $C\bar{1}r_2$-$C\bar{1}s_2$ and $C\bar{1}$. FEBS Lett. 141, pp. 19–24.
1365. Law, S.K.A. & Levine, R.P. (1977). Interaction between the third complement protein and cell surface macromolecules. Proc. Natl. Acad. Sci. USA 74, pp. 2701–2705.
1366. Law, S.K.A., Lichtenberg, N.A. & Levine, R.P. (1979). Evidence for an ester linkage between the labile binding site of C3b and receptive surfaces. J. Immunol. 123, pp. 1388–1394.
1367. Pangburn, M.K., Schreiber, R.D. & Müller-Eberhard, H.J. (1981). Formation of the initial C3 convertase of the alternative complement pathway. Acquisition of C3b-like activities by spontaneous hydrolysis of the putative thiolester in native C3. J. Exp. Med. 154, pp. 856–862.
1368. Sim, R.B., Twose, T.M., Paterson, D.S. & Sim, E. (1981). The covalent binding reaction of complement component C3. Biochem. J. 193, pp. 115–127.
1369. Vogt, W, Schmidt, G., van Buthlar, B. & Dieminger, I. (1978). A new function of the activated third component of complement: Binding to C5, an essential step for C5 activation. Immunology 34, pp. 29–40.
1370. Isenman, D.E., Podack, E.R. & Cooper, N.R. (1980). The interaction of C5 with C3b in free solution: A sufficient condition for cleavage by fluid phase C3/C5 convertase. J. Immunol. 124, pp. 326–331.

1371. Takata, Y., Kinoshita, T., Kozono, H., Takeda, J., Tanaka, E., Hong, K. & Inoue, K. (1987). Covalent association of C3b with C4b within C5 convertase of the classical complement pathway. J. Exp. Med. 165, pp. 1494–1507.
1372. Volanakis, J.E., Schrohenloher, R.E. & Stroud, R.M. (1977). Human factor $\overline{\mathrm{D}}$ of the alternative complement pathway: Purification and characterization. J. Immunol. 119, pp. 337–342.
1373. Lesavre, P.H. & Müller-Eberhard, H.J. (1978). Mechanism of action of factor $\overline{\mathrm{D}}$ of the alternative complement pathway. J. Exp. Med. 148, pp. 1498–1509.
1374. Johnson, D.M.A., Gagnon, J. & Reid, K.B.M. (1984). Amino acid sequence of human factor $\overline{\mathrm{D}}$ of the complement system. FEBS Lett. 166, pp. 347–351.
1375. Nieman, M.A., Brown, A.S., Bennett, J.C. & Volanakis, J.E. (1984). Amino acid sequence of human $\overline{\mathrm{D}}$ of the alternative complement pathway. Biochemistry 23, pp. 2482–2486.
1376. Mole, J.E. & Anderson, J.K. (1987). Cloning the cDNA for complement factor $\overline{\mathrm{D}}$: Evidence for the existence of a zymogen for the serum enzyme. Complement 4, Abstract 191.
1377. Fearon, D.T, Austen, K.F. & Ruddy, S. (1974). Properdin factor $\overline{\mathrm{D}}$: Characterization of its active site and isolation of the precursor form. J. Exp. Med. 139, pp. 355–366.
1378. Lesavre, P.H., Hugli, T.E., Esser, A.F. & Müller-Eberhard, H.J. (1979). The alternative pathway C3/C5 convertase: Chemical basis of factor B activation. J. Immunol. 123, pp. 529–534.
1379. Davis, A.E., Zalut, C., Rosen, F.S. & Alper, C.A. (1979). Human factor $\overline{\mathrm{D}}$ of the alternative complement pathway. Physicochemical characteristics and N-terminal amino acid sequence. Biochemistry 18, pp. 5082–5087.
1380. Johnson, D.M.A., Gagnon, J. & Reid, K.B.M. (1980). Factor $\overline{\mathrm{D}}$ of the alternative pathway of human complement. Biochem. J. 187, pp. 863–874.
1381. Rosen, B.S., Cook, K.S., Yaglom, J., Groves, D.L., Volanakis, J.E., Damm, D., White, T. & Spiegelman, B.M. (1989). Adipsin and complement factor $\overline{\mathrm{D}}$ activity: An immune-related defect in obesity. Science 244, pp. 1483–1487.
1382. Cook, K.S., Groves, D.L., Min, H.Y. & Spiegelman, B.M. (1985). A developmentally regulated mRNA from 3T3 adipocytes encodes a novel serine protease homologue. Proc. Natl. Acad. Sci. USA 82, pp. 6480–6484.
1383. Curman, B., Sandberg-Trägårdh, L. & Peterson, P.A. (1977). Chemical characterization of human factor B of the alternate pathway of complement activation. Biochemistry 16, pp. 5368–5375.
1384. Gagnon, J. & Christie, D.L. (1983). Amino acid sequence of the Bb fragment from complement factor B. Biochem. J. 209, pp. 51–60.
1385. Christie, D.L. & Gagnon, J. (1983). Amino acid sequence of the Bb fragment from complement factor B. Sequence of the major cyanogen bromide-cleavage peptide (CB-II) and completion of the sequence of the Bb fragment. Biochem. J. 209, pp. 61–70.
1386. Sackstein, R., Colten, H.R. & Woods, D.E. (1983). Phylogenetic conservation of a class III major histocompatability complex antigen, factor B. J. Biol. Chem. 258, pp. 14693–14699.
1387. Campbell, R.D. & Porter, R.R. (1983). Molecular cloning and characterization of the gene coding for human complement protein B. Proc. Natl. Acad. Sci. USA 80, pp. 4464–4468.
1388. Campbell, R.D., Bentley, D.R. & Morley, B.J. (1984). The factor B and C2 genes. Phil. Trans. Royal Soc. Lond. B 306, pp. 367–378.
1389. Campbell, R.D., Morley, B.J., Sargent, C.A. & Janjua, N.J. (1985). Molecular basis for allelic variation at the factor B locus. Complement 2, Abstract 36.
1390. Christie, D.L. & Gagnon, J. (1982). Isolation, characterization and N-terminal sequences of the CNBr-cleavage peptides from human complement factor B. Biochem. J. 201, pp. 555–567.
1391. Wu, L., Morley, B.J. & Campbell, R.D. (1987). Cell specific expression of the human complement protein factor B gene: Evidence for the role of two distinct 5′-flanking elements. Cell 48, pp. 331–342.
1392. Morel, Y., Bristow, J., Gitelman, S.E. & Miller, W.L. (1989). Transcript encoded on the opposite strand of the human steroid 21-hydroxylase/complement component C4 gene locus. Proc. Natl. Acad. Sci. USA 86, pp. 6582–6586.

1393. Nicol, P.A.E. & Lachman, P.J. (1973). The alternative pathway of complement activation. The role of C3 and its inactivator (KAF). Immunology 24, pp. 259–275.
1394. Lachman, P.J. & Halwachs, L. (1975). The influence of C3b inactivator (KAF) concentration on the ability of serum to support complement activation. Clin. Exp. Immunol. 21, pp. 109–114.
1395. Pangburn, M.K. & Müller-Eberhard, H.J. (1980). Relation of a putative thiolester bond in C3 to activation of the alternative pathway and the binding of C3b to biological targets of complement. J. Exp. Med. 152, pp. 1102–1114.
1396. Hack, C.E., Paardekooper, J., Smeenk, R.J.T., Abbink, J., Eerenberg, A.J.M. & Nuijens, J.H. (1988). Disruption of the internal thioester bond in the third component of complement (C3) results in the exposure of neodeterminants also present on activation products of C3. J. Immunol. 141, pp. 1602–1609.
1397. Pryzdial, E.L.G. & Isenman, D.E. (1986). A reexamination of the role of magnesium in the human alternative pathway of complement. Mol. Immunol. 23, pp. 87–96.
1398. Müller-Eberhard, H.J. & Götze, O. (1972). C3 proactivator convertase and its mode of action. J. Exp. Med. 135, pp. 1003–1008.
1399. Hunsicker, L.G., Ruddy, S. & Austen K.F. (1973). Alternate complement pathway: Factors involved in cobra venom factor (CoVF) activation of the third component of complement (C3). J. Immunol. 110, pp. 128–138.
1400. Vogt, W., Dames, W., Schmidt, G. & Dieminger, L. (1977). Complement activation by the properdin system: Formation of a stoichiometric, C3 cleaving complex of properdin factor B with C3b. Immunochemistry 14, pp. 201–205.
1401. Smith, C.A., Vogel, C.-W. & Müller-Eberhard, H.J. (1982). Ultrastructure of cobra venom factor-dependent C3/C5 convertase and its zymogen, factor B of human complement. J. Biol. Chem. 257, pp. 9879–9882.
1402. Smith, C.A., Vogel, C.-W. & Müller-Eberhard, H.J. (1984). MHC class III products: An electron microscopic study of the C3 convertase of human complement. J. Exp. Med. 159, pp. 324–329.
1403. Pryzdial, E.L.G. & Isenman, D.E. (1987). Alternative complement pathway activation fragment Ba binds to C3b. J. Biol. Chem. 262, pp. 1519–1525.
1404. Lambris, J.D. & Müller-Eberhard, H.J. (1984). Isolation and characterization of a 33,000 Dalton fragment of complement factor B with catalytic and C3b binding activity. J. Biol. Chem. 259, pp. 12685–12690.
1405. Kinoshita, T., Takata, Y., Kozono, H., Takeda, J., Hong, K. & Inoue, K. (1988). C5 convertase of the alternative complement pathway: Covalent linkage between two molecules within the trimolecular complex enzyme. J. Immunol. 141, pp. 3895–3901.
1406. Tack, B.F., Morris, S.C. & Prahl, J.W. (1979). Fifth component of human complement: Purification from plasma and polypeptide chain structure. Biochemistry 18, pp. 1490–1497.
1407. DiScipio, R.G., Smith, C.A., Müller-Eberhard, H.J. & Hugli, T.E. (1983). The activation of human complement component C5 by fluid phase C5 convertase. J. Biol. Chem. 258, pp. 10629–10636.
1408. Fernandez, H.N. & Hugli, T.E. (1978). Primary structural analysis of the polypeptide portion of human C5a anaphylatoxin. J. Biol. Chem. 253, pp. 6955–6964.
1409. Gerard, C. & Hugli, T.E. (1980). Amino acid sequence of anaphylatoxin from the fifth component of porcine complement. J. Biol. Chem. 255, pp. 4710–4715.
1410. Gennaro, R., Simonic, T., Negri, A., Motola, C., Secchi, C., Ronchi, S. & Romeo, D. (1986). C5a fragment of bovine complement. Eur. J. Biochem. 155, pp. 77–86.
1411. Cui, L.-X., Ferreri, K. & Hugli, T.E. (1985). Characterization of rat C5a, a uniquely active spasmogen. Complement 2, pp. 18–19.
1412. Wetsel, R.A., Ogata, R.T. & Tack, B.F. (1987). Primary structure of the fifth component of complement. Biochemistry 26, pp. 737–743.
1413. Lundwall, Å., Wetsel, R.A., Kristensen, T., Whitehead, A.S., Woods, D.E., Ogden, R.C., Colten, H.R. & Tack, B.F. (1985). Isolation and sequence analysis of a cDNA clone encoding the fifth complement component. J. Biol. Chem. 260, pp. 2108–2112.
1414. Wetsel, R.A., Lemons, R.S., Le Beau, M.M., Barnum, S.R., Noack, D. & Tack, B.F. (1988). Molecular analysis of human complement component C5: Localization of the structural gene to chromosome 9. Biochemistry 27, pp. 1474–1482.
1415. Jeremiah, S.J., West, L.F., Davis, M., Povey, S., Carritt, B. & Fey, G.H. (1988). The

assignment of the human gene coding for complement C5 to chromosome 9q22–9q33. Ann. Hum. Genet. 52, pp. 111–116.
1416. Podack, E.R., Kolb, W.P. & Müller-Eberhard, H.J. (1976). Purification of the sixth and seventh component of human complement without loss of hemolytic activity. J. Immunol. 116, pp. 263–269.
1417. DiScipio, R.G. & Gagnon, J. (1982). Characterization of human complement components C6 and C7. Mol. Immunol. 19, pp. 1425–1431.
1418. Chakravarti, D.N., Chakravarti, B., Parra, C.A. & Müller-Eberhard, H.J. (1989). Structural homology of complement protein C6 with other channel-forming proteins of complement. Proc. Natl. Acad. Sci. USA 86, pp. 2799–2803.
1419. Lachman, P.J. & Hobart, M.J. (1978). A further 'complement super-gene'. J. Immunol. 120, pp. 1781–1782.
1420. Lachman, P.J., Hobart, M.J. & Woo, P. (1978). Combined genetic deficiency of C6 and C7 in man. Clin. Exp. Immunol. 33, pp. 193–203.
1421. Tschopp, J., Masson, D. & Stanley, K.K. (1986). Structural/functional similarity between proteins involved in complement- and cytotoxic-T-lymphocyte-mediated cytolysis. Nature 322, pp. 831–834.
1422. Kolb, W.P. & Müller-Eberhard, H.J. (1976). The membrane attack mechanism of complement: The three polypeptide chain structure of the eighth component (C8). J. Exp. Med. 143, pp. 1131–1139.
1423. Steckel, E.W., York, R.G., Monahan, J.B. & Sodetz, J.M. (1980). The eighth component of complement. J. Biol. Chem. 255, pp. 11997–12005.
1424. Alper, C.A., Marcus, D., Raum, D., Petersen, B.H. & Spira, T.J. (1983). Genetic polymorphism in C8 β-chains. J. Clin. Invest. 72, pp. 1526–1531.
1425. Rogde, S., Olaisen, B., Gedde-Dahl, T. & Teisberg, P. (1986). The C8A and C8B loci are closely linked on chromosome 1. Ann. Hum. Genet. 50, pp. 139–144.
1426. Tedesco, F., Densen, P., Villa, M.A., Petersen, B.H. & Sirchia, G. (1983). Two types of dysfunctional eighth component of complement (C8) molecules in C8 deficiency in man. J. Clin. Invest. 71, pp. 183–191.
1427. Tedesco, F., Villa, M.A., Densen, P. & Sirchia, G. (1983). β-Chain deficiency in three patients with dysfunctional C8 molecules. Mol. Immunol. 20, pp. 47–51.
1428. Rao, A.G., Howard, O.M.Z., Ng, S.C., Whitehead, A.S., Colten, H.R. & Sodetz, J.M. (1987). Complementary DNA and derived amino acid sequence of the α subunit of human complement protein C8: Evidence for the existence of a separate α subunit messenger RNA. Biochemistry 26, pp. 3556–3564.
1429. Südhof, T.C., Goldstein, J.L., Brown, M.S. & Russel, D.W. (1985). The LDL receptor gene: A mosaic of exons shared with different proteins. Science 228, pp. 815–822.
1430. Tschopp, J. & Mollnes, T.-E. (1986). Antigenic crossreactivity of the α-subunit of complement component C8 with the cysteine-rich domain shared by complement component C9 and low density lipoprotein receptor. Proc. Natl. Acad. Sci. USA 83, pp. 4223–4227.
1431. Haefliger, J.-A., Tschopp, J., Nardelli, D., Wahli, W., Kocher, H.-P., Tosi, M. & Stanley, K.K. (1987). Complementary DNA cloning of complement C8β and its sequence homology to C9. Biochemistry 26, pp. 3551–3556.
1432. Howard, O.M.Z., Rao, A.G. & Sodetz, J.M. (1987). Complementary DNA and derived amino acid sequence of the β-subunit of human complement protein C8: Identification of a close structural and ancestral relationship to the α-subunit and C9. Biochemistry 26, pp. 3565–3570.
1433. Ng, S.C., Rao, A.G., Howard, O.M.Z. & Sodetz, J.M. (1987). The eighth component of human complement: Evidence that it is an oligomeric serum protein assembled from products of three different genes. Biochemistry 26, pp. 5229–5233.
1434. Haefliger, J.-A., Jenne, D., Stanley, K.K. & Tschopp, J. (1987). Structural homology of human complement component C8γ and plasma protein HC: Identity of the cysteine bond pattern. Biochem. Biophys. Res. Commun. 149, pp. 750–754.
1435. Hunt, L.T., Elzanowski, A. & Barker, W.C. (1987). The homology of complement factor C8 gamma and alpha-1-microglobulin. Biochem. Biophys. Res. Commun. 149, pp. 282–288.
1436. Luzio, J.P. & Stanley, K.K. (1988). Sequence homology of complement C8γ chain with

α_1-microglobulin and its implications for C8 structure and function. Mol. Immunol. 25, pp. 513–516.

1437. Ng, S. & Sodetz, J.M. (1987). Biosynthesis of C8 by hepatocytes. J. Immunol. 139, pp. 3021–3027.
1438. Densen, P. & McRill, C.M. (1988). Differential functional expression of the C8 subunits. J. Immunol. 141, pp. 2674–2679.
1439. Biesecker, G. & Müller-Eberhard, H.J. (1980). The ninth component of human complement: Purification and physicochemical characterization. J. Immunol. 124, pp. 1291–1296.
1440. DiScipio, R.G., Gehring, M.R., Podack, E.R., Kan, C.C., Hugli, T.E. & Fey, G.H. (1984). Nucleotide sequence of cDNA and derived amino acid sequence of human complement component C9. Proc. Natl. Acad. Sci. USA 81, pp. 7298–7302.
1441. Stanley, K.K., Kocher, H.P., Luzio, J.P., Jackson, P. & Tschopp, J. (1985). The sequence and topology of human complement component C9. EMBO J. 4, pp. 375–382.
1442. Marazziti, D., Eggertsen, G., Fey, G.H. & Stanley, K.K. (1988). Relationship between the gene and protein structure in human complement component C9. Biochemistry 27, pp. 6529–6534.
1443. DiScipio, R.G. & Hugli, T.E. (1985). The architecture of complement component C9 and poly(C9). J. Biol. Chem. 260, pp. 14802–14809.
1444. Biesecker, G., Gerard, C. & Hugli, T.E. (1982). An amphiphilic structure of the ninth component of human complement. J. Biol. Chem. 257, pp. 2584–2590.
1445. Abbott, C., West, L., Povey, S., Jeremiah, S., Murad, Z., DiScipio, R. & Fey, G.H. (1989). The gene for human complement component C9 mapped to chromosome 5 by polymerase chain reaction. Genomics 4, pp. 606–609.
1446. Rogne, S., Myklebost, O., Stanley, K. & Geurts van Kessel, A. (1989). The gene for human complement C9 is on chromosome 5. Genomics 5, pp. 149–152.
1447. Stanley, K.K. & Herz, J. (1987). Topological mapping of complement component C9 by recombinant DNA techniques suggests a novel mechanism for its insertion into target membranes. EMBO J. 6, pp. 1951–1957.
1448. Cooper, N.R. & Müller-Eberhard, H.J. (1970). The reaction mechanism of human C5 in immune hemolysis. J. Exp. Med. 132, pp. 775–793.
1449. Podack, E.R., Kolb, W.P. & Müller-Eberhard, H.J. (1978). The C5b–6 complex: Formation, isolation and inhibition of its activity by lipoprotein and the S-protein of human serum. J. Immunol. 120, pp. 1841–1848.
1450. Yamamoto, K. & Gewurz, H. (1978). The complex of C5b and C6: Isolation, characterization and identification of a modified form of C5b consisting of three polypeptides. J. Immunol. 120, pp. 2008–2014.
1451. Hammer, C.H., Abramovitz, A.S. & Mayer, M.M. (1976). A new activity of complement component C3: Cell-bound C3b potentiates lysis of erythrocytes by C5b–6 and terminal components. J. Immunol. 117, pp. 830–834.
1452. Podack, E.R., Biesecker, G., Kolb, W.P. & Müller-Eberhard, H.J. (1978). The C5b–6 complex: Reaction with C7, C8, C9. J. Immunol. 121, pp. 484–490.
1453. Preissner, K.T., Podack, E.R. & Müller-Eberhard, H.J. (1985). The membrane attack complex of complement: Relation of C7 to the metastable membrane binding site of the intermediate complex C5b–7. J. Immunol. 135, pp. 445–451.
1454. Preissner, K.T., Podack, E.R. & Müller-Eberhard, H.J. (1985). Self-association of the seventh component of human complement (C7): Dimerization and polymerization. J. Immunol. 135, pp. 452–458.
1455. Monahan, J.B. & Sodetz, J.M. (1980). Binding of the eighth component of complement to the soluble cytolytic complex is mediated by its β subunit. J. Biol. Chem. 255, pp. 10579–10582.
1456. Monahan, J.B. & Sodetz, J.M. (1981). Role of the β subunit in interaction of the eighth component of human complement with the membrane-bound cytolytic complex. J. Biol. Chem. 256, pp. 3258–3262.
1457. Steward, J.L., Kolb, W.P. & Sodetz, J.M. (1987). Evidence that C5b recognizes and mediates C8 incorporation into the cytolytic complex of complement. J. Immunol. 139, pp. 1960–1964.
1458. Brickner, A. & Sodetz, J.M. (1984). Function of subunits within the eighth component

of human complement: Selective removal of the γ chain reveals it has no direct role in cytolysis. Biochemistry 23, pp. 832–837.
1459. Brickner, A. & Sodetz, J.M. (1985). Functional domains on the α-subunit of the eighth component of human complement: Identification and characterization of a distinct binding site for the γ chain. Biochemistry 24, pp. 4603–4606.
1460. Steward, J.L. & Sodetz, J.M. (1985). Analysis of the specific association of the eighth and ninth components of human complement: Identification of a direct role for the α subunit of C8. Biochemistry 24, pp. 4598–4602.
1461. Steckel, E.W. & Sodetz, J.M. (1983). Evidence of direct insertion of terminal complement proteins into cell membrane bilayers during cytolysis. J. Biol. Chem. 258, pp. 4318–4324.
1462. Yoden, A., Moriyama, T., Inoue, K. & Inai, S. (1988). The role of the C9b domain in the binding of C9 molecules to EAC1–8 defined by monoclonal antibodies to C9. J. Immunol. 140, pp. 2317–2321.
1463. Tschopp, J., Podack, E.R. & Müller-Eberhard, H.J. (1982). Ultrastructure of the membrane attack complex of complement: Detection of the tetramolecular C9-polymerizing complex C5b–8. Proc. Natl. Acad. Sci. USA 79, pp. 7474–7478.
1464. Tschopp, J., Podack, E.R. & Müller-Eberhard, H.J. (1985). The membrane attack complex of complement: C5b–8 complex as accelerator of C9 polymerization. J. Immunol. 134, pp. 495–499.
1465. Ramm, L.E., Whitlow, M.B. & Mayer, M.M. (1982). Size of the transmembrane channels produced by complement protein C5b–8. J. Immunol. 129, pp. 1143–1146.
1466. Dankert, J.R. & Esser, A.F. (1985). Proteolytic modification of human complement protein C9: Loss of poly(C9) and circular lesion formation without impairment of function. Proc. Natl. Acad. Sci. USA 82, pp. 2128–2132.
1467. Morgan, B.P., Patel, A.K. & Campbell, A.K. (1987). The ring-like classical complement lesion is not the functional pore of the membrane attack complex. Biochem. Soc. Trans. 15, pp. 659–660.
1468. Campbell, A.K., Daw, R.A., Hallet, M.B. & Luzio, J.P. (1981). Direct measurement of the increase in intracellular free calcium ion concentration in response to action of complement. Biochem. J. 194, pp. 551–560.
1469. Malinski, J.A. & Nelsestuen, G.L. (1989). Membrane permeability to macromolecules mediated by the membrane attack complex. Biochemistry 28, pp. 61–70.
1470. Haupt, H., Heimburger, N., Kranz, T. & Schwick, H.G. (1970). Ein Beitrag zur Isolierung und Charakterisierung des $C\bar{1}$-Inaktivators aus Humanplasma. Eur. J. Biochem. 17, pp. 254–261.
1471. Tosi, M., Duponchel, C., Bourgarel, P., Colomb, M. & Meo, T. (1986). Molecular cloning of human $C\bar{1}$ inhibitor: Sequence homologies with α_1-antitrypsin and other members of the serpins superfamily. Gene 42, pp. 265–272.
1472. Davis III, A.E, Whithehead, A.S., Harrison, R.A., Dauphinais, A., Bruns, G.A.P., Cicardi, M. & Rosen, F.S. (1986). Human inhibitor of the first component of complement, $C\bar{1}$: Characterization of cDNA clones and localization of the gene to chromosome 11. Proc. Natl. Acad. Sci. USA 83, pp. 3161–3165.
1473. Que, B.G. & Petra, P.H. (1986). Isolation and analysis of a cDNA coding for human $C\bar{1}$ inhibitor. Biochem. Biophys. Res. Commun. 137, pp. 620–625.
1474. Rauth, G., Schumaker, G., Buckel, P. & Müller-Esterl, W. (1988). Molecular cloning of the cDNA coding for human $C\bar{1}$ inhibitor. Protein Seq. Data Anal. 1, pp. 251–257.
1475. Carter, P.E., Dunbar, B. & Fothergill, J.E. (1988). Genomic and cDNA cloning of the human $C\bar{1}$ inhibitor. Eur. J. Biochem. 173, pp. 163–169.
1476. Skriver, K., Radziejewska, E., Silberman, J.A., Donaldson, V.H. & Bock, S.C.(1989). CpG mutations in the reactive site of human $C\bar{1}$ inhibitor. J. Biol. Chem. 264, pp. 3066–3071.
1477. Salvesen, G.S., Catanese, J.J., Kress, L.F. & Travis, J. (1985). Primary structure of the reactive site of human $C\bar{1}$-inhibitor. J. Biol. Chem. 260, pp. 2432–2436.
1478. Aulak, K.S., Pemberton, P.A., Rosen, F.S., Carrell, R.W., Lachman, P.J. & Harrison, R.A. (1988). Dysfunctional $C\bar{1}$-inhibitor (At), isolated from a type II hereditary-angio-oedema plasma, contains a P1 'reactive centre' (Arg444 → His) mutation. Biochem. J. 253, pp. 615–618.

1479. Skriver, K., Wikoff, W., Stoppa-Lyonnet, D., Donaldson, V.H. & Bock, S.C. (1989). High rate mutational inactivation in serpins with P1-arginine residues encoded by CGX triplets. Submitted.
1480. Skriver, K., Wikoff, W.R., Abildgaard, U., Hultin, M.A., Kaplan, A.P. & Bock, S.C. (1989). Sites flanking serpin reactive centre determine inhibitor/substrate status and level in circulation. Submitted.
1481. Levy, N.J., Ramesh, M., Cicardi, M., Harrison, R.A. & Davis, A.E. (1989). Type II hereditary angioneurotic edema that may result from a single nucleotide change in the codon for alanine-436 in the C$\overline{1}$ inhibitor gene. Proc. Natl. Acad. Sci. USA, 87, pp. 265–268.
1482. Ariga, T., Igarashi, T., Ramesh, N., Parad, D., Cicardi, M. & Davis III, A.E. (1989). Type I C$\overline{1}$ inhibitor deficiency with a small messenger RNA resulting from deletion of one exon. J. Clin. Invest. 83, pp. 1888–1893.
1483. Sim, R.B., Reboul, A., Arlaud, G.J., Villiers, C.L. & Colomb, M.G. (1979). Interaction of ^{125}I-labelled complement subcomponents C$\overline{1}$r and C$\overline{1}$s with protease inhibitors in plasma. FEBS Lett. 97, pp. 111–115.
1484. Arlaud, G.J., Reboul, A., Sim, R.B. & Colomb, M.G. (1979). Interaction of C$\overline{1}$-inhibitor with the C$\overline{1}$r and C$\overline{1}$s subcomponents in human C$\overline{1}$. Biochim. Biophys. Acta 576, pp. 151–162.
1485. Ziccardi, R.J. & Cooper, N.R. (1979). Active disassembly of the first complement component, C$\overline{1}$, by C$\overline{1}$ inactivator. J. Immunol. 123, pp. 788–792.
1486. Sim, R.B., Arlaud, G.J. & Colomb, M.G. (1980). Kinetics of reaction of human C$\overline{1}$-inhibitor with the human complement system proteases C$\overline{1}$r and C$\overline{1}$s. Biochim. Biophys. Acta 612, pp. 433–449.
1487. Ziccardi, R.J. (1982). A new role for C$\overline{1}$-inhibitor in homeostasis: Control of activation of the first component of human complement. J. Immunol. 128, pp. 2505–2508.
1488. Ziccardi, R.J. (1986). Control of C1 activation by nascent C3b and C4b: A mechanism of feedback inhibiton. J. Immunol. 136, pp. 3378–3383.
1489. Whaley, K. & Ruddy, S. (1976). Modulation of the alternative complement pathway by β1H globulin. J. Exp. Med. 144, pp. 11471163.
1490. Kristensen, T., Wetsel, R.A. & Tack, B.F. (1986). Structural analysis of human complement protein H: Homology with C4b binding protein, β_2-glycoprotein I and the Ba fragment of B. J. Immunol. 136, pp. 3407–3411.
1491. Ripoche, J., Day, A.J., Willis, A.C., Belt, K.T., Campbell, R.D. & Sim, R.B. (1986). Partial characterization of human complement factor H by protein and cDNA sequencing: Homology with other complement and non-complement proteins. Biosci. Rep. 6, pp. 65–72.
1492. Schulz, T.F., Schwäble, W., Stanley, K.K., Weiss, E. & Dierich, M.P. (1986). Human complement factor H: Isolation of cDNA clones and partial cDNA sequence of the 38-kDa tryptic fragment containing the binding site for C3b. Eur. J. Immunol. 16, pp. 1351–1355.
1493. Vik, D.P., Keeney, J.B., Muñoz-Cánoves, P., Chaplin, D.D. & Tack, B.F. (1988). Structure of murine complement factor H gene. J. Biol. Chem. 263, pp. 16720–16724.
1494. de Cordoba, S.R., Lublin, D.M., Rubinstein, P. & Atkinson, J.P. (1985). Human genes for three complement components that regulate the activation of C3 are tightly linked. J. Exp. Med. 161, pp. 1189–1195.
1495. Weis, J.H., Morton, C.C., Bruns, G.A.P., Weis, J.J., Klickstein, L.B., Wong, W.W. & Fearon, D.T. (1987). A complement receptor locus: Genes encoding C3b/C4b receptor and C3d/Epstein-Barr virus receptor map to 1q32. J. Immunol. 138, pp. 312–315.
1496. Hing, S., Day, A.J., Linton, S.J., Ripoche, J., Sim, R.B., Reid, K.B.M. & Solomon, E. (1988). Assignment of complement components C4 binding protein (C4BP) and factor H (FH) to chromosome 1q, using cDNA probes. Ann. Hum. Genet. 52, pp. 117–122.
1497. Scharfstein, J., Ferreira, A., Gigli, I. & Nussenzweig, V. (1978). Human C4-binding protein. J. Exp. Med. 148, pp. 207–222.
1498. Reid, K.B.M. & Gagnon, J. (1982). Human C4b-binding protein: N-terminal amino acid sequence analysis and limited proteolysis by trypsin. FEBS Lett. 137, pp. 75–79.
1499. Hillarp, A. & Dahlbäck, B. (1988). Novel subunit in the C4b-binding protein required for protein S binding. J. Biol. Chem. 263, pp. 12759–12764.

1500. Chung, L.P., Gagnon, J. & Reid, K.B.M. (1985). Amino acid sequence studies of human C4b-binding protein: N-terminal sequence analysis and alignment of the fragments produced by limited proteolysis with chymotrypsin and the peptides produced by cyanogen bromide treatment. Mol. Immunol. 22, pp. 427–435.
1501. Lintin, S.J., Lewin, A. & Reid, K.B.M. (1987). Studies on the structure of the human C4b-binding protein gene and characterization of the protein signal sequence. Complement 4, Abstract 166.
1502. Lintin, S.J. & Reid, K.B.M. (1986). Studies on the structure of the human C4b-binding protein gene. FEBS Lett. 204, pp. 77–81.
1503. Dahlbäck, B., Smith, C.A. & Müller-Eberhard, H.J. (1983). Visualization of human C4b-binding protein and its complexes with vitamin K-dependent protein S and complement protein C4b. Proc. Natl. Acad. Sci. USA 80, pp. 3461–3465.
1504. Dahlbäck, B. & Müller-Eberhard, H.J. (1984). Ultrastructure of C4b-binding protein fragments formed by limited proteolysis using chymotrypsin. J. Biol. Chem. 259, pp. 11631–11634.
1505. Perkins, S.J., Chung, L.P. & Reid, K.B.M. (1986). Unusual ultrastructure of complement-component-C4b-binding protein of human complement by synchroton X-ray scattering and hydrodynamic analysis. Biochem. J. 233, pp. 799–807.
1506. Nagasawa, S., Unno, H., Ichihara, C., Koyama, J. & Koide, T. (1983). Human C4b-binding protein, C4bp. FEBS Lett. 164, pp. 135–138.
1507. Janatova, J., Reid, K.B.M. & Willis, A.C. (1989). Disulfide bonds are localized within short consensus repeat units of complement regulatory proteins: C4b-binding protein. Biochemistry 28, pp. 4754–4761.
1508. Kaidoh, T., Natsuume-Sakai, S. & Takahashi, M. (1981). Murine C4-binding protein: A rapid purification method by affinity chromotography. J. Immunol. 126, pp. 463–467.
1509. Barnum, S.R., Kristensen, T., Chaplin, D.D., Seldin, M.F. & Tack, B.F. (1989). Molecular analysis of the murine C4b-binding protein gene. Chromosome assignment and partial gene organization. Biochemistry 28, pp. 8312–8317.
1510. Suzuki, K. & Nishioka, J. (1988). Binding site for vitamin K-dependent protein S on complement C4-binding protein. J. Biol. Chem. 263, pp. 17034–17039.
1511. Dahlbäck, B. & Hildebrand, B. (1983). Degradation of human complement component C4b in the presence of the C4b-binding protein-protein S complex. Biochem. J. 209, pp. 857–863.
1512. Nagasawa, S., Mizuguchi, K., Ichihari, C. & Koyama, J. (1982). Limited chymotryptic cleavage of human C4-binding protein: Isolation of a carbohydrate-containing core domain and an active fragment. J. Biochem. (Tokyo) 92, pp. 1329–1332.
1513. Hillarp, A. & Dahlbäck, B. (1987). The protein S binding site localized to the central core of C4b-binding protein. J. Biol. Chem. 262, pp. 11300–11307.
1514. Lublin, D.M. & Atkinson, J.P. (1989). Decay-accelerating factor: Biochemistry, molecular biology and function. Ann. Rev. Immunol. 7, pp. 35–58.
1515. Nicholson-Weller, A., Burge, J., Fearon, D.T., Weller, P.F. & Austen, K.F. (1982). Isolation of a human erythrocyte membrane glycoprotein with decay-accelerating activity for C3 convertases of the complement system. J. Immunol. 129, pp. 184–189.
1516. Medof, M.E., Walter, E.I., Roberts, W.L., Haas, R. & Rosenberry, T.L. (1986). Decay accelerating factor of complement is anchored to cells by a C-terminal glycolipid. Biochemistry 25, pp. 6740–6747.
1517. Davitz, M.A., Low, M.G. & Nussenzweig, V. (1986). Release of decay-accelerating factor (DAF) from the cell membrane by phosphatidyl inositol-specific phospholipase C (PIPLC). J. Exp. Med. 163, pp. 1150–1161.
1518. Hoffmann, E.M. (1969). Inhibition of complement by a substance isolated from human erythrocytes. Immunochemistry 6, pp. 391–403.
1519. Kinoshita, T., Medof, M.E., Silber, R. & Nussenzweig, V. (1985). Distribution of decay-accelerating factor in the peripheral blood of normal individuals and patients with paroxymal nocturnal hemoglobinuria. J. Exp. Med. 162, pp. 75–92.
1520. Nicholson-Weller, A., March, J.P., Rosen, C.E., Spicer, D.B. & Austen, K.F. (1985). Surface membrane expression by human blood leukocytes and platelets of decay-accelerating factor, a regulatory protein of the complement system. Blood 65, pp. 1237–1244.

1521. Asch, A.S., Kinoshita, T., Jaffe, E.A. & Nussenzweig, V. (1986). Decay-accelerating factor is present on cultured human umbilical vein endothelial cells. J. Exp. Med. 163, pp. 221–226.
1522. Lublin, D.M., Krsek-Stapels, J., Pangburn, M.K. & Atkinson, J.P. (1986). Biosynthesis and glycosylation of the human complement regulatory protein decay-accelerating factor. J. Immunol. 137, pp. 1629–1635.
1523. Lublin, D.M., Lemons, R.S., Lebeau, M.M., Holers, V.M., Tykocinski, M.G., Medof, M.E. & Atkinson, J.P. (1987). The gene encoding decay-accelerating factor (DAF) is located in the complement regulatory locus on the long arm of chromosome 1. J. Exp. Med. 165, pp. 1731–1736.
1524. Rey-Campos, J., Rubinstein, P. & de Cordoba, S.R. (1987). Decay-accelerating factor. J. Exp. Med. 166, pp. 246–252.
1525. Caras I.W., Weddell, G.N., Davitz, M.A., Nussenzweig, V. & Martin Jr., D.W. (1987). Signal for attachment of a phospholipid membrane anchor in decay accelerating factor. Science 238, pp. 1280–1283.
1526. Tykocinski, M.L., Shu, H.K., Ayers, D.J., Walter, E.I., Getty, R.R., Groger, R.K., Hauer, C.A. & Medof, M.E. (1988). Glycolipid reanchoring of T-lymphocyte surface antigen CD8 using the 3′ end sequence of decay-accelerating factor's mRNA. Proc. Natl. Acad. Sci. USA 85, pp. 3555–3559.
1527. Caras, I.W. & Weddell, G.N. (1989). Signal peptide for protein secretion directing glycophospholipid membrane anchor attachment. Science 243, pp. 1196–1198.
1528. Caras, I.W., Weddell, G.N. & Williams, S.R. (1989). Analysis of the signal for attachment of a glycophospholipid membrane anchor. J. Cell Biol. 108, pp. 1387–1396.
1529. Reddy, P., Caras, I. & Krieger, M. (1989). Effects of O-linked glycosylation on the cell surface expression and stability of decay-accelerating factor, a glycophospholipid-anchored membrane protein. J. Biol. Chem. 264, pp. 17329–17336.
1530. Medof, M.E., Walter, E.I., Rutgers, J.L., Knowles, D.M. & Nussenzweig, V. (1985). Identification of the complement decay-accelerating factor (DAF) on epithelium and glandular cells and in body fluids. J. Exp. Med. 165, pp. 848–864.
1531. Fearon, D.T. (1979). Regulation of the amplification of C3 convertase of human complement by an inhibitory protein isolated from human erythrocyte membrane. Proc. Natl. Acad. Sci. USA 76, pp. 5867–5871.
1532. Fearon, D.T. (1980). Identification of the membrane glycoprotein that is the C3b receptor of the human erythrocyte, polymorphonuclear leukocyte, and monocyte. J. Exp. Med. 152, pp. 20–30.
1533. Dykman, T.R., Cole, J., Iida, K. & Atkinson, J.P. (1983). Polymorphism of human erythrocyte C3b/C4b receptor. Proc. Natl. Acad. Sci. USA 80, pp. 1698–1702.
1534. Dykman, T.R., Hatch, J.A. & Atkinson, J.P. (1984). Polymorphism of the human C3b/C4b receptor: Identification of a third allele and analysis of receptor phenotypes in families and patients with systemic lupus erythematosus. J. Exp. Med. 159, pp. 691–703.
1535. Dykman, T.R., Hatch, J.A., Aqua, M.S. & Atkinson, J.P. (1985). Polymorphism of the C3b/C4b (CR1) receptor: Characterization of a fourth allele. J. Immunol. 136, pp. 1787–1789.
1536. Lublin, D.M., Griffith, R.C. & Atkinson, J.P. (1986). Influence of glycosylation on allelic and cell-specific M_r variation, receptor processing, and ligand binding of the human complement C3b/C4b receptor. J. Biol. Chem. 261, pp. 5736–5744.
1537. Holers, V.M., Chaplin, D.D., Leykam, J.F., Gruner, B.A., Kuman, V. & Atkinson, J.P. (1987). Human complement receptor C3b/C4b receptor (CR1) mRNA polymorphism that correlates with the CR1 allelic molecular weight polymorphism. Proc. Natl. Acad. Sci. USA 84, pp. 2459–2463.
1538. Sim, R.B. (1985). Large-scale isolation of complement receptor type 1 (CR1) from human erythrocytes. Biochem. J. 232, pp. 883–889.
1539. Wong, W.W., Klickstein, L.B., Smith, J.A., Weis, J.H. & Fearon, D.T. (1985). Identification of a partial cDNA clone for the human receptor for complement fragments C3b/C4b. Proc. Natl. Acad. Sci. USA 82, pp. 7711–7715.
1540. Wong, W.W., Cahill, J.M., Rosen, M.D., Kennedy, C.A., Bonaccio, E.T., Morris, M.J., Wilson, J.G., Klickstein, L.B. & Fearon, D.T. (1989). Structure of the human CR1 gene. J. Exp. Med. 169, pp. 847–863.
1541. Cole, J.L., Housely Jr., G.A., Dykman, T.R., MacDermott, P. & Atkinson, J.P. (1985).

Identification of an additional class of C3-binding membrane proteins of human peripheral blood leukocytes and cell lines. Proc. Natl. Acad. Sci. USA 82, pp. 859–863.
1542. Seya, T., Ballard, L., Bora, N., McNearney, T. & Atkinson, J.P. (1987). Membrane cofactor protein (MCP or gp45-70): A distinct complement regulatory protein with a wide tissue distribution. Complement 4, abs 268.
1543. Ballard, L., Seya, T., Teckman, J., Lublin, D.M. & Atkinson, J.P. (1987). A polymorphism of the complement regulatory protein MCP (membrane cofactor protein or gp45-70). J. Immunol. 138, pp. 3850–3855.
1544. Ballard, L.L., Bora, N.S., Yu, G.H. & Atkinson, J.P. (1988). Biochemical characterization of membrane cofactor protein of the complement system. J. Immunol. 141, pp. 3923–3929.
1545. Gigli, I., Fujita, T. & Nussenzweig, V. (1979). Modulation of the classical pathway C3 convertase by plasma proteins C4 binding protein and C3b inactivator. Proc. Natl. Acad. Sci. USA 76, pp. 6596–6600.
1546. Nagasawa, S., Ischihara, C. & Stroud, R.M. (1980). Cleavage of C4b by C3b inactivator: production of a nicked form of C4b, C4b′, as an intermediate cleavage product of C4b by C3b inactivator. J. Immunol. 125, pp. 578–582.
1547. Medof, M.E., Kinoshita, T. & Nussenzweig, V. (1985). Inhibition of complement activation on the surface of cells after incorporation of decay-accelerating factor (DAF) into their membranes. J. Exp. Med. 160, pp. 1558–1578.
1548. Iida, K. & Nussenzweig, V. (1981). Complement receptor is an inhibitor of the complement cascade. J. Exp. Med. 153, pp. 1138–1150.
1549. Weiler, J.M., Daha, M.R., Austen, K.F. & Fearon D.T. (1976). Control of the amplification convertase of complement by the plasma protein βH1. Proc. Natl. Acad. Sci. USA 73, pp. 3268–3272.
1550. Fujita, T., Inoue, T., Ogawa, K., Iida, K. & Tamura, N. (1987). The mechanism of action of decay-accelerating factor (DAF). J. Exp. Med. 166, pp. 1221–1228.
1551. Pangburn, M.K., Schreiber, R.D. & Müller-Eberhard, H.J. (1977). Human complement C3b inactivator: Isolation, characterization, and demonstration of an absolute requirement for the serum protein β1H for cleavage of C3b and C4b in solution. J. Exp. Med. 146, pp. 257–270.
1552. Fearon, D.T. (1977). Purification of C3b inactivator and demonstration of its two polypeptide chain structure. J. Immunol. 119, pp. 1248–1252.
1553. Goldberger, G., Arnaout, M.A., Kay, R., Rits, M. & Colten, H.R. (1984). Biosynthesis and postsynthetic processing of human C3b/C4b inactivator (factor I) in three heptoma cell lines. J. Biol. Chem. 259, pp. 6492–6497.
1554. Davis III, A.E. (1981). The C3b inactivator of the human complement system: Homology with serine proteases. FEBS Lett. 134, pp. 143–150.
1555. Yuan, J.-M., Hsiung, L-M. & Gagnon, J. (1986). CNBr cleavage of the light chain of human complement factor I and alignment of the fragments. Biochem. J. 233, pp. 339–345.
1556. Catterall, C.F., Lyons, A., Sim, R.B., Day, A.J. & Harris, T.J.R. (1987). Characterization of the primary amino acid sequence of human complement component Factor I from an analysis of cDNA clones. Biochem. J. 242, pp. 849–856.
1557. Fujita, T., Gigli, I. & Nussenzweig, V. (1978). Human C4 binding protein. II. Role in proteolysis of C4b by C3b-inactivator. J. Exp. Med. 148, pp. 1044–1051.
1558. Fujita, T. & Nussenzweig, V. (1979). The role of C4-binding protein and β1H in proteolysis of C3b and C4b. J. Exp. Med. 150, pp. 267–276.
1559. Seya, T., Turner, J.R. & Atkinson, J.P. (1986). Purification and characterization of a membrane protein (gp45-70) that is a cofactor for cleavage of C3b and C4b. J. Exp. Med. 163, pp. 837–855.
1560. Harrison, R.A. & Lachman, P.J. (1980). The physiological breakdown of the third component of human complement. Mol. Immunol. 17, pp. 9–20.
1561. Davis III, A.E. & Harrison, R.A. (1982). Structural characterization of factor I mediated cleavage of the third component of complement. Biochemistry 21, pp. 5745–5749.
1562. Lambris, J.D., Avila, D., Becherer, J.D. & Müller-Eberhard, H.J. (1988). A discontinuous factor H binding site in the third component of complement as delineated

by synthetic peptides. J. Biol. Chem. 263, pp. 12147–12150.
1563. Becherer, J.D. & Lambris, J.D. (1988). Identification of the C3b receptor-binding domain in third component of complement. J. Biol. Chem. 263, pp. 14586–14591.
1564. Alsenz, J., Lambris, J.D., Schulz, T.F. & Dierich, M.P. (1984). Localization of the complement-component-C3b-binding site and the cofactor activity for factor I in the 38 kDa tryptic fragment of factor H. Biochem. J. 224, pp. 389–398.
1565. Chung, L.P. & Reid, K.B.M. (1985). Structural and functional studies on C4b-binding protein, a regulatory component of the human complement system. Biosci. Rep. 5, pp. 855–865.
1566. Pillemer, L., Blum, L., Lepow, I.H., Ross, O.A., Todd, E.W. & Wardlaw, A.C. (1954). The properdin system and immunity. I. Demonstration and isolation of a new serum protein, properdin, and its role in immune phenomena. Science 120, pp. 279–285.
1567. Smith, C.A., Pangburn, M.K., Vogel, C.-W. & Müller-Eberhard, H.J. (1984). Molecular architecture of human properdin, a positive regulator of the alternative pathway of complement. J. Biol. Chem. 259, pp. 4582–4588.
1568. Farries, T.C., Finch, J.T., Lachman, P.J. & Harrison, R.A. (1987). Resolution and analysis of 'native' and 'activated' properdin. Biochem. J. 243, pp. 507–517.
1569. Pangburn, M.K. (1989). Analysis of the natural polymeric forms of human properdin and their functions in complement activation. J. Immunol. 142, pp. 202–207.
1570. Reid, K.B.M. & Gagnon, J. (1981). Amino acid sequence studies of human properdin – N-terminal sequence analysis and alignment of the fragments produced by limited proteolysis with trypsin and the peptides produced by cyanogen bromide treatment. Mol. Immunol. 18, pp. 949–959.
1571. Medicus, R.G., Esser, A.F., Fernandez, H.N. & Müller-Eberhard, H.J. (1980). Native and activated properdin: Interconvertibility and identity of amino- and carboxy-terminal sequences. J. Immunol. 124, pp. 602–606.
1572. Minta, J.A. & Lepow, I.H. (1974). Studies on the subunit structure of human properdin. Immunochemistry 11, pp. 361–368.
1573. Goundis, D. & Reid, K.B.M. (1988). Properdin, the terminal complement components, thrombospondin and the circumsporozoite protein of malaria parasites contain similar sequence motifs. Nature 335, pp. 82–85.
1574. Goonewardena, P., Sjöholm, A.G., Nilsson, L.-Å & Petterson, U. (1988). Linkage analysis of the properdin deficiency gene: Suggestion of a locus in the proximal part of the short arm of the X chromosome. Genomics 2, pp. 115–118.
1575. Goundis, D., Holt, S.M., Boyd, Y. & Reid, K.B.M. (1989). Localization of the properdin structural locus to Xp11.23–Xp21.1. Genomics 5, pp. 56–60.
1576. Fearon, D.T. & Austen, K.F. (1975). Properdin: Binding to C3b and stabilization of the C3b-dependent C3 convertase. J. Exp. Med. 142, pp. 856–863.
1577. Medicus, R.G., Götze, O. & Müller-Eberhard, H.J. (1976). Alternative pathway of complement: Recruitment of precursor properdin by the labile C3/C5 convertase and the potentiation of the pathway. J. Exp. Med. 144, pp. 1076–1093.
1578. Daoudaki, M.E., Becherer, J.D. & Lambris, J.D.(1988). A 34-amino-acid peptide of the third component of complement mediates properdin binding. J. Immunol. 140, pp. 1577–1580.
1579. Podack, E.R. & Müller-Eberhard, H.J. (1979). Isolation of human S-protein, an inhibitor of the membrane attack complex of complement. J. Biol. Chem. 254, pp. 9908–9914.
1580. Dahlbäck, B. & Podack, E.R. (1985). Characterization of human S-protein, an inhibitor of the membrane attack complex of complement. Demonstration of a free reactive thiol group. Biochemistry 24, pp. 2368–2374.
1581. Jenne, D. & Stanley, K.K. (1985). Molecular cloning of S-protein, a link between complement, coagulation and cell-substrate adhesion. EMBO J. 4, pp. 3153–3157.
1582. Suzuki, S., Oldberg, Å., Hayman, E.G., Pierschbacher, M.D. & Ruoslahti, E. (1985). Complete amino acid sequence of human vitronectin deduced from cDNA. Similarity of cell attachment sites in vitronectin and fibronectin. EMBO J. 4, pp. 2519–2524.
1583. Jenne, D. & Stanley, K.K. (1987). Nucleotide sequence and organization of the human S-protein gene: Repeating peptide motifs in the 'pexin' family and a model for their evolution. Biochemistry 26, pp. 6735–6742.

1584. Fryklund, L. & Sievertsson, H. (1978). Primary structure of somatomedin B. FEBS Lett. 87, pp. 55–60.
1585. McGuire, E.A., Peacock, M.E., Inhorn, R.C., Siegel, N.R. & Tollefsen, D.M. (1988). Phosphorylation of vitronectin by a protein kinase in human plasma. J. Biol. Chem. 263, pp. 1942–1945.
1586. Jenne, D., Hille, A., Stanley, K.K. & Huttner, W.B. (1989). Sulfation of two tyrosine-residues in human complement S-protein (vitronectin). Eur. J. Biochem. 185, pp. 391–395.
1587. Stanley, K.K. (1986). Homology with hemopexin suggests a possible scavenging function for S-protein/vitronectin. FEBS Lett. 199, pp. 249–253.
1588. Podack, E.R., Kolb, W.P. & Müller-Eberhard, H.J. (1977). The C5b–7 complex: Formation, isolation, properties and subunit composition. J. Immunol. 119, pp. 2024–2028.
1589. Kolb, W.P. & Müller-Eberhard, H.J. (1975). The membrane attack mechanism of complement. J. Exp. Med. 141, pp. 724–735.
1590. Preissner, K.T., Podack, E.R. & Müller-Eberhard, H.J. (1983). Localization of the S-protein within the inhibitor complexes SC5b–7, SC5b–8, and SC5b–9 of human complement. Immunobiology 164, p. 286.
1591. Podack, E.R., Preissner, K.T. & Müller-Eberhard, H.J. (1984). Inhibition of C9 polymerization within the SC5b–9 complex of complement by S-protein. Acta Pathol. Microbiol. Immunol. Scand. Sect. C, 284 Suppl., pp. 89–96.
1592. Kinoshita, T., Hong, K. & Inoue, K. (1981). Soluble C5b–9 complex of guinea pig complement: Demonstration of its heterogeneity and the mechanism of its C9 hemolytic activity as transfer of reversibly bound C9 molecules from the complex. Mol. Immunol. 18, pp. 423–431.
1593. Tschopp, J., Masson, D., Schäfer, S., Peitsch, M. & Preissner, K.T. (1988). The heparin binding domain of S-protein/vitronectin binds to complement components C7, C8, and C9 and perforin from cytolytic T-cells and inhibits their lytic activities. Biochemistry 27, pp. 4103–4109.
1594. Preissner, K.T., Wassmuth, R. & Müller-Berghaus, G. (1985). Physicochemical characterization of human S-protein and its function in the blood coagulation system. Biochem. J. 231, pp. 349–355.
1595. Preissner, K.T. & Müller-Berghaus, G. (1986). S-protein modulates the heparin-catalyzed inhibition of thrombin by antithrombin III. Eur. J. Biochem. 156, pp. 645–650.
1596. Podack, E.R., Dahlbäck, B. & Griffin, J.H. (1986). Interaction of S-protein of complement with thrombin and antithrombin III during coagulation. J. Biol. Chem. 261, pp. 7387–7392.
1597. Suzuki, S., Pierschbacher, M.D., Hayman, E.G., Nguyen K., Öhgren, Y. & Ruoslahti, E. (1984). Domain structure of vitronectin. J. Biol. Chem. 259, pp. 15307–15314.
1598. Preissner, K.T. & Müller-Berghaus, G. (1987). Neutralization and binding of heparin by S protein/vitronectin in the inhibition of factor Xa by antithrombin III. J. Biol. Chem. 262, pp. 12247–12253.
1599. Preissner, K.T., Zwicker, L. & Müller-Berghaus, G. (1987). Formation, characterization and detection of a ternary complex between S-protein, thrombin and antithrombin III in serum. Biochem. J. 243, pp. 105–111.
1600. Jenne, D., Hugo, F. & Bhakdi, S. (1985). Interaction of complement S-protein with thrombin-antithrombin complexes: A role for S-protein in haemostasis. Thromb. Res. 38, pp. 401–412.
1601. Murphy, B.F., Kirszbaum, L., Walker, I.D. & d'Apice, A.J.F. (1988). SP-40,40, a newly identified normal human serum protein found in the SC5b–9 complex of complement and in the immune deposits in glomerulonephritis. J. Clin. Invest. 81, pp. 1858–1864.
1602. Kirszbaum, L., Sharpe, J.A., Murphy, B., d'Apice, A.J.F., Classon, B., Hudson, P. & Walker, I.D. (1989). Molecular cloning and characterization of the novel, human complement-associated protein, SP-40,40: A link between the complement and reproductive systems. EMBO J. 8, pp. 711–718.
1603. Jenne, D. & Tschopp, J. (1989). Molecular structure and functional characterization of a human complement cytolysis inhibitor found in blood and seminal plasma: Identity to sulfated glycoprotein 2, a constituent of rat testis fluid. Proc. Natl. Acad. Sci. USA 86,

pp. 7123–7127.
1604. Zalman, L.S., Wood, L.M. & Müller-Eberhard, H.J. (1986). Isolation of a human erythrocyte membrane protein capable of inhibiting expression of homologous complement transmembrane channels. Proc. Natl. Acad. Sci. USA 83, pp. 6975–6979.
1605. Schönermark, S., Rauterberg, E.W., Shin, M.L., Löke, S., Roelcke, D. & Hänsch, G.M. (1986). Homologous species restriction in lysis of human erythrocytes: A membrane-derived protein with C8-binding capacity functions as an inhibitor. J. Immunol. 136, pp. 1772–1776.
1606. Hänsch, G.M., Weller, P.F. & Nicholson-Weller, A. (1988). Release of C8 binding protein (C8bp) from the cell membrane by phosphatidyl inositol-specific phospholipase C. Blood 72, pp. 1089–1092.
1607. Schönermark, S., Filsinger, S., Berger, B. & Hänsch, M. (1988). The C8-binding protein of human erythrocytes: Interaction with the components of the complement-attack phase. Immunology 63, pp. 585–590.
1608. Hugli, T.E. (1984). Structure and function of the anaphylatoxins. Springer Semin. Immunopath. 7, pp. 193–219.
1609. Corbin, N.C. & Hugli, T.E. (1976). The primary structure of porcine C3a anaphylatoxin. J. Immunol. 117, pp. 990–995.
1610. Jacobs, J.W., Rubin, J., Hugli, T.E., Bogardt, R.A., Mariz, I.K., Daniels, J.S., Daughaday, W.H. & Bradshaw, R.A. (1978). Purification, characterization, and amino acid sequence of rat anaphylatoxin (C3a). Biochemistry 17, pp. 5031–5038.
1611. Gerard, N.P., Lively, M.O. & Gerard, C. (1988). Amino acid sequence of guinea pig C3a anaphylatoxin. Protein Seq. Data Anal. 1, pp. 473–478.
1612. Smith, M.A., Gerrie, L.M., Dunbar, B. & Fothergill, J.E. (1982). Primary structure of bovine complement activation fragment C4a, the third anaphylatoxin. Purification and complete amino acid sequence. Biochem. J. 207, pp. 253–260.
1613. Cui, L., Ferreri, K. & Hugli, T.E. (1988). Structural characterization of the C4a anaphylatoxin from rat. Mol. Immunol. 25, pp. 663–671.
1614. Greer, J. (1985). Model structure for the inflammatory protein C5a. Science 228, pp. 1055–1060.
1615. Zuiderweg, E.R.P., Nettesheim, D.G., Mollison, K.W. & Carter, G.W. (1989). Tertiary structure of human complement component C5a in solution from nuclear magnetic resonance data. Biochemistry 28, pp. 172–185.
1616. Hugli, T.E. & Erickson, B.W. (1977). Synthetic peptides with the biological activities and specificity of human C3a anaphylatoxin. Proc. Natl. Acad. Sci. USA 74, pp. 1826–1830.
1617. Caporale, L.H., Tippett, P.S., Erickson, B.W. & Hugli, T.E. (1980). The active site of C3a anaphylatoxin. J. Biol. Chem. 255, pp. 10758–10763.
1618. Lu, Z.-X., Fok, K.-F., Erickson, B.W. & Hugli, T.E. (1984). Conformational analysis of COOH-terminal segments of C3a. J. Biol. Chem. 259, pp. 7367–7370.
1619. Hugli, T.E., Kawahara, M.S., Unson, C.G., Molinor, R.L. & Erickson, B.W. (1983). The active site of human C4a anaphylatoxins. Mol. Immunol. 20, pp. 637–645.
1620. Chenoweth, D.E., Erickson, B.W. & Hugli, T.E. (1979). Human C5a-related synthetic peptides as neutrophilic chemotactic factors. Biochem. Biophys. Res. Commun. 86, pp. 227–234.
1621. Chenoweth, D.E. & Hugli, T.E. (1980). Human C5a and C5a analogs as probes of the neutrophilic C5a receptor. Mol. Immunol. 17, pp. 151–161.
1622. Mollison, K.W., Mandecki, W., Zuiderweg, E.R.P., Fayer, L., Fey, T.A., Krause, R.A., Conway, R.G., Miller, L., Edalji, R.P., Shallcross, M.A., Lane, B., Fox, J.L., Greer, J. & Carter, G.W. (1989). Identification of receptor-binding residues in the inflammatory complement protein C5a by site-directed mutagenesis. Proc. Natl. Acad. Sci. USA 86, pp. 292–296.
1623. Chenoweth, D.E. & Hugli, T.E. (1978). Demonstration of a specific C5a receptor on intact human polymorphonuclear leukocytes. Proc. Natl. Acad. Sci. USA 75, pp. 3943–3947.
1624. Chenoweth, D.E., Goodman, M.G. & Weigle, W.O. (1982). Demonstration of a specific receptor for human C5a anaphylatoxin on murine macrophages. J. Exp. Med. 156, pp. 68–78.
1625. Shin, H.S., Snyderman, R., Friedman, E., Mellors, A. & Mayer, M.M. (1968).

Chemotactic and anaphylatoxic fragment cleaved from the fifth component of guinea pig complement. Science 162, pp. 361–363.
1626. Ward, P.A. & Newman, L.J. (1969). A neutrophil chemotactic factor from human C5. J. Immunol. 102, pp. 93–99.
1627. Kay, A.B., Shin, H.S. & Austen K.F. (1973). Selective attraction of eosinophil chemotactic factor of anaphylaxis (ECF-A) and a fragment cleaved from the fifth component of complement (C5a). Immunology 24, pp. 969–974.
1628. Lett-Brown, M.A., Boetcher, D.A. & Leonard, E.J. (1976). Chemotactic responses of normal human basophils to C5a and to lymphocyte-derived chemotactic factor. J. Immunol. 117, pp. 246–252.
1629. Snyderman, R., Shin, H.S. & Hausman, M.H. (1971). A chemotactic factor for mononuclear leukocytes. Proc. Soc. Exp. Biol. Med. 138, pp. 387–390.
1630. O'Flaherty, J.T. & Ward, P.A. (1978). Leukocyte aggregation induced by chemotactic factors. Inflammation 3, pp. 177–194.
1631. Craddock, P.R., Hammerschmidt, D., Whith, J.G., Dalmasso, A.P. & Jacobs, H.S. (1977). Complement (C5a)-induced granulocyte aggregation *in vitro*: A possible mechanism of complement-mediated leukostasis and leukopenia. J. Clin. Invest. 60, pp. 260–264.
1632. Hammerschmidt, D.E., Harris, P., Wayland, J.H. & Jacob, J.S. (1978). Intravascular granulocyte (PMN) aggregation in live animals: A complement (C) derived mechanism of ischemia. Blood 52 (Suppl. 1), p. 125.
1633. Goetzl, E.J. & Austen K.F. (1974). Stimulation of human neutrophil leukocyte aerobic glucose metabolism by purified chemotactic factors. J. Clin. Invest. 53, pp. 591–599.
1634. Goldstein, I.M., Feit, F. & Weissman, G. (1975). Enhancement of nitroblue tetrazolium dye reduction by leukocytes exposed to a component of complement in the absence of phagocytosis. J. Immunol. 114, pp. 516–518.
1635. Dahinden, C.A., Fehr, J. & Hugli, T.E. (1983). Role of cell surface contact on the kinetics of superoxide production by granulocytes. J. Clin. Invest. 72, pp. 113–121.
1636. Huey, R., Bloor, C.M., Kawahara, M.S. & Hugli, T.E. (1983). Potentiation of the anaphylatoxins *in vivo* using an inhibitor of serum carboxypeptidase N (SCPN). Am. J. Pathol. 112, pp. 48–60.
1637. Bokish, V.A. & Müller-Eberhard, H.J. (1970). Anaphylatoxin inactivator of human plasma; its isolation and characterization as a carboxypeptidase. J. Clin. Invest. 49, pp. 2427–2436.
1638. Oshima, G., Kato, J. & Erdös, E.G. (1974). Subunits of human plasma carboxypeptidase N (kininase I; anaphylatoxin inactivator). Biochim. Biophys. Acta 365, pp. 344–348.
1639. Oshima, G., Kato, J. & Erdös, E.G. (1975). Plasma carboxypeptidase N, subunits and characteristics. Arch. Biochem. Biophys. 170, pp. 132–138.
1640. Plummer Jr., T.H. & Hurwitz, M.Y. (1978). Human plasma carboxypeptidase N. J. Biol. Chem. 253, pp. 3907–3912.
1641. Skidgel, R.A., Bennett, C.D., Schilling, J.W., Tan, F., Weerasinghe, D.K. & Erdös, E.G. (1988). Amino acid sequence of the N-terminus and selected peptides of the active subunit of human plasma carboxypeptidase N: Comparison with other carboxypeptidases. Biochem. Biophys. Res. Commun. 154, pp. 1323–1329.
1642. Gebhard, W., Schube, M. & Eulitz, M. (1989). cDNA cloning and complete primary structure of the small, active subunit of human carboxypeptidase N (kininase 1). Eur. J. Biochem. 178, pp. 603–607.
1643. Tan, F., Weerasinghe, D.K., Skidgel, R.A., Tamei, H., Kaul, R.K., Roninson, I.B., Schilling, J.W. & Erdös, E.G. (1990). The deduced protein sequence of the human carboxypeptidase N high molecular weight subunit reveals the presence of leucine-rich tandem repeats. J. Biol. Chem. 265, pp. 13–19.
1644. Weis, J.J., Tedder, T.F. & Fearon, D.T. (1984). Identification of a 145,000 M_r membrane protein as the C3d receptor (CR2) of human B lymphocytes. Proc. Natl. Acad. Sci. USA 81, pp. 881–885.
1645. Weis, J.J. & Fearon, D.T. (1985). The identification of N-linked oligosaccharides on the human CR2/Epstein-Barr virus receptor and their function in receptor metabolism, plasma membrane expression, and ligand binding. J. Biol. Chem. 260, pp. 13824–13830.
1646. Weis, J.J., Fearon D.T., Klickstein, L.B., Wong, W.W., Richards, S.A., de Bruyn Kops,

A., Smith, J.A. & Weis, J.H. (1986). Identification of a partial clone for the C3d/Epstein-Barr virus receptor of human B lymphocytes: Homology with the receptor for fragments C3b and C4b of the third and fourth components of complement. Proc. Natl. Acad. Sci. USA 83, pp. 5639–5643.

1647. Toothaker, L.E., Henjes, A.J. & Weis, J.J. (1989). Variability of CR2 gene products is due to alternative exon usage and different CR2 alleles. J. Immunol. 142, pp. 3668–3675.
1648. Lambris, J.D., Ganu, V.S., Hirani, S. & Müller-Eberhard, H.J. (1985). Mapping of the C3d receptor (CR2)-binding site and a neoantigenic site in the C3d domain of the third component of complement. Proc. Natl. Acad. Sci. USA 82, pp. 4235–4239.
1649. Mitomo, K., Fujita, T. & Iida, K. (1987). Functional and antigenic properties of complement receptor type 2, CR2. J. Exp. Med. 165, pp. 1424–1429.
1650. Arnaout, M.A., Gupta, S.K., Pierce, M.W. & Tenen, D.G. (1988). Amino acid sequence of the alpha subunit of human leukocyte adhesion receptor Mo1 (complement receptor type 3). J. Cell Biol. 106, pp. 2153–2158.
1651. Corbi, A.L., Kishimoto, T.K., Miller, L.J. & Springer, T.A. (1988). The human leukocyte adhesion glycoprotein Mac-1 (complement receptor type 3, CD11b) α subunit. J. Biol. Chem. 263, pp. 12403–12411.
1652. Hickstein, D.D., Hickey, M.J., Ozols, J., Baker, D.M., Back, A.L. & Roth, G.J. (1989). cDNA sequence for the aM subunit of the human neutrophile adherence receptor indicates homology to integrin α subunits. Proc. Natl. Acad. Sci. USA 86, pp. 257–261.
1653. Arnaout, M.A., Remold-O'Donnel, E., Pierce, M.W., Harris, P. & Tenen, D.G. (1988). Molecular cloning of the α subunit of human and guinea pig leukocyte adhesion glycoprotein Mo1: Chromosomal location and homology to the α subunits of integrins. Proc. Natl. Acad. Sci. USA 85, pp. 2776–2780.
1654. Corbi, A.L., Larson, R.S., Kishimoto, T.K., Springer, T.A. & Morton, C.C. (1988). Chromosomal location of the genes encoding the leukocyte adhesion receptors LFA-1, Mac1 and p150,95, J. Exp. Med. 167, pp. 1597–1607.
1655. Law, S.K.A., Gagnon, J., Hildreth, J.E.K., Wells, C.E., Willis, A.C. & Wong, A.J. (1987). The primary structure of the β-subunit of the cell surface adhesion glycoproteins LFA-1, CR3 and p150,95 and its relationship to the fibronectin receptor. EMBO J. 6, pp. 915–919.
1656. Kishimoto, T.K., O'Connor, K., Lee, A., Roberts, T.M. & Springer, T.A. (1987). Cloning of the β subunit of the leukocyte adhesion proteins: Homology to an extracellular matrix receptor defines a novel supergene family. Cell 48, pp. 681–690.
1657. Solomon, E., Palmer, R.W., Hing, S. & Law, S.K.A. (1988). Regional localization of CD18, the β-subunit of the cell surface adhesion molecule LFA-1, on human chromosome 21 by in situ hybridization. Ann. Hum. Genet. 52, pp. 123–128.
1658. Wright, S.D., Reddy, P.A., Jong, M.T.C. & Erickson, B.W. (1989). C3bi receptor (complement receptor type 3) recognizes a region of complement protein C3 containing the sequence Arg-Gly-Asp. Proc. Natl. Acad. Sci. USA 84, pp. 1965–1968.
1659. Kaplan, A.P. (1983). Hageman factor-dependent pathways: Mechanism of initiation and bradykinin formation. Fed. Proc. 42, pp. 3123–3127.
1660. Regoli, D. (1987). Kinins. Br. Med. Bull. 43, pp. 270–284.
1661. Proud, D. & Kaplan, A.P. (1988). Kinin formation: Mechanisms and role in inflammatory disorders. Ann. Rev. Immunol. 6, pp. 49–83.
1662. Baenziger, J.U. (1984). The oligosaccharides of plasma glycoproteins: Synthesis, structure, and function. In: The Plasma Proteins IV, (Putnam, F. W., Ed.), pp. 271–315, Academic Press, New York.
1663. Briggs, M.S. & Gierasch, L.M. (1986). Molecular mechanism of protein secretion. The role of the signal sequence. Adv. Protein Chem. 38, pp. 109–180.
1664. Walter, P. & Lingappa, V.R. (1986). Mechanism of protein translocation across the endoplasmic reticulum membranes. Ann. Rev. Cell Biol. 2, pp. 499–516.
1665. Robinson, A. & Austen, B. (1987). The role of topogeneic sequences in the movement of proteins through membranes. Biochem. J. 246, pp. 249–261.
1666. Singer, S.J., Maher, P.A. & Yaffe, M.P. (1987). On the translocation of proteins across membranes. Proc. Natl. Acad. Sci. USA 84, pp. 1015–1019.
1667. Singer, S.J., Maher, P.A. & Yaffe, M.P. (1987). On the transfer of integral proteins into membranes. Proc. Natl. Acad. Sci. USA 84, pp. 1960–1964.

1668. Rapoport, T.A. (1985). Protein translocation across and insertion into membranes. CRC Crit. Rev. Biochem. 20, pp. 73–137.
1669. Verner, K. & Schatz, G. (1988). Protein translocation across membranes. Science 241, pp. 1307–1313.
1670. Dam, H. (1935). The antihaemorrhagic vitamin of the chick. Biochem. J. 29, pp. 1273–1285.
1671. Suttie, J.W. (1985). Vitamin K-dependent carboxylase. Ann. Rev. Biochem. 54, pp. 459–477.
1672. Hubbard, B.R., Ulrich, M.M.W., Jacobs, M., Vermeer, C., Walsh, C., Furie, B. & Furie, B.C. (1989). Vitamin K-dependent carboxylase: Affinity purification from bovine liver using a synthetic propeptide containing the γ-carboxylation recognition site. Proc. Natl. Acad. Sci. USA 86, pp. 6893–6897.
1673. Hubbard, B.R., Jacobs, M., Ulrich, M.M.W., Walsh, C., Furie, B. & Furie, B.C. (1989). Vitamin K-dependent carboxylation. J. Biol. Chem. 264, pp. 14145–14150.
1674. Jorgensen, M.J., Cantor, A.B., Furie, B.C., Brown, C.L., Shoemaker, C.B. & Furie, B. (1987). Recognition site directing vitamin K-dependent γ-carboxylation resides on the propeptide of factor IX. Cell 48, pp. 185–191.
1675. Rabiet, M.-J., Jorgensen, M.J., Furie, B. & Furie, B.C. (1987).Effect of propeptide mutations on post-translational processing of factor IX. J. Biol. Chem. 262, pp. 14895–14898.
1676. Galeffi, P. & Brownlee, G.G. (1987). The propeptide region of clotting factor IX is a signal for a vitamin K dependent carboxylase: Evidence from protein engineering of amino acid–4. Nucl. Acids Res. 15, pp. 9505–9513.
1677. Suttie, J.W., Hoskins, J.A., Engelke, J., Hopfgartner, A., Ehrlich, H., Bang, N.U., Belagaje, R.M., Schoner, B. & Long, G.L. (1987). Vitamin K-dependent carboxylase: Possible role of substrate 'propeptide' as an intracellular recognition site. Proc. Natl. Acad. Sci. USA 84, pp. 634–637.
1678. Knobloch, J.E. & Suttie, J.W. (1987). Vitamin K–dependent carboxylase. J. Biol. Chem. 262, pp. 15334–15337.
1679. Foster, D.C., Rudinski, M.S., Schach, B.G., Berkner, K.L., Kumar, A.A., Hagen, F.S., Sprecher, C.A., Insley, M.Y. & Davie, E.W. (1987). Propeptide of human protein C is necesssary for γ-carboxylation. Biochemistry 26, pp. 7003–7011.
1680. Sugo, T., Persson, U. & Stenflo, J. (1985). Protein C in bovine plasma after warfarin treatment. J. Biol. Chem. 260, pp. 10453–10457.
1681. Stenflo, J. & Fernlund, P. (1984). β-Hydroxyaspartic acid in vitamin K-dependent plasma proteins from scorbutic and warfarin-treated guinea pigs. FEBS Lett. 168, pp. 287–292.
1682. Walsh, P.N. (1972). The role of platelets in the contact phase of blood coagulation. Br. J. Haematol. 22, pp. 237–254.
1683. Walsh, P.N. (1974). Platelet coagulant activities and hemostasis: A hypothesis. Blood 43, pp. 597–605.
1684. Majerus, P.W. & Miletich, J.P. (1978). Relationship between platelets and coagulation factors in hemostasis. Ann. Rev. Med. 29, pp. 41–49.
1685. Greengard, J.S. & Griffin, J.H. (1984). Receptors for high molecular weight kininogen on stimulated washed platelets. Biochemistry 23, pp. 6863–6869.
1686. Gustafson, E.J., Schutsky, D., Knight, L.C. & Schmaier, A.H. (1986). High molecular weight kininogen binds to unstimulated platelets. J. Clin. Invest. 78, pp. 310–318.
1687. Schmaier, A.H., Zuckerberg, A., Silverberg, C., Kuchibhotla, J., Tuszynski, G.P. & Colman, R.W. (1983). High-molecular weight kininogen. J. Clin. Invest. 71, pp. 1477–1489.
1688. Schmaier, A.H., Smith, P.M., Purdon, A.D., White, J.G. & Colman, R.W. (1986). High molecular weight kininogen: Localization in the unstimulated and activated platelet and activation by a platelet calpain(s). Blood 67, pp. 119–130.
1689. Greengard, J.S., Heeb, M.J., Ersdal, E., Walsh, P.N. & Griffin, J.H. (1986). Binding of coagulation factor XI to washed human platelets. Biochemistry 25, pp. 3884–3890.
1690. Sinha, D., Seaman, F.S., Koshy, A., Knight, L.C. & Walsh, P.N. (1984). Blood coagulation factor XIa binds specifically to a site on activated human platelets distinct from that for factor XI. J. Clin. Invest. 73, pp. 1550–1556.

1691. Walsh, P.N. & Griffin, J.H. (1981). Contributions of human platelets to the proteolytic activation of blood coagulation factors XII and XI. Blood 57, pp. 106–118.
1692. Walsh, P.N., Sinha, D., Koshy, A., Seaman, F.S. & Bradford, H. (1986). Functional characterization of platelet-bound factor XIa: Retention of factor XIa activity on the platelet surface. Blood 68, pp. 225–230.
1693. Soons, H., Janssen-Claessen, T., Hemker, H.C. & Tans, G. (1986). The effect of platelets in the activation of human blood coagulation factor IX by factor XIa. Blood 68, pp. 140–148.
1694. Walsh, P.N., Sinha, D., Kueppers, F., Seaman, F.S. & Blankstein, K.B. (1987). Regulation of factor XIa activity by platelets and α_1-proteinase inhibitor. J. Clin. Invest. 80, pp. 1578–1586.
1695. Nachman, R.L. & Jaffe, A.E. (1975). Subcellular platelet factor VIII and von Willebrand factor. J. Exp. Med. 141, pp. 1101–1113.
1696. Ahmad, S.S., Rawala-Sheikh, R. & Walsh, P.N. (1989). Comparative interactions of factor IX and factor IXa with human platelets. J. Biol. Chem. 264, pp. 3244–3251.
1697. Rosing, J., van Rijn, J.L.M.L., Bevers, E.M., van Dieijen, G., Comfurius, P. & Zwaal, R.F.A. (1985). The role of activated human platelets in prothrombin and factor X activation. Blood 65, pp. 319–332.
1698. Breederveld, K., Giddings, J.C., Ten Cate, J.W. & Bloom, A.L. (1975). The localization of factor V within normal human platelets and the demonstration of the platelet factor V antigen in congenital factor V deficiency. Br. J. Haematol. 29, pp. 405–412.
1699. Chesney, C.M., Pifer, D. & Colman, R.W. (1981). Subcellular localization and secretion of factor V from human platelets. Proc. Natl. Acad. Sci. USA 78, pp. 5180–5184.
1700. Tracy, P.B., Peterson, J.M., Nesheim, M.E., McDuffie, F.C. & Mann, K.G. (1979). Interaction of coagulation factor V and Va with platelets. J. Biol. Chem. 254, pp. 10354–10361.
1701. Comp, P.C. & Esmon, C.T. (1979). Activated protein C inhibits platelet prothrombin-converting activity. Blood 54, pp. 1272–1281.
1702. Dahlbäck, B. & Stenflo, J. (1980). Inhibitory effect of activated protein C on activation of prothrombin by platelet-bound factor Xa. Eur. J. Biochem. 107, pp. 331–335.
1703. Harris, K.W. & Esmon, C.T. (1985). Protein S is required for bovine platelets to support activated protein C binding and activity. J. Biol. Chem. 260, pp. 2007–2010.
1704. Bevers, E.M., Comfurius, P., van Rijn, J.L.M.L., Hemker, H.C. & Zwaal, R.F.A. (1982). Generation of prothrombin-converting activity and the exposure of phosphatidylserine at the outer surface of platelets. Eur. J. Biochem. 122, pp. 429–436.
1705. Zwaal, R.F.A., Comfurius, P. & van Deenen, L.L.M. (1977). Membrane asymmetry and blood coagulation. Nature 268, pp. 358–360.
1706. Sims, P.J., Wiedmer, T., Esmon, C.T., Weiss, H.J. & Shattil, S.J. (1989). Assembly of the platelet prothrombinase complex is linked to vesiculation of the platelet plasma membrane. J. Biol. Chem. 264, pp. 17049–17057.
1707. Harmon, J.T. & Jamieson, G.A. (1986). The glycocalicin portion of glycoprotein Ib expresses both high and moderate affinity receptor sites for thrombin. J. Biol. Chem. 261, pp. 13224–13229.
1708. Harmon, J.T. & Jamieson, G.A. (1986). Activation of platelets by α-thrombin is a receptor-mediated event. J. Biol. Chem. 261, pp. 15928–15933.
1709. Altieri, D.C. & Edgington, T.S. (1988). The saturable high affinity association of factor X to ADP-stimulated monocytes defines a novel function of the Mac-1 receptor. J. Biol. Chem. 263, pp. 7007–7015.
1710. Altieri, D.C., Morrissey, J.H. & Edgington, T.S. (1988). Adhesive receptor Mac-1 coordinates the activation of factor X on stimulated cells of monocytic and myeloid differentiation: An alternative initiation of the coagulation protease cascade. Proc. Natl. Acad. Sci. USA 85, pp. 7462–7466.
1711. van Iwaarden, F., de Grodt, P.G. & Bouma, B.N. (1988). The binding of high molecular weight kininogen to cultured human endothelial cells. J. Biol. Chem. 263, pp. 4698–4703.
1712. Schmaier, A.H., Kuo, A., Lundberg, D., Murray, S. & Cines, D.B. (1988). The expression of high molecular weight kininogen on human umbilical vein endothelial cells. J. Biol. Chem. 263, pp. 16327–16333.

1713. Stern, D.M., Drillings, M., Nossel, H.L., Hurlet-Jensen, A., LaGamma, K.S. & Owen, J. (1983). Binding of factors IX and IXa to cultured vascular endothelial cells. Proc. Natl. Acad. Sci. USA 80, pp. 4119–4123.
1714. Stern, D.M., Nawroth, P.P., Kisiel, W., Vehar, G. & Esmon, C.T. (1985). The binding of factor IXa to cultured bovine aortic endothelial cells. J. Biol. Chem. 260, pp. 6717–6722.
1715. Heimark, R.L. & Schwartz, S.M. (1983). Binding of coagulation factors IX and X to the endothelial cell surface. Biochem. Biophys. Res. Commun. 111, pp. 723–731.
1716. Rimon, S., Melamed, R., Savion, N., Scott, T., Nawroth, P.P. & Stern, D.M. (1987). Identification of a factor IX/IXa binding protein on the endothelial cell surface. J. Biol. Chem. 262, pp. 6023–6031.
1717. Ryan, J., Wolitzky, B., Heimer, B., Lambrose, T., Felix, A., Tam., J.P., Huang, L.H., Nawroth, P., Wilner, G., Kisiel, W., Nelsestuen, G.L. & Stern, D.M. (1989). Structural determinants of the factor IX molecule mediating interaction with the endothelial cell binding site are distinct from those involved in phospholipid binding. J. Biol. Chem. 264, pp. 20283–20287.
1718. Stern, D.M., Nawroth, P.P., Kisiel, W., Handley, D., Drillings, M. & Bartos, J. (1984). A coagulation pathway on bovine aortic segments leading to generation of factor Xa and thrombin. J. Clin. Invest. 74, pp. 1910–1921.
1719. Stern, D.M., Drillings, M., Kisiel, W., Nawroth, P., Nossel, H.L. & LaGamma, K.S. (1984). Activation of factor IX bound to cultured bovine aortic endothelial cells. Proc. Natl. Acad. Sci. USA 81, pp. 913–917.
1720. Stern, D., Nawroth, P., Handley, D. & Kisiel, W. (1985). An endothelial cell-dependent pathway of coagulation. Proc. Natl. Acad. Sci. USA 82, pp. 2523–2527.
1721. Nelsestuen, G.L. (1984). Calcium function in blood coagulation. Metal Ions 17, pp. 353–380.
1722. Sakariassen, K.S., Bolhuis, P.A. & Sixma, J.J. (1979). Human blood platelet adhesion to artery subendothelium is mediated by factor VIII-von Willebrand factor bound to the subendothelium. Nature 279, pp. 636–638.
1723. Nyman, D. (1977). Interaction of collagen with the factor VIII antigen activity-von Willebrand factor complex. Thromb. Res. 11, pp. 433–438.
1724. Scott, D.M., Griffin, B., Pepper, D.S. & Barnes, M.J. (1981). The binding of purified factor VIII/von Willebrand factor to collagens of differing type and form. Thromb. Res. 24, pp. 467–472.
1725. Morton, L.F., Griffin, B., Pepper, D.S. & Barnes, M.J. (1983). The interaction between collagens and factor VIII/von Willebrand factor: Investigation of the structural requirements for interaction. Thromb. Res. 32, pp. 545–556.
1726. Houdijk, W.P.M., Sakariassen, K.S., Nievelstein, P.F.E.M. & Sixma, J.J. (1985). Role of factor VIII-von Willebrand factor and fibronectin in the interaction of platelets in flowing blood with monomeric and fibrillar collagen types I and III. J. Clin. Invest. 75, pp. 531–540.
1727. Fauvel, F., Grant, M.E., Legrand, Y.J., Souchon, H., Tobelem, G., Jackson, D.S. & Caen, J.P. (1983). Interaction of blood platelets with a microfibrillar extract from adult bovine aorta: Requirement for von Willebrand factor. Proc. Natl. Acad. Sci. USA 80, pp. 551–554.
1728. Wagner, D.D., Urban-Pickering, M. & Marden, V.J. (1984). Von Willebrand protein binds to extracellular matrices independently of collagen. Proc. Natl. Acad. Sci. USA 81, pp. 471–475.
1729. Coller, B.S., Peerschke, E.I., Scudder, L.E. & Sullivan, C.A. (1983). Studies with a murine monoclonal antibody that abolishes ristocetin-induced binding of von Willebrand factor to platelets: Additional evidence in support of Gp Ib as the platelet receptor for von Willebrand factor. Blood 61, pp. 99–110.
1730. Ruggeri, Z.M., Bader, R. & de Marco, L. (1982). Glanzmann thrombastenia: Deficient binding of von Willebrand factor to thrombin-stimulated platelets. Proc. Natl. Acad. Sci. USA 79, pp. 6038–6041.
1731. Ruggeri, Z.M., de Marco, L., Gatti, L., Bader, R. & Montgomery, R.R. (1983). Platelets have more than one binding site for von Willebrand factor. J. Clin. Invest. 72, pp. 1–12.

1732. Gralnick, H.R., Williams, S.B. & Coller, B.S. (1984). Fibrinogen competes with von Willebrand factor for binding to the glycoprotein IIb/IIIa complex when platelets are stimulated with thrombin. Blood 64, pp. 797–800.
1733. Zucker, M.B., Mosesson, M.W., Broekman, M.J. & Kaplan, K.L. (1979). Release of platelet fibronectin (cold-insoluble globulin) from alpha granules induced by thrombin or collagen; lack of requirement for plasma fibronectin in ADP-induced platelet aggregation. Blood 54, pp. 8–12.
1734. Nachman, R.L. & Leung, L.L.K. (1982). Complex formation of platelet membrane glycoprotein IIb and IIIa with fibrinogen. J. Clin. Invest. 69, pp. 263–269.
1735. Coller, B.S., Peerschke, E.I., Scudder, L.E. & Sullivan, C.A. (1983). A murine monoclonal antiboby that completely blocks the binding of fibrinogen to platelets produce a thrombastenic-like state in normal platelets and binds to glycoprotein IIb and/or IIIa. J. Clin. Invest. 72, pp. 325–338.
1736. Bennett, J.S., Hoxie, J.A., Leitman, S.F., Vilaire, G. & Cines, D.G. (1983). Inhibition of fibrinogen binding to stimulated human platelets by a monoclonal antibody. Proc. Natl. Acad. Sci. USA 80, pp. 2417–2421.
1737. Schullek, J., Jordan, J. & Montgomery, R.D. (1984). Interaction of von Willebrand factor with human platelets in the plasma milieu. J. Clin. Invest. 73, pp. 421–428.
1738. Piétu, G., Cherel, G., Marguerie, G. & Meyer, D. (1984). Inhibition of von Willebrand factor-platelet interaction by fibrinogen. Nature 308, pp. 648–649.
1739. Lawler, J. (1986). The structural and functional properties of thrombospondin. Blood 67, pp. 1197–1209.
1740. Leung, L.L.K. (1984). Role of thrombospondin in platelet aggregation. J. Clin. Invest. 74, pp. 1764–1772.
1741. Leung, L.L.K. & Nachman, R.L. (1982). Complex formation of platelet thrombospondin with fibrinogen. J. Clin. Invest. 70, pp. 542–549.
1742. Pierschbacher, M.D. & Ruoslahti, E. (1984). Cell attachment activity of fibronectin can be duplicated by small synthetic fragments of the molecule. Nature 309, pp. 30–33.
1743. Haverstick, D.M., Cowan, J.F., Yamada, K.M. & Sontoro, S.A. (1985). Inhibition of platelet adhesion to fibronectin, fibrinogen, and von Willebrand factor substrates by a synthetic tetrapeptide derived from the cell-binding domain of fibronectin. Blood 66, pp. 946–952.
1744. Plow, E.F., Piersbacher, M.D., Ruoslahti, E., Marguerie, G.A. & Ginsberg, M.H. (1985). The effect of Arg-Gly-Asp-containing peptides on fibrinogen and von Willebrand factor binding to platelets. Proc. Natl. Acad. Sci. USA 82, pp. 8037–8061.
1745. Pytela, R., Piersbacher, M.D., Ginsberg, M.H., Plow, E.F. & Ruoslahti, E. (1986). Platelet membrane glycoprotein IIb/IIIa: Member of a family of Arg-Gly-Asp-specific adhesion receptors. Science 231, pp. 1559–1562.
1746. Lam., S.C.-T., Plow, E.F., Smith, M.A., Andrieux, A., Ryckwaert, J.-J., Marguerie, G., & Ginsberg, M. (1987). Evidence that arginyl-glycyl-aspartate peptides and fibrinogen γ chain peptides share a common binding site on platelets. J. Biol. Chem. 262, pp. 947–950.
1747. Santoro, S.A. & Lawing Jr., W.J. (1987). Competition for related but nonidentical binding sites on the glycoprotein IIb-IIIa complex by peptides derived from platelet adhesive proteins. Cell 48, pp. 867–873.
1748. Plow, E.F., Pierschbacher, M.D., Ruoslahti, E., Marguerie, G. & Ginsberg, M.H. (1987). Arginyl-glycyl-aspartic acid sequences and fibrinogen binding to platelets. Blood 70, pp. 110–115.
1749. Bennett, J.S., Shattil, S.J., Power, J.W. & Gartner, T.K. (1988). Interaction of fibrinogen with its platelet receptor. J. Biol. Chem. 263, pp. 12948–12953.
1750. Hawiger, J., Kloczewiak, M., Bednarek, M.A. & Timmons, S. (1989). Platelet receptor recognition domains in the α chain of human fibrinogen: Structure-function analysis. Biochemistry 28, pp. 2909–2914.
1751. Kloczewiak, M., Timmons, S., Bednarek, M.A., Sakon, M. & Hawiger, J. (1989). Platelet receptor recognition domain on the γ chain of human fibrinogen and its synthetic peptide analogues. Biochemistry 28, pp. 2915–2919.
1752. Timmons, S., Bednarek, M.A., Kloczewiak, M. & Hawiger, J. (1989). Antiplatelet 'hybrid' peptides analogous to receptor recognition domains on γ and α chains of

human fibrinogen. Biochemistry 28, pp. 2919–2913.
1753. Andrieux, A., Hudry-Clergeon, G., Ryckwaert, J.-J., Chapel, A., Ginsberg, M.H., Plow, E.F. & Marguerie, G. (1989). Amino acid sequences in fibrinogen mediating its interaction with its platelet receptor, GPIIbIIIa. J. Biol. Chem. 264, pp. 9258–9265.
1754. Girma, J.-P., Meyer, D., Verweij, C.L., Pannekoek, H. & Sixma, J.J. (1987). Structure-function relationship of human von Willebrand factor. Blood 70, pp. 605–611.
1755. Ruggeri, Z.M. & Zimmerman, T.S. (1987). von Willebrand factor and von Willebrand disease. Blood 70, pp. 895–904.
1756. Titani, K. & Walsh, K.A. (1988). Human von Willebrand factor: The molecular glue of platelet plugs. TIBS 13, pp. 94–97.
1757. Shelton-Inloes, B.B., Titani, K. & Sadler, J.E. (1986). cDNA sequences for human von Willebrand factor reveal five types of repeated domains and five possible protein sequence polymorphisms. Biochemistry 25, pp. 3164–3171.
1758. Titani, K., Kumar, S., Takio, K., Ericsson, L.E., Wade, R.D., Asida, K., Walsh, K.A., Chopek, M.W., Sadler, J.E. & Fujikawa, K. (1986). Amino acid sequence of human von Willebrand factor. Biochemistry 25, pp. 3171–3184.
1759. Verweij, C.L., Diergaarden, P.J., Hart, M. & Pannekoek, H. (1986). Full-length von Willebrand factor (vWF) cDNA encodes highly repetitive protein considerably longer than the mature vWF subunit. EMBO J. 5, pp. 1839–1847.
1760. Bonthron, D., Orr, E.C., Mitsock, L.M., Ginsburg, D., Handin, R.I. & Orkin, S.H. (1986). Nucleotide sequence of pre-pro-von Willebrand factor cDNA. Nucl. Acids Res. 14, pp. 7125–7127.
1761. Ginsburg, D., Handin, R.I., Bonthron, D.T., Donlon, T.A., Bruns, G.A.P., Latt, S.A. & Orkin, S.H. (1985). Human von Willebrand factor (vWF): Isolation of complementary DNA (cDNA) clones and chromosomal location. Science 228, pp. 1401–1406.
1762. VerWeij, C.L., de Vries, C.J.M., Distel, B., van Zonnenfeld, A.-J., van Kessel, A.G., van Mourik, J.A. & Pannekoek, H. (1985). Construction of cDNA coding for human von Willebrand factor using antibody probes for colony-screening and mapping of the chromosomal gene. Nucl. Acids Res. 13, pp. 4699–4717.
1763. Collins, C.J., Underdahl, J.P., Levene, R.B., Ravera, C.P., Morin, M.J., Dombalagian, M.J., Ricca, G., Livingston, D.M. & Lynch, D.C. (1987). Molecular cloning of the human gene for von Willebrand factor and identification of the transcription initiation site. Proc. Natl. Acad. Sci. USA 84, pp. 4393–4397.
1764. Bonthron, D. & Orkin, S.H. (1988). The human von Willebrand factor gene. Eur. J. Biochem. 171, pp. 51–57.
1765. Mancuso, D.J., Tuley, E.A., Westfield, L.A., Worrall, N.K., Shelton-Inloes, B.B., Sorace, J.M., Alevy, Y.G. & Sadler, J.E. (1989). Structure of the gene for human von Willebrand factor. J. Biol. Chem. 264, pp. 19514–19527.
1766. Fay, P.J., Kawai, Y., Wagner, D.D., Ginsburg, D., Bothron, D., Ohlsson-Wilhelm, D.M., Chavin, S.I., Abraham, G.N., Handin, R.I., Orkin, S.H., Montgomery, R.R. & Marder, V.J. (1986). Propolypeptide of von Willebrand factor circulates in blood and is identical to von Willebrand antigen II. Science 232, pp. 995–998.
1767. Sadler, J.E., Shelton-Inloes, B.B., Sorace, J.M., Harlan, J., Titani, K. & Davie, E.W. (1985). Cloning and characterization of two cDNAs coding for human von Willebrand factor. Proc. Natl. Acad. Sci. USA 82, pp. 6394–6398.
1768. Verweij, C.L., Hart, M. & Pannekoek, H. (1987). Expression of variant von Willebrand factor (vWF) cDNA in heterologous cells: Requirement of the pro-polypeptide in vWF multimer formation. EMBO J. 6, pp. 2885–2890.
1769. Wise, R.J., Pittman, D.D., Handin, R.I., Kaufman, R.J. & Orkin, S.H. (1988). The propeptide of von Willebrand factor independently mediates the assembly of von Willebrand multimers. Cell 52, pp. 229–236.
1770. Marti, T., Rösselet, S.J., Titani, K. & Walsh, K.A. (1987). Identification of disulfide-bridged substructures within human von Willebrand factor. Biochemistry 26, pp. 8099–8109.
1771. Foster, P.A., Fulcher, C.A., Marti, T., Titani, K. & Zimmerman, T.S. (1987). A major factor VIII binding domain resides within the amino-terminal 272 amino acid residues of von Willebrand factor. J. Biol. Chem. 262, pp. 8443–8446.
1772. Takahashi, Y., Kalafatis, M., Girma, J.-P., Sewerin, K., Anderson, L.-O. & Meyer, D.

(1987). Localization of a factor VIII binding domain on a 34 kilodalton fragment of the N-terminal portion of von Willebrand factor. Blood 70, pp. 1679–1682.

1773. Piétu, G., Ribba, A.S., Meulien, P. & Meyer, D. (1989). Epitope localization within the 106 N-terminal amino acids of von Willebrand factor (vWF) of a monoclonal antibidy to vWF which inhibits its binding to factor VIII. Thromb. Haemostas. 62, Abstract 665.

1774. Bahou, W.F., Ginsburg, D., Sikkink, R., Litwiller, R. & Fass, D.N. (1989). A monoclonal antibody to von Willebrand factor (vWF) inhibits factor VIII binding. J. Clin. Invest. 84, pp. 56–61.

1775. Fujimura, Y., Titani, K., Holland, L.Z., Russell, S.R., Roberts, J.R., Elder, J.H., Ruggeri, Z.M. & Zimmerman, T.S. (1986). von Willebrand factor. A reduced and alkylated 52/48 kDa fragment beginning at amino acid residue 449 contains the domain interacting with platelet glycoprotein Ib. J. Biol. Chem. 261, pp. 381–385.

1776. Mohri, H., Fujimura, Y., Shima, M., Yoshioka, A., Houghten, R.A., Ruggeri, Z.M. & Zimmerman, T.S. (1988). Structure of the von Willebrand factor domain interacting with glycoprotein Ib. J. Biol. Chem. 263, pp. 17901–17904.

1777. Berliner, S., Niiya, K., Roberts, J.R., Houghten, R.A. & Ruggeri, Z.M. (1988). Generation and characterization of peptide-specific antibodies that inhibit von Willebrand factor binding to glycoprotein IIb-IIIa without interacting with other adhesive molecules. J. Biol. Chem. 263, pp. 7500–7505.

1778. Roth, G.J., Titani, K., Hoyer, L.W. & Hickey, M.J. (1986). Localization of binding sites within human von Willebrand factor for monomeric type III collagen. Biochemistry 25, pp. 8357–8361.

1779. Pareti, F.I., Fujimura, Y., Dent, J.A., Holland, L.Z., Zimmerman, T.S. & Ruggeri, Z.M. (1986). Isolation and characterization of a collagen binding domain in human von Willebrand factor. J. Biol. Chem. 261, pp. 15310–15315.

1780. Pareti, F.I., Niiya, K., McPherson, J.M. & Ruggeri, Z.M. (1987). Isolation and characterization of two domains of human von Willebrand factor that interact with fibrillar collagen types I and III. J. Biol. Chem. 262, pp. 13835–13841.

1781. Kalafatis, M., Takahashi, Y., Grima, J.-P. & Meyer, D. (1987). Localization of a collagen-interactive domain of human von Willebrand factor between amino acid residues Gly911 and Glu1365. Blood 70, pp. 1577–1583.

1782. Fujimura, Y., Titani, K., Holland, L.Z., Roberts, J.R., Kostel, P., Ruggeri, Z.M. & Zimmerman, T.S. (1987). A heparin-binding domain of human von Willebrand factor. Characterization and localization to a tryptic fragment extending from amino acid residues Val449 to Lys728. J. Biol. Chem. 262, pp. 1734–1739.

1783. Ruoslahti, E. (1988). Fibronectin and its receptors. Ann. Rev. Biochem. 57, pp. 375–413.

1784. Skorstengaard, K., Jensen, M.S., Petersen, T.E. & Magnusson, S. (1985). Purification and complete primary structures of the heparin-, cell-, and DNA-binding domains of bovine plasma fibronectin. Eur. J. Biochem. 154, pp. 15–29.

1785. Skorstengaard, K., Jensen, M.S., Sahl, P., Petersen, T.E. & Magnusson, S. (1986). Complete primary structure of bovine plasma fibronectin. Eur. J. Biochem. 161, pp. 441–453.

1786. Kornblihtt, A.R., Umezawa, K., Vibe-Pedersen, K. & Baralle, F.E. (1985). Primary structure of human fibronectin: Differential splicing may generate at least 10 polypeptides from a single gene. EMBO J. 4, pp. 1755–1759.

1787. Frazier, W.A. (1987). Thrombospondin: A modular adhesive glycoprotein of platelets and nucleated cells. J. Cell Biol. 105, pp. 625–632.

1788. Lawler, J. & Hynes, R.O. (1986). The structure of human thrombospondin, an adhesive glycoprotein with multiple calcium binding sites and homologies with several different proteins. J. Cell. Biol. 103, pp. 1635–1648.

1789. Hennesy, S.W., Frazier, B.A., Kim, D.D., Deckwerth, T.L., Baumgartel, D.M., Rotwein, P. & Frazier, W.A. (1989). Complete thrombospondin mRNA sequence includes potential regulatory sites in the 3' untranslated region. J. Cell Biol. 108, pp. 729–736.

1790. Titani, K., Takio, K., Handa, M. & Ruggeri, Z.M. (1987). Amino acid sequence of the von Willebrand factor-binding domain of platelet membrane glycoprotein Ib. Proc. Natl. Acad. Sci. USA 84, pp. 6610–6614.

1791. Lopez, J.A., Chung, D.W., Fujikawa, K., Hagen, F.S., Papayannopoulou, T. & Roth, G.J. (1987). Cloning of the α-chain of human platelet glycoprotein Ib: A transmembrane protein with homology to leucine-rich α_2-glycoprotein. Proc. Natl. Acad. Sci. USA 84, pp. 5615–5619.
1792. Tsuji, T., Tsunehisa, S., Watanabe, Y., Yamamoto, K., Tohyama, H. & Osawa, T. (1983). The carbohydrate moiety of human platelet glycocalicin. J. Biol. Chem. 258, pp. 6335–6339.
1793. Takahashi, N., Takahashi, Y. & Putnam, F.W. (1985). Periodicity of leucine and tandem repetition of a 24-amino-acid segment in the primary structure of leucine-rich α_2-glycoprotein of human serum. Proc. Natl. Acad. Sci. USA 82, pp. 1906–1910.
1794. Lopez, J.A., Chung, D.W., Fujikawa, K., Hagen, F.S., Davie, E.W. & Roth, G.J. (1988). The α and β chains of human platelet glycoprotein Ib are both transmembrane proteins containing a leucine-rich amino acid sequence. Proc. Natl. Acad. Sci. USA 85, pp. 2135–2139.
1795. Muszbek, L. & Laposata, M. (1989). Glycoprotein Ib and glycoprotein IX in human platelets are acylated with palmitic acid through thioester linkages. J. Biol. Chem. 264, pp. 9716–9719.
1796. Wyler, B., Bienz, D., Clementson, K.J. & Luscher, E.F. (1986). Glycoprotein Ib_β is the only phosphorylated major membrane glycoprotein in human platelets. Biochem. J. 234, pp. 373–379.
1797. Wardell, M.R., Reynolds, C.C., Berndt, M.C., Wallace, R.W. & Fox, J.E.B. (1989). Platelet glycoprotein IB_β is phosphorylated on serine 166 by cyclic AMP-dependent protein kinase. J. Biol. Chem. 264, pp. 15656–15661.
1798. Vicente, V., Kostel, P.J. & Ruggeri, Z.M. (1988). Isolation and functional characterization of the von Willebrand factor-binding domain located between residues His^1-Arg^{293} of the α-chain of glycoprotein Ib. J. Biol. Chem. 263, pp. 18473–18479.
1799. Vicente, V., Houghten, R.A. & Ruggeri, Z.M. (1990). Identification of a site in the α chain of platelet glycoprotein Ib that participates in von Willebrand factor binding. J. Biol. Chem. 265, pp. 274–280.
1800. Hickey, M.J., Williams, S.A. & Roth, G.J. (1989). Human platelet glycoprotein IX: An adhesive prototype of leucine-rich glycoproteins with flank-center-flank structures. Proc. Natl. Acad. Sci. USA 86, pp. 6773–6777.
1801. Poncz, M., Eisman, R., Heidenreich, R., Silver, S.M., Vilaire, G., Surrey, S., Schwartz, E. & Bennett, J.S. (1987). Structure of the platelet membrane glycoprotein IIb. J. Biol. Chem. 262, pp. 8476–8482.
1802. Loftus, J.C., Plow, E.F., Frelinger III, A.L., D'Souza, S.E., Dixon, D., Lacy, J., Sorge, J. & Ginsberg, M.H. (1987). Molecular cloning and chemical synthesis of a region of platelet glycoprotein IIb involved in adhesive function. Proc. Natl. Acad. Sci. USA 84, pp. 7114–7118.
1803. Bray, P.F., Rosa, J.-P., Johnston, G.I., Shiu, D.T., Cook, R.G., Lau, C., Kan, Y.W., McEver, R.P. & Shuman, M.A. (1987). Platelet glycoprotein IIb. J. Clin. Invest. 80, pp. 1812–1817.
1804. Uzan, G., Frachet, P., Lajmanovich, A., Prandini, M.-H., Denarier, E., Duperray, A., Loftus, J., Ginsberg, M., Plow, E. & Marguerie, G. (1988). cDNA clones for human platelet GPIIb corresponding to mRNA from megakaryocytes and HEL cells. Eur. J. Biochem. 171, pp. 87–93.
1805. Ginsberg, M.H., Loftus, J., Ryckwaert, J., Pierschbacher, M., Pytela, R., Ruoslahti, E. & Plow, E.F. (1987). Immunochemical and aminoterminal sequence comparison of two cytoadhesins indicates they contain similar or identical β subunits and distinct α subunits. J. Biol. Chem. 262, pp. 5437–5440.
1806. Charo, I.F., Fitzgerald, L.A., Steiner, B., Rall Jr., S.C., Bekeart, L.S. & Phillips, D.R. (1986). Platelet glycoproteins IIb and IIIa: Evidence for a family of immunologically and structurally related glycoproteins in mammalian cells. Proc. Natl. Acad. Sci. USA 83, pp. 8351–8356.
1807. Hiraiwa, A., Matsukage, A., Shiku, H., Takahashi, T., Naito, K. & Yamada, K. (1987). Purification and partial amino acid sequence of human platelet membrane glycoproteins IIb and IIIa. Blood 69, pp. 560–564.
1808. Sosnoski, D.M., Emanuel, B.S., Hawkins, A.L., van Tuinen, P., Ledbetter, D. H., Nussbaum, R.L., Kaos, K.-F., Schwartz, E., Phillips, D., Bennett, J.S., Fitzgerald, L. &

Poncz, M. (1988). Chromosome localization of the genes for the vitronectin and fibronectin receptors α subunits and for platelet glycoproteins IIb and IIIa. J. Clin. Invest. 81, pp. 1993–1998.

1809. Bray, P.F., Barsh, G., Rosa, J.-P., Lug, X.Y., Magenis, E. & Shuman, M.A. (1988). Physical linkage of the genes for platelet membrane glycoproteins IIb and IIIa. Proc. Natl. Acad. Sci. USA 85, pp. 8683–8687.
1810. Calvete, J.J., Alvarez, M.V., Rivas, G., Hew, C.-L., Henschen, A. & Gonzáles-Rodríguez, J. (1989). Interchain and intrachain disulphide bonds in human platelet glycoprotein IIb. Biochem. J. 261, pp. 551–560.
1811. Loftus, J.C., Plow, E.F., Jennings, L.K. & Ginsberg, M.H. (1988). Alternative proteolytic processing of platelet membrane glycoprotein IIb. J. Biol. Chem. 263, pp. 11025–11028.
1812. Watterson, D.M., Sharief, F. & Vanaman, T.C. (1980). The complete amino acid sequence of the Ca^{2+}-dependent modulator protein (calmodulin) of bovine brain. J. Biol. Chem. 255, pp. 962–975.
1813. Collins, J.H., Greaser, M.L., Potter, J.D. & Horn, M.J. (1977). Determination of the amino acid sequence of troponin C from rabbit skeletal muscle. J. Biol. Chem. 252, pp. 6356–6362.
1814. Gariépy, J. & Hodges, R.S. (1983). Primary sequence analysis and folding behavior of EF hands in relation to the mechanism of action of troponin C and calmodulin. FEBS Lett. 160, pp. 1–6.
1815. Kretsinger, R.H. & Nockolos, C.E. (1973). Carp muscle calcium-binding protein. J. Biol. Chem. 248, pp. 3313–3326.
1816. Kunicki, T.J., Pidard, D., Rosa, J.P. & Nurden, A.T. (1981). The formation of calcium-dependent complexes of platelet membrane glycoproteins IIb and IIIa in solution as determined by crossed immunoelectrophoresis. Blood 58, pp. 268–278.
1817. Fujimura, K. & Phillips, D.R. (1983). Calcium cation regulation of glycoprotein IIb-IIIa complex formation in platelet plasma membrane. J. Biol. Chem. 262, pp. 14080–14086.
1818. Jennings, L.K. & Phillips, D.R. (1982). Purification of glycoproteins IIb and IIIa from human platelet membranes and characterization of a calcium-dependent glycoprotein IIb-IIIa complex. J. Biol. Chem. 257, pp. 10458–10466.
1819. Calvete, J.J., Henschen, A. & Gonzáles Rodríguez, J. (1989). Complete localization of the intrachain disulphide bonds and the N-glycosylation points in the α-subunit of human platelet glycoprotein IIb. Biochem. J. 261, pp. 561–568.
1820. Fitzgerald, L.A., Steiner, B., Rall Jr., S.C., Lo, S. & Phillips, D.R. (1987). Protein sequence of endothelial glycoprotein IIIa derived from a cDNA clone. J. Biol. Chem. 262, pp. 3936–3939.
1821. Rosa, J.-P., Bray, P.F., Gaynet, O., Johnston, G.I., Cook, R.G., Jackson, K.W., Shuman, M.A. & McEver, R.P. (1988). Cloning of glycoprotein IIIa from human erythroleukaemia cells and localization of the gene to human chromosome 17. Blood 72, pp. 593–600.
1822. Zimrin, A.B., Eisman, R., Vilaire, G., Schwartz, E., Bennett, J.S. & Poncz, M. (1988). Structure of glycoprotein IIIa. J. Clin. Invest. 81, pp. 1470–1475.
1823. van Kuppevelt, T.H.M.S.M., Languing, L.R., Gailit, J.O., Suzuki, S. & Ruoslahti, E. (1989). An alternative cytoplasmic domain of the integrin β_3 subunit. Proc. Natl. Acad. Sci. USA 86, pp. 5415–5418.
1824. Hynes, R.O. (1987). Integrins: A family of cell surface receptors. Cell 48, pp. 549–554.
1825. Ruoslahti, E. & Pierschbacher, M.D. (1987). New perspectives in cell adhesion: RGD and integrins. Science 238, pp. 491–497.
1826. Phillips, D.R., Charo, I.F., Praise, L.V. & Fitzgerald, L.A. (1988). The platelet membrane glycoprotein IIb-IIIa complex. Blood 71, pp. 831–843.
1827. Tamkun, J.W., DeSimone, D.W., Fonda, D., Patel, R.S., Buck, G., Horwitz, A.F. & Hynes, R.O. (1986). Structure of integrin, a glycoprotein involved in the transmembrane linkage between fibronectin and actin. Cell 46, pp. 271–282.
1828. Argraves, W.S., Suzuki, S., Arai, H., Thompson, K., Pierschbacher, M. & Ruoslahti, E. (1987). Amino acid sequence of the human fibronectin receptor. J. Cell Biol. 105, pp. 1183–1190.
1829. Zhang, Y., Saison, M., Spaepen, M., De Strooper, B., Van Leuven, F., David, G., Van

den Berghe, H. & Cassiman, J.-J. (1988). Mapping of human fibronectin receptor β subunit gene to chromosome 10. Somat. Cell Mol. Genet. 14, pp. 99–104.

1830. De Simone, D.W. & Hynes, R.O. (1988). Xenopus laevis integrins. J. Biol. Chem. 263, pp. 5333–5340.

1831. Takada, Y. & Hemler, M.E. (1989). The primary structure of the VLA-2/collagen receptor $\alpha 2$ subunit (platelet GPIa): Homology to other integrins and the presence of a possible collagen-binding domain. J. Cell Biol. 109, pp. 397–407.

1832. Larson, R.S., Corbi, A.L., Berman, L. & Springer, T. (1989). Primary structure of the leukocyte function-associated molecule-1 α subunit: An integrin with an embedded domain defining a protein superfamily. J. Cell Biol. 108, pp. 703–712.

1833. Corbi, A.L., Miller, L.J., O'Connor, K., Larson, R.S. & Springer, T.A. (1987). cDNA cloning and complete primary structure of the α subunit of a leukocyte adhesion glycoprotein, p150,95. EMBO J. 6, pp. 4023–4028.

1834. Pytela, R., Pierschbacher, M.D. & Ruoslahti, E. (1985). A 125/115 kDa cell surface receptor specific for vitronectin interacts with the arginine-glycine-aspartic acid adhesion sequence derived from fibronectin. Proc. Natl. Acad. Sci. USA 82, pp. 5766–5770.

1835. Suzuki, S., Argraves, W.S., Arai, H., Languino, L.R., Pierschbacher, M.D. & Ruoslahti, E. (1987). Amino acid sequence of the vitronectin receptor α subunit and comparative expression of adhesion receptor mRNAs. J. Biol. Chem. 262, pp. 14080–14086.

1836. Suzuki, S., Argraves, W.S., Pytela, R., Arai, H., Krusius, T., Pierschbacher, M.D. & Ruoslahti, E. (1986). cDNA and amino acid sequence of the cell adhesion protein receptor recognizing vitronectin reveal a transmembrane domain and homologies with other adhesion protein receptors. Proc. Natl. Acad. Sci. USA 83, pp. 8614–8618.

1837. Fitzgerald, L.A., Poncz, M., Steiner, B., Rall Jr., S.C., Bennett, J.S. & Phillips, D.R. (1987). Comparison of cDNA-derived protein sequences of the human fibronectin and vitronectin receptor α-subunits and platelet glycoprotein IIb. Biochemistry 26, pp. 8158–8165.

1838. Fitzgerald, L.A., Charo, I.F. & Phillips, D.R. (1985). Human and bovine endothelial cells synthesize membrane proteins similar to human platelet glycoprotein IIb and IIIa. J. Biol. Chem. 260, pp. 10893–10896.

1839. Leeksma, O.C., Zandbergen-Spaargaren, J., Giltay, J.C. & van Mourik, J.A. (1986). Cultured human endothelial cells synthesize a plasma membrane protein complex immunologically related to the platelet glycoprotein IIb/IIIa complex. Blood 67, pp.1176–1180.

1840. Huber, R. & Carrell, R.W. (1989). Implications of the three-dimensional structure of α_1-antitrypsin for structure and function of serpins. Biochemistry 28, pp. 8951–8966.

1841. Ragg, H. (1986). A new member of the plasma protease inhibitor gene family. Nucl. Acids Res. 14, pp. 1073–1088.

1842. Blinder, M.A., Marasa, J.C., Reynolds, C.H., Deaven, L.L. & Tollefsen, D. (1988). Heparin cofactor II: cDNA sequence, chromosome localization, restriction fragment length polymorphism, and expression in *Eschericia coli*. Biochemistry 27, pp. 752–759.

1843. Kurachi, K., Chandra, T., Degen, S.J.F., White, T.T., Marchioro, T.L., Woo, S.L.C. & Davie, E.W. (1981). Cloning and sequence of cDNA coding for α_1-antitrypsin. Proc. Natl. Acad. Sci. USA 78, pp. 6826–6830.

1844. Hill, R.E., Shaw, P.H., Boyd, P.A., Baumann, H. & Hastie, N.D. (1984). Plasma proteinase inhibitors in mouse and man: Divergence within the reactive centre region. Nature 311, pp. 175–177.

1845. Chao, S., Chai, K.X., Chao, L. & Chao, J. (1990). Molecular cloning and primary structure of rat α_1-antitrypsin. Biochemistry 29, pp. 323–329.

1846. Chandra, T., Stackhouse, R., Kidd, V.J., Robson, K.J.H. & Woo, S.L.C. (1983). Sequence homology between human α_1-antichymotrypsin, α_1-antitrypsin, and antithrombin III. Biochemistry 22, pp. 5055–5061.

1847. Kageyama, R., Ohkubo, H. & Nakanishi, S. (1984). Primary structure of human preangiotensinogen deduced from the cloned cDNA sequence. Biochemistry 23, pp. 3603–3609.

1848. Ohkubo, H., Kageyama, R., Ujihara, M., Hirose, T., Inayama, S. & Nakanishi, S.

(1983). Cloning and sequence analysis of cDNA for rat angiotensinogen. Proc. Natl. Acad. Sci. USA 80, pp. 2196–2200.
1849. Flink, I.L., Bailey, T.J., Gustafson, T.A., Markham, B.E. & Morkin, E. (1986). Complete amino acid sequence of human thyroxine-binding globulin deduced from cloned DNA: Close homology to the serine antiproteases. Proc. Natl. Acad. Sci. USA 83, pp. 7708–7712.
1850. Hammond, G.L., Smith, C.L., Goping, I.S., Underhill, D.A., Harley, M.J., Reventos, J., Musto, N.A., Gunsalus, G.L. & Bardin, C.W. (1987). Primary structure of human corticosteroid binding globulin, deduced from hepatic and pulmonary cDNAs, exhibits homology with serine protease inhibitors. Proc. Natl. Acad. Sci. USA 84, pp. 5153–5157.
1851. Yoon, J.-B., Towle, H.C. & Seelig, S. (1987). Growth hormone induces two mRNA species of the serine protease inhibitor gene family in rat liver. J. Biol. Chem. 262, pp. 4284–4289.
1852. Upton, C., Carrell, R.W. & McFadden, G. (1986). A novel member of the serpin superfamily is encoded on a circular plasmid-like DNA species isolated from rabbit cells. FEBS Lett. 207, pp. 115–120.
1853. McReynolds, L., O'Malley, B.W., Nisbet, A.D., Fothergill, J.E., Givol, D., Robertson, M. & Brownlee, G.G. (1978). Sequence of chicken ovalbumin mRNA. Nature 273, pp. 723–730.
1854. Heilig, R., Muraskowsky, R., Kloepfer, C. & Mandel, J.L. (1982). The ovalbumin gene family: Complete sequence and structure of the Y gene. Nucl. Acids Res. 10, pp. 4363–4382.
1855. Heilig, R., Perrin, F., Gannon, F., Mandel, J.L. & Chambon, P. (1980). The ovalbumin gene family: Structure of the X gene and evolution of duplicated split genes. Cell 20, pp. 625–637.
1856. Hejgaard, J., Rasmussen, S.K., Brandt, A. & Svendsen, I. (1985). Sequence homology between barley endosperm protein Z and protease inhibitors of the α_1-antitrypsin family. FEBS Lett. 180, pp. 89–94.
1857. Pickup, D.J., Ink, B.S., Hu, W., Ray, C.A. & Joklik, W.K. (1986). Hemorrhage in lesions caused by cowpox virus is induced by a viral protein that is related to plasma protein inhibitors of serine proteases. Proc. Natl. Acad. Sci. USA 83, pp. 7698–7702.
1858. Kanost, M.R., Prasad, S.V. & Wells, M.A. (1989). Primary structure of a member of the serpin superfamily of proteinase inhibitors from an insect, Manduca sexta. J. Biol. Chem. 264, pp. 965–972.
1859. Bao, J., Sifers, R.N., Kidd, V.J., Ledley, F.D. & Woo, S.L.C. (1987). Molecular evolution of serpins: Homologous structure of the human α_1-antichymotrypsin and α_1-antitrypsin gene. Biochemistry 26, pp. 7755–7759.
1860. Ragg, H. & Preibisch, G. (1988). Structure and expression of the gene coding for the human serpin hLS2. J. Biol. Chem. 263, pp. 12129–12134.
1861. Tanaka, T, Ohkubo, H. & Nakanishi, S. (1984). Common structural organization of the angiotensinogen and the α_1-antitrypsin gene. J. Biol. Chem. 259, pp. 8063–8065.
1862. Woo, S.L.C., Beattie, W.G., Catterall, J.F., Dugaiczyk, A., Staden, R., Brownlee, G.G. & O'Malley, B.W. (1981). Complete nucleotide sequence of the chicken chromosomal ovalbumin gene and its biological significance. Biochemistry 20, pp. 6437–6446.
1863. Carrell, R.W. (1986). α_1-antitrypsin: Molecular pathology, leukocytes, and tissue damage. J. Clin. Invest. 78, pp. 1427–1431.
1864. Rabin, M., Watson, M., Kidd, V., Woo, S.L.C., Breg, W.R. & Ruddle, F.H. (1986). Regional location of α_1-antichymotrypsin and α_1-antitrypsin genes on human chromosome 14. Somat. Cell Mol. Genet. 12, pp. 209–214.
1865. Perlino, E., Cortese, R. & Ciliberto, G. (1987). The human α_1-antitrypsin gene is transcribed from two different promoters in macrophages and hepatocytes. EMBO J. 6, pp. 2767–2771.
1866. Johnson, D. & Travis, J. (1978). Structural evidence for methionine at the reactive site of human α_1-proteinase inhibitor. J. Biol. Chem. 253, pp. 7142–7144.
1867. Loebermann, H., Tokuoka, R., Diesenhofer, J. & Huber, R. (1984). Human α_1-proteinase inhibitor. J. Mol. Biol. 177, pp. 531–556.
1868. Nukiwa, T., Brantly, M., Ogushi, F., Fells, G., Satoh, K., Stier, L., Courtney, M. &

Crystal, R.G. (1987). Characterization of the M1(Ala213) type of α_1-antitrypsin, a newly recognized common 'normal' α_1-antitrypsin haplotype. Biochemistry 26, pp. 5259–5267.

1869. Brantly, M., Nukiwa, T. & Crystal, R.G. (1988). Molecular basis of alpha-1-antitrypsin deficiency. Am. J. Med. 84 (Suppl. 6A), pp. 13–31.

1870. Brennan, S.O. & Carrell, R.W. (1986). α_1-Antitrypsin Christchurch; 363 Glu → Lys: Mutation at the P'_5 position does not affect inhibitory activity. Biochim. Biophys. Acta 873, pp. 13–19.

1871. Jeppsson, J.-O. & Laurell, C.-B. (1988). The amino acid substitution of human α_1-antitrypsin M_3, X and Z. FEBS Lett. 231, pp. 327–330.

1872. Nukiwa, T., Satoh, K., Brantly, M., Ogushi, F., Fells, G.A., Courtney, M. & Crystal, R.G. (1986). Identification of a second mutation in the protein-coding sequence of the Z-type alpha-1-antitrypsin gene. J. Biol. Chem. 261, pp. 15989–15994.

1873. Takahashi, H., Nukiwa, T., Satoh, K., Ogushi, F., Brantly, M., Fells, G., Stier, L., Courtney, M. & Crystal, R.G. (1988). Characterization of the gene and protein of the α_1-antitrypsin 'deficiency' allele $M_{Procida}$. J. Biol. Chem. 263, pp. 15528–15534.

1874. Curiel, D.T., Holmes, M.D., Okayama, H., Brantly, M.L., Vogelmeier, C., Travis, W.D., Stier, L.E., Perks, W.H. & Crystal, R.G. (1989). Molecular basis of the liver and lung disease associated with the α_1-antitrypsin deficiency allele M_{Malton}. J. Biol. Chem. 264, pp. 13938–13945.

1875. Owen, M.C., Brennan, S.O., Lewis, J.H. & Carrell, R.W. (1983). Mutation of antitrypsin to antithrombin. N. Eng. J. Med. 309, pp. 694–698.

1876. Hofker, M.H., Nukiwa, T. van Paassen, H.M.B., Nelen, M., Franis, R.R., Klasen, E.C. & Crystal, R.G. (1987). A Pro → Leu substitution in codon 369 in the α_1-antitrypsin deficiency variant Pi $M_{Heerlen}$. Am. J. Hum. Genet. 41, Abstract 220.

1877. Nukiwa, T., Takahashi, H., Brantly, M., Courtney, M. & Crystal, R.G. (1987). α_1-Antitrypsin $Null_{Granite\ Falls}$, a nonexpression α_1-antitrypsin gene associated with a frameshift to stop mutation in a coding exon. J. Biol. Chem. 262, pp. 11999–12004.

1878. Satoh, K., Nukiwa, T., Brantly, M., Garver Jr, R.I., Hofker, M., Courtney, M. & Crystal, R.G. (1988). Emphysema associated with complete absence of α_1-antitrypsin in serum and the homozygous inheritance of stop codon in α_1-antitrypsin coding exon. Am. J. Hum. Genet. 42, pp. 77–83.

1879. Sifers, R.N., Brashears-Macatee, S., Kidd, V.J., Muensch, H. & Woo, S.L.C. (1988). A frameshift mutation results in a truncated α_1-antitrypsin that is retained within the rough endoplasmic reticulum. J. Biol. Chem. 263, pp. 7330–7335.

1880. Curiel, D., Brantly, M., Curiel, E., Stier, L. & Crystal, R.G. (1989). α_1-Antitrypsin deficiency caused by the α_1-antitrypsin $Null_{Mattawa}$ gene. J. Clin. Invest. 83, pp. 1144–1152.

1881. Schapira, M., Ramus, M.-A., Jallat, S., Carvallo, D. & Courtney, M. (1985). Recombinant α_1-antitrypsin Pittsburg (Met^{358} → Arg) is a potent inhibitor of plasma kallikrein and activated factor XII fragment. J. Clin. Invest. 76, pp. 635–637.

1882. Scott, C.F., Carrell, R.W., Glaser, C.B., Kueppers, F., Lewis, J.H. & Colman, R.W. (1986). Alpha-1-antitrypsin-Pittsburg. J. Clin. Invest. 77, pp. 631–634.

1883. Morii, M. & Travis, J. (1983). Amino acid sequence at the reactive site of human α_1-antichymotrypsin. J. Biol. Chem. 258, pp. 12749–12752.

1884. Tollefsen, D.M., Majerus, D.W. & Blank, M.K. (1982). Heparin cofactor II. J. Biol. Chem. 257, pp. 2162–2169.

1885. Tollefsen, D.M., Pestka, C.A. & Monafo, W.J. (1983). Activation of heparin cofactor II by dermatan sulfate. J. Biol. Chem. 258, pp. 6713–6716.

1886. Hortin, G., Tollefsen, D.M. & Strauss, A.W. (1986). Identification of two sites of sulfation of human heparin cofactor II. J. Biol. Chem. 261, pp. 15827–15880.

1887. Griffith, M.J., Noyes, C.M. & Church, F.C. (1985). Reactive site peptide structural similarity between heparin cofactor II and antithrombin III. J. Biol. Chem. 260, pp. 2218–2225.

1888. Hortin, G.L., Tollefsen, D.M. & Benutto, B.M. (1989). Antithrombin activity of a peptide corresponding to residues 54–75 of heparin cofactor II. J. Biol. Chem. 264, pp. 13979–13982.

1889. Blinder, M.A., Andersson, T.R., Abildgaard, U. & Tollefsen, D.M. (1989). Heparin cofactor II_{Oslo}. J. Biol. Chem. 264, pp. 5128–5133.

1890. Blinder, M.A. & Tollefsen, D.M. (1990). Site-directed mutagenesis of arginine103 and lysine185 in the proposed glycosaminoglycan-binding site of heparin cofactor II. J. Biol. Chem. 265, pp. 286–291.
1891. Trent, J.M., Flink, I.L., Morkin, E., Van Tuinen, P. & Ledbetter, D.H. (1987). Localization of the human thyroxine-binding globulin gene to the long arm of the X-chromosome (Xq21–22). Am. J. Hum. Genet. 41, pp. 428–435.
1892. Takeda, K., Mori, Y., Sobieszczyk, S., Seo, H., Dick, M., Watson, F., Flink, I.L., Seino, S., Bell, G.I. & Refetoff, S. (1989). Sequence of the variant thyroxine-binding globulin of Australian Aborigines. J. Clin. Invest. 83, pp. 1344–1348.
1893. Mori, Y., Refetoff, S., Seino, S., Flink, I.L. & Murata, Y. (1986). N. Eng. J. Med. 314, p. 694 (abs).
1894. Kunapuli, S.P. & Kumar, A. (1986). Difference in the nucleotide sequence of human angiotensinogen cDNA. Nucl. Acids Res. 14, p. 7509.
1895. Tewksbury, D.A., Dart, R.A. & Travis, J. (1981). The amino terminal amino acid sequence of human angiotensinogen. Biochem. Biophys. Res. Commun. 99, pp. 1311–1315.
1896. Campdell, D.J. (1987). Circulating and tissue angiotensin systems. J. Clin. Invest. 79, pp. 1–6.
1897. Doolittle, R.F. (1985). The genealogy of some recently evolved vertebrate proteins. TIBS 10, pp. 233–237.
1898. Patthy, L. (1985). Evolution of the proteases of blood coagulation and fibrinolysis by assembly from modules. Cell 41, pp. 657–663.
1899. Patthy, L. (1987). Intron-dependent evolution: Preferred types of exons and introns. FEBS Lett. 214, pp. 1–7.
1900. Patthy, L. (1987). Detecting homology of distantly related proteins with consensus sequences. J. Mol. Biol. 198, pp. 567–577.
1901. Doolittle, R.F. & Fang, D.F. (1987). Reconstructing the evolution of vertebrate blood coagulation from a consideration of the amino acid sequences of clotting proteins. CSH Symp. Quant. Biol. 52, pp. 869–874.
1902. Blake, C.C.F., Harlos, K. & Holland, S.K. (1987). Exon and domain evolution in the proenzymes of blood coagulation and fibrinolysis. CSH Symp. Quant. Biol. 52, pp. 925–931.
1903. Patthy, L. (1988). Detecting distant homologies of mosaic proteins. J. Mol. Biol. 202, pp. 689–696.
1904. Grubb, A. & Löfberg, H. (1982). Human γ-trace, a basic microprotein: Amino acid sequence and presence in the adenohypophysis. Proc. Natl. Acad. Sci. USA 79, pp. 3024–3027.
1905. Machleidt, W., Borchart, U., Fritz, H., Brzin, J., Ritonja, A. & Turk, V. (1983). Primary structure of stefin, a cytosolic inhibitor of cysteine proteinases from human polymorphonuclear granulocytes. Hoppe-Seyler's Z. Physiol. Chem. 364, pp. 1481–1486.
1906. Ritonja, A., Machleidt, W. & Barrett, A. (1985). Amino acid sequence of the intracellular cysteine proteinase inhibitor cystatin B from human liver. Biochem. Biophys. Res. Commun. 131, pp. 1187–1192.
1907. Hirado, M., Tsunasawa, S., Sakiyama, F., Niiobe, M. & Fujii, S. (1985). Complete amino acid sequence of bovine colostrum low-M_r cysteine proteinase inhibitor. FEBS Lett. 186, pp. 41–45.
1908. Takio, K., Kominami, E., Wakamatsu, N., Katunuma, N. & Titani, K. (1983). Amino acid sequence of rat liver thiol proteinase inhibitor. Biochem. Biophys. Res. Commun. 115, pp. 902–908.
1909. Takio, K., Kominami, E., Bando, Y., Katunuma, N. & Titani, K. (1984). Amino acid sequence of rat epidermal thiol proteinase inhibitor. Biochem. Biophys. Res. Commun. 121, pp. 149–154.
1910. Poser, J.W., Esch, F.S., Ling, N.S. & Price, P.A. (1980). Isolation and sequence of the vitamin K–dependent protein from human bone. J. Biol. Chem. 255, pp. 8685–8691.
1911. Celeste, A.J., Rosen, V., Buecker, J.L., Kriz., R., Wang, E.A. & Wozney, J.M. (1986). Isolation of the human gene for bone Gla protein utilizing mouse and rat cDNA clones. EMBO J. 5, pp. 1885–1890.
1912. Price, P.A. & Williamson, M.K. (1985). Primary structure of bovine matrix Gla protein,

a new vitamin K-dependent bone protein. J. Biol. Chem. 260, pp. 14971–14975.

1913. Kiefer, M.C., Bauer, D.M., Young, D., Hermsen, K.M., Masiarz, F.R. & Barr, P.J. (1988). The cDNA and derived amino acid sequence for human and bovine matrix Gla protein. Nucl. Acids Res. 16, p. 5213.

1914. Pennica, D., Kohr, W.J., Kuang, W.-J., Glaister, D., Aggarwal, B.B., Chen, E.Y. & Goeddel, D.V. (1987). Identification of human uromodulin as the Tamm-Horsefall urinary glycoprotein. Science 236, pp. 83–88.

1915. Nielsen, L.B. & Petersen, T.E. (1989). Personal communication.

1916. Højrup, P., Jensen, M.S. & Petersen, T.E. (1985). Amino acid sequence of bovine protein Z: A vitamin K-dependent serine protease homologue. FEBS Lett. 184, pp. 333–338.

1917. Gray, A., Dull, T.J. & Ulrich, A. (1983). Nucleotide sequence of epidermal growth factor cDNA predicts a 128,000-molecular weight protein precursor. Nature 303, pp. 722–725.

1918. Scott, J., Urdea, M., Quiroga, M., Sanchez-Pescador, R., Fong, N., Selby, M., Rutter, W.J. & Bell, G.I. (1983). Structure of a mouse submaxillary messenger RNA encoding epidermal growth factor and seven related proteins. Science 221, pp. 236–240.

Index